The Human Body
in Health & Disease

Ruth L. Memmler, MD

Professor Emeritus
Life Sciences
formerly Coordinator
Health, Life Sciences, Nursing
East Los Angeles College
Los Angeles, California

•

Barbara Janson Cohen, BA, MSEd.

Assistant Professor
Delaware County Community College
Media, Pennsylvania

•

Dena Lin Wood, RN, MS

Staff Nurse
Verdugo Hills Visiting Nurse Association
Glendale, California

**Illustrated by Janice A. Schwegler, MS, CMI
and
Anthony Ravielli**

The Human Body in Health & Disease

Eighth Edition

Lippincott
Philadelphia • New York

Acquisitions Editor: Andrew Allen
Editorial Assistant: Laura Dover
Project Editor: Barbara Ryalls
Design Coordinator: Doug Smock
Interior Designer: Susan Hess Blaker
Cover Designer: Lawrence R. Didona
Production Manager: Helen Ewan
Production Coordinator: Nannette Winski
Indexer: Alberta Morrison

8th Edition

Library of Congress Cataloging-in-Publication Data

Memmler, Ruth Lundeen.
    The human body in health and disease / Ruth L. Memmler, Barbara
Janson Cohen, Dena Lin Wood, illustrations by Janice A. Schwegler and Anthony Ravielli.
    —8th ed.
        p.   cm.
    Includes bibliographical references and index.
    ISBN 0-397-55175-4 Softcover   ISBN 0-397-55174-6 Casebound
    1. Human physiology.   2. Physiology, Pathological.   3. Human
anatomy.   I. Cohen, Barbara J.   II. Wood, Dena Lin.   III. Title.
    [DNLM: 1. Anatomy.   2. Pathology.   3. Physiology.   QT 200 M533h
1996]
QP34.5.M48   1996
612—dc20
DNLM/DLC
for Library of Congress   95-17706
                                                                                       CIP

♾ This Paper Meets the Requirements of ANSI/NISO Z39.48-1992 (Permanence of Paper)

The material contained in this volume was submitted as previously unpublished material, except in the instances in which credit has been given to the source from which some of the illustrative material was derived.

Any procedure or practice described in this book should be applied by the health-care practitioner under appropriate supervision in accordance with professional standards of care used with regard to the unique circumstances that apply in each practice situation. Care has been taken to confirm the accuracy of information presented and to describe generally accepted practices. However, the authors, editors, and publisher cannot accept any responsibility for errors or omissions or for any consequences from application of the information in this book and make no warranty, express or implied, with respect to the contents of the book.

The authors and publisher have exerted every effort to ensure that drug selection and dosage set forth in this text are in accordance with current recommendations and practice at the time of publication. However, in view of ongoing research, changes in government regulations, and the constant flow of information relating to drug therapy and drug reactions, the reader is urged to check the package insert for each drug for any change in indications and dosage and for added warnings and precautions. This is particularly important when the recommended agent is a new or infrequently employed drug.

Materials appearing in this book prepared by individuals as part of their official duties as U.S. Government employees are not covered by the above-mentioned copyright.

9 8 7 6 5 4 3

Preface

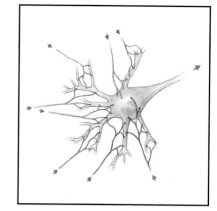

The Human Body in Health and Disease, Eighth Edition, traces the organization of the body from the single cell to the coordinated whole. A major theme is the interaction of all body systems for the maintenance of a stable internal state, a condition termed *homeostasis*. Each chapter on the body systems also contains a discussion of conditions that may upset this balance to produce disease.

▶ Organization

1. Chapters with a common theme are grouped into units. These include units on basic body structure, disease and the role of the skin in protecting against disease, the systems involved in movement and support, coordination and control, circulation, obtaining and using energy, and reproduction.
2. In this edition, the previous chapter on reproduction now has been divided into two chapters, one on the male and female reproductive systems and a second on development and birth. This reduces the size of the chapters and gives greater flexibility in planning the curriculum.

▶ Aids to Learning

1. Each chapter begins with Behavioral Objectives that state the learning goals for the chapter.
2. Selected Key Terms are listed at the start of each chapter.
3. New terms are emphasized in the text with ***boldface italic*** type. Each new term is followed by a phonetic pronunciation.
4. Abundant illustrations are included to clarify and expand on the text. In many chapters information is summarized in charts to organize the material and aid in review.
5. Each chapter is followed by a Summary outline for review.
6. Questions for Study and Review based on the text appear at the end of each chapter.
7. There is a complete Glossary with pronunciations.

8. Appendices summarize information from the text and give additional related material.
9. A section on Medical Terminology shows how scientific words are built from prefixes, roots, and suffixes. Learning these word parts will help the student remember terms and understand the meaning of new terms.

New with This Edition

1. More information on physiology (body functions) has been included. This appears in chapters on the muscular system, nervous system, urinary system, and others.
2. The number of illustrations has been increased considerably. In addition, many of the original illustrations have been redrawn for greater clarity. Photographs have been added, including photographs of tissues seen under the microscope.
3. Each chapter includes special interest boxes which expand on the text or add related information. One box in each chapter centers on normal structure and function and the other on disease conditions or clinical practice. Each type of box is signified with a logo:

 represents a box focusing on normal function and

represents a box with a clinical focus.

4. An appendix on laboratory values has been added. It includes studies on urine, blood cells, and blood chemistry.

Supplements

1. Study Guide
The study guide that accompanies **The Human Body in Health and Disease** has been revised and expanded. New illustrations have been included for labeling. New forms of questions have been added, including short essay questions.

2. Instructor's Manual
The Instructor's Manual includes:
—A test file with questions similar to those in the text and workbook. This test file will also be provided in computerized form as described below.
—Answers to the Workbook exercises
—Suggestions for teaching and for additional studies

3. Computerized Test Bank
With this edition, a Computerized Test Bank is available which allows instructors to select, modify, or add questions as required.

4. Color Transparencies
A set of color transparencies of key illustrations in the text is available for classroom instruction.

Guide to Pronunciation

The stressed syllable in each word is shown with capital letters. The vowel pronunciations are as follows:
Any vowel at the end of a syllable is given as a long sound:

a as in say
e as in be
i as in nice
o as in go
u as in true

A vowel followed by a consonant and the letter e (as in *rate*) also is given a long pronunciation.
Any vowel followed by a consonant receives a short pronunciation:

a as in absent
e as in end
i as in bin
o as in not
u as in up

In some cases, the letter h is added to a syllable to make a short vowel sound.

Acknowledgments

The authors are grateful to all those who helped in the preparation of this eighth edition of *The Human Body in Health and Disease.* Major changes have been made in the art program for this edition. We thank Janice Schwegler, the artist for this edition, for her talent, creativity, and patience. For supervising the art program and designing the book we owe many thanks to Susan Blaker. We also wish to thank Dana Morse Bittus for giving generously of her time and expertise in preparing the photomicrographs.

We are indebted to those who reviewed the text. Margaret M. Nowack evaluated the entire manuscript; Beverly M. Kovanda gave advice on laboratory tests. Above all, we thank the many instructors nationwide who have sent reviews in response to user surveys. We value their opinions and use their suggestions in striving to improve this text.

At Lippincott-Raven Publishers we thank our editor, Andrew Allen, who sees each edition through from start to finish. We extend our appreciation for the great effort, support, and attention to detail shown by the Editorial Assistant, Laura Dover. Finally, we thank the Project Editor for this edition, Barbara Ryalls.

Brief Contents

UNIT **I**

The Body as a Whole 1

1   Introduction to the Human Body   3
2   Chemistry, Matter, and Life   15
3   Cells and Their Functions   25
4   Tissues, Glands, and Membranes   39

UNIT **II**

Disease and the First Line of Defense . . 53

5   Disease and Disease-Producing
    Organisms   55
6   The Skin in Health and Disease   73

UNIT **III**

Movement and Support 83

7   The Skeleton: Bones and Joints   85
8   The Muscular System   115

UNIT **IV**

Coordination and Control 133

9   The Nervous System: The Spinal Cord
    and Spinal Nerves   135
10 The Nervous System: The Brain and Cranial
    Nerves   151

11 The Sensory System **167**

12 The Endocrine System:
Glands and Hormones **185**

UNIT **V**

Circulation and Body Defense **199**

13 The Blood **201**

14 The Heart and Heart Disease **219**

15 Blood Vessels and Blood Circulation **237**

16 The Lymphatic System and Lymphoid Tissue **261**

17 Body Defenses, Immunity, and Vaccines **271**

Unit **VI**

Energy: Supply and Use **283**

18 Respiration **285**

19 Digestion **301**

20 Metabolism, Nutrition, and Body Temperature **321**

21 Body Fluids **333**

22 The Urinary System **339**

UNIT **VII**

Perpetuating Life **357**

23 The Male and Female Reproductive Systems **359**

24 Development and Birth **379**

25 Heredity and Hereditary Diseases **391**

Suggestions for Further Study **401**
Glossary **403**
Medical Terminology **417**
Appendices **423**
Index **435**

Contents

UNIT **I**

The Body as a Whole 1

1 Introduction to the Human Body   3

Studies of the Human Body   4

Body Systems   4

Body Processes   5

Directions in the Body   5

Body Cavities   7

The Metric System   8

2 Chemistry, Matter, and Life   15

What Is Chemistry?   16

A Look at Atoms   16

Elements, Molecules, and Compounds   16

Chemical Bonds   19

Acids, Bases, and Buffers   20

Radioactivity   21

The Chemistry of Living Matter   22

3 Cells and Their Functions   25

Structure of the Cell   26

The Organelles   26

Cell Functions   29

Cells and Cancer   35

4 Tissues, Glands, and Membranes   39

Tissue Classification   40

Epithelial Tissue   40

Connective Tissue   42

Muscle Tissue   44

Nervous Tissue   **46**

Membranes   **46**

Benign and Malignant Tumors   **48**

UNIT **II**

Disease and the First Line of Defense   53

5   Disease and Disease-Producing Organisms   55

What Is Disease?   **56**

The Study of Disease   **56**

The Microorganisms   **58**

Parasitic Worms   **63**

Microbial Control   **65**

Chemotherapy   **67**

Laboratory Identification of Pathogens   **68**

6   The Skin in Health and Disease   73

Structure of the Integumentary System   **74**

Appendages of the Skin   **75**

Functions of the Skin   **76**

Observation of the Skin   **77**

Care of the Skin   **79**

Skin Diseases   **80**

UNIT **III**

Movement and Support   83

7   The Skeleton: Bones and Joints   85

The Bones   **86**

The Joints   **104**

8   The Muscular System   115

Characteristics of Skeletal Muscle   **116**

The Mechanics of Muscle Movement   **119**

Skeletal Muscle Groups   **123**

Effects of Aging on Muscles   **129**

Muscular Disorders   **129**

UNIT **IV**

Coordination and Control 133

9 The Nervous System: The Spinal Cord and Spinal Nerves 135

The Nervous System as a Whole **136**

Neurons and Their Functions **136**

The Spinal Cord **141**

Spinal Nerves **144**

The Autonomic Nervous System **144**

10 The Nervous System: The Brain and Cranial Nerves 151

The Brain and its Protective Structures **152**

Divisions of the Brain **156**

Brain Studies **160**

Disorders of the Brain **161**

Cranial Nerves **163**

11 The Sensory System 167

Senses and Sensory Mechanisms **168**

The Eye **168**

The Ear **175**

Other Special Sense Organs **179**

General Senses **179**

12 The Endocrine System: Glands and Hormones 185

Functions of Hormones **186**

The Endocrine Glands and Their Hormones **186**

Other Hormone-Producing Tissues **193**

Hormones and Treatment **195**

Hormones and Stress **196**

UNIT **V**

Circulation and Body Defense 199

13 The Blood 201

Functions of the Blood **202**

Blood Constituents **202**

Hemostasis **207**

Blood Typing and Transfusions **207**

Use of Blood and Blood Components **208**

Blood Disorders **210**

Blood Studies **213**

14   The Heart and Heart Disease   219

Circulation and the Heart   **220**

Structure of the Heart   **220**

Function of the Heart   **223**

Heart Disease   **227**

Prevention of Heart Ailments   **230**

Instruments Used in Heart Studies   **231**

Treatment of Heart Disease   **231**

15   Blood Vessels and Blood Circulation   237

Blood Vessels   **238**

Names of Systemic Arteries   **239**

Names of Systemic Veins   **244**

The Physiology of Circulation   **248**

Pulse and Blood Pressure   **252**

Disorders Involving the Blood Vessels   **253**

16   The Lymphatic System and Lymphoid Tissue   261

The Lymphatic System   **262**

Lymphoid Tissue   **264**

The Reticuloendothelial System   **267**

Disorders of the Lymphatic System and Lymphoid
Tissue   **267**

17   Body Defenses, Immunity, and Vaccines   271

The Body's Defenses Against Disease   **272**

The Occurrence of Infection   **272**

Nonspecific Defenses   **272**

Immunity   **274**

Disorders of the Immune System   **279**

Transplants and the Rejection Syndrome   **280**

Unit **VI**

Energy: Supply and Use **283**

18   Respiration   285

The Respiratory System   **286**

The Lung Cavities   **290**

Physiology of Respiration   **290**

Disorders of the Respiratory System   **294**

Special Equipment for Respiratory Tract Treatments   **298**

19   Digestion   301

What the Digestive System Does   **302**

The Accessory Structures   **313**

The Process of Digestion   **315**

20   Metabolism, Nutrition, and Body Temperature   321

Metabolism   **322**

Some Practical Aspects of Nutrition   **323**

Body Temperature   **325**

21   Body Fluids   333

Fluid Compartments   **334**

Intake and Output of Water   **334**

Fluid Therapy   **337**

22   The Urinary System   339

Organs of the Urinary System   **340**

The Kidneys   **340**

The Ureters   **349**

The Urinary Bladder   **349**

The Urethra   **350**

Disorders of the Urinary System   **350**

The Urine   **353**

UNIT **VII**

Perpetuating Life   357

23   The Male and Female Reproductive Systems   359

The Male Reproductive System   **360**

Hormonal Control of Male Reproduction   **363**

Disorders of the Male Reprodutive System   **364**

The Female Reproductive System   **365**

The Menstrual Cycle   **368**

Disorders of the Female Reproductive System   **370**

Menopause   **374**

Contraception   **376**

24   Development and Birth   379

Pregnancy   **380**

Childbirth   **383**

The Mammary Glands and Lactation   **385**

Disorders of Pregnancy   **385**

25 Heredity and Hereditary Diseases   391

Genes and Chromosomes   **392**

Hereditary Traits   **393**

Genetic Diseases   **393**

Treatment and Prevention of Genetic Diseases   **396**

Suggestions for Further Study   401

Glossary   403

Medical Terminology   417

Appendix 1—Common Metric Measurements   423

Appendix 2—Celsius–Fahrenheit Temperature Conversion   424

Appendix 3—Laboratory Tests   425

Table 1—Routine Urinalysis   **425**

Table 2—Complete Blood Count (CBC)   **426**

Table 3—Blood Chemistry Tests   **427**

Appendix 4—Typical Disease Conditions and Causative Organisms   429

Table 1—Bacterial Diseases   **429**

Table 2—Fungal Diseases   **431**

Table 3—Viral Diseases   **432**

Table 4—Protozoan Diseases   **433**

Index   435

Special Interest Boxes

1 Introduction to the Human Body
A Map of the Body   **7**
Imaging the Body   **10**

2 Chemistry, Matter, and Life
Trace Elements   **22**
Enzymes   **23**

3 Cells and Their Functions
Cell Junctions   **29**
Lysosomes   **29**

4 Tissues, Glands, and Membranes
Collagen: The Body's Scaffolding   **44**
Wound Healing   **44**

5 Disease and Disease-Producing Organisms
The CDC   **57**
Antibiotic Resistance   **68**

6 The Skin in Health and Disease
Vitamin D   **77**
Burns   **80**

7 The Skeleton: Bones and Joints
The Timetable of Bone Formation   **89**
Joint Replacement   **111**

8 The Muscular System
Rigor Mortis   **119**
Anabolic Steroids   **120**

Topics of interest are highlighted in boxes within each chapter. One box, indicated by an open book logo, contains information on basic structure and function of the body. Another, shown by a test tube logo, centers on a clinical topic.

9   The Nervous System: The Spinal Cord and Spinal Nerves
Dermatomes   **145**
Cell Receptors—Getting the Message   **147**

10   The Nervous System: The Brain and Cranial Nerves
Sleep   **159**
Magnetic Resonance Imaging   **162**

11   The Sensory System
Mechanoreceptors   **168**
Referred Pain   **182**

12   The Endocrine System: Glands and Hormones
Growth Hormone   **189**
Seasonal Affective Disorder   **195**

13   The Blood
Hemoglobin   **204**
Reticulocytes   **211**

14   The Heart and Heart Disease
Cardiac Reserve   **225**
What Are All Those DLs?   **230**

15   Blood Vessels and Blood Circulation
Blood Vessels in the Fetus   **239**
Special Methods for Measuring Blood Pressure   **253**

16   The Lymphatic System and Lymphoid Tissue
How Lymphatic Capillaries Work   **264**
Cancer and the Lymphatic System   **268**

17   Body Defenses, Immunity, and Vaccines
Antibodies   **276**
AIDS Update   **280**

18   Respiration
The Air We Breathe   **292**
Liquid Ventilation   **298**

19   Digestion
Surface Area and Transport   **311**
Hepatitis B   **314**

20   Metabolism, Nutrition, and Body Temperature
Food as Good Medicine   **325**
Obesity   **327**

21 **Body Fluids**
How Much to Drink   **335**
Sodium and Potassium Imbalances   **337**

22 **The Urinary System**
The Transport Maximum   **347**
Angiotensin   **351**

23 **The Male and Female Reproductive Systems**
In Vitro Fertilization   **364**
STDs   **374**

24 **Development and Birth**
The Origin of Tissues   **383**
The Use of Ultrasound in Obstetrics   **387**

25 **Heredity and Hereditary Diseases**
The Human Genome Project   **392**
Correction of Genetic Disorders   **396**

The Human Body
in Health & Disease

UNIT I

The Body as a Whole

This unit presents the basic levels of organization within the human body. Included are brief descriptions of the systems, which are made of groups of organs. Organs are made of tissues, which in turn are made of the smallest units, called *cells*. A short survey of chemistry, which deals with the composition of all matter and is important for the understanding of human physiology, is incorporated in this unit. These chapters should prepare the student for the more detailed study of individual body systems in the units that follow.

CHAPTER 1

Introduction to the Human Body

Selected Key Terms

The following terms are defined in the Glossary:

anabolism

anatomy

ATP

catabolism

cell

gram

homeostasis

liter

metabolism

meter

negative feedback

organ

pathology

physiology

system

tissue

Behavioral Objectives

After careful study of this chapter, you should be able to:

- Define the terms *anatomy, physiology,* and *pathology*
- Describe the organization of the body from cells to the whole organism
- List 10 body systems and give the general function of each
- Define *metabolism* and name the two phases of metabolism
- Differentiate between extracellular and intracellular fluids
- Briefly explain the role of ATP in the body
- Define *homeostasis* and *negative feedback*
- List and define the main directional terms for the body
- List and define the three planes of division of the body
- Name the subdivisions of the dorsal and ventral cavities
- Name the basic metric units of length, weight, and volume
- Name and locate the nine regions of the abdomen
- Define the metric prefixes *kilo-, centi-, milli-,* and *micro-*

Memmler, RL., Cohen, BJ, Wood, DL. *THE HUMAN BODY IN HEALTH AND DISEASE,* 8/e, © 1996 Lippincott-Raven Publishers.

❱ Studies of the Human Body

Everyone is interested in the human body and how it is affected by disease. Articles on health and medicine appear daily in newspapers and magazines. More and more, ordinary people not specifically trained in science are being asked to make decisions on scientific matters that affect society.

The scientific term for the study of body structure is *anatomy* (ah-NAT-o-me). Part of this word means "to cut," because one of the many ways to learn about the human body is to cut it apart, or *dissect* it. *Physiology* (fiz-e-OL-o-je) is the term for the study of how the body functions, and the two sciences are closely related. They form the basis for all medical practice. Anything that upsets the normal structure or working of the body is considered a *disease* and is studied as the science of *pathology* (pah-THOL-o-je). It is hoped that from these studies you will gain an appreciation for the design and balance of the human body and for living organisms in general.

All living things are organized from very simple levels to more complex levels. Living matter begins with simple chemicals. These chemicals are formed into the complex substances that make living *cells*—the basic units of all life. Specialized groups of cells form *tissues,* and tissues may function together as *organs.* Organs functioning together for the same general purpose make up organ *systems,* which maintain the body.

❱ Body Systems

The body systems have been listed as 9, 10, or 11 in number, depending on how much detail one wishes to include. Here is one such list:

1. The *skeletal system.* The basic framework of the body is a system of over 200 bones and the joints between them, collectively known as the *skeleton.*
2. The *muscular system.* Body movements result from the action of the skeletal muscles, which are attached to the bones. Other types of muscles are present in the walls of body organs such as the intestine and the heart.
3. The *circulatory system.* The heart and blood vessels make up the system that pumps blood to all

the body tissues, bringing with it nutrients, oxygen, and other substances and carrying away waste materials. Lymphatic vessels play an important auxiliary role in circulation.

4. The *digestive system.* This system comprises all the organs that are involved with taking in food and converting it into substances that body cells can use. Examples of these organs are the mouth, esophagus, stomach, intestine, liver, and pancreas.
5. The *respiratory system.* This system includes the lungs and the passages leading to the lungs. The purpose of this system is to take in air and conduct it to the areas designed for gas exchange. Oxygen passes from the air into the blood and is carried to all tissues by the circulatory system. In like manner, carbon dioxide, as a gaseous waste product, is taken from the tissues to be expelled from the body.
6. The *integumentary* (in-teg-u-MEN-tar-e) *system.* The word *integument* (in-TEG-u-ment) means skin. The skin with its appendages is considered to be a separate body system. The structures associated with the skin include the hair, the nails, and the sweat and oil glands.
7. The *urinary system.* This system is also called the *excretory system.* Its main components are the kidneys, the ureters, the bladder, and the urethra. Its chief purpose is to rid the body of waste products and excess water. (Note that other waste products are removed by the digestive and respiratory systems and by the skin).
8. The *nervous system.* The brain, the spinal cord, and the nerves make up this very complex system by which the body is controlled and coordinated. The organs of special sense (such as the eyes, ears, taste buds, and organs of smell), together with the receptors for touch and other senses, receive stimuli from the outside world. These stimuli are converted into impulses that are transmitted to the brain. The brain directs the body's responses to these messages and also to messages coming from within the body. Such higher functions as memory and reasoning also occur in the brain.
9. The *endocrine* (EN-do-krin) *system.* The scattered organs known as *endocrine glands* are grouped together because they share a similar function. All produce special substances called

hormones, which regulate such body activities as growth, food utilization within the cells, and reproduction. Examples of endocrine glands are the thyroid and the pituitary glands.

10. The *reproductive system.* This system includes the external sex organs and all related internal structures that are concerned with the production of offspring.

Body Processes

All the life-sustaining reactions that go on within the body systems together make up *metabolism* (meh-TAB-o-lizm). Metabolism can be divided into two types of activities:

1. In *catabolism* (kah-TAB-o-lizm), complex substances, including the nutrients from digested foods, are broken down into simpler compounds with the release of energy.
2. In *anabolism* (ah-NAB-o-lizm), simple compounds are used to manufacture complex materials needed for growth, function, and repair of tissues.

The energy obtained from the breakdown of nutrients is used to form a compound often described as the "energy currency" of the cell. It has the long name of *adenosine triphosphate* (ah-DEN-o-sene tri-FOS-fate), but is commonly abbreviated *ATP.* Chapter 20 has more information on metabolism and ATP.

It is important to recognize that our bodies are composed of large amounts of fluids. Certain kinds of fluids bathe the cells, carry nutrient substances to and from the cells, and transport the nutrients into and out of the cells. This group of fluids is called *extracellular fluid* because it includes all body fluids outside the cells. A second type of fluid, *intracellular fluid,* is contained within the cells. Extracellular and intracellular fluids account for approximately 60% of an adult's weight.

Homeostasis

The purpose of metabolism is to maintain a state of balance within the organism, an important characteristic of all living things. Such conditions as body temperature, the composition of body fluids, heart rate, respiration rate, and blood pressure must be kept within set limits to maintain health. This steady state within the organism is called *homeostasis* (ho-me-o-STA-sis), which literally means "staying (stasis) the same (homeo)."

Homeostasis is maintained by the feedback of information. Sensors throughout the body monitor internal conditions and bring them back to normal when they shift, much as a thermostat regulates the temperature of a house to a set level. Since each change, up or down, must be reversed to restore the norm, this mechanism is described as *negative feedback.* The main controllers in this self-regulation are the nervous and endocrine systems. You will find many examples of negative feedback throughout this book.

Directions in the Body

Because it would be awkward and inaccurate to speak of bandaging the "southwest part" of the chest, a number of terms are used to designate position and directions in the body. They refer to the body in the *anatomic position*—upright with face front, arms at the sides with palms forward, and feet parallel. The main directional terms are as follows (Fig. 1-1):

1. *Superior* is a term meaning above, or in a higher position. Its opposite, *inferior,* means below, or lower. The heart, for example, is superior to the intestine.
2. *Ventral* and *anterior* mean the same thing in humans: located toward the belly surface or front of the body. Their corresponding opposites, *dorsal* and *posterior,* refer to locations nearer the back.
3. *Cranial* means nearer to the head; *caudal,* nearer to the sacral region of the spinal column (*i.e.,* where the tail is located in lower animals).
4. *Medial* means nearer to an imaginary plane that passes through the midline of the body, dividing it into left and right portions. *Lateral,* its opposite, means farther away from the midline, toward the side.
5. *Proximal* means nearer the origin of a structure; *distal,* farther from that point. For example, the part of your thumb where it joins your hand is its proximal region; the tip of the thumb is its distal region.

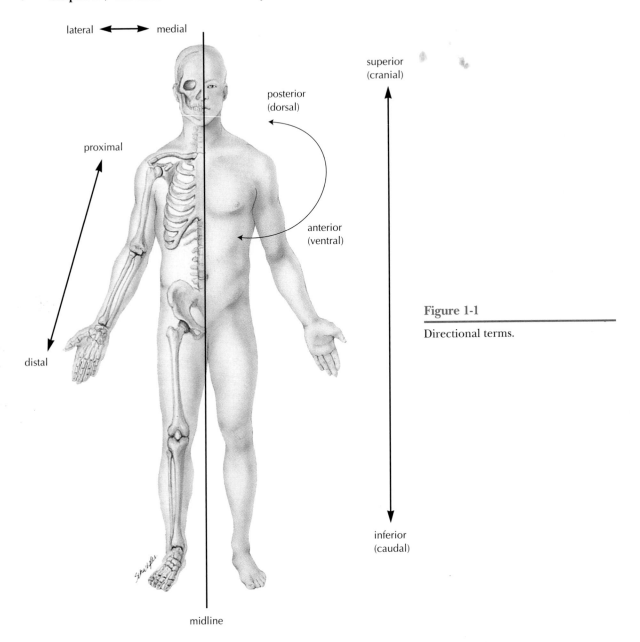

Figure 1-1

Directional terms.

Planes of Division

For convenience in visualizing the spatial relations of various body structures to each other, anatomists have divided the body into three imaginary planes, each of which is a cut through the body in a different direction (Fig. 1-2).

1. The *sagittal* (SAJ-ih-tal) *plane.* If you were to cut the body in two from front to back, separating it into right and left portions, the sections you would see would be sagittal sections. A cut exactly down the midline of the body, separating it into equal right and left halves, is a *midsagittal* section.

2. The *frontal plane.* If the cut were made in line with the ears and then down the middle of the body, creating a front and a rear portion, you would see

frontal
plane

sagittal
plane

transverse
plane

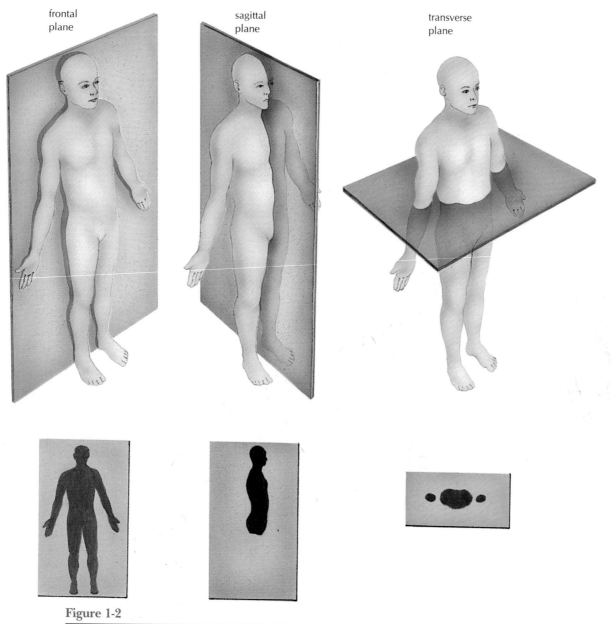

Figure 1-2

Planes of division.

a front (anterior or ventral) section and a rear (posterior or dorsal) section. Another name for this plane is *coronal plane.*

3. The ***transverse plane.*** If the cut were made horizontally, across the other two planes, it would divide the body into an upper (superior) part and a lower (inferior) part. There could be many such cross sections, each of which would be on a transverse plane, also called a *horizontal plane.*

▶ Body Cavities

The body contains a few large internal spaces, or ***cavities,*** within which various organs are located.

A Map of the Body

The body's surface landmarks provide a good introduction to body structure. These landmarks are also useful for orientation in surgery and other clinical procedures. For example, the lower tip of the sternum, the xiphoid process, is used as a reference point in the administration of CPR (cardiopulmonary resuscitation).

Feel for some of the other superficial markings to learn how your body is organized. The joint between the mandible and the temporal bone of the skull (the temporomandibular, or TM, joint) can be felt in front of the ear canal as you move your lower jaw up and down. The pulse of the facial artery can be felt in a small depression about 1 inch in front of the angle of the lower jaw. The pulse of the carotid artery can be felt at the side of the neck.

Touch behind your neck for the spine of the 7th cervical vertebra, the prominent vertebra at the base of the neck. At the distal end of the forearm , you can feel the pulse of the radial artery at the lateral surface of the anterior wrist.

In the torso, you can feel the ribs and the lower margin of the rib cage. By placing your hands on your hips, you can feel the crest of the hip bone. When it hurts to sit on a hard chair, you become aware of the lower bone of the pelvic girdle on each side.

Many other features can be seen and felt at the surface to tell us about the structure and the condition of the body.

There are two groups of cavities: *dorsal* and *ventral* (Fig. 1-3).

Dorsal Cavity

The dorsal body cavity has two regions: the *cranial cavity,* containing the brain, and the *spinal canal,* enclosing the spinal cord. These two areas form one continuous space.

Ventral Cavity

The ventral cavity is much larger than the dorsal cavity. Unlike the dorsal cavity, it is not continuous, but is separated by the *diaphragm* (DI-ah-fram), a muscle used in breathing. The *thoracic* (tho-RAS-ik)

cavity is located above the diaphragm. Its contents include the heart, the lungs, and the large blood vessels that join the heart. The heart is contained in the pericardial sac; the lungs are in the pleural sacs (Fig. 1-4). The *abdominopelvic* (ab-dom-ih-no-PEL-vik) *cavity* is located below the diaphragm. This space is subdivided into two regions. The uppermost *abdominal* cavity contains the stomach, most of the intestine, the kidneys, the liver, the gallbladder, the pancreas, and the spleen. The lower portion, set off by an imaginary line across the top of the hip bones, is the *pelvic cavity.* This cavity contains the urinary bladder, the rectum, and the internal parts of the reproductive system.

REGIONS OF THE ABDOMEN

Because the abdomen is so large, it is helpful to divide it for reference into nine regions (see Fig. 1-4). The three central regions are the *epigastric* (ep-ih-GAS-trik) *region,* located just below the breastbone; the *umbilical* (um-BIL-ih-kal) *region* around the umbilicus (um-BIL-ih-kus), commonly called the *navel;* and the *hypogastric* (hi-po-GAS-trik) *region,* the lowest of all the midline regions. At each side are the right and left *hypochondriac* (hi-po-KON-dre-ak) *regions,* just below the ribs; then the right and left *lumbar regions;* and finally the right and left *iliac,* or *inguinal* (IN-gwih-nal), *regions.* A simpler but less precise division into four quadrants is sometimes used. These are the right upper quadrant (RUQ), left upper quadrant (LUQ), right lower quadrant (RLQ), and left lower quadrant (LLQ) (Fig. 1-5).

▶ The Metric System

Now that we have set the stage for further study of the body and its structure and processes, a look at the metric system is in order, since this is the system used for all scientific measurements. The drug industry and the health care industry already have converted to the metric system, so anyone who plans a career in health care should be acquainted with metrics.

To use the metric system easily and correctly may require a bit of effort, as is often the case with any new idea. Actually, you already know something about the metric system because our monetary sys-

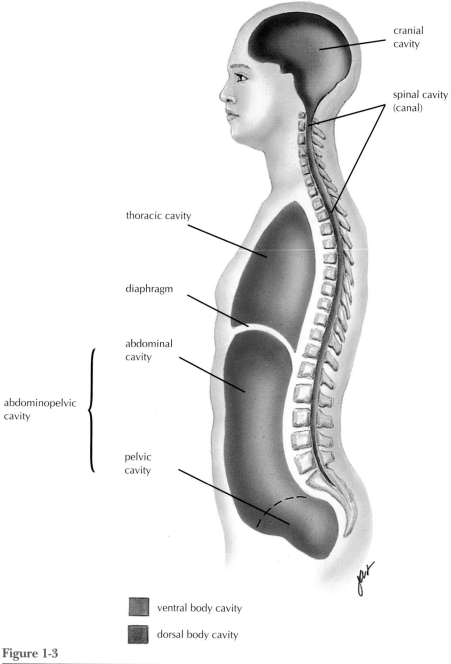

cranial
cavity

spinal cavity
(canal)

thoracic cavity

diaphragm

abdominal
cavity

abdominopelvic
cavity

pelvic
cavity

ventral body cavity

dorsal body cavity

Figure 1-3

Side view of the body cavities.

tem is similar to it in that both are decimal systems. One hundred cents equal one dollar; one hundred centimeters equal one meter. The decimal system is based on multiples of the number 10, which are indicated by prefixes:

kilo = 1000

centi = 1/100

milli = 1/1000

micro = 1/1,000,000

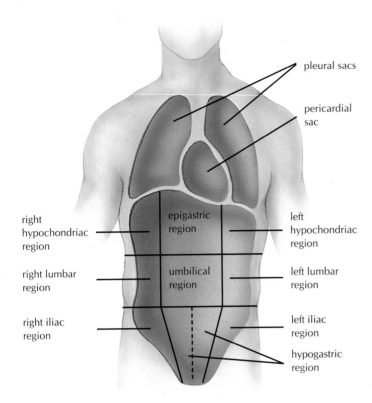

right
hypochondriac
region

epigastric
region

left
hypochondriac
region

right lumbar
region

umbilical
region

left lumbar
region

right iliac
region

left iliac
region

hypogastric
region

pleural sacs

pericardial
sac

Figure 1-4

Front view of the thoracic cavity and the nine regions of the abdomen.

The basic unit of length in the metric system is the *meter*. Thus, 1 kilometer is equal to 1000 meters. A centimeter is 1/100 of a meter; stated another way, there are 100 centimeters in 1 meter. A changeover to the metric system has been slow in coming to the United States. Often measurements on packages, bottles, and yard goods are now given according to both scales. In this text, equivalents in the more familiar units of inches, feet, and so forth are included along with the metric units for comparison. There are 2.5 centimeters (cm) or 25 millimeters (mm) in 1 inch, as shown in Figure 1-6. Some equivalents that may help you to appreciate the size of various body parts are as follows:

1 mm = 0.04 inch, or 1 inch = 25 mm
1 cm = 0.4 inch, or 1 inch = 2.5 cm
1 m = 3.3 feet, or 1 foot = 30 cm

The same prefixes used for linear measurements are used for weights and volumes. The *gram* is the standard unit of weight. Thirty grams are approximately equal to 1 ounce, and 1 kilogram to 2.2 pounds. Drug dosages are usually stated in grams or

Imaging the Body

Diagnostic imaging techniques are based on the physical properties of the human body and its response to forces such as ionizing radiation, magnetic fields, and sound waves. With diagnostic imaging, it is possible to view the interior of the body without resorting to exploratory surgery.

The oldest and most familiar method is radiography (ra-de-OG-rah-fe), which uses x-rays (a form of radiation) to produce a picture on a special type of film. The dark areas indicate where the beam passed through the body and exposed the film. The light areas show where the beam was blocked or deflected. Structures clearly visible on x-ray films are dense structures such as bones. Structures not as dense as bone will also deflect the rays to a smaller extent. Sometimes an opaque material is added to show outlines of soft tissue, as in a "barium swallow." X-rays are damaging to body tissues, but with modern equipment extremely low dosages can be used.

Figure 1-6

Comparison of centimeters and inches.

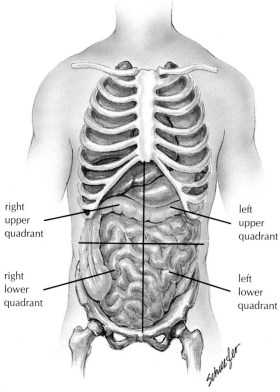

right
upper
quadrant

left
upper
quadrant

right
lower
quadrant

left
lower
quadrant

Figure 1-5

Quadrants of the abdomen showing the organs within each quadrant.

milligrams. One thousand milligrams equal 1 gram; a 500-milligram (mg) dose would be the equivalent of 0.5 gram (g), and 250 mg are equal to 0.25 g.

The dosages of liquid medications are given in units of volume. The standard metric measurement for volume is the *liter* (LE-ter). There are 1000 milliliters (mL) in a liter. A liter is slightly greater than a quart, a liter being equal to 1.06 quarts. For smaller quantities, the milliliter is used most of the time. There are 5 mL in a teaspoon and 15 mL in a tablespoon. A fluid ounce contains 30 mL.

The Celsius (centigrade) temperature scale, now in use by most countries and by scientists in this country, is discussed in Chapter 20. A chart of all the common metric measurements and their equivalents is shown in Appendix 1. A Celsius–Fahrenheit temperature conversion scale appears in Appendix 2.

SUMMARY

I. Studies of the human body
  A. Anatomy—study of structure
  B. Physiology—study of function
  C. Pathology—study of disease

II. Organization
  A. Levels—cell, tissue, organ, organ system
  B. Body systems
    1. Skeletal system—support
    2. Muscular system—movement
    3. Circulatory system (includes lymphatic system)—transport
    4. Digestive system—intake and breakdown of food
    5. Respiratory system—intake of oxygen and release of carbon dioxide
    6. Integumentary system—skin and associated structures (appendages)
    7. Urinary system—elimination of waste and water
    8. Nervous system—receipt of stimuli and control of responses
    9. Endocrine system—production of hormones for regulation of growth, metabolism, reproduction
    10. Reproductive system—production of offspring

III. Body processes
  A. Metabolism—all the chemical reactions needed to sustain life

1. Catabolism—breakdown of complex substances into simpler substances
2. Anabolism—building of body materials

B. ATP (adenosine triphosphate)—energy compound of cells

C. Body fluids
1. Extracellular—outside the cells
2. Intracellular—inside the cells

D. Homeostasis—steady state of body conditions; maintained by negative feedback

IV. Body positions, directions, planes

A. Anatomic position—upright, palms forward

B. Directions
1. Superior—above or higher; inferior—below or lower
2. Ventral (anterior)—toward belly or front surface; dorsal (posterior)—nearer to back surface
3. Cranial—nearer to head; caudal—nearer to sacrum
4. Medial—toward midline; lateral—toward side
5. Proximal—nearer to point of origin; distal—farther from point of origin

C. Planes of division
1. Sagittal—from front to back, dividing the body into left and right parts
   a. Midsagittal—exactly down the midline
2. Frontal (coronal)—from left to right, dividing the body into anterior and posterior parts
3. Transverse—horizontally, dividing the body into superior and inferior parts

V. Body cavities

A. Dorsal—contains cranial and spinal cavities for brain and spinal cord

B. Ventral
1. Thoracic—chest cavity containing heart and lungs and divided from abdominal cavity by diaphragm
2. Abdominopelvic
   a. Abdominal—upper region containing stomach, most of intestine, kidneys, liver, spleen, etc.
   b. Pelvic—lower region containing reproductive organs, urinary bladder, rectum
   c. Abdomen divided for reference into nine regions: epigastric, umbilical, hypogastric, right and left hypochondriac, right and left lumbar, right and left iliac (inguinal)

VI. Metric system—based on multiples of 10

A. Basic units
1. Meter—length
2. Liter—volume
3. Gram—weight

B. Prefixes—indicate multiples of 10
1. Kilo—1000 times
2. Centi—1/100th (0.01)
3. Milli—1/1000th (0.001)
4. Micro—1/1,000,000 (0.000001)

C. Temperature—measured in Celsius (centigrade) scale

QUESTIONS FOR STUDY AND REVIEW

1. List three types of study of the human body and define each.
2. Define *cell, tissue, organ,* and *system.*
3. List 10 body systems and briefly describe the function of each.
4. Name and define the two phases of *metabolism.*
5. What is ATP?
6. Define the term *homeostasis* and give two examples. How is homeostasis maintained?
7. Stand in the anatomic position.
8. List the opposite term for each of the following body directions: *superior, ventral, anterior, cranial, medial,* and *proximal.* Define each term and its opposite.
9. Describe the location of the elbow in relation to the wrist; in relation to the shoulder.
10. Describe the location of the stomach in relation to the urinary bladder; in relation to the lungs.

11. Describe the location of the ears in relation to the nose.
12. What are the three main body planes? Explain the division each makes.
13. Name the cavities within the dorsal and ventral cavities. Name one organ found in each.
14. Make a rough sketch of the abdomen and label the nine divisions.
15. Name and locate the four quadrants of the abdomen.
16. Why should you learn the metric system? What are its advantages?

CHAPTER 2

Chemistry, Matter, and Life

Selected Key Terms

The following terms are defined in the Glossary:

acid

atom

base

buffer

carbohydrate

compound

covalent bond

electrolyte

electron

element

enzyme

ion

lipid

molecule

neutron

pH

protein

proton

solution

suspension

Behavioral Objectives

After careful study of this chapter, you should be able to:

- Describe the structure of an atom
- Differentiate between atoms and molecules
- Define the atomic number of an atom
- Differentiate between elements and compounds and give several examples of each
- Explain why water is so important to the body
- Define *mixture*; list the three types of mixtures and give two examples of each
- Differentiate between ionic and covalent bonds
- Define the terms *acid, base,* and *salt*
- Explain how the numbers on the pH scale relate to acidity and alkalinity
- Define *buffer* and explain why buffers are important in the body
- Define *radioactivity* and cite several examples of how radioactive substances are used in medicine
- List three characteristics of organic compounds
- Name the three main types of organic compounds and the building blocks of each
- Define *enzyme*; describe how enzymes work

Memmler, RL, Cohen, BJ, Wood, DL. *THE HUMAN BODY IN HEALTH AND DISEASE,* 8/e,
© 1996 Lippincott-Raven Publishers.

What Is Chemistry?

Great strides toward an understanding of living organisms, including the human being, have come to us through *chemistry,* the science that deals with the composition of matter. Knowledge of chemistry and chemical changes helps us understand the normal and abnormal functioning of the body and its parts. The digestion of food in the intestinal tract, the production of urine by the kidneys, the regularity of breathing—all body processes—are based on chemical principles. Chemistry also is important in *microbiology* (the study of microscopic plants and animals) and *pharmacology* (the study of drugs). The various solutions that are used to cleanse the skin before a surgical operation are chemicals, as are aspirin, penicillin, and all other drugs used in treating disease. To provide some understanding of the importance of chemistry in the life sciences, this chapter briefly describes *atoms* and *molecules, elements, compounds,* and *mixtures,* which are fundamental forms of matter.

A Look at Atoms

Atoms are small particles that form the building blocks of matter, the smallest complete units of which all matter is made. To visualize the size of an atom, one can think of placing millions of them on the sharpened end of a pencil and still have room for many more. Everything about us, everything we can see and touch, is made of atoms—the food we eat, the atmosphere, the water in the ocean, the smoke coming out of a chimney.

Despite the fact that the atom is such a tiny particle, it has been carefully studied and has been found to have a definite structure. At the center of the atom is a nucleus, which contains positively charged electric particles called *protons* and noncharged particles called *neutrons.* Outside the nucleus, in regions called *orbitals,* are negatively charged particles called *electrons* (Fig. 2-1).

The protons and electrons always are equal in number in any atom. Collectively, they are responsible for all the atom's characteristics. The neutrons and protons are tightly bound in the nucleus, contributing nearly all the atom's weight. The positively charged protons keep the negatively charged electrons in the orbital area around the nucleus by means of the opposite charges the particles possess.

Positively (+) charged protons attract negatively (−) charged electrons. The electrons contribute important chemical characteristics to the atom.

Most atoms have several orbitals of electrons. However, each orbital can hold only two electrons. The orbitals are arranged into energy levels. Energy levels are identified by their distance from the nucleus of the atom. The first energy level, the one closest to the nucleus, is composed of one orbital. The second energy level, the next in distance away from the nucleus, can have four orbitals. Since each orbital contains two electrons, the second energy level has the capacity to hold eight electrons. The electrons farthest away from the nucleus are the particles that give the atom its chemical characteristics.

If the outermost energy level has more than four electrons but fewer than its capacity of eight, the atom normally completes this level by gaining electrons. Such an atom is called a *nonmetal.* The oxygen atom illustrated in Figure 2-1 has six electrons in its second, or outermost, level. When oxygen enters into chemical reactions, its chemical behavior is to gain two electrons. The oxygen atom then has two more electrons than protons. If the outermost shell has fewer than four electrons, the atom normally loses those electrons to attain a complete outer energy level. Such an atom is called a *metal.*

Elements, Molecules, and Compounds

Atoms are the fundamental units that make up the chemical *elements* from which all matter is made. An element is a substance that cannot be decomposed—that is, changed into something else—by physical means (*e.g.,* by the use of heat, pressure, or electricity). Examples of elements include various gases, such as hydrogen, oxygen, and nitrogen; liquids, such as the mercury used in thermometers and blood pressure instruments; and many solids, such as iron, aluminum, gold, silver, and carbon. Graphite (the so-called "lead" in a pencil), coal, charcoal, and diamonds are examples of the element carbon. The entire universe is made up of about 105 elements.

Elements can be identified by their names, their symbols, or their atomic numbers. Chemists use symbols as a shorthand method to name elements. The atomic number of an element is equal to the num-

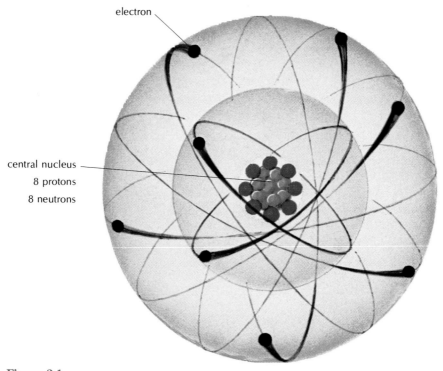

electron

central nucleus
8 protons
8 neutrons

Figure 2-1

Representation of the oxygen atom. Eight protons and eight neutrons are tightly bound
in the central nucleus, around which the eight electrons revolve.

ber of protons that are present in the nucleus. Because the number of protons is equal to the number
of electrons in an atom, the atomic number also
identifies the number of electrons whirling about
the nucleus. Table 2-1 lists some elements found in
the human body along with their functions.

When, on the basis of electron structure, two or
more atoms unite, a ***molecule*** is formed. A molecule
can be made of like atoms—the oxygen molecule is
made of two identical atoms—but more often it is
made of two or more different atoms. For example,
a molecule of water (H_2O) contains 1 atom of oxygen (O) and 2 atoms of hydrogen (H) (Fig. 2-2).
Substances that contain molecules formed by the
union of two or more different atoms are called ***compounds.*** Compounds may be made of a few elements
in a simple combination; for example, the gas carbon monoxide (CO) contains 1 atom of carbon (C)
and 1 atom of oxygen (O). Alternatively, compounds may be very complex; some proteins, for
example, have thousands of atoms.

It is interesting to observe how different a compound is from any of its constituents. For example,

a molecule of liquid water is formed from oxygen
and hydrogen, both of which are gases. Another example is a crystal sugar, glucose ($C_6H_{12}O_6$). Its constituents include 12 atoms of the gas hydrogen,
6 atoms of the gas oxygen, and 6 atoms of the solid
element carbon. In this case, the component gases
and the solid carbon do not in any way resemble the
glucose.

More About Water

Water is the most abundant compound in the
body. No plant or animal, including the human, can
live very long without it. Water is of critical importance in all physiologic processes in body tissues.
Water carries substances to and from the cells and
makes possible the essential processes of absorption,
exchange, secretion, and excretion. What are some
of the properties of water that make it such an ideal
medium for living cells?

1. Water can dissolve many different substances in
   large amounts. For this reason it is called the ***universal solvent.*** All the materials needed by the body,

TABLE 2-1
Some Common Chemical Elements*

Name	Symbol	Function
Oxygen	O	Part of water; needed to metabolize nutrients for energy
Carbon	C	Basis of all organic compounds; in carbon dioxide, the waste gas of metabolism
Hydrogen	H	Part of water; participates in energy metabolism, acid–base balance
Nitrogen	N	Present in all proteins, ATP (the energy compound), and nucleic acids (DNA and RNA)
Calcium	Ca	Builds bones and teeth; needed for muscle contraction, nerve impulse conduction, and blood clotting
Phosphorus	P	Active ingredient in the energy-storing compound ATP; builds bones and teeth; in cell membrane and nucleic acids
Potassium	K	Nerve impulse conduction; muscle contraction; water balance and acid–base balance
Sulfur	S	Part of many proteins
Sodium	Na	Water balance; nerve impulse conduction; muscle contraction
Chlorine	Cl	Water balance; acid–base balance; in stomach acid
Iron	Fe	Part of hemoglobin, the compound that carries oxygen in red blood cells

*The elements are listed in decreasing order by weight in the body.

such as gases, minerals, and nutrients, dissolve in water to be carried from place to place.

2. Water is very stable as a liquid at ordinary temperatures. Water does not freeze until the temperature drops to 0°C (32°F) and does not boil until the temperature reaches 100°C (212°F). This stability provides a constant environment for body cells. Water can also be used to distribute heat throughout the body and to cool the body by evaporation of sweat from the body surface.

3. Water participates in chemical reactions in the body. It is needed directly in the process of digestion and in many of the metabolic reactions that occur in the cells.

Mixtures, Solutions, and Suspensions

Not all elements or compounds combine chemically when brought together. The air we breathe every day is a mixture of gases, largely nitrogen, oxygen, and carbon dioxide, along with smaller percentages of other substances. The constituents in the air maintain their identity, although the proportions of each may vary. Blood plasma is also a mixture in which the various components maintain their identity. The many valuable compounds in the plasma remain separate entities with their own properties. Such combinations are called ***mixtures***— blends of two or more substances. A mixture, such as salt water, in which the component substances remain evenly distributed is called a ***solution.*** The dissolving substance, in this case water, is the ***solvent;*** the substance dissolved, in this case salt, is the ***solute.*** Solutions of glucose, salts, or both in water are used for intravenous fluid treatments.

In some mixtures, the material distributed in the solvent settles out unless the mixture is constantly

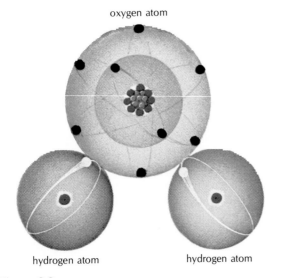

oxygen atom

hydrogen atom   hydrogen atom

Figure 2-2

Molecule of water.

shaken; this type of mixture is called a *suspension.* Settling occurs in a suspension because the particles in the mixture are large and heavy. Examples of suspensions are milk of magnesia, india ink, and, in the body, red blood cells suspended in blood plasma.

One other type of mixture is of importance in the body. In a *colloidal suspension,* the particles do not dissolve but remain distributed in the solvent because they are so small. The fluid that fills the cells (cytoplasm) is a colloidal suspension, as is blood plasma.

▶ Chemical Bonds

Ionic Bonds and Electrolytes

When discussing the structure of the atom, we mentioned the positively charged (+) protons that are located in the nucleus, with the corresponding number of negatively charged (−) electrons found in the surrounding space and neutralizing the protons. If a single electron were removed from the sodium atom, it would leave one proton not neutralized, and the atom would have a positive charge (Na^+). This actually happens during a chemical change. Similarly, atoms can gain electrons so that there are more electrons than protons. Chlorine, which has seven electrons in its outermost energy

level, tends to gain one electron to fill the level to its capacity. Such an atom of chlorine is negatively charged (Cl^-) (Fig. 2-3). An atom with a positive or negative charge is called an *ion.* An ion that is positively charged is a *cation,* whereas a negatively charged ion is an *anion.*

Let us imagine a sodium atom coming in contact with a chlorine atom. The chlorine atom gains an electron from the sodium atom, and the two newly formed ions (Na^+ and Cl^-), because of their opposite charges, which attract each other, cling together and produce the compound sodium chloride, ordinary table salt. A bond formed by this method of electron transfer is called an *ionic bond* (Fig. 2-4).

In the fluids and cells of the body, ions make it possible for materials to be altered, broken down, and recombined to form new substances. Calcium ions (Ca^{++}) are necessary for the clotting of blood, the contraction of muscle, and the health of bone tissue. Bicarbonate ions (HCO_3^-) are required for the regulation of acidity and alkalinity of the tissues. The stable condition of the normal organism, homeostasis, is influenced by ions.

Compounds that form ions whenever they are in solution are called *electrolytes* (e-LEK-tro-lites). Electrolytes are responsible for the acidity and the alka-

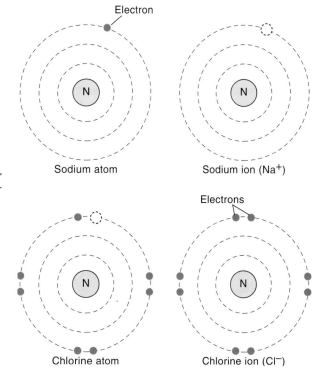

Figure 2-3

Formation of Na^+ cation and Cl^- anion. Only the electrons in the outermost energy level are shown.

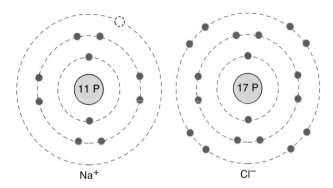

Na⁺   Cl⁻

Figure 2-4

A sodium ion with 11 protons in the nucleus and 10 electrons in orbitals is attracted to a chlorine ion with 17 protons in the nucleus and 18 electrons in orbitals to form the compound sodium chloride.

linity of solutions. They also include a variety of mineral salts, such as sodium and potassium chloride. Electrolytes must be present in exactly the right quantities in the fluid within the cell (intracellular) and outside the cell (extracellular), or there will be very damaging effects, preventing the cells in the body from functioning properly.

Because ions are charged particles, electrolyte solutions can conduct an electric current. (In practice, the term *electrolytes* is also used to refer to the ions themselves in body fluids.) Records of electric currents in tissues are valuable indications of the functioning or malfunctioning of tissues and organs. The *electrocardiogram* (e-lek-tro-KAR-de-o-gram) and the *electroencephalogram* (e-lek-tro-en-SEF-ah-lo-gram) are graphic tracings of the electric currents generated by the heart muscle and the brain, respectively (see Chaps. 10 and 14).

Covalent Bonds

Although many chemical compounds are formed by ionic bonds, a much larger number are formed by another type of chemical bond. This type of bond involves not the exchange of electrons but a sharing of electrons between the atoms in the molecule. The electrons orbit around both of the atoms in the bond to make both of them stable. The electrons may be equally shared, as in the case of a hydrogen molecule (H_2), or they may be held closer to one atom than the other, as in the case of water (H_2O), shown in Figure 2-2. These bonds are called *covalent bonds.* Covalently bonded compounds do not conduct an electric current in solution. Carbon, the element that is the basis of organic chemistry, forms covalent bonds. Thus, the compounds that are characteristic of living things are covalently bonded compounds. These bonds may involve the sharing of one, two, or three pairs of electrons between atoms.

▶ Acids, Bases, and Buffers

An *acid* is a chemical substance capable of donating a hydrogen ion (H^+) to another substance. A common example is hydrochloric acid, the acid found in stomach juices:

HCl	→	H⁺	+	Cl⁻
hydrochloric acid		hydrogen ion		chlorine ion

A *base* is a chemical substance, usually containing a hydroxide ion (OH^-), that can accept a hydrogen ion. Sodium hydroxide, which releases hydroxide ion in solution, is an example of a base:

NaOH	→	Na⁺	+	OH⁻
sodium hydroxide		sodium ion		hydroxide ion

A reaction between an acid and a base produces a *salt,* such as sodium chloride:

$$HCl + NaOH \rightarrow NaCl + H_2O$$

The greater the concentration of hydrogen ions in a solution, the greater is the acidity of that solution. As the concentration of hydrogen ions becomes less than it is in pure water, the more alkaline (basic) the solution becomes. Acidity is indicated by pH units, which represent the concentration of hydrogen ions in a solution. These units are listed on a scale from 0 to 14, with 0 being the most acidic and 14 being the most basic (Fig. 2-5).

A pH of 7.0 is neutral, having an equal number of hydrogen and hydroxide ions. Each pH unit on the scale represents a 10-fold change in the number of hydrogen and hydroxide ions present. A solution registering 5.0 on the scale has 10 times the number of hydrogen ions as a solution that registers 6.0. A solution registering 9.0 has one tenth the number of

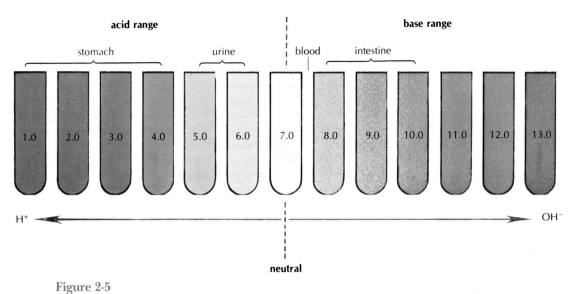

Figure 2-5

The pH scale measures degree of acidity or alkalinity.

hydrogen ions and 10 times the number of hydroxide ions as one registering 8.0. Thus, the lower the pH rating, the greater is the acidity, and the higher the pH, the greater is the alkalinity. Blood is only slightly alkaline, with a pH range of 7.35 to 7.45, or nearly neutral.

A delicate balance exists in the acidity or alkalinity of body fluids. If a person is to remain healthy, these chemical characteristics must remain within narrow limits. The substances responsible for the balanced chemical state are the acids, bases, and buffers. *Buffers* form a chemical system that prevents sharp changes in hydrogen ion concentration and thus maintains a relatively constant pH. Buffers are very important in maintaining the stability of body fluids.

▶ Radioactivity

No discussion of atoms is complete without reference to the part some play in the diagnosis and treatment of disease. Atoms of an element may exist in several forms, called *isotopes.* These forms are alike in their chemical reactions but different in weight (*e.g.,* heavy oxygen and regular oxygen). The greater weight of the heavier isotopes is due to the presence of one or more extra neutrons in the nucleus. Some isotopes are stable and maintain a constant charac-

ter. Others disintegrate (fall apart) as they give off small particles and rays; these are said to be *radioactive.* Radioactive elements may occur naturally, as is the case with such very heavy isotopes as radium and uranium. Others may be produced artificially from nonradioactive elements such as iodine and gold by bombardment (smashing) of the atoms in special machines.

The rays given off by some radioactive elements have the ability to penetrate and destroy tissues and so are used in the treatment of cancer. Radiation therapy is often given by means of special machines (such as linear accelerators) that are able to release particles to destroy tumors. The sensitivity of the younger cells and the dividing cells in a growing cancer allows selective destruction of the abnormal cells with minimal damage to normal tissues. Modern radiation instruments produce tremendous amounts of energy (in the multimillion electron-volt range) and yet can destroy deep-seated cancers without causing serious skin reactions.

Radioactive isotopes, such as cobalt 60, in the form of pellets, may be sealed in stainless steel cylinders. These cylinders are mounted on arms or cranes that permit the proper alignment for directing the beams through a porthole to the area to be treated.

In the form of needles, seeds, or tubes, implants containing radioactive isotopes are widely used in the treatment of many types of cancer.

In addition to its therapeutic values, irradiation is extensively used in diagnosis. X-rays penetrate tissues and produce an impression of their interior on a photographic plate. Radioactive iodine and other "tracers" taken orally or injected into the bloodstream are used to diagnose abnormalities of several body organs. Rigid precautions must be followed by hospital personnel to protect themselves and the patient when using irradiation in diagnosis or therapy because the rays can destroy both healthy and diseased tissues.

▶ The Chemistry of Living Matter

Of the 100 or more elements that have been found in nature, only a relatively small number, mostly the lighter ones, are important components of living cells. Hydrogen, oxygen, carbon, and nitrogen are the elements that make up about 99% of the cells. Calcium, sodium, potassium, phosphorus, sulfur, chlorine, and magnesium are the seven elements that make up most of the remaining 1% of tissue elements. A number of others are present in trace amounts, such as iron, copper, iodine, and fluorine.

The chemical compounds that characterize living things are called *organic compounds.* All of these contain the element *carbon.* Since carbon can combine with a variety of different elements and can even bond to other carbon atoms to form long chains, most organic compounds consist of large, complex molecules. The starch found in potatoes, the fat in the tissue under the skin, and many drugs are examples of organic compounds. These large molecules are often formed from simpler molecules called *building blocks,* which bond together in long chains. The main types of organic compounds are carbohydrates, fats, and proteins. All three contain carbon, hydrogen, and oxygen as their main ingredients.

Organic Compounds

Carbohydrates are the simple sugars, called *monosaccharides* (mon-o-SAK-ah-rides), or molecules made from simple sugars linked together. Carbohydrates in the form of sugars and starches are important sources of energy in the diet. Examples of carbohydrates in the body are the glucose that circulates in the blood as food for the cells and a storage form of glucose called *glycogen* (GLI-ko-jen).

Trace Elements

Trace elements are required in very small amounts by the body but are absolutely essential for health. Many are used in enzymes and vitamins.

Iodine is needed for the manufacture of thyroid hormones. Selenium helps to prevent damage to cells caused by the chemical process of oxidation. A deficiency of selenium leads to liver and muscular disorders. Growth is impaired if the body lacks chromium. This element works with insulin, the hormone that helps the cells to take up glucose from the blood. Copper is needed for the proper use of iron, and cobalt is a part of vitamin B_{12} needed for blood cell formation. Zinc is used in energy metabolism and for the transport of carbon dioxide, the waste gas of metabolism.

Other elements that are needed in trace amounts include silicon, fluorine, tin, and manganese. More information on trace elements may be discovered as nutrition studies continue.

Fats, or *lipids,* are also stored for energy in the body. In addition, they provide insulation and protection for the body organs. Fats are made from a substance called *glycerol* (glycerine) in combination with fatty acids. A group of complex lipids containing phosphorus, called *phospholipids* (fos-fo-LIP-ids), is important in the body. Among other things, phospholipids make up a major part of the membrane around living cells.

All proteins contain, in addition to carbon, hydrogen, and oxygen, the element *nitrogen.* They may also contain sulfur and phosphorus. Proteins are composed of building blocks called *amino* (ah-ME-no) *acids.* Although there are only about twenty different amino acids found in the body, a vast number of proteins can be made by linking them together in different combinations. Proteins are the structural materials of the body, found in muscle, bone, and connective tissue. They also make up the pigments that give hair, eyes, and skin their color. It is protein that makes each individual physically distinct from others. All three categories of organic compounds, in addition to minerals and vitamins, must be taken in as part of a normal diet. These compounds will be discussed further in Chapters 19 and 20.

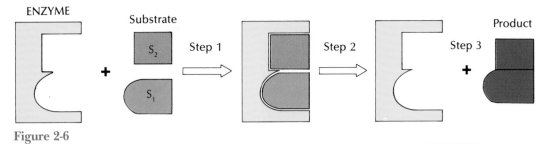

Figure 2-6

Diagram of enzyme action. The enzyme combines with substance 1 (S_1) and substance 2 (S_2). When a new product is formed, the enzyme is released unchanged.

Enzymes

 Enzymes enable metabolic reactions to occur at a fast enough rate to maintain life. These catalysts are needed because body conditions are mild. More extreme measures to speed chemical reactions, such as increased temperature or pressure, cannot be applied without damaging cells. Many different enzymes are needed, because each one works only on a specific substance and does only one specific chemical job.

The production of each enzyme is controlled by a gene, and in this way the genes govern all cell activities. Many hereditary disorders result from defects in the genes that control specific enzymes. These include PKU (phenylketonuria) and Tay Sachs disease. A common example of enzyme deficiency occurs in people who are described as "lactose intolerant." These people lack the enzyme lactase used to digest the sugar in milk (lactose). Eating dairy products causes cramps, diarrhea, and gas. Someone with lactose intolerance must avoid milk products, buy lactose-free milk, or add lactase to their milk.

Enzymes

An important group of proteins is the *enzymes*. Enzymes function as *catalysts* in the hundreds of reactions that occur in metabolism. A catalyst is a substance that speeds up the rate of a reaction but is not changed or used up in that reaction. Because they are constantly reused, only very small amounts of enzymes are needed. Each chemical reaction in the body requires a specific enzyme, so enzymes really control the metabolism of the cells. Some of the vitamins and minerals needed in the diet are important because they make up parts of enzymes.

In its action, the shape of the enzyme is important. It must match the shape of the substance or substances with which the enzyme combines in much the same way as a key fits a lock. This so-called "lock and key" mechanism is illustrated in Figure 2-6.

Harsh conditions, such as extremes of temperature or pH, can alter the shape of an enzyme and stop its action. Such an event is always harmful to the cells.

You can usually recognize the names of enzymes because, with few exceptions, they end with the suffix *-ase*. Examples are amylase, lipase, and oxidase. The first part of the name usually refers to the substance acted on or the type of reaction involved.

SUMMARY

I. **Atoms**—basic units of matter that make up the various elements on earth
  A. Structure
    1. Protons—positively charged particles in the nucleus
    2. Neutrons—neutral particles in the nucleus
    3. Electrons—negatively charged particles in energy levels around the nucleus
II. **Molecules**—combinations of two or more atoms
  A. Compounds—combinations of different atoms (*e.g.*, water)
III. **Mixtures**—blends of two or more substances

A. Solution—substance (solute) remains evenly distributed in solvent (*e.g.,* salt in water)

B. Suspension—material settles out of mixture on standing (*e.g.,* red cells in blood plasma)

C. Colloidal suspension—particles do not dissolve but remain suspended (*e.g.,* cytoplasm)

IV. Chemical bonds

A. Ionic bonds—atoms exchange electrons to become stable

B. Covalent bonds—atoms share electrons to become stable

V. Types of compounds

A. Acid—donates hydrogen ions

B. Base—accepts hydrogen ions

C. Salt—formed by reaction between acid and base

VI. pH

A. Measure of acidity or alkalinity of a solution

B. Scale goes from 0 to 14; 7 is neutral

C. Buffer—maintains constant pH of a solution

VII. Radioactive elements (isotopes)

A. Diagnosis

B. Cancer therapy

C. X-ray films

VIII. Organic compounds—all contain carbon

A. Examples

1. Carbohydrates

2. Lipids (*e.g.,* fats)

3. Proteins (*e.g.,* enzymes–organic catalysts)

QUESTIONS FOR STUDY AND REVIEW

1. Define *chemistry* and tell something about what is included in this study.
2. Define an atom and describe the basic structure of an atom.
3. Define *molecule, element, compound,* and *mixture.*
4. What is an element and what are some examples of elements?
5. Why is water so important to life?
6. How does a mixture differ from a compound? Give two examples of each.
7. You dissolve a teaspoon of sugar in a cup of tea. Which is the solute? the solvent?
8. What are ions and how are they related to electrolytes? What are some examples of ions and of what importance are they in the body?
9. Explain how covalent bonds are formed. Give two examples of covalently bonded compounds.
10. What would a pH reading of 6.5 indicate about a solution? 8.5? 7.0?
11. What is meant by *radioactivity* and what are some of its practical uses?
12. What are organic compounds and what element is found in all of them?
13. What elements are found in the largest amounts in living cells?
14. What are the building blocks of carbohydrates? fats? proteins?
15. What are enzymes? What type of organic compound are they?
16. Why is the shape of an enzyme important to its action?

CHAPTER 3

Cells and Their Functions

Selected Key Terms

The following terms are defined in the Glossary:

active transport

cancer

cell membrane

centriole

chromosome

cytoplasm

diffusion

DNA

interphase

isotonic

micrometer

mitochondria

mitosis

mutation

nucleotide

nucleus

organelle

osmosis

phagocytosis

ribosome

RNA

Behavioral Objectives

After careful study of this chapter, you should be able to:

- List three types of microscopes used to study cells
- Describe the function and composition of cytoplasm
- Name and describe the main organelles in the cell
- Give the composition, location, and function of DNA in the cell
- Give the composition, location, and function of RNA in the cell
- Explain briefly how cells make proteins
- Name and briefly describe the stages in cell division
- Define six methods by which substances enter and leave cells
- Explain what will happen if cells are placed in solutions with the same or different concentrations than the cell fluids
- Define *cancer*
- List several risk factors in cancer

Memmler, RL, Cohen, BJ, Wood, DL. *THE HUMAN BODY IN HEALTH AND DISEASE*, 8/e, © 1996 Lippincott-Raven Publishers.

The cell is the basic unit of all life. It is the simplest structure that shows all the characteristics of life: growth, metabolism, responsiveness, reproduction, and homeostasis. In fact, it is possible for a single cell to live independently of other cells. Examples of such independent cells are some of the organisms that produce disease and other microscopic organisms. All the activities of the human body, which is composed of millions of cells, result from the activities of individual cells. Any materials produced within the body are produced by cells.

The scientific study of cells began some 350 years ago with the invention by Anton van Leeuwenhoek of the microscope. In time, this single-lens microscope was replaced by the modern compound light microscope, which has two sets of lenses. This is the type in use in most laboratories. A great boon to cell biologists has been the development of the *transmission electron microscope,* which by a combination of magnification and enlargement of the resulting image affords magnification to 1 million times or more (Fig. 3-1). Another type, the *scanning electron microscope,* does not magnify as much (×250,000) but gives a three-dimensional picture of an object.

Before they are examined under the microscope, cells and tissues usually are colored with special dyes called *stains* to aid in viewing. These stains produce the variety of colors you see when looking at pictures of cells and tissues taken under a microscope. The metric unit used for microscopic measurements is the *micrometer* (MI-kro-me-ter). This unit is 1/1000 of a millimeter and is symbolized with the Greek letter for m as µm.

▶ Structure of the Cell

Just as people may look different but still have certain features in common—two eyes, a nose, and a mouth, for example—all cells share certain characteristics. A "typical" animal cell is shown in Fig. 3-2.

The outer covering of the cell is the *cell membrane,* also called the *plasma membrane* (Fig. 3-3). This membrane is composed of two layers of lipid molecules in which float a variety of different proteins and a small amount of carbohydrates. The proteins may act as receptors, that is, points of attachment for materials coming to the cell in the blood; they may also act as carriers, shuttling materials into or out of the cell. Some form channels through which selected substances can pass. The cell membrane is

very important in regulating what can enter and leave the cell.

The main substance that fills the cell and holds the cell contents is the *cytoplasm* (SI-to-plazm). This is a suspension of nutrients, minerals, enzymes, and other specialized materials in water. Although the composition of the cytoplasm has been analyzed, no one has been able to produce it in a laboratory; there must be something about the organization of these substances that has not yet been discovered.

▶ The Organelles

Just as the body has different organs to carry out special functions, the cell contains specialized subdivisions that perform different tasks. These structures are called *organelles,* which means "little organs." Each of these will be examined in turn (see Fig. 3-2).

The largest organelle is the *nucleus* (NU-kle-us). This is often called the *control center* of the cell because it contains the genetic material, which governs all the activities of the cell, including cell reproduction. Within the nucleus is a smaller globule called the *nucleolus* (nu-KLE-o-lus), which means "little nucleus." Its functions are not entirely understood, but it is believed to act in the manufacture of proteins within the cell. The actual formation of proteins occurs on small bodies called *ribosomes* (RI-bo-somes). These are attached to a network of membranes throughout the cell called the *endoplasmic reticulum* (en-do-PLAS-mik re-TIK-u-lum). The name literally means "network" (reticulum) "within the cytoplasm" (endoplasmic), but for ease it is almost always called simply the *ER.*

Fairly large and important organelles are the *mitochondria* (mi-to-KON-dre-ah). These are round or bean-shaped structures with folded membranes on the inside. Within the mitochondria, food is converted to energy for the cell in the form of ATP. These are the "power plants" of the cell. Active cells have large numbers of mitochondria. Other organelles in a typical cell include the *centrioles,* rod-shaped bodies near the nucleus that function in cell division; *lysosomes* (LI-so-somes), which contain digestive enzymes; and the *Golgi* (GOL-je) *apparatus,* which formulates special substances, such as mucus, released from cells. All these cell structures are summarized in Table 3-1 for easy study.

Although the basic structure of all body cells is the same, individual cells may vary widely in size,

(Text continues on page 29)

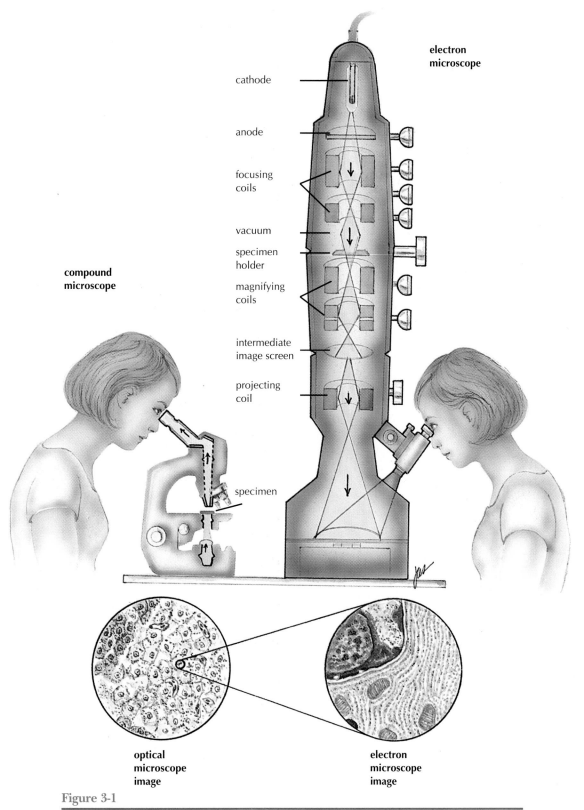

electron
microscope

cathode

anode

focusing
coils

vacuum

specimen
holder

magnifying
coils

intermediate
image screen

projecting
coil

compound
microscope

specimen

optical
microscope
image

electron
microscope
image

Figure 3-1

A simplified comparison of an optical (compound) microscope and an electron microscope.

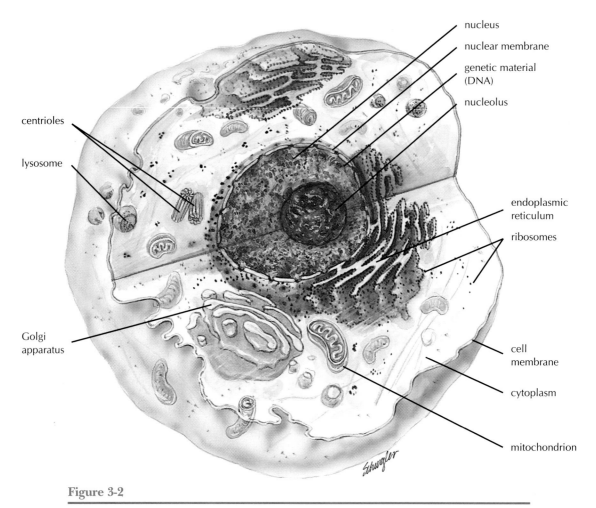

centrioles

lysosome

Golgi
apparatus

nucleus

nuclear membrane

genetic material
(DNA)

nucleolus

endoplasmic
reticulum

ribosomes

cell
membrane

cytoplasm

mitochondrion

Figure 3-2

Diagram of a typical animal cell showing the main organelles.

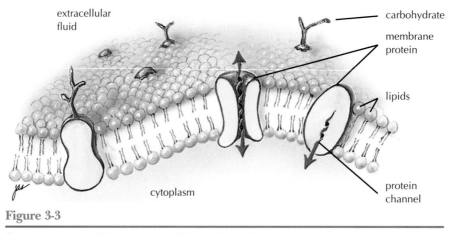

extracellular
fluid

carbohydrate

membrane
protein

lipids

protein
channel

cytoplasm

Figure 3-3

Current concept of the structure of the cell membrane.

Cell Junctions

Some cells, such as blood cells, are not in direct contact with other cells. Most cells, however, are held in contact with other cells. What keeps cells together? First, the cell membranes contain complex substances that are sticky and tend to bind cells to each other. Second, there are special intercellular junctions, which may vary according to the functions of the cells.

A *tight junction* is formed when the membranes of adjacent cells actually become fused. This type of junction is found in the lining of the digestive tract and acts to keep digestive juices and other harmful substances from damaging the organs. One substance that can seep through the tight junctions of the stomach and damage the deeper tissues is alcohol.

Connections can form between adjacent cells where tissues are under great stress, for example in the skin. Protein filaments extend from the cells, bridging the space between the membranes and anchoring cells to each other. These junctions are called *desmosomes* (DES-mo-somz).

Finally, *gap junctions* allow materials to pass between cells. Small tubes made of protein connect the cells and act as channels for exchange of chemicals. These are found where cells need to communicate easily, as in the nervous system.

shape, and composition according to the function of each. In size they may range from the 7 µm of a red blood cell to the 200 µm or more of a muscle cell. A neuron with its long fibers is very different in appearance from a muscle cell or from a transparent cell in the clear lens of the eye. Most human cells have all the organelles described, but these may vary in number. Some cells have small, hairlike projections from the surface called *cilia* (SIL-e-ah), which wave to create movement around the cell. Examples are the cells that line the passageways of the respiratory and reproductive tracts. A long, whiplike extension from the cell is a *flagellum* (flah-JEL-lum). Each human sperm cell has a flagellum that is used for locomotion. Thus, each cell is specialized for its particular function.

▶ Cell Functions

Protein Synthesis

Since protein molecules play an indispensable part in the body's activities, we need to identify the cellular substances that direct the production of protein. In the cytoplasm and in the nucleus are chemicals called *nucleic* (nu-KLE-ik) *acids.* The two nucleic acids important in protein production are *deoxyribonucleic* (de-ok-se-RI-bo-nu-kle-ik) *acid,* or *DNA,* and *ribonucleic* (RI-bo-nu-kle-ik) *acid,* abbreviated *RNA.* Both of these are large, complex molecules composed of subunits called *nucleotides* (NU-kle-o-tides) (Fig. 3-4). There are four nucleotides in DNA and four in RNA, but only three of these are common to both.

DNA

DNA molecules are found mostly in the nucleus of the cell, where they make up the *chromosomes* (KRO-mo-somes), dark-staining, threadlike bodies. Looking at Figure 3-4, you can see that DNA exists as a double strand. The two strands are matched according to the pattern of the nucleotides. The adenine (A) nucleotide always pairs with the thymine (T) nucleotide; the cytosine (C) nucleotide

Lysosomes

Lysosomes are small sacs in the cytoplasm of the cell that break down waste and foreign matter. They are most apparent in phagocytes, cells that engulf impurities in tissues and body fluids.

Lysosomes contain digestive enzymes that would be harmful if they escaped into the cell. Normally, these enzymes are held safely within the membrane of the lysosome. However, if the membrane breaks or weakens, the cell begins to digest itself in a process called *autolysis* (aw-TOL-ih-sis), literally "self destruction." Some investigators believe that autolysis is involved in cell aging and in a group of disorders described as *autoimmune diseases,* in which the body develops an immune response to its own tissues. Rheumatoid arthritis, a disorder of the joints, is one example of such a disease.

TABLE 3-1
Cell Structures

Name	Description	Function
Cell membrane	Outer layer of the cell; composed mainly of lipids and proteins	Limits the cell; regulates what enters and leaves the cell
Cytoplasm	Colloidal suspension that fills cell	Holds cell contents
Nucleus	Large, dark-staining body near the center of the cell; composed of DNA and proteins	Contains the chromosomes with the genes (the hereditary material that directs all cell activities)
Nucleolus	Small body in the nucleus; composed of RNA, DNA, and protein	Needed for protein manufacture
Endoplasmic reticulum (ER)	Network of membranes in the cytoplasm	Used for storage and transport; holds ribosomes
Ribosomes	Small bodies attached to the ER; composed of RNA and protein	Manufacture proteins
Mitochondria	Large organelles with folded membranes inside	Convert energy from nutrients into ATP
Golgi apparatus	Layers of membranes	Put together special substances such as mucus
Lysosomes	Small sacs of digestive enzymes	Digest substances within the cell
Centrioles	Rod-shaped bodies (usually 2) near the nucleus	Help separate the chromosomes in cell division
Cilia	Short, hairlike projections from the cell	Create movement around the cell
Flagellum	Long, whiplike extension from the cell	Moves the cell

always pairs with the guanine (G) nucleotide. The two strands of DNA are held together by weak bonds. The doubled strands are then coiled into a spiral, giving DNA the descriptive name of the *double helix.*

Specific regions of the DNA in the chromosomes make up the **genes,** the hereditary factors of each cell. It is the genes that carry the messages for the development of particular inherited characteristics, such as brown eyes or curly hair. Their message is actually contained in the pattern of the four nucleotides in the DNA. These code for the amino acids in the various cellular proteins. DNA is considered the master blueprint of the cell, and, because it contains the DNA, the nucleus is considered the cell's control center.

THE ROLE OF RNA

A blueprint is only a map. The directions it illustrates must be translated into appropriate actions. RNA is the substance needed for this step. RNA is much like DNA except that it is a single strand and has one nucleotide that is different from those in

DNA. By pairing of nucleotides, RNA picks up the message of the DNA, which has by now broken its weak bonds and uncoiled into single strands. This *messenger RNA* travels into the cytoplasm to the ribosomes attached to the ER. Ribosomes are the organelles responsible for protein manufacture. They are so named because they are composed mainly of RNA. At the ribosomes the genetic message is decoded to produce chains of amino acids, the building blocks of proteins.

Cell Division

To ensure that every cell in the body has similar genetic information, cell reproduction occurs by a very precise division process (Fig. 3-5). In this process of **mitosis** (mi-TO-sis), each original parent cell divides to form two identical daughter cells. Before mitosis can occur, however, the genetic information in the parent cell must be doubled. This duplication occurs during the period between mitoses that is termed **interphase.** While DNA is uncoiled from its double-stranded form, each strand

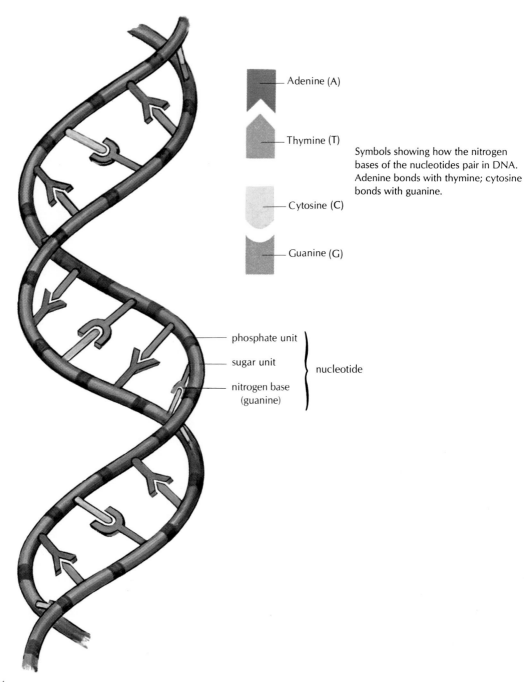

Adenine (A)

Thymine (T)

Symbols showing how the nitrogen bases of the nucleotides pair in DNA. Adenine bonds with thymine; cytosine bonds with guanine.

Cytosine (C)

Guanine (G)

phosphate unit

sugar unit

nitrogen base (guanine)

} nucleotide

Figure 3-4

Schematic representation of the basic structure of a DNA molecule. Each structural unit, or nucleotide, consists of a phosphorus-containing unit and a sugar unit to which is attached a nitrogen base. There are four different nucleotides. Their arrangement "spells out" the genetic instructions that control all activities of the cell. The symbols in the upper right show how the nucleotides pair in DNA.

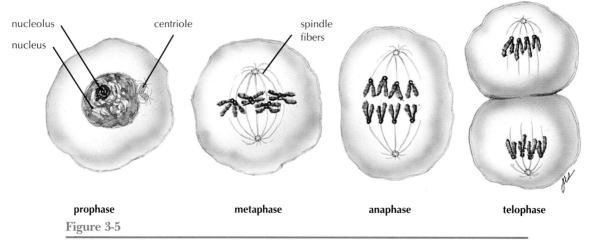

prophase   **metaphase**   **anaphase**   **telophase**

Figure 3-5

The stages of mitosis. The cell shown is for illustration only. It is not a human cell, which has 46 chromosomes.

duplicates itself according to the pattern of the nucleotides.

Mitosis itself is described as occurring in four stages, during which distinct changes can be seen in the dividing cell. In *prophase* (PRO-faze), the doubled strands return to their spiral organization and become visible under the microscope as dark, threadlike chromosomes. Meanwhile, in the cytoplasm, the two centrioles move to opposite ends of the cell, trailing very thin threadlike substances that form a structure resembling a spindle stretched across the cell. In *metaphase* (MET-ah-faze), the chromosomes line up across the threadlike spindle. Then, in *anaphase* (AN-ah-faze), the duplicated chromosomes separate and begin to move toward opposite ends of the cell. As mitosis continues into *telophase* (TEL-o-faze), the nuclear area becomes pinched in the middle until two nuclear regions have formed. A similar change occurs in the cell membrane, making the cell resemble a dumbbell. The midsection between the two halves of the dumbbell becomes progressively smaller until, finally, the cell splits in two. There are now two identical daughter cells, which are themselves identical to, but smaller than, the parent cell. The two new cells will grow and mature and continue to carry out life functions.

During mitosis, all the organelles, except those needed for the division process, temporarily disappear. After the cell splits, these organelles reappear in each daughter cell. Also at this time, the centrioles usually duplicate in preparation for the next cell division.

Body cells differ in the rate at which they reproduce. Some, such as nerve cells and muscle cells, stop dividing at some point in development and are not replaced if they die. They remain in interphase. Others, such as skin cells, multiply rapidly to replace cells destroyed by injury, disease, or natural wear-and-tear. As a person ages, characteristic changes in the overall activity of his or her body cells take place. One example of these is the slowing down of repair processes. A bone fracture, for example, takes considerably longer to heal in an aged person than in a young one.

Movement of Substances Across the Cell Membrane

To function, cells require nutrients. How to receive this nourishment would seem to be a problem, since cells are surrounded by a protective membrane. However, if a cell is bathed in a liquid containing dissolved nutrient materials, an interesting thing happens: the liquid with the dissolved nutrient particles passes through the cell membrane. Not only do these molecules pass in, but waste products pass out of the cell in the opposite direction, enabling the cell to perform the function of elimination. The membrane also keeps valuable proteins and other substances from leaving the cell and prevents the admission of undesirable substances. For this reason, the cell membrane is classified as a *semipermeable* (sem-e-PER-me-ah-bl) membrane. It is permeable or passable to some molecules but impassable to others.

Water, a tiny molecule, is always able to penetrate the membrane with ease. In contrast, other molecules cannot pass through because they are too large, even though they are readily soluble. It is by the process of digestion that large molecules are split into smaller molecules so that they can travel through the membrane. For example, sucrose (table sugar) is converted to glucose, a smaller molecule that is a major source of cellular energy.

Various physical processes are responsible for exchanges through cell membranes. Some of these processes are:

1. *Diffusion,* the constant movement of molecules from a region of relatively higher concentration to one of lower concentration. Molecules, especially those in solution, tend to spread throughout an area until they are equally concentrated in all parts of a container (Figs. 3-6 and 3-7). When substances diffuse through a membrane, such as the intact cell membrane, passage is limited to those particles small enough to pass through spaces in the membrane. A large-scale example is shown in Figure 3-8.

2. *Osmosis.* Of all substances, water moves most rapidly through the cell membrane. The term *osmosis* means specifically the diffusion of water through a semipermeable membrane. The water molecules move, as expected, from an area where they are in higher number to an area where they are in lower number. That is, they move from a more dilute solution into a more concentrated solution. The tendency of a solution to pull water into it is called the *osmotic pressure* of the solution. This force is directly related to concentration: the higher the concentration of a solution, the greater is its tendency to pull water into it.

3. *Filtration,* the passage of water containing dissolved materials through a membrane as a result of a mechanical ("pushing") force on one side (Fig. 3-9). An example of filtration in the human body is the formation of urine in the microscopic functional units of the kidney (see Chap. 22).

Figure 3-6

Diffusion of gaseous molecules throughout a given space. When a solution vaporizes, the molecules tend to spread throughout the area. Common examples are the diffusion of perfumes, oils, and cleaning solutions.

Figure 3-7

Diffusion of a solid in a liquid. Here, a solid diffuses into water. The molecules of the solid tend to spread evenly throughout the water.

The three processes just described do not require cellular energy. They depend on the natural energy of the molecules for movement. The next three processes to be described do require cellular energy.

4. *Active transport.* Often molecules move into or out of a living cell in a direction opposite to the way they would normally flow by diffusion. That is, they move from an area where they are in rela-

tively lower concentration to an area where they are in higher concentration. This movement, because it is against the natural flow, requires energy in the form of ATP. It also requires proteins in the cell membrane that act as *carriers* for the molecules. This process, called *active transport,* is a function of the living cell membrane. It allows the cell to take in what it needs from the surrounding fluids and to release materials from the cell. Because the cell membrane can carry on active trans-

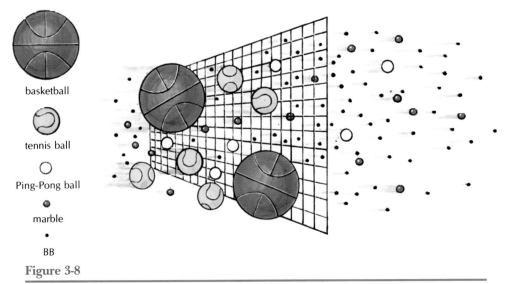

basketball

tennis ball

Ping-Pong ball

marble

BB

Figure 3-8

Diffusion through a semipermeable membrane. In this example, large objects (basketballs and tennis balls) cannot pass through the net. In the human body, large particles (proteins and blood cells) do not pass through intact membranes, such as the walls of the blood capillaries.

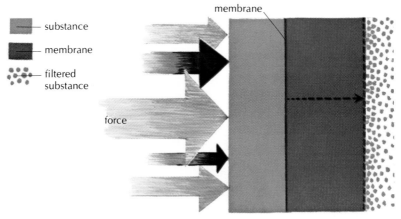

- substance
- membrane
- filtered substance

membrane

force

Figure 3-9

Filtration. A mechanical force pushes a substance through a membrane.

port, it is most accurately described as **selectively permeable.** It regulates what can enter and leave the cell based on the needs of the cell.

5. **Phagocytosis** (fag-o-si-TO-sis) is the engulfing of relatively large particles by the cell membrane and the movement of these particles into the cell. Certain white blood cells carry out phagocytosis to rid the body of foreign material and dead cells.

6. In **pinocytosis** (pi-no-si-TO-sis), droplets of fluid are engulfed by the cell membrane. This is a way for large protein molecules in suspension to travel into the cell. The word *pinocytosis* means "cell drinking."

HOW OSMOSIS AFFECTS CELLS

As stated earlier, water moves very easily through the cell membrane. Therefore, for a normal fluid balance to be maintained, all cells must be kept in solutions that have the same concentration of molecules as the fluids within the cell—the intracellular fluids (Fig. 3-10). Such equivalent solutions are described as **isotonic.** Tissue fluids and blood plasma are isotonic for body cells. Manufactured solutions that are isotonic for the cells and can thus be used to replace body fluids include 0.9% salt or **normal saline** and 5% dextrose (glucose).

A solution that is less concentrated than the intracellular fluid is described as **hypotonic.** A cell placed in a hypotonic solution draws water in, swells, and may burst. When this occurs to a red blood cell, the cell is said to **hemolyze** (HE-mo-lize). If a cell is placed in a **hypertonic** solution, which is more concentrated than the cell fluids, it loses water to the surrounding fluids and shrinks. Osmosis affects the total amount and distribution of body fluids, as discussed in Chapter 21.

▶ Cells and Cancer

Through the study of cells we gain some insight into the laws of growth. For reasons that remain obscure, cells develop various forms, and those that are of the same kind congregate to form one of the basic tissues. These tissues, in turn, become the specialized organs. In the early stages of the body's development, the cells multiply rapidly, and hence the body grows until a point of maximum growth has been reached. From here on, cell division is not so rapid. It continues, however, at a rate sufficient to replace cells that for one reason or another have become damaged and discarded. In this manner, the tissues are maintained, and every cell and formation of cells has its purpose.

Occasionally, some change, or **mutation,** occurs in the genetic material (DNA) of a cell. If such a cell does not die naturally or get destroyed by the immune system, it may begin to multiply out of control and spread to other tissues, producing **cancer.** Cancer cells form tumors, which interfere with normal functions, crowding out normal cells and robbing them of nutrients. There is more information on the various types of tumors in Chapter 4.

The causes of cancer are very complex, involving the interaction of factors in the cell and the environment. Because cancer may take a long time to develop, it is often difficult to identify its cause or

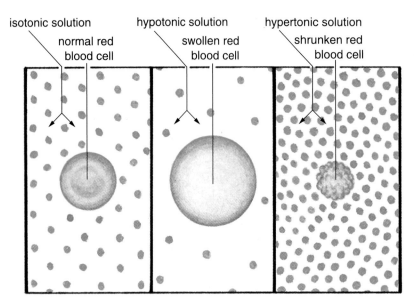

isotonic solution
normal red blood cell

hypotonic solution
swollen red blood cell

hypertonic solution
shrunken red blood cell

Figure 3-10

Osmosis. Water molecules moving through a red blood cell membrane in three different concentrations of fluid. *(Left)* The normal saline solution has a concentration nearly the same as that inside the cell, and water molecules move into and out of the cell at the same rate. *(Center)* The dilute solution causes the cell to swell and eventually hemolyze (burst) because of the large number of water molecules moving into the cell. *(Right)* The concentrated solution causes the water molecules to move out of the cell, leaving it shrunken.

causes. Certain forces increase the chances of developing the disease and are considered risk factors. These include:

1. ***Heredity.*** Certain types of cancer occur more frequently in some families than in others, indicating that there is some inherited predisposition to the development of cancer.
2. ***Chemicals.*** Certain industrial and environmental chemicals are known to increase the risk of cancer. Any chemical that causes cancer is called a ***carcinogen*** (kar-SIN-o-jen). The most common carcinogens in our society are those present in cigarette smoke. Carcinogens are also present, both naturally and as additives, in foods. Certain drugs may be carcinogenic.
3. ***Ionizing radiation.*** Certain types of radiation can produce damage to cell DNA that may lead to

cancer. These include x-rays, rays from radioactive substances, and ultraviolet rays. For example, the ultraviolet rays received from exposure to the sun are very harmful to the skin.

4. ***Physical irritation.*** Continued irritation, such as the intake of hot foods or the contact of a hot pipestem on the lip, increases cell division and thus increases the chance of mutation.
5. ***Diet.*** It has been shown that diets high in fats and total calories are associated with an increased occurrence of certain forms of cancer. A general lack of fiber and insufficient amounts of certain fruits and vegetables in the diet can leave one susceptible to cancers of the digestive tract.
6. ***Viruses*** have been implicated in cancers of the liver, the blood (leukemias), and lymphatic tissues (lymphomas).

SUMMARY

I. Microscopes
  A. Types
    1. Compound light microscope
    2. Transmission electron microscope—magnifies up to 1 million times
    3. Scanning electron microscope—gives three-dimensional image
  B. Stains—dyes used to aid in viewing cells under the microscope

  C. Micrometer—metric unit commonly used for microscopic measurements
II. Cell structure
  A. Cell membrane—regulates what enters and leaves cell
  B. Cytoplasm—colloidal suspension that holds organelles
  C. Organelles—subdivisions that carry out special functions

1. Nucleus, nucleolus, ribosomes, centrioles
2. Mitochondria, ER, lysosomes, Golgi apparatus
3. Cilia, flagellum

III. Cell functions

A. Protein synthesis—carried out by nucleic acids
  1. DNA
     a. Double strand of nucleotides
     b. Located in the nucleus
     c. Carries the genetic message
  2. RNA
     a. Single strand of nucleotides
     b. Located in the cytoplasm
     c. Translates DNA message into proteins
     d. Ribosomes—site of protein synthesis
B. Cell division
  1. Duplication of chromosomes during interphase
  2. Mitosis—separation of chromosomes and division into two identical daughter cells
  3. Four stages—prophase, metaphase, anaphase, telophase
C. Movement of substances across cell membrane
  1. Do not require cell energy
     a. Diffusion—molecules move from area of higher concentration to area of lower concentration
     b. Osmosis—diffusion of water through semipermeable membrane
     c. Filtration—movement of materials through cell membrane under mechanical force
  2. Require cell energy
     a. Active transport—movement of molecules from area of lower concentration to area of higher concentration
     b. Phagocytosis—engulfing of large particles by cell membrane
     c. Pinocytosis—intake of droplets of fluid
D. Effect of osmosis on cells
  1. Isotonic solution—same concentration as cell fluids; cell remains the same
  2. Hypotonic solution—lower concentration than cell fluids; cell swells
  3. Hypertonic solution—higher concentration than cell fluids; cell shrinks

IV. Cells and cancer

A. Mutation—change in DNA
B. Risk factors
  1. Heredity
  2. Chemicals—carcinogens
  3. Ionizing radiation
  4. Physical irritation
  5. Diet
  6. Viruses

QUESTIONS FOR STUDY AND REVIEW

1. Why is the study of cells so important in the study of the body?
2. Define the term *organelle*. List 10 organelles found in cells and give the function of each.
3. Compare DNA and RNA with respect to location in the cell and composition.
4. What are chromosomes? Where are they located in the cell and what is their function?
5. Explain the role of each of the following in protein synthesis: DNA, nucleotide, RNA, ribosomes.
6. Name the stage between one mitosis and the next.
7. Name the process of cell division. Name the stages of cell division and describe what happens during each.
8. Explain why it is necessary to reduce large molecules to smaller molecules by the process of digestion.
9. List and define six methods by which materials cross the cell membrane. Which of these requires cellular energy?
10. What substance moves most rapidly through the cell membrane?
11. Why is the cell membrane described as selectively permeable?
12. What is meant by the term *isotonic*? Name four isotonic solutions.
13. What will happen to a red blood cell placed in a 5.0% salt solution? in distilled water?
14. Define *mutation*.
15. List six risk factors associated with cancer.

CHAPTER 4

Tissues, Glands, and Membranes

Selected Key Terms
The following terms are defined in the Glossary:

adipose

areolar

benign

cartilage

collagen

endocrine

epithelium

exocrine

fascia

histology

malignant

membrane

metastasis

mucosa

myelin

neoplasm

neuroglia

neuron

serosa

Behavioral Objectives
After careful study of this chapter, you should be able to:

- Name the four main groups of tissues and give the location and general characteristics of each
- Describe the difference between exocrine and endocrine glands and give examples of each
- Give examples of soft, fibrous, hard, and liquid connective tissues
- Name four types of membranes and give the location and functions of each
- Describe two types of epithelial membranes
- Explain the difference between benign and malignant tumors and give several examples of each type
- List four methods of diagnosing cancer
- List four methods of treating cancer

Memmler, RL, Cohen, BJ, Wood, DL. *THE HUMAN BODY IN HEALTH AND DISEASE*, 8/e,
© 1996 Lippincott-Raven Publishers.

39

Tissues are groups of cells similar in structure, arranged in a characteristic pattern, and specialized for the performance of specific tasks. The study of tissues is known as *histology* (his-TOL-o-je). The tissues in our bodies might be compared with the different materials used to construct a building. Think for a moment of the great variety of materials used according to need—wood, stone, steel, plaster, insulation, and so forth. Each of these has different properties, but together they contribute to the building as a whole. The same may be said of tissues in the body.

Tissue Classification

The four main groups of tissue are the following:

1. *Epithelial* (ep-ih-THE-le-al) *tissue* covers surfaces, lines cavities, and forms glands.
2. *Connective tissue* supports and forms the framework of all parts of the body.
3. *Muscle tissue* contracts and produces movement.
4. *Nervous tissue* conducts nerve impulses.

This chapter concentrates mainly on epithelial and connective tissues; the other types of tissues receive more attention in later chapters.

Epithelial Tissue

Epithelial tissue, or *epithelium* (ep-ih-THE-le-um), forms a protective covering for the body and all the organs. It is the main tissue of the outer layer of the skin. It forms the lining of the intestinal tract, the respiratory and urinary passages, the blood vessels, the uterus, and other body cavities.

Epithelium has many forms and many purposes, and the cells of which it is composed vary accordingly. Epithelial tissue is classified according to the shape and the arrangement of its cells (Fig. 4-1). The cells may be

1. *Squamous* (SKWA-mus)—flat and irregular
2. *Cuboidal*—square
3. *Columnar*—long and narrow.

They may be arranged in a single layer, described as *simple,* or in many layers, termed *stratified.* Thus, a single layer of flat, irregular cells would be described as *simple squamous epithelium,* whereas tissue with many layers of these same cells would be described as *stratified squamous epithelium.*

The cells of some kinds of epithelium produce secretions, such as *mucus* (MU-kus) (a clear, sticky fluid), digestive juices, sweat, and other substances. The digestive tract is lined with a special kind of epithelium, the cells of which not only produce secretions but also are designed to absorb digested foods. The air that we breathe passes over yet another form of epithelium that lines the respiratory tract. This lining secretes mucus and is provided with tiny hairlike projections called *cilia.* Together, the mucus and the cilia help trap bits of dust and other foreign particles that could otherwise reach the lungs and damage them.

Some organs, such as the urinary bladder, must vary a great deal in size during the course of their work; for this purpose there is a special wrinkled, crepe-like tissue, called *transitional epithelium,* which is capable of great expansion yet will return to its original form once tension is relaxed—as when, in this case, the bladder is emptied. Certain areas of the epithelium that form the outer layer of the skin are capable of modifying themselves for greater strength whenever they are subjected to unusual wear and tear; the growth of calluses is a good example of this.

Epithelium repairs itself very quickly if it is injured. If, for example, there is a cut, the cells near and around the wound immediately form daughter cells, which grow until the cut is closed. Epithelial tissue reproduces frequently in areas of the body subject to normal wear and tear, such as the skin, the inside of the mouth, and the lining of the intestinal tract.

Glands

The active cells of many glands are epithelial tissue. A gland is a group of cells specialized to produce a substance that is sent out to other parts of the body. These secretions are manufactured from blood constituents.

Glands are divided into two categories:

1. *Exocrine* (EK-so-krin) *glands* have ducts or tubes to carry the secretion from the gland to another organ, to a body cavity, or to the body surface. These *external secretions* are effective in a limited area near their origins. Examples include the digestive juices, the secretions from the sebaceous (oil) glands of the skin, and tears from the lacrimal glands. These secretions are discussed in the chapters on specific systems.

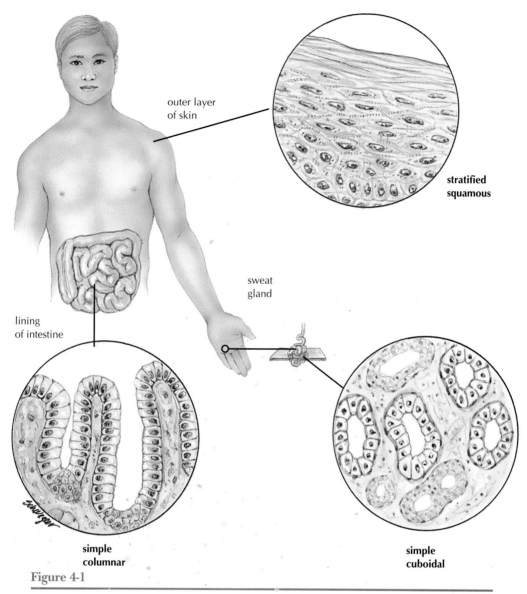

Figure 4-1

Three types of epithelium.

2. ***Endocrine*** (EN-do-krin) ***glands*** depend on blood flowing through the gland to carry the secretion to another organ. These secretions, called ***hormones,*** have specific effects on other tissues. Endocrine glands are the so-called "ductless glands."

Exocrine glands vary in design from very simple depressions resembling tiny dimples to more complex structures. Simple tubelike glands are found in the stomach wall and in the intestinal lining. Complex glands composed of treelike groups of ducts are found in the liver, the pancreas, and the salivary glands. Most glands are made largely of epithelial tissue with a framework of connective tissue. There may be a tough connective tissue capsule (a fibrous envelope) enclosing the gland, with extensions into the organ that form partitions. Between the partitions are groups of cells that unite to form lobes.

Endocrine glands produce ***internal secretions,*** which are carried to all parts of the body by the blood. These substances often affect tissues at a considerable distance from the point of origin. Endocrine glands, because they secrete directly into the bloodstream, have an extensive blood vessel network. The organs believed to have the richest blood supply in the body are the tiny adrenal glands located near the upper part of the kidneys. Some glands, such as the pancreas, have both endocrine and exocrine functions and therefore have both ducts and a rich blood supply. Hormones and the glands that produce them are discussed in Chapter 12.

▶ Connective Tissue

The supporting fabric of all parts of the body is connective tissue (Fig. 4-2). This is so extensive and widely distributed that if we were able to dissolve all the tissues except connective tissue, we would still be able to recognize the contours of the parts and the organs of the entire body.

Connective tissue has large amounts of nonliving material, called *intercellular material*, between the cells. This material contains varying amounts of water, fibers, and hard minerals. Connective tissue may be classified simply according to its degree of hardness:

1. Soft connective tissue—loosely held together with semi-liquid material between the cells; includes adipose (fat) tissue and loose (areolar) connective tissue
2. Fibrous connective tissue—Most connective tissue contains fibers, but this type is densely packed with fibers. Examples are ligaments, tendons, and fascia.
3. Hard connective tissue—has a very firm consistency, as in cartilage, or is hardened, as in bone, by minerals in the intercellular material.
4. Liquid connective tissue—Blood and lymph (the fluid that circulates in the lymphatic system) are examples of liquid connective tissues. Cells are suspended in a fluid environment.

Soft Connective Tissue

The loose, or *areolar* (ah-RE-o-lar), form of connective tissue is found in membranes around vessels and organs, between muscles, and under the skin. It is the most common type of connective tissue in the body. It contains cells and fibers in a very loose, jellylike background material. *Adipose* (AD-ih-pose) *tissue* contains cells that are able to store large amounts of fat. This tissue is used as a reserve of energy for

the body. Fat also serves as a heat insulator and as padding for various structures, such as organs and joints.

Fibrous Connective Tissue

Fibrous connective tissue is very dense and has large numbers of fibers that give it strength and flexibility. The main type of fiber in this and other connective tissues is *collagen* (KOL-ah-jen), a flexible white protein. Some tissues, such as the tissue that makes up the vocal cords, consist mainly of elastic fibers that allow the tissue to stretch. This type of connective tissue serves as a binding between organs and also as a framework for some organs. Particularly strong forms make up the tough *capsules* around certain organs, such as the kidneys, liver, and glands, the deep fascia around muscles, and the *periosteum* (per-e-OS-te-um) around bones. If the fibers in the connective tissue are all arranged in the same direction, like the strands of a cable, the tissue can pull in one direction. Examples are the cordlike *tendons,* which connect muscles to bones, and the *ligaments,* which connect bones to other bones.

FASCIA

The word *fascia* (FASH-e-ah) means "band"; hence, fascial membranes are bands or sheets that support the organs and hold them in place. An example of a fascia is the continuous sheet of tissue that underlies the skin. This contains fat (adipose tissue) and is called the *superficial fascia. Superficial* refers to a surface; the superficial fascia is closer than any other kind of fascia to the surface of the body.

As we penetrate more deeply into the body, we find examples of the *deep fascia,* which contains no fat and has many different purposes. Deep fascia covers and protects the muscle tissue with coverings known as *muscle sheaths.* The blood vessels and the nerves are sheathed with fascia; the brain and the spinal cord are encased in a multilayered covering called the *meninges* (men-IN-jeze). In addition, fascia serves to anchor muscle tissue to structures such as the bones.

TISSUE REPAIR

Like epithelial tissue, fibrous connective tissue can repair itself easily. A large, gaping wound requires a correspondingly large growth of this new connective tissue; such new growth is called *scar tissue.* Excess production of collagen in the formation

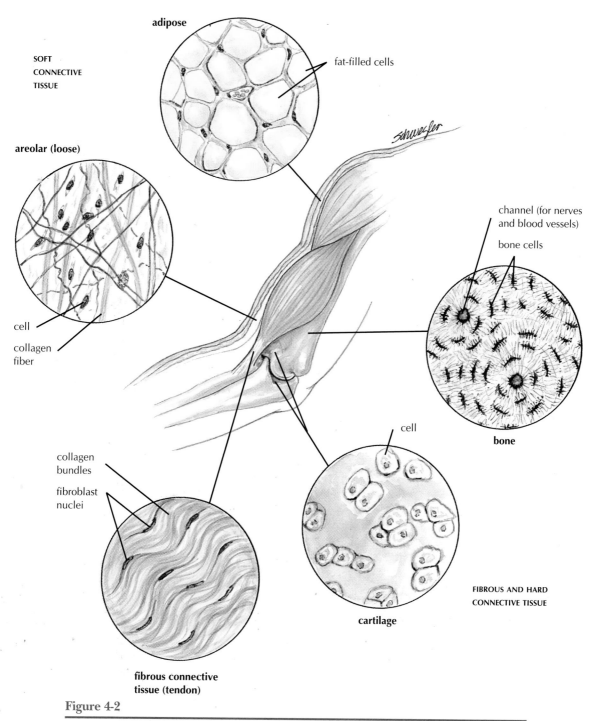

SOFT CONNECTIVE TISSUE

adipose

fat-filled cells

areolar (loose)

cell

collagen fiber

channel (for nerves and blood vessels)

bone cells

bone

cell

collagen bundles

fibroblast nuclei

FIBROUS AND HARD CONNECTIVE TISSUE

fibrous connective tissue (tendon)

cartilage

Figure 4-2

Connective tissue.

Collagen: The Body's Scaffolding

The most abundant protein in the body, making up about 25% of total protein, is the tough, flexible, white material known as *collagen*. Its name comes from a Greek word meaning "glue," indicating that it is the main structural substance in the body. Collagen is the major ingredient in all connective tissue.

The arrangement of the fibers in collagen gives this substance its different properties. It may form a fibrous connective tissue, as in the heart valves, a cord of great strength, as in tendons and ligaments, or a tough, transparent tissue, as in the cornea of the eye. It is collagen that gives strength and resilience to the skin.

The varied properties of collagen are evident in the preparation of a gelatin dessert. Gelatin is a collagen extract made by boiling animal bones and other connective tissue. It is a viscous liquid in hot water but forms a semisolid gel on cooling.

of a scar may result in the development of *keloids* (KE-loyds), sharply raised areas on the surface of the skin. These are not dangerous but may be removed for the sake of appearance. The process of repair includes stages in which new blood vessels are formed in the wound, followed by the growth of the scar tissue. Overdevelopment of the blood vessels in the early stages of repair may lead to the formation of excess tissue. Normally, however, the blood vessels are gradually replaced by white fibrous connective tissue, which forms the scar. Suturing (sewing) the edges of a clean wound together, as is done in the case of operative wounds, decreases the amount of scar tissue needed and hence reduces the size of the resulting scar. Such scar tissue may be stronger than the original tissue.

Hard Connective Tissue

The hard connective tissues, which as the name suggests are more solid than the other groups, include cartilage and bone. A common form of cartilage is the tough, elastic, translucent material popularly called *gristle*. This and other forms of cartilage are found in such places as between the segments of the spine and at the ends of the long bones. In these positions, cartilage acts as a shock absorber and as a

bearing surface that reduces friction between moving parts. Cartilage is found in other structures, including the nose, the ear, and parts of the larynx, or "voice box." Except in joint cavities, cartilage is covered by a layer of fibrous connective tissue called *perichondrium* (per-e-KON-dre-um).

The tissue of which bones are made, called *osseous* (OS-e-us) *tissue*, is very much like cartilage in its cellular structure. In fact, the skeleton of the fetus in the early stages of development is made almost entirely of cartilage. This tissue gradually becomes impregnated with the calcium salts that make bone characteristically solid and hard. Within the bones are nerves, blood vessels, bone-forming cells, and a special form of tissue, bone marrow, in which blood cells are manufactured.

The liquid connective tissues, blood and lymph, are discussed in Chapters 13 and 16, respectively.

▶ Muscle Tissue

Muscle tissue is designed to produce movement by a forcible contraction. The cells of muscle tissue are threadlike and thus are called *muscle fibers*. If a piece of well-cooked meat is pulled apart, small groups of

Wound Healing

The recovery of tissue after injury involves two processes: the regeneration of tissue and the growth of fibrous tissue to form a scar.

Tissue repair begins after blood has clotted and a scab has formed at the surface of the wound. From damaged capillaries, new vessels branch and grow into the injured tissue. Fibroblasts (cells that produce fibers) manufacture collagen to close the gap made by the wound. This material, formed early in the repair process, will become the scar of the wound.

While tissue is organizing below the surface, the upper layer of the skin, the epidermis, is regenerating between the scab and the underlying fibrous tissue. This layer finally is completely restored, releasing the scab. The underlying scar tissue may then continue to show at the surface as a white line. Scar tissue is strong, but it is not as flexible as normal tissue and does not function like the tissue it replaces.

these muscle fibers may be seen. Muscle tissue is usually classified as follows (Fig. 4-3):

1. ***Skeletal muscle,*** which combines with connective tissue structures such as tendons (discussed later along with the skeleton) to provide for movement of the body. This type of tissue is also known as ***voluntary muscle,*** since it can be made to contract by an act of will. In other words, in theory at least, any of your skeletal muscles can be made to contract as you want them to.

   The next two groups of muscle tissue are known as ***involuntary muscle*** because they typically contract independently of the will. In fact, most of the time we do not think of their actions at all. These are:

2. ***Cardiac muscle,*** which forms the bulk of the heart wall and is known also as ***myocardium*** (mi-o-KAR-de-um). This is the muscle that produces the regular contractions known as *heartbeats.*

3. ***Smooth muscle,*** known also as ***visceral muscle,*** which forms the walls of the ***viscera*** (VIS-er-ah), or organs of the ventral body cavities (with the exception of the heart). Some examples of visceral muscles are those that move food and waste materials along the digestive tract. Visceral muscles

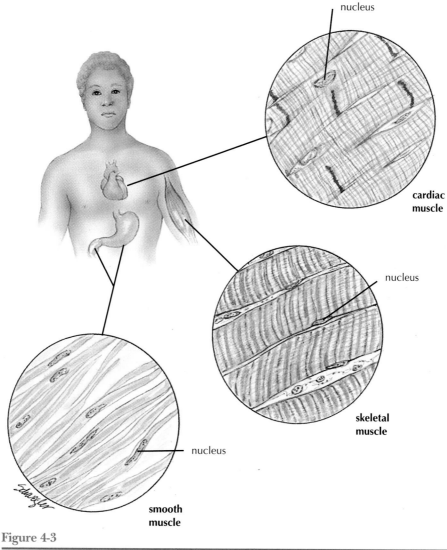

nucleus

cardiac
muscle

nucleus

skeletal
muscle

nucleus

smooth
muscle

Figure 4-3

Muscle tissue.

are found in other kinds of structures as well. Many tubular structures contain them, such as the blood vessels and the tubes that carry urine from the kidneys. Even certain structures at the base of body hair have this type of muscle. When these muscles contract, the skin condition we call *gooseflesh* results. Other types of visceral muscles are discussed under the various body systems.

Muscle tissue, like nervous tissue, repairs itself only with difficulty or not at all once an injury has been sustained. When injured, muscle tissue is frequently replaced with connective tissue.

▶ Nervous Tissue

The human body is made up of countless structures, both large and small, each of which contributes something to the action of the whole organism. This aggregation of structures might be compared to an army. For all the members of the army to work together, there must be a central coordinating and order-giving agency somewhere; otherwise chaos would ensue. In the body this central agency is the **brain.** Each structure of the body is in direct communication with the brain by means of its own set of telephone wires, called **nerves.** The nerves from even the most remote parts of the body all come together and form a great trunk cable called the **spinal cord,** which in turn leads directly into the central switchboard of the brain. Here, messages come in and orders go out 24 hours a day. This entire communication system, brain and all, is made of nervous tissue.

The basic unit of nervous tissue is the **neuron** (NU-ron), or nerve cell (Fig. 4-4). A neuron consists of a nerve cell body plus small branches, like those of a tree, called *fibers.* One type of fiber, the **dendrite** (DEN-drite), carries nerve impulses, or messages, to the nerve cell body. A single fiber, the **axon** (AK-son), carries impulses away from the nerve cell body. Neurons may be quite long; their fibers can extend for several feet. A nerve is a bundle of nerve cell fibers held together with connective tissue.

Just as telephone wires are insulated to keep them from being short-circuited, some axons are insulated and protected by a coating of material called **myelin** (MI-eh-lin). Groups of myelinated fibers form "white matter," so called because of the color of the myelin, which is very much like fat in appearance and consistency. Not all neurons have myelin, however; some axons are unmyelinated, as are all dendrites and all cell bodies. These areas appear gray in color. Since the outer layer of the brain has large collections of cell bodies and unmyelinated fibers, the brain is popularly termed *gray matter.*

Nervous tissue is supported by ordinary connective tissue everywhere except in the brain and spinal cord. Here, the supporting cells are **neuroglia** (nu-ROG-le-ah), which have a protective function as well. Neuroglia, also called *glial cells,* are named from the Greek word *glia* meaning "glue." There is a more detailed discussion of nervous tissue in Chapter 9.

▶ Membranes

Membranes are thin sheets of tissue. Their properties vary: some are fragile, others tough; some are transparent, others opaque (*i.e.,* they cannot be seen through). Membranes may cover a surface, may serve as dividing partitions, may line hollow organs and body cavities, or may anchor various organs. They may contain cells that secrete lubricants to ease the movement of organs such as the heart and the movement of joints.

Epithelial Membranes
Epithelial membrane is so named because its outer surface is made of epithelium. Underneath, however, there is a layer of connective tissue that strengthens the membrane, and in some cases there is a thin layer of smooth muscle under that. Epithelial membranes may be divided into two subgroups:

1. **Serous** (SE-rus) **membranes** line the walls of body cavities and are folded back onto the surface of internal organs, forming the outermost layer of many of them.
2. **Mucous** (MU-kus) **membranes** line tubes and other spaces that open to the outside of the body.

Epithelial membranes are made of closely crowded active cells that manufacture lubricants and protect the deeper tissues from invasion by microorganisms. Serous membranes secrete a thin, watery lubricant. Mucous membranes produce a rather thick and sticky substance called **mucus.** (Note that the adjective *mucous* contains an "o," whereas the noun *mucus* does not.)

The special serous membrane lining a closed cavity or covering an organ is called the **serosa** (se-RO-sah). In referring to the mucous membrane of a particular part, the noun **mucosa** (mu-KO-sah) is used.

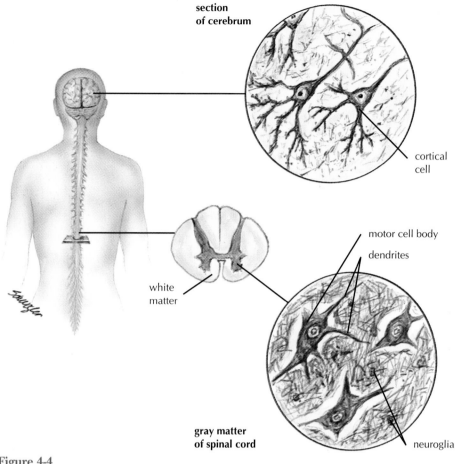

section
of cerebrum

cortical
cell

motor cell body

dendrites

white
matter

gray matter
of spinal cord

neuroglia

Figure 4-4

Nervous tissue.

SEROUS MEMBRANES

Serous membranes, unlike mucous membranes, usually do not communicate with the outside of the body. This group lines the closed ventral body cavities. There are three serous membranes:

1. The *pleurae* (PLU-re) or *pleuras* (PLU-rahs) line the thoracic cavity and cover each lung.
2. The *pericardium* (per-ih-KAR-de-um) is a sac that encloses the heart. It fits into a space in the chest between the two lungs.
3. The *peritoneum* (per-ih-to-NE-um) is the largest serous membrane. It lines the walls of the abdominal cavity, covers the organs of the abdomen, and forms supporting and protective structures within the abdomen (see Fig. 19-3).

The epithelium covering serous membranes is of a special kind called *mesothelium* (mes-o-THE-le-um). The mesothelium is smooth and glistening and is lubricated so that movements of the organs can take place with a minimum of friction.

Serous membranes are so arranged that one portion forms the lining of the closed cavity while another part covers the surface of the organs. The serous membrane attached to the wall of a cavity or sac is known as the *parietal* (pah-RI-eh-tal) *layer;* the word *parietal* refers to a wall. Parietal pleura lines the chest wall, and parietal pericardium lines the sac that encloses the heart. Because internal organs are called *viscera,* the membrane attached to the organs is the *visceral layer.* On the surface of the heart is visceral pericardium, and each lung surface is covered by visceral pleura.

Areas lined with serous membrane are examples of *potential spaces* because it is *possible* for a space to exist. Normally the membrane surfaces are in direct contact and a minimal amount of lubricant is present. Only if substances accumulate between the membranes, as when inflammation causes the production of excessive amounts of fluid, is there an actual space.

MUCOUS MEMBRANES

Mucous membranes form extensive continuous linings in the digestive, the respiratory, the urinary, and the reproductive systems, all of which are connected with the outside of the body. They vary somewhat in both structure and function. The cells that line the nasal cavities and most parts of the respiratory tract are supplied with tiny, hairlike extensions called *cilia*, which have already been mentioned in describing cells. The microscopic cilia move in a wavelike manner that forces the secretions outward away from the deeper parts of the lungs. In this way, foreign particles such as bacteria and dust become trapped in the sticky mucus and are prevented from causing harm. Ciliated epithelium is also found in certain tubes of both the male and the female reproductive systems.

The mucous membranes that line the digestive tract have special functions. For example, the mucous membrane of the stomach serves to protect the deeper tissues from the action of powerful digestive juices. If for some reason a portion of this membrane is injured, these juices begin to digest a part of the stomach itself—as in the case of peptic ulcers. Mucous membranes located farther along in the digestive system are designed to absorb food materials, which are then transported to all the cells of the body. Other mucous membranes are discussed in more detail later.

Other Membranes

Another epithelial membrane is the skin, or *cutaneous* (ku-TA-ne-us) *membrane*, described fully in Chapter 6. *Synovial* (sin-O-ve-al) *membranes* are thin connective tissue membranes that line the joint cavities. They secrete a lubricating fluid that reduces friction between the ends of bones, thus permitting free movement of the joints.

With this we conclude our brief introduction to membranes. As we study each system in turn, other membranes will be encountered. They may have unfamiliar names, but they will be either epithelial or connective tissue membranes, and we will become familiar with their general locations; in short, they will be easy to recognize and remember.

Membranes and Disease

We are all familiar with a number of diseases that directly affect membranes. These range all the way from the common cold, which is an inflammation of the mucosa of the nasal passages, to the sometimes fatal condition known as *peritonitis,* an infection of the peritoneum, which can follow rupture of the appendix.

Membranes can act as pathways along which disease may spread. In general, epithelial membranes seem to have more resistance to infections than do layers made of connective tissue. However, lowered resistance may allow the transmission of infection along any membrane. For example, infections may travel along the lining of the tubes of the reproductive system into the urinary system in males. In females, an infection may travel up the tubes and spaces of the reproductive system into the peritoneal cavity (see Figs. 23-1 and 23-12).

Sometimes fibrous membranes form planes of division extending in such a way that infection in one area is prevented from reaching another space. In other cases a vertical plane that separates two areas may seem to encourage the travel of bacteria either upward or downward. An infection of the tonsils, for example, may travel down to the chest along fibrous membranes.

The connective tissue or collagen diseases, such as *systemic lupus erythematosus* (LU-pus er-ih-them-ah-TO-sus) (SLE) and *rheumatoid arthritis,* may affect many parts of the body, because collagen is the major intercellular protein in solid connective tissue. In systemic lupus erythematosus, serous membranes such as the pleura, pericardium, and peritoneum are often involved. In rheumatoid arthritis, the synovial membrane becomes inflamed and swollen, and the cartilage in the joints is gradually replaced with fibrous connective tissue.

▶ Benign and Malignant Tumors

For a variety of reasons, the normal pattern of cell and tissue growth may be broken by an upstart formation of cells having no purpose whatsoever in the body. Any abnormal growth of cells is called a *tumor,* or *neoplasm.* If the tumor is confined to a local area and does not spread, it is called a *benign* (be-NINE) tumor. If the tumor spreads to neighboring tissues or to distant parts of the body, it is called a *malignant* (mah-LIG-nant) tumor. The general term for any type of malignant tumor is *cancer*. The process of tumor cell spread is called *metastasis* (meh-TAS-tah-sis).

Tumors are found in all kinds of tissue, but they occur most frequently in those tissues that repair themselves most quickly, specifically epithelium and connective tissue, in that order.

Benign Tumors

Benign tumors, theoretically at least, are not dangerous in themselves; that is, they do not spread. Their cells stick together, and often they are encapsulated, that is, surrounded by a containing membrane. Benign tumors grow as a single mass within a tissue, lending themselves neatly to complete surgical removal. Of course, some benign tumors can be quite harmful; they may grow within an organ, increase in size, and cause considerable mechanical damage. A benign tumor of the brain, for example, can kill a person just as a malignant one can, since it grows in an enclosed area and compresses vital brain tissue. Some examples of benign tumors are given below (note that most of the names end in *-oma*, which means "tumor").

1. *Papilloma* (pap-ih-LO-mah), a tumor that grows in epithelium as a projecting mass. One example is a wart.
2. *Adenoma* (ad-eh-NO-mah), an epithelial tumor that grows in and about the glands (*adeno-* means "gland").
3. *Lipoma* (lip-O-mah), a connective tissue tumor originating in fatty (adipose) tissue.
4. *Osteoma* (os-te-O-mah), a connective tissue tumor that originates in the bones.
5. *Myoma* (mi-O-mah), a tumor of muscle tissue. Rare in voluntary muscle, it is common in some types of involuntary muscle, particularly in the uterus (womb). When found in the uterus, however, it is ordinarily called a *fibroid*.
6. *Angioma* (an-je-O-mah), a tumor that usually is composed of small blood or lymphatic vessels; an example is a birthmark.
7. *Nevus* (NE-vus), a small skin tumor of one of a variety of tissues. Some nevi are better known as moles; some are angiomas. Ordinarily, these tumors are harmless, but they can become malignant.
8. *Chondroma* (kon-DRO-mah), a tumor of cartilage cells that may remain within the cartilage or develop on the surface, as in the joints.

Malignant Tumors

Malignant tumors, unlike benign tumors, can cause death no matter where they occur. The word *cancer* means "crab," and this is descriptive: a cancer sends out clawlike extensions into neighboring tissue. A cancer also spreads "seeds," which plant themselves in other parts of the body. These seeds are, of course, cancer cells, and they are transported everywhere by either the blood or the lymph (another circulating fluid). When the cancer cells reach their destination, they immediately form new (secondary) growths, or *metastases* (meh-TAS-tah-seze). Malignant tumors, moreover, grow much more rapidly than benign ones.

Malignant tumors generally are classified in two main categories according to the type of tissue in which they originate:

1. *Carcinoma* (kar-sih-NO-mah). This type of cancer originates in epithelium and is by far the most common form of cancer. Common sites of carcinoma are the skin, mouth, lung, breast, stomach, colon, prostate, and uterus. Carcinomas are usually spread by the lymphatic system (see Chap. 16).
2. *Sarcoma* (sar-KO-mah). These are cancers of connective tissue of all kinds and hence may be found anywhere in the body. Their cells are usually spread by the blood stream, and they often form secondary growths in the lungs.

There are other types of malignant tumors. One that originates in a birthmark or a mole is called a *melanoma* (mel-ah-NO-mah). A *neuroma* (nu-RO-mah) is a tumor that arises from a nerve. Because nervous tissue does not multiply throughout life, however, it is rarely involved in cancer. Usually, a tumor of the nervous system originates in the connective tissue of the brain or spinal cord and is called a *glioma* (gli-O-mah). A malignant neoplasm of lymphatic tissue is called a *lymphoma* (lim-FO-mah).

Symptoms of Cancer

Everyone should be familiar with certain signs that may be indicative of early cancer, so that these signs can be reported immediately, before the condition spreads. It is unfortunate that early cancer is painless; otherwise, cancer would not be the problem that it is. Early symptoms may include unaccountable loss of weight, unusual bleeding or discharge, persistent indigestion, chronic hoarseness or cough, changes in the color or size of moles, a sore that does not heal in a reasonable time, the presence of an unusual lump, and the presence of white patches inside the mouth or white spots on the tongue.

Detection and Treatment of Cancer

Cancer diagnosis is becoming increasingly specific, leading to more precise methods of treatment. The diagnostic methods listed below include some advanced techniques for showing the size and location of tumors and other abnormalities.

1. *Biopsy* (BI-op-se) is the removal of living tissue for the purpose of microscopic examination of the cells.
2. *Ultrasound* is the use of reflected high-frequency sound waves to differentiate various kinds of tissue.
3. *Computed tomography* (*CT*) is the use of x-rays to produce a cross-sectional picture of body parts, such as the brain (see Fig. 10-9).
4. *Magnetic resonance imaging* (*MRI*) is the use of magnetic fields and radio waves to show changes in soft tissues.

These studies are used for a process called *staging,* which means the classification of a tumor based on size and extent of invasion. Staging helps the physician select appropriate treatment and predict the outcome of the disease.

Benign tumors usually can be removed completely by surgery; malignant tumors cannot be treated so easily, for a number of reasons. If cancerous tissue is removed surgically, there is always the probability that a few hidden cells will be left behind to grow anew. If the cells have spread to distant parts of the body, there is little that anyone can do. Sometimes surgery is preceded by or followed by radiation. Radiation therapy is administered by x-ray machines or by the placement of radioactive seeds within the involved organ. Radiation destroys the more rapidly dividing cancer cells while causing less damage to the more slowly dividing normal cells. Methods are being developed that will permit more accurate focusing of the beam of radiation and thus reduce the damage to normal body structures.

Drugs used for the treatment of cancer include those that act selectively on tumor cells. Known as *antineoplastic* (an-ti-ne-o-PLAS-tik) *agents,* these drugs are most effective when used in combination. The treatment of cancer with antineoplastic agents is one form of *chemotherapy* (ke-mo-THER-ah-pe) (see Chap. 5). Certain types of leukemia and various cancers of the lymphatic system are often effectively treated by this means. Research continues to develop new drugs and more effective drug combinations.

Based on the concept that the immune system normally rids the body of wayward cancerous cells, techniques of *immunotherapy* (ih-mu-no-THER-ah-pe) are being tested. The goal of these efforts is to strengthen each patient's immune system and enable it to destroy malignant cells. Work with these methods continues.

The *laser* (LA-zer), a device that produces a very highly concentrated and intense beam of light, may sometimes successfully destroy the tumor or may be employed as a cutting device for removing the growth. Important advantages of the laser are its ability to coagulate blood so that bleeding is largely prevented and the capacity to direct a narrow beam of light accurately to attack harmful cells and avoid normal ones.

SUMMARY

I. Tissue—groups of cells specialized for a specific task
  A. Epithelial tissue—covers surfaces, lines cavities
    1. Squamous, cuboidal, and columnar cells
    2. Simple or stratified cell layers
    3. Glands
      a. Exocrine
        (1) Secrete through ducts
        (2) Produce external secretions
      b. Endocrine
        (1) Secrete into bloodstream
        (2) Produce internal secretions—hormones
  B. Connective tissue—supports, binds, forms framework of body

1. Soft—jellylike intercellular material
    a. Adipose—stores fat
    b. Areolar (loose)
2. Fibrous—Dense tissue with collagenous or elastic fibers between cells
    a. Tendons—attach muscle to bone
    b. Ligaments—connect bones
    c. Capsules—around organs
    d. Fascia—bands or sheets that support organs
3. Hard—firm and solid
    a. Cartilage—found at ends of bones, nose, outer ear, trachea, etc.
    b. Bone—contains mineral salts
4. Liquid
    a. Blood
    b. Lymph
C. Muscle tissue—contracts to produce movement
1. Skeletal muscle—voluntary; moves skeleton
2. Cardiac muscle—forms the heart
3. Smooth muscle—involuntary; forms visceral organs
D. Nervous tissue
    l. Neuron—nerve cell
        a. Cell body—contains nucleus
        b. Dendrite—fiber carrying impulses toward cell body
        c. Axon—fiber carrying impulses away from cell body
    2. Myelin—fatty material that insulates some axons
        a. Myelinated fibers—make up white matter
        b. Unmyelinated cells and fibers—make up gray matter
    b. Neuroglia—support and protect nervous tissue

II. **Membrane**—thin sheet of tissue
    A. Main types
        1. Epithelial membrane—outer layer epithelium
            a. Serous membrane—secretes watery fluid
                (1) Parietal layer—lines body cavity
                (2) Visceral layer—covers internal organs
            b. Mucous membrane
                (1) Secretes mucus
                (2) Lines tube or space that opens to the outside (*e.g.,* respiratory, digestive, reproductive tracts)
        2. Cutaneous membrane—skin
        3. Synovial membrane—lines a joint
    B. Membranes and disease
III. **Tumor** (neoplasm)—uncontrolled growth of cells
    A. Types
        1. Benign—localized
        2. Malignant—invades tissue and spreads to other parts of the body (metastasizes)
    B. Symptoms—weight loss, bleeding, persistent indigestion, hoarseness or cough, change in mole, lump, nonhealing sore
    C. Detection
        1. Biopsy (study of tissue), ultrasound, CT, MRI
        2. Staging—classification based on size of tumor and extent of invasion
    D. Treatment
        1. Surgical removal
        2. Radiation
        3. Chemotherapy—drugs
        4. Immunotherapy

QUESTIONS FOR STUDY AND REVIEW

1. Define *tissue*. Give a few general characteristics of tissues.
2. Define *epithelium* and give two examples.
3. Describe the difference between endocrine and exocrine glands and give examples of each.
4. Define *connective tissue*. Name the main kinds of connective tissue and give an example of each.
5. What kinds of fibers are found in connective tissue?
6. Name three kinds of muscle tissue and give an example of each.
7. What is the difference between voluntary and involuntary muscle?

8. What is the main purpose of nervous tissue? What is its basic structural unit called?

9. Define *myelin*. Where is it found?

10. Compare how easily the four basic groups of tissues repair themselves.

11. What are some general characteristics of epithelial membranes?

12. Name two types of epithelial membrane. Give their general characteristics and two examples of each.

13. Name and describe the two layers of serous membranes.

14. Give two examples of fascia.

15. What is an infection of the peritoneum called? What is a possible cause of this condition?

16. Name two diseases of membranes.

17. What is the relation between membranes and disease?

18. What is a tumor? In what kinds of tissue are tumors most commonly found?

19. What is the difference between a benign and a malignant tumor?

20. Name four examples of benign tumors and tell where each is found.

21. Name the two categories of malignant tumor. In what kinds of tissue are these tumors found? Which kind is the most common?

22. Name some early symptoms of cancer.

23. In what ways is cancer diagnosed?

24. In what ways is cancer treated?

25. What is staging and how is it used?

UNIT II

Disease and the First Line of Defense

The two chapters in this unit incorporate a discussion of deviations from the normal, which are the basis for disease. One chapter is devoted largely to the most common causes of disease. These are the microorganisms, including bacteria, viruses and protozoa, and larger organisms such as worms. The most important defense against the multitude of causes of disease is the skin, the first line of defense. The skin, as well as being classified as a system, is the largest organ of the body. Its properties and functions are discussed in this unit.

CHAPTER 5

Disease and Disease-Producing Organisms

Selected Key Terms
The following terms are
defined in the Glossary:

acute

asepsis

chemotherapy

chronic

diagnosis

disease

epidemic

etiology

microorganism

pathogen

pathophysiology

prognosis

sign

spore

sterilization

symptom

systemic

toxin

Behavioral Objectives

After careful study of this chapter, you should be able to:

- Define disease and list seven causes of disease
- List eight predisposing causes of disease
- Differentiate between acute and chronic disease
- List four types of organisms studied in microbiology and give the characteristics of each
- List some diseases caused by each type of microorganism
- Describe the three types of bacteria according to shape
- List several diseases in humans caused by worms
- Describe how infections may be spread
- Describe several public health measures taken to prevent the spread of disease
- Differentiate *sterilization, disinfection,* and *antisepsis*
- Describe universal precautions
- Define *chemotherapy*
- Describe several methods used to identify microorganisms in the laboratory

Memmler, RL, Cohen, BJ, Wood, DL. *THE HUMAN BODY IN HEALTH AND DISEASE,* 8/e,
© 1996 Lippincott-Raven Publishers

What Is Disease?

Disease may be defined as the abnormal state in which part or all of the body is not properly adjusted or is not capable of carrying on all its required functions. There are marked variations in the extent of disease and in its effect on the person. Disease can have a number of direct causes, such as the following:

1. *Disease-producing organisms.* Some of these are discussed in this chapter. Organisms are believed to play a part in at least half of all human illness.
2. *Malnutrition.* This is a lack of essential vitamins, minerals, proteins, or other substances required for normal life processes.
3. *Physical agents.* These include excessive heat or cold, and injuries that cause cuts, fractures, or crushing damage to tissues.
4. *Chemicals.* Some chemicals that may be poisonous or otherwise injurious if present in excess, are lead compounds (in paints), carbolic acid (in certain antiseptic solutions), and household and industrial poisons.
5. *Birth defects.* Abnormalities of structure and function that are present at birth are termed *congenital.* Such abnormalities may be inherited (passed on by the parents through their reproductive cells) or acquired during the process of development within the mother's uterus (womb) (see Chap. 24).
6. *Degenerative processes.* Degeneration (de-jen-er-A-shun) means "breaking down." With aging, tissue degenerates and thus becomes less active and less capable of performing its normal functions. Such deterioration may be caused by continuous infection, by repeated minor injuries to tissues, or by the normal "wear and tear" of life. Thus, degeneration is an anticipated result of aging.
7. *Neoplasms.* The word *neoplasm* means "new growth" and refers to cancer and other types of tumors (see Chap. 4)

Other factors that enter into the production of a disease are known as *predisposing causes.* Although a predisposing cause may not in itself give rise to a disease, it increases the probability of a person's becoming ill. Examples of predisposing causes include the following:

1. *Age.* As stated earlier, the degenerative processes of aging can be a direct cause of disease. But a person's age at any time can be a predisposing factor. For instance, measles is more common in children than in adults. Other diseases may appear most commonly in young adults or people in middle years.
2. *Sex.* Certain diseases are more characteristic of one sex than the other. Men are more susceptible to early heart disease, whereas women are more likely to develop diabetes.
3. *Heredity.* Some individuals inherit a "tendency" to acquire certain diseases—particularly diabetes and many allergies.
4. *Living conditions and habits.* Persons who habitually drive themselves to exhaustion, do not get enough sleep, or pay little attention to diet are highly vulnerable to the onslaught of disease. Overcrowding and poor sanitation invite epidemics. The abuse of drugs, alcohol, and tobacco also can lower vitality and predispose to disease.
5. *Occupation.* A number of conditions are classified as "occupational diseases." For instance, coal miners have been susceptible to lung damage caused by the constant inhalation of coal dust. Asbestos and other industrial substances have been associated with increased occurrence of cancer.
6. *Physical exposure.* Undue chilling of all or part of the body, or prolonged exposure to heat, can lower the body's resistance to disease.
7. *Preexisting illness.* Any preexisting illness, even the common cold, increases one's chances of contracting another disease.
8. *Psychogenic influences.* *Psycho* refers to the mind, *genic* to origin. Some physical disturbances are due either directly or indirectly to emotional upsets caused by conditions of stress and anxiety in daily living. Headaches and so-called "nervous indigestion" are examples.

The Study of Disease

The modern approach to the study of disease emphasizes the close relation of the pathologic and physiologic aspects and the need to understand the fundamentals of each in treating any body disorder. The term used for this combined study in medical science is *pathophysiology.*

Underlying the basic medical sciences are the still more fundamental disciplines of physics and chemistry. A knowledge of both of these is essential to any real understanding of the life processes.

Disease Terminology

The study of the cause of any disease, or the theory of its origin, is *etiology* (e-te-OL-o-je). Any study of a disease usually includes some indication of *incidence,* which means its range of occurrence and its tendency to affect certain groups of individuals more than others. Information about its geographic distribution and its tendency to appear in one sex, age group, or race more or less frequently than another is usually included in a presentation on disease incidence.

Diseases are often classified on the basis of severity and duration as:

1. *Acute.* These are relatively severe but usually last a short time.
2. *Chronic.* These are often less severe but likely to be continuous or recurring for long periods.
3. *Subacute.* These are intermediate between acute and chronic, not being as severe as acute infections nor as long lasting as chronic disorders.

Another term used in describing certain diseases is *idiopathic* (id-e-o-PATH-ik), which means "self-originating" or "without known cause."

A *communicable* disease is one that can be transmitted from one person to another. If many people in a given region acquire a certain disease at the same time, that disease is said to be *epidemic.* If a given disease is found to a lesser extent but continuously in a particular region, the disease is *endemic* to that area. A disease that is prevalent throughout an entire country or continent, or the whole world, is said to be *pandemic.*

DIAGNOSIS, TREATMENT, PROGNOSIS, AND PREVENTION

To treat a patient, a physician obviously must first reach a conclusion as to the nature of the illness—that is, make a *diagnosis.* To do this he or she must know the *symptoms,* which are the conditions of disease noted by the patient, and the *signs,* which are the objective manifestations the physician or other health care professional can observe. Sometimes a characteristic group of symptoms and signs accompanies a given disease. Such a group is called a *syndrome* (SIN-drome). Frequently, the physician uses laboratory tests to help establish the diagnosis. A *prognosis* (prog-NO-sis) is a prediction of the probable outcome of a disease based on the condition of the patient and the physician's knowledge about the disease.

Nurses play an extremely valuable role in this process by observing closely for signs, collecting and organizing information from the patient about his or her symptoms, and then reporting this information to the physician. Once a patient's disorder is known, the physician prescribes a course of treatment, known as *therapy.* Specific measures in a course of treatment are carried out by the nurse and other health care providers under the physician's orders.

In recent years, physicians, nurses, and other health care workers have taken on increasing responsibilities in *prevention.* Throughout most of medical history, the physician's aim has been to cure patients of existing diseases. However, the modern concept of prevention seeks to stop disease before it actually happens—to keep people well through the promotion of health. A vast number of organizations exist for this purpose, ranging from the World Health Organization (WHO) on an international level down to local private and community health

The CDC

In any article on infectious diseases in the United States, the initials CDC are likely to appear. This federal agency, the full name of which is the Centers for Disease Control and Prevention, is located in Atlanta, Georgia. Since its establishment in 1946, it has traced the origins and spread of infectious diseases and has gradually expanded its role to include the prevention of injury and disease.

The CDC participated in the fight against polio, which has virtually been eliminated in the United States, and joined the World Health Organization in efforts to eradicate smallpox worldwide. More recently, it has led in efforts to study and treat toxic shock syndrome, Legionnaire's disease, AIDS, and tuberculosis. In 1993, researchers at the CDC, working with local health agencies, rapidly identified the strain of hantavirus that caused a serious and often fatal pulmonary disease in persons in the Southwest.

The CDC employs approximately 6500 people working in more than 170 occupations. Many are assigned to various state, federal, and foreign locations. Here they investigate health problems, conduct research, and promote public health policies. Their stated goal is "healthy people in a healthy world—through prevention."

programs. A rapidly growing responsibility of the nursing professional and of other health occupations is educating individual patients toward the maintenance of total health, physical and mental.

Modes of Infection

The predominant cause of disease in humans is the invasion of the body by disease-producing *microorganisms* (mi-kro-OR-gan-izms). The word *organism* means "anything having life"; *micro* means "small." Hence, a microorganism is a tiny living thing, too small to be seen by the naked eye. Other terms for microorganism are *microbe* and, more popularly, *germ*. A microbe, or any other organism, that lives on or within a living *host* and at the host's expense is called a *parasite*.

Although most microorganisms are harmless to humans, and many are beneficial, a few types cause illness; that is, they are *pathogenic* (path-o-JEN-ic). Any disease-causing organism is a *pathogen* (PATH-o-jen). If the body is invaded by pathogens, with adverse effects, the condition is called an *infection*. If the infection is restricted to a relatively small area of the body, it is *local*. A generalized, or *systemic* (sis-TEM-ik), infection is one in which the whole body is affected. Systemic infections usually are spread by the blood. An infection that takes hold because the host has been weakened by disease is described as an *opportunistic infection*. People with depressed immune systems, such as those with AIDS, become infected with organisms that are ordinarily harmless.

MODES OF TRANSMISSION

Microorganisms may be transmitted from an infected human, insect, or animal host to a susceptible human being; this transfer may be by direct or indirect contact. For example, infected human hosts may transfer their microorganisms to other individuals through direct personal contact such as shaking hands, kissing, or having sexual intercourse. Indirect contact involves touching objects that have been contaminated by an infected person. For example, microorganisms may be transferred indirectly through bedding, toys, food, and dishes. Also, insects may deposit infectious material on food, skin, or clothing. Pets may be the source of a number of infections.

PORTAL OF ENTRY AND EXIT

There are several avenues through which microorganisms may enter the body: the skin, respiratory tract, and digestive system, as well as the urinary and reproductive systems. These portals of entry may also serve as exit routes. For example, discharges from the respiratory and intestinal tracts may spread infection through contamination of air, through contamination of hands, and through contamination of food and water supplies. (Microbial control is discussed later in this chapter.)

▶ The Microorganisms

Microorganisms are simple, usually single-celled forms of life (Fig. 5-1). The study of these microscopic organisms is *microbiology* (mi-kro-bi-OL-o-je). Other sciences have grown up within the science of microbiology, and each has become a specialty in itself. Some examples of these more specialized sciences include the following:

1. *Bacteriology* (bak-te-re-OL-o-je) is the study of bacteria, both beneficial and disease-producing. It includes the study of rickettsias and chlamydias, which are extremely small bacteria that multiply within living cells.
2. *Mycology* (my-KOL-o-je) is the study of fungi, which include yeasts and molds.
3. *Virology* (vi-ROL-o-je) is the study of viruses, extremely small infectious agents that can multiply only within living cells.
4. *Protozoology* (pro-to-zo-OL-o-je) is the study of single-celled animals called *protozoa*. Although the term *parasitology* (par-ah-si-TOL-o-je) is the study of parasites in general, in practice it usually refers to the study of protozoa and worms (helminths).

Despite the fact that this book is concerned with pathogenic forms, most microorganisms are harmless to humans and are absolutely essential to the continuation of all life on earth. It is through the actions of microorganisms that dead animals and plants are decomposed and transformed into substances that enrich the soil. Sewage is rendered harmless by microorganisms. Several groups of bacteria transform the nitrogen of the air into a form usable by plants, a process called *nitrogen fixation.* Farmers take advantage of this by allowing a field to lie fallow (untilled) so that the nitrogen of its soil can be replenished. Certain bacteria and fungi produce the antibiotics that make our lives safer. Others produce the fermented products that make our lives more enjoyable, such as beer, wine, cheeses, and yogurt.

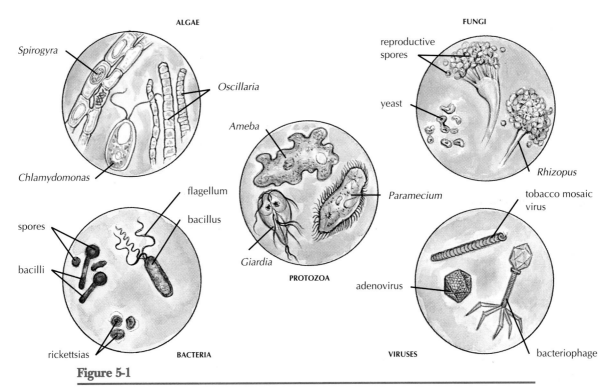

Figure 5-1

Some examples of microorganisms.

Bacteria

We have a population (flora) of microorganisms that normally grows on and within our bodies. We live in balance with these organisms, and they prevent the growth of other harmful varieties. However, some microorganisms that are normally harmless may become pathogenic for persons who are in a weakened state as a result, for example, of disease, injury, or malnutrition.

Bacteria

Bacteria are one-celled organisms that are among the most primitive forms of life on earth. They can be seen only with a microscope; from 10 to 1000 bacteria (depending on the species) would, if lined up, span a pinhead. Staining of the cells with dyes helps make their structures more visible and reveals information about their properties.

Bacteria are found everywhere: in soil, in hot springs, in polar ice, and on and within plants and animals. Their requirements for water, nutrients, oxygen, temperature, and other factors vary widely according to species. Some are capable of carrying out photosynthesis, like green plants; others must take in organic nutrients, as do animals. Some, de-

scribed as ***anaerobic*** (an-air-O-bik), can grow in the absence of oxygen; others, called ***aerobic*** (air-O-bik), require oxygen. Some bacteria can produce ***spores,*** resistant forms that can tolerate long periods of dryness or other adverse conditions. Because these spores are easily airborne and are resistant to ordinary methods of disinfection, pathogenic organisms that form spores are particularly dangerous. Some bacteria are capable of swimming rapidly about by themselves by means of threadlike appendages called ***flagella*** (flah-JEL-ah).

Bacteria compose the largest group of pathogens. Not surprisingly, these pathogenic bacteria are most at home within the "climate" of the human body. When living conditions are ideal, the organisms reproduce by binary fission (simple cell division) with unbelievable rapidity. If they succeed in overcoming the body's natural defenses, they can cause damage in two ways: by producing poisons, or ***toxins,*** and by entering the body tissues and growing within them. Table 1 of Appendix 4 lists some typical pathogenic bacteria and the diseases they cause.

There are so many different types of bacteria that their classification is complicated. For our purposes, a convenient and simple grouping is based on the shape and arrangement of these organisms as seen with a microscope (Figs. 5-2 and 5-3):

1. *Rod-shaped cells—bacilli* (bah-SIL-i). These cells are straight and slender. Some are cigar-shaped, with tapering ends. Typical bacillary diseases include tetanus, diphtheria, tuberculosis, typhoid fever, and Legionnaire's disease.
2. *Spherical cells—cocci* (KOK-si). These cells are round and are seen in characteristic arrangements. Those that are in pairs are called *diplococci* (*diplo-* means "double"). Those that are arranged in chains, like a string of beads, are called *streptococci* (*strepto-* means "chain"). A third group, seen in large clusters, are known as *staphylococci* (staf-ih-lo-KOK-si) (*staphylo-* means "bunch of grapes"). Among the diseases caused by diplococci are gonorrhea and meningitis; streptococci and

staphylococci are responsible for a wide variety of infections, including pneumonia, rheumatic fever, and scarlet fever.

3. *Curved rods.* One type, which has only a slight curvature, like a comma, is called *vibrio* (VIB-re-o). Cholera is caused by a vibrio. Another form, which resembles a corkscrew, is known as *spirillum* (spi-RIL-um). (The plural is *spirilla.*) Bacteria very similar to the spirilla, but capable of waving and twisting motions, are called *spirochetes* (SPI-ro-ketes). The most serious and widespread spirochetal infection is syphilis. In syphilis, the spirochetes enter the body at the point of contact, usually through the genital skin or mucous membranes. They then travel to the bloodstream and thus set up a systemic infection. (See Table 1 in Appendix 4 for a summary of the three stages of syphilis.)

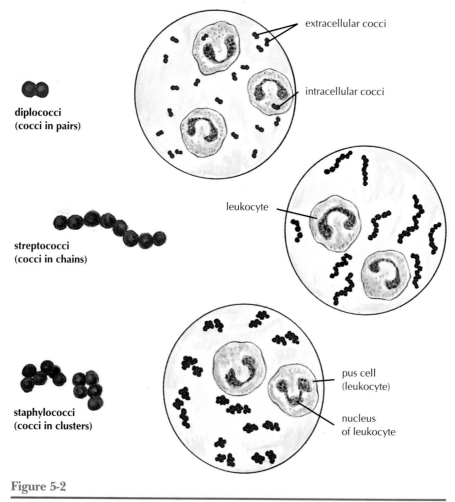

Figure 5-2

Spherical bacteria (Gram stained).

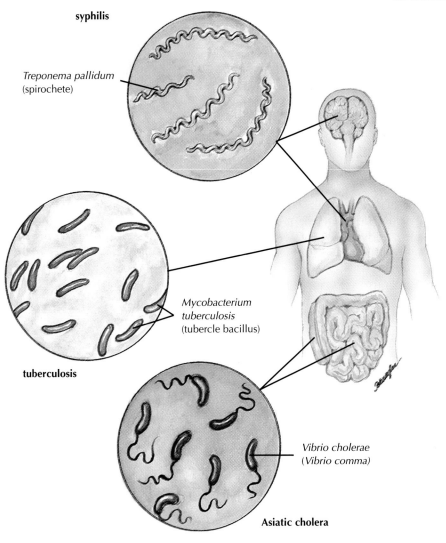

syphilis

Treponema pallidum
(spirochete)

*Mycobacterium
tuberculosis*
(tubercle bacillus)

tuberculosis

Vibrio cholerae
(*Vibrio comma*)

Asiatic cholera

Figure 5-3

Rod-shaped and curved bacteria. The illustrations show areas of the body invaded by these pathogens.

A spirochete is also responsible for Lyme disease, which has increased in the United States since it first appeared in the early 1960s. People who walk in or near woods are advised to wear white protective clothing that covers their ankles. They should examine their bodies for the freckle-sized ticks that carry the disease.

4. ***Rickettsias*** (rih-KET-se-ahs) and the ***chlamydias*** (klah-MID-e-ahs) (sometimes called *rickettsiae* and *chlamydiae*) are classified as bacteria, although they are considerably smaller. These microorganisms can exist only inside living cells. Because they exist at the expense of their hosts, they are parasites, and because they can grow only within living cells, they are referred to as *obligate intracellular parasites*.

The rickettsias are the cause of a number of serious diseases in humans, such as typhus and Rocky Mountain spotted fever. In almost every instance, these organisms are transmitted through the bites of insects, such as lice, ticks, and fleas. A few common rickettsial diseases are listed in Table 1 of Appendix 4.

The chlamydias are smaller than the rickettsias. They are the causative organisms in trachoma (a serious eye infection that ultimately causes blindness), parrot fever or psittacosis, the sexually transmitted disease lymphogranuloma venereum, and some respiratory diseases (see Appendix 4, Table 1).

Fungi

The true *fungi* (FUN-ji) are a large group of simple plantlike organisms. Only a very few types are pathogenic. Although fungi are much larger and more complicated than bacteria, they are still a simple form of life. They differ from the higher plants in that they lack the green pigment chlorophyll, which enables most plants to use the energy of sunlight to manufacture food. Like bacteria, fungi grow best in dark, damp places. Fungi reproduce in several ways, including by simple cell division and by production of large numbers of reproductive spores. Single-celled forms of fungi are generally referred to as *yeasts;* the fuzzy, filamentous forms are called *molds.*

Familiar examples of fungi are mushrooms, puffballs, bread molds, and the yeasts used in baking and brewing. Diseases caused by fungi are called *mycotic* (mi-KOT-ik) infections (*myco-* means "fungus"). Examples of these are athlete's foot and ringworm. *Tinea capitis* (TIN-e-ah KAP-ih-tis), which involves the scalp, and *tinea corporis* (kor-PO-ris), which may be found almost anywhere on the nonhairy parts of the body, are common types of ringworm. One yeast-like fungus that may infect a weakened host is *Candida.* This is a normal inhabitant of the mouth and digestive tract that may produce skin lesions, an oral infection called *thrush,* digestive upset, or inflammation of the vaginal tract (vaginitis) as a secondary infection. Although few systemic diseases are caused by fungi, some are very dangerous, and all are difficult to cure. Pneumonia can be caused by the inhalation of fungal spores contained in dust particles. Table 2 in Appendix 4 is a list of typical fungal diseases.

Viruses

Although bacteria seem small, they are enormous in comparison with *viruses.* Viruses are comparable in size to large molecules, but unlike other molecules, they contain genetic material and are able to reproduce. Viruses are so tiny that they are invisible with a light microscope; they can be seen only with an electron microscope. Because of their very small size and the difficulties associated with growing them in the laboratory, viruses were not studied with much success until the middle of this century.

Viruses are the smallest known infectious agents. They have some of the fundamental properties of living matter, but they are not cellular and they have no enzyme system. Like the rickettsias and the chlamydias, they can grow only within living cells—they are obligate intracellular parasites; unlike these, however, the viruses are not usually susceptible to antibiotics.

There is no universally accepted classification of viruses. For our purposes it is appropriate to think of them in relation to the diseases they cause. There are a considerable number of them—measles, poliomyelitis, hepatitis, chickenpox, and the common cold, to name a few. *AIDS* (acquired immunodeficiency syndrome) is a very serious viral disease discussed in Chapter 17. The virus that causes AIDS and other representative viruses are listed in Table 3 in Appendix 4.

Protozoa

With the *protozoa* (pro-to-ZO-ah), we come to the only group of microbes that can be described as animal-like. Although protozoa are one-celled, like bacteria, they are much larger.

Protozoa are found all over the world in the soil and in almost any body of water from moist grass to mud puddles to the sea. There are four main divisions of protozoa:

1. *Amebas* (ah-ME-bas). An ameba is an irregular blob of cytoplasm that propels itself by extending part of its cell (a pseudopod, or "false foot") and then flowing into the extension. Amebic dysentery is caused by a pathogen of this group.
2. *Ciliates* (SIL-e-ates). This type of protozoon is covered with tiny hairs called *cilia* that produce a wave motion to propel the organism.
3. *Flagellates* (FLAJ-eh-lates). These organisms are propelled by long, whiplike filaments called *flagella.* One of this group, a *trypanosome* (tri-PAN-o-some), causes African sleeping sickness (Fig. 5-4).
4. *Sporozoa* (spor-o-ZO-ah). Unlike other protozoa, sporozoa cannot propel themselves. They are obligate parasites, unable to grow outside a host. Malaria is caused by members of the genus *Plasmodium* (plaz-MO-de-um). These protozoa, carried by a type of mosquito, cause much serious

illness in the tropics, resulting in up to 3 million deaths each year.

Two other members of this group have emerged as opportunistic infections in people suffering from AIDS or those who are otherwise weakened. *Pneumocystis carinii* causes a previously rare pneumonia. *Cryptosporidium* causes severe and prolonged diarrhea.

Figure 5-4 illustrates some of the pathogenic protozoa and their portals of entry. Table 4 in Appendix 4 presents a list of typical pathogenic protozoa with the diseases they cause.

Parasitic Worms

Many species of worms, also referred to as **helminths**, are parasitic by nature and select the human organism as their host. The study of worms, particularly parasitic ones, is called **helminthology** (hel-min-THOL-o-je). Whereas invasion by any form of organism is usually called an *infection*, the presence of parasitic worms in the body also can be termed an **infestation** (Fig. 5-5). A microscope is required for one to see the eggs or larval forms of most worm infestations.

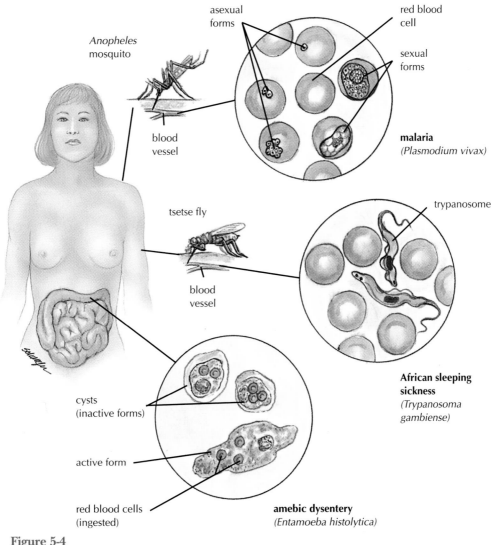

Figure 5-4

Pathogenic protozoa.

Roundworms
INTESTINAL ROUNDWORMS

Many human parasitic worms are classified as roundworms, one of the most common of which is the large worm **ascaris** (AS-kah-ris). This worm is prevalent in many parts of Asia, where it is found mostly in larval form. In the United States, it is found especially frequently in children (ages 4–12 years) of the rural South. *Ascaris* is a long, white-yellow worm pointed at both ends. It may infest the lungs or the intestines, producing intestinal obstruction if present in large numbers. The eggs produced by the adult worms are very resistant and can live in soil during either freezing or hot, dry weather and cannot be destroyed even by strong antiseptics. The embryo worms develop within the eggs deposited with excreta in the soil, and later reach the digestive system of a victim by means of contaminated food. Discovery of this condition may be made by a routine stool examination.

PINWORMS

Another fairly common infestation, particularly in children, is the seat worm, or **pinworm** (*Enterobius vermicularis*), which is also very hard to control and eliminate. The worms average 12 mm (somewhat less than 1/2 inch) in length and live in the large intestine. The adult female moves outside the vicinity of the anus to lay its thousands of eggs. These eggs are often transferred by a child's fingers from the itching anal area to the mouth. In the digestive system of the host, the eggs develop to form new adult worms, and thus a new infestation is begun. The child also may infect others by this means. In addition, pinworm eggs that are expelled from the body constitute a hazard, since they may live in the external environment for several months. Patience and every precaution, with careful attention to medical instructions, are necessary to rid the patient of the worms. Washing the hands, keeping the fingernails clean, and avoiding finger sucking are all essential.

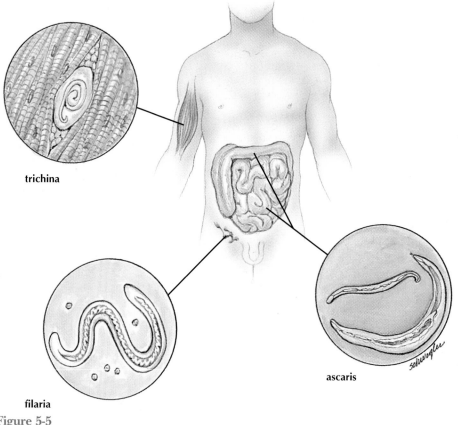

trichina

filaria

ascaris

Figure 5-5

Common parasitic worms.

HOOKWORMS

Hookworms are parasites that live in the small intestine. They are dangerous because they suck blood from the host, causing such a severe anemia (blood deficiency) that the victim becomes sluggish, both physically and mentally. Most victims become susceptible to various chronic infections because of extremely reduced resistance following great and continuous blood loss. Hookworms lay thousands of eggs, which are distributed in the soil by contaminated excreta. The eggs develop into small larvae, which are able to penetrate the intact skin of bare feet. They enter the blood and, by way of the circulating fluids, the lungs, and the upper respiratory tract, finally reach the digestive system. Prevention of this infestation is best accomplished by the proper disposal of excreta, attention to sanitation, and the wearing of shoes in areas where the soil is contaminated.

OTHER ROUNDWORMS

Most roundworms are transmitted by excreta; however, the small *trichina* (trik-I-nah), found mainly in pork, is an exception. These tiny, round worms become enclosed in cysts, or sacs, inside the muscles of rats, pigs, and humans. If undercooked pork is eaten, the cysts in it are dissolved by the host's digestive juices, and the tiny worms mature and travel to the host's muscles, where they again become encased. This disease is known as *trichinosis* (trik-ih-NO-sis).

The tiny, threadlike worm that causes *filariasis* (fil-ah-RI-ah-sis) is transmitted by such biting insects as flies and mosquitoes. The worms grow in large numbers, causing various body disturbances. If the lymphatic vessels become clogged by them, a condition called *elephantiasis* (el-eh-fan-TI-ah-sis) results, in which the lower extremities and the scrotum may become tremendously enlarged. Filariasis is most common in tropical and subtropical lands, such as southern Asia and many of the South Pacific islands.

Flatworms

Some flatworms resemble long ribbons, whereas others have the shape of a leaf. Tapeworms may grow in the intestinal tract to a length of from 1.5 meters to 15 meters (5–50 feet). They are spread by infected, improperly cooked meats, including beef, pork, and fish. Like that of most intestinal worm parasites, the flatworm's reproductive system is very highly developed, so that each worm produces an enormous number of eggs, which then may contaminate food, water, and soil. Leaf-shaped flatworms, known as *flukes,* may invade various parts of the body, including the blood, lungs, liver, and intestine.

▶ Microbial Control

The Spread of Microorganisms

There is scarcely a place on earth that is naturally free of microorganisms. One exception is the interior of normal body tissue. However, body surfaces and passageways leading to the outside of the body—such as the mouth, throat, nasal cavities, and large intestine—harbor an abundance of both harmless and pathogenic microbes. As is explained in Chapter 17, the body has natural defenses against these organisms. If these natural defenses are sound, a person may harbor many microbes without ill effect. However, if that person's resistance becomes lowered, an infection can result.

Microbes are spread about through an innumerable variety of means. The simplest way is by person-to-person contact. The more crowded the living conditions, the greater are the chances of epidemics. The atmosphere is a carrier of microorganisms. Although microbes cannot fly, the dust of the air is alive with them. In close quarters, the atmosphere is further contaminated by bacteria-laden droplets discharged by sneezing, coughing, and even normal conversation. Pathogens also are spread by such pests as rats, mice, fleas, lice, flies, and mosquitoes. Microbial growth is further abetted by a prevalence of dirt and a lack of sunlight. In poor areas throughout the world, there is often a combination of crowded conditions and poor sanitation. Many inhabitants do not receive immunizations and have lowered resistance because of poor nutrition and other undesirable health practices. As a result, epidemics are apt to begin in these districts.

Microbes and Public Health

All societies establish and enforce measures designed to protect the health of their populations. Most of these practices are concerned with preventing the spread of infectious organisms. A few examples of fundamental public health considerations are listed below:

1. ***Sewage and garbage disposal.*** In times past, when people disposed of the household "slops" by the simple expedient of throwing them out the window, great epidemics were inevitable. Modern practice is to divert sewage into a processing plant in which harmless microbes are put to work destroying the pathogens. The resulting noninfectious "sludge" makes excellent fertilizer.
2. ***Purification of the water supply.*** Drinking water that has become polluted with untreated sewage may be contaminated with such dangerous pathogens as typhoid bacilli, the viruses of polio and hepatitis, and dysentery amebas. The municipal water supply usually is purified by a filtering process, and a close and constant watch is kept on its microbial population. Industrial and chemical wastes, such as asbestos fibers, acids and detergents from homes and from industry, and pesticides used in agriculture, complicate the problem of obtaining pure drinking water.
3. ***Prevention of food contamination.*** Various national, state, and local laws seek to prevent outbreaks of disease through contaminated food. Not only can certain animal diseases (*e.g.,* tuberculosis and tularemia) be passed on, but food is a natural breeding place for many dangerous pathogens. Organisms that cause food poisoning are the botulism bacillus (*Clostridium botulinum*) that grows in improperly canned foods, so-called "staph" (*Staphylococcus aureus*), and species of *Salmonella* transmitted in eggs, poultry, and dairy products. Recently, a variety of the normally harmless intestinal bacillus *Escherichia coli* has caused outbreaks of food poisoning from undercooked meat. For further information, see Table 1 of Appendix 4.

   Most cities have sanitary regulations requiring, among other things, compulsory periodic inspection of food-handling establishments.
4. ***Milk pasteurization.*** Milk is rendered free of pathogens by pasteurization, a process in which the milk is heated to 63°C (145°F) for 30 minutes and then allowed to cool rapidly before being bottled. Sometimes slightly higher temperatures are used for a much shorter time with satisfactory results. The entire pasteurization process, including the cooling and bottling, is accomplished in a closed system, without any exposure to air. Pasteurized milk still contains microbes, but no harmful ones. Pasteurization is also used to preserve other beverages and dairy products.

Aseptic Methods

In the practice of medicine, surgery, nursing, and other health fields, specialized procedures are performed for the purpose of reducing to a minimum the influence of pathogenic organisms. The word *sepsis* means "poisoning due to pathogens"; ***asepsis*** (a-SEP-sis) is its opposite—a condition in which no pathogens are present. Procedures that are designed to kill, remove, or prevent the growth of microbes are called *aseptic methods.*

There are a number of terms designating aseptic practices. These are often confused with one another. Some of the more commonly used terms and their definitions are as follows (Fig. 5-6):

1. ***Sterilization.*** To sterilize an object means to kill *every* living microorganism on it. In operating rooms and delivery rooms especially, as much of the environment as possible is kept sterile, including the gowns worn by operating room personnel and the instruments used. The usual sterilization agent is live steam under pressure in an ***autoclave.*** Dry heat is also used. Ethylene oxide, a gas, is used to sterilize supplies and equipment not able to withstand high temperatures. Most pathogens can be killed by exposure to boiling water for 4 minutes. However, the time and temperature required to ensure the destruction of all spore-forming organisms in sterilization are much greater than those required to kill most pathogens.
2. ***Disinfection.*** Disinfection refers to any measure that kills all pathogens (except spores) but does not necessarily kill all harmless microbes. Most ***disinfectants*** (disinfecting agents) are chemicals that can be applied directly to nonliving surfaces. Examples are chlorine compounds, such as household bleach, and phenol (carbolic acid). Two other terms for bacteria-killing agents, synonymous with ***disinfectant,*** are ***bactericide*** and ***germicide,*** agents that kill bacteria and germs.
3. ***Antisepsis.*** This term refers to any process in which pathogens are not necessarily killed but are prevented from multiplying, a state called ***bacteriostasis*** (bak-te-re-o-STA-sis). (*Stasis* means "steady state.") ***Antiseptics*** are less powerful than disinfectants and are safe to use on living tis-

Figure 5-6

Aseptic methods. (**A**) Sterilization. (**B**) Disinfection. (**C**) Antisepsis.

sues. An antiseptic may be a dilute form of a disinfectant.

UNIVERSAL PRECAUTIONS

Universal precautions have been introduced as a consequence of the prevalence of certain incurable blood-borne diseases—notably hepatitis and AIDS. The health care worker (and everyone else) is encouraged to observe universal blood and body fluid precautions. Universal precautions apply to fluids such as blood, semen, abdominal fluids, amniotic fluids, or any tissues or secretions that may contain these fluids. The health care worker is advised to protect herself or himself as necessary with gloves, gowns, and protective goggles. *Handwashing should be thorough and frequent. Needles are never recapped.* Protection of broken skin is of utmost importance. All waste and laundry from health care facilities is treated as if contaminated.

▶ Chemotherapy

Chemotherapy (ke-mo-THER-ah-pe) means the treatment of a disease by the administration of a chemical agent. The definition has come to include the treatment of disease by any natural or artificial (synthetic) substance.

Antibiotics

An *antibiotic* is a chemical substance produced by living cells. It has the power to kill or arrest the growth of pathogenic microorganisms by upsetting vital chemical processes within them. Antibiotics that are relatively nontoxic to the host are used for the treatment of infectious diseases. Most antibiotics are derived from molds and soil bacteria. Penicillin, the first widely used antibiotic, is made from a common blue mold, *Penicillium*. Another large group of antibiotics is produced by *Cephalosporium*. The many compounds of this family can be recognized by the "cephal" in the name.

Although the development of antibiotics has been of incalculable benefit to humanity, it has given rise to serious complications. One danger is that of opportunistic infections. It may well be that coexisting in the body with disease-causing bacteria is a second type of disease organism, such as a fungus. Up to the time of administration of antibiotics, the fungus is of no danger to the body because its growth is suppressed by the bacteria. However, if antibiotics eliminate the natural enemy of the fungus without affecting the fungus itself, there is nothing to prevent the fungus from growing unrestrainedly and setting up a new infection that is very difficult to cure.

Another danger in the use of antibiotics is the development of allergies (immunologic reactions) to these substances. This complication can have very dangerous consequences.

Finally, the widespread use of antibiotics has resulted in the natural evolution of strains of pathogens that are resistant to such medications. One of the greatest problems in hospitals today is the prevalence of antibiotic-resistant pathogens, including certain streptococci, staphylococci, and bacilli. These pathogens may cause serious infections that are unresponsive to chemotherapy. About 5% of acute care hospital patients contract one or more of these infections. Patients who are elderly or severely debilitated are most susceptible to these *nosocomial* (nos-o-KO-me-al) diseases (hospital-acquired diseases).

Antineoplastic Agents

Antineoplastic agents are a group of chemotherapeutic drugs extensively employed to treat cancers. These agents are toxic to tumor cells but are also damaging to normal cells and should be administered by persons who understand the complications caused by these drugs. Because of their weakened state, such patients are subject to the development of opportunistic infections.

▶ Laboratory Identification of Pathogens

The nurse, physician, or laboratory worker may obtain specimens from patients to identify bacteria and other organisms. Specimens most frequently studied are blood, spinal fluid, feces, urine, and sputum, and swabbings from other areas. Swabs are used to collect specimens from the nose, throat, eyes, and cervix, as well as from ulcers or other infected areas.

Staining Techniques

There are so many different kinds of bacteria requiring identification that the laboratory must use a number of procedures for determining which organisms are present in the material obtained from a patient. One of the most frequently used methods

Antibiotic Resistance

Since their introduction in the 1940s and 1950s, antibiotics have been considered miracle drugs. The first, penicillin, saved many lives on the battlefields of World War II. Over the years, however, enthusiasm for these drugs became too great. Not only were they used in the treatment of human diseases, but they became widely used in agriculture as well, to protect farm animals from infections that cut down on productivity.

The result was predictable. Populations of microorganisms, because they multiply so rapidly, gradually have been selected for resistance to many antibiotics. The organisms fight back by blocking entrance of the drugs, by producing enzymes that inactivate the drugs, by pumping the drugs out of the cell, or by producing alternate compounds that are unaffected by the drugs. As fast as pharmaceutical companies alter the drugs, bacteria find new strategies to evade them.

Already, some common organisms such as strains of staphylococci, pneumococci, *E. coli* (an intestinal organism), and tuberculosis are resistant to most antibiotics. To make matters worse, bacteria resistant to a specific antibiotic can give this resistance directly to other organisms by transfer of genetic material.

for beginning the process of identification involves the application of colored dyes, known as *stains,* to a thin smear of the specimen on a glass slide.

The most commonly used staining procedure is known as the *Gram stain* (see Fig. 5-2). A bluish-purple dye (such as crystal violet) is applied, and then a weak solution of iodine is added. This causes a colorfast combination within certain organisms so that washing with alcohol does not remove the dye. These bacteria are said to be *gram positive* and appear bluish-purple under the microscope. Examples are the pathogenic staphylococci and streptococci; the cocci that cause certain types of pneumonia; and the bacilli that produce diphtheria, tetanus, and anthrax. Other organisms are said to be *gram negative* because the coloring can be removed from them by the use of a solvent. These are then stained for visibility, usually with a red dye. Examples of gram-negative organisms are the diplococci that cause gonorrhea and epidemic meningitis and the bacilli that produce typhoid fever, influenza, and one type of dysentery. The colon bacillus (*E. coli*) normally found in the bowel is also gram negative, as is the cholera vibrio.

Another stain used to identify organisms is the ***acid-fast stain***. After being stained with a reddish dye (carbolfuchsin), the smear is treated with acid. Most bacteria quickly lose their stain upon application of the acid, but the organisms that cause tuberculosis and leprosy remain colored. Such organisms are said to be *acid-fast*.

A few organisms, such as the spirochetes of syphilis and the rickettsias, do not stain with any of the commonly used dyes. Special staining techniques must be used to identify these organisms.

Other Methods

In addition to the various staining procedures, laboratory techniques for identifying bacteria include: (1) growing cells in cultures, a process using substances called ***media*** (such as nutrient broth or agar) that bacteria can use as food; (2) studying the ability of bacteria to act on (ferment) various carbohydrates (sugars); (3) observing reactions to various test chemicals; (4) inoculating animals and analyzing their reactions to the injections; and (5) studying bacteria by serologic (immunologic) tests based on the antigen–antibody reaction (see Chap. 17).

These are only a few of the many laboratory procedures that play a vital part in the process of diagnosing disease.

SUMMARY

I. Causes of disease
  A. Direct—disease-producing organisms, malnutrition, physical and chemical agents, congenital and inherited abnormalities, degeneration, neoplasms
  B. Indirect (predisposing causes)—age, sex, heredity, living conditions and habits, occupation, physical exposure, preexisting illness, psychogenic influences
II. Study of disease—pathophysiology
  A. Terminology
    1. Etiology—study of causation
    2. Incidence—range of occurrence
    3. Disease description
      a. Acute—severe, of short duration
      b. Chronic—less severe, of long duration
      c. Subacute—intermediate between acute and chronic
      d. Idiopathic—of unknown cause
      e. Communicable—transmissible
      f. Epidemic—widespread in a given region
      g. Endemic—characteristic of a given region
      h. Pandemic—prevalent throughout an entire country or the world
      i. Opportunistic—appearing in a weakened host
      j. Nosocomial—hospital-acquired
    4. Diagnosis—determination of the nature of the illness
      a. Symptom—change in body function felt by the patient
      b. Sign—change in body function observable by others
      c. Syndrome—characteristic group of signs and symptoms
    5. Prognosis—prediction of probable outcome of disease
    6. Therapy—course of treatment
    7. Prevention—removal of potential causes of disease
  B. Infection—invasion of body by microorganisms (pathogens)
    1. Parasite—organism that lives on or within a host at host's expense
    2. Modes of transmission—direct and indirect
    3. Portals of entry and exit—skin; respiratory, digestive, and reproductive systems
III. Microorganisms—single-celled organisms visible only with a microscope
  A. Microbiology—the study of microorganisms
    1. Bacteriology—study of bacteria
    2. Mycology—study of fungi

3. Virology—study of viruses
4. Protozoology—study of protozoa
B. Bacteria
  1. Bacilli—straight rods; may produce spores (resistant forms)
  2. Cocci—spheres
    a. Diplococci—pairs
    b. Streptococci—chains
    c. Staphylococci—clusters
  3. Curved rods
    a. Vibrios—comma shaped
    b. Spirilla—corkscrew or wavy
    c. Spirochetes—flexible spirals
  4. Modified bacteria—obligate intracellular parasites
    a. Rickettsias
    b. Chlamydias
C. Fungi—simple, plantlike organisms including yeasts and molds
D. Viruses—smallest infectious agents; obligate intracellular parasites
E. Protozoa—single-celled, animal-like organisms, including amebas, ciliates, flagellates, sporozoa
IV. Parasitic worms—helminths
A. Roundworms—ascaris, pinworms, trichinas, filarias, hookworms
B. Flatworms—tapeworms, flukes

V. Microbial control
A. Spread of microorganisms
B. Public health measures—sewage disposal, water purification, food inspection, milk pasteurization
C. Aseptic methods
  1. Sterilization—total removal of organisms
  2. Disinfection—destruction of all pathogens except spores
  3. Antisepsis—pathogens prevented from multiplying (bacteriostasis)
D. Universal precautions—used on assumption that all body fluids have potential for transmission of disease
VI. Chemotherapy—drug treatment
A. Antibiotics—used for nontoxic treatment of infection
B. Antineoplastics—used for treatment of cancer
VII. Laboratory identification of pathogens
A. Cultivation of organisms
B. Staining of cells
  1. Gram stain—most commonly used stain
  2. Acid-fast stain—used to identify tuberculosis
C. Other tests

QUESTIONS FOR STUDY AND REVIEW

1. What is disease? List five direct and five indirect causes of disease.
2. Explain the difference between the terms in each of the following pairs:
  a. *etiology* and *incidence*
  b. *epidemic* and *endemic*
  c. *diagnosis* and *prognosis*
  d. *acute* and *chronic*
  e. *idiopathic* and *communicable*
  f. *symptom* and *sign*
  g. *pathogen* and *parasite*
  h. *aerobic* and *anaerobic*
3. What are the three characteristic shapes of bacterial cells? Name a typical disease caused by each group.
4. Name the portals of entry for disease.
5. What are spores? Why are they important in the study of disease?
6. In what ways are microorganisms beneficial to humans?
7. How do the rickettsias and the chlamydias differ from other bacteria in size and living habits?
8. Name two diseases caused by rickettsias and two due to chlamydias.
9. What is the typical mode of transmission of rickettsial infections?
10. Name two types of pathogenic fungi and one common fungal infection.
11. List four viral diseases. What does the acronym (abbreviation) *AIDS* mean?
12. What microbial group is described as animal-like? Name three diseases these microbes cause.

13. What are the most common ways by which disease organisms are spread?
14. What measures do communities take to prevent outbreaks of disease?
15. Define the term *asepsis*.
16. Compare the terms *sterilization, disinfection,* and *antisepsis.* How is each accomplished?
17. Why are universal precautions followed? What measures are included in the use of universal precautions?
18. Define *chemotherapy.* Name two types of chemotherapeutic agents.
19. What are some of the disadvantages of the use of antibiotics?
20. Who would be subject to a nosocomial infection? to an opportunistic infection?
21. What are laboratory stains and how are they used?
22. Give examples of gram-positive, gram-negative, and acid-fast organisms.

Selected Key Terms

The following terms are defined in the Glossary:

- dermatitis
- dermis
- epidermis
- erythema
- integument
- keratin
- lesion
- melanin
- sebaceous
- sebum
- subcutaneous
- sudoriferous

Behavioral Objectives

After careful study of this chapter, you should be able to:

- Describe the layers of the integumentary system
- Describe the location and function of the appendages of the skin
- List the main functions of the skin
- Summarize the information to be gained by observation of the skin
- List the main diseases of the skin

Memmler, RL., Cohen, BJ, Wood, DL. *THE HUMAN BODY IN HEALTH AND DISEASE*, 8/e, © 1996 Lippincott-Raven Publishers

The skin is often considered to be merely a membrane covering the body. However, in both structure and function the skin assumes more complex properties. The skin and its appendages are easily observed, giving us clues to the functioning and health of other body systems. The skin is classified in the following three ways:

1. It may be called an enveloping *membrane* because it is a thin layer of tissue that covers the entire body.
2. It may be referred to as an *organ* (the largest one, in fact) because it contains several kinds of tissue, including epithelial, connective, and nervous tissues.
3. It is most properly known as the *integumentary* (in-teg-u-MEN-tar-e) *system* because it includes glands, vessels, nerves, and a subcutaneous (sub-ku-TA-ne-us) layer that work together as a body system. The name is from the word *integument* (in-TEG-u-ment), which means "covering." The term *cutaneous* (ku-TA-ne-us) also refers to the skin.

▶ Structure of the Integumentary System

The integumentary system consists of the skin and the subcutaneous (under the skin) layer, which contains structures extending from the skin. The skin itself consists of two main layers, different from each other in structure and function (Fig. 6-1). These are:

1. The *epidermis* (ep-ih-DER-mis), or outermost layer, which is subdivided into *strata* (STRA-tah), or layers, and is made entirely of epithelial cells with no blood vessels.
2. The *dermis,* or true skin, which has a framework of connective tissue and contains many blood vessels, nerve endings, and glands.

Epidermis

The epidermis is the surface layer of the skin, the outermost cells of which are constantly lost through wear and tear. Since there are no blood vessels in the epidermis, the only living cells are in its deepest layer, the **stratum germinativum** (jer-min-a-TI-vum), where nourishment is provided by capillaries in the underlying dermis. The cells in this layer are con-

stantly dividing and producing daughter cells, which are pushed upward toward the surface. As the surface cells die from the gradual loss of nourishment, they undergo changes. Mainly, they ·develop large amounts of a protein called **keratin** (KER-ah-tin), which serves to thicken and protect the skin.

By the time epidermal cells reach the surface, they have become flat and horny, forming the uppermost layer of the epidermis, the **stratum corneum** (KOR-ne-um). Between the stratum germinativum and the stratum corneum there are additional layers that vary with the thickness of the skin. Cells in the deepest layer of the epidermis produce **melanin** (MEL-ah-nin), a dark pigment that colors the skin; irregular patches of melanin are called *freckles*. Ridges and grooves in the skin of the fingers, palms, toes, and soles form unchanging patterns determined by heredity. These ridges actually serve to prevent slipping, but because they are unique to each individual, fingerprints and footprints are excellent means of identification. The ridges are due to elevations and depressions in the epidermis and the dermis. The deep surface of the epidermis is accurately molded upon the outer part of the dermis, which has raised and depressed areas.

Dermis

The **dermis,** or **corium** (KO-re-um), the so-called "true skin," has a framework of elastic connective tissue and is well supplied with blood vessels and nerves. The thickness of the dermis and epidermis varies. Some areas, such as the soles of the feet and the palms of the hands, are covered with very thick layers of skin, whereas others, such as the eyelids, are covered with very thin and delicate layers. Most of the appendages of the skin, including the sweat glands, the oil glands, and the hair, are located in the dermis and may extend into the subcutaneous layer. Because of its elasticity, the skin can stretch, even dramatically as in pregnancy, with little damage.

Subcutaneous Layer

The dermis rests on the subcutaneous layer, sometimes referred to as the *superficial fascia,* which connects the skin to the surface muscles. This layer consists of elastic and fibrous connective tissue and adipose (fat) tissue. The fat serves as insulation and as a reserve store for energy. Continuous bundles of

nerve endings

pore (opening of sweat gland)

stratum corneum

stratum germinativum (growing layer)

papilla

epidermis

dermis (corium)

sebaceous (oil) gland

muscle

sudoriferous (sweat) gland

subcutaneous layer

fibrous connective tissue

adipose (fat) cells

blood vessels

nerve

hair follicle

pressure receptor

Figure 6-1

Cross section of the skin.

elastic fibers connect the subcutaneous tissue with the dermis, so there is no clear boundary between the two. The major blood vessels that supply the skin run through the subcutaneous layer. The blood vessels that are concerned with temperature regulation are located here. Some of the appendages of the skin, such as sweat glands and hair roots, extend into the subcutaneous layer. The subcutaneous tissues are rich in nerves and nerve endings, including those that supply the dermis and epidermis. The thickness of the subcutaneous layer varies in different parts of the body, the layer being at its thinnest on the eyelids and at its thickest on the abdomen.

Figure 6-2 is a photograph of the skin and its appendages as seen through a microscope.

▶ Appendages of the Skin

Sweat Glands

The *sudoriferous* (su-do-RIF-er-us) *glands,* or sweat glands, are coiled, tubelike structures located in the dermis and the subcutaneous tissue. Each gland has an excretory tube that extends to the surface and opens at a pore. The slant at which the excretory tube joins the skin serves as a valve. Sudoriferous glands function to regulate body temperature through the evaporation of sweat from the body surface. Sweat consists of water with small amounts of mineral salts and other substances.

Another type of sweat gland is located mainly in the armpits and the groin area. These glands release their secretions through the hair follicles in re-hair follicle   sebaceous gland

epidermis

dermis

subcutaneous
adipose tissue

sweat gland

Figure 6-2

Microscopic view of thin skin showing tissue layers and some appendages. (Adapted from Cormack DH: Essential Histology, Plate 12-1B. Philadelphia, JB Lippincott, 1993)

sponse to emotional stress and sexual stimulation. The secretions contain cellular material that is broken down by bacteria, producing body odor.

Several types of glands associated with the skin are modified sweat glands. These are the ***ceruminous*** (seh-RU-min-us) ***glands*** in the ear canal that produce ear wax, or ***cerumen,*** the ***ciliary*** (SIL-e-er-e) ***glands*** at the edges of the eyelids, and the ***mammary glands.***

Sebaceous Glands

The ***sebaceous*** (se-BA-shus) ***glands*** are saclike in structure, and their oily secretion, ***sebum*** (SE-bum), lubricates the skin and hair and prevents drying. The ducts of the sebaceous glands open into the hair follicles.

Babies are born with a covering produced by these glands that resembles cream cheese; this secretion is called the ***vernix caseosa*** (VER-niks ka-se-O-sah).

Blackheads consist of a mixture of dried sebum and keratin that may collect at the openings of the sebaceous glands. If these glands become infected, pimples result. If sebaceous glands become blocked by accumulated sebum, a sac of this secretion may form and gradually increase in size. These sacs are referred to as ***sebaceous cysts.*** Usually it is not difficult to remove such tumorlike cysts by surgery.

Hair and Nails

Almost all of the body is covered with hair, which in most areas is very soft and fine. Hair is composed mainly of keratin and is not living. Each hair develops within a sheath called a ***follicle,*** and new hair is formed from cells at the bottom of the follicles (see Fig. 6-1). Attached to most hair follicles is a thin band of involuntary muscle. When this muscle contracts, the hair is raised, forming "goose bumps" on the skin. As it contracts, the muscle presses on the sebaceous gland associated with the hair follicle, causing the release of sebum.

Nails are protective structures made of hard keratin produced by cells that originate in the outer layer of the epidermis (stratum corneum). New cells form continuously at the proximal end of the nail in an area called the ***nail root.*** Nails of both the toes and the fingers are affected by general health.

Changes in nails, including abnormal color, thickness, shape, or texture (*e.g.,* grooves or splitting) occur in chronic diseases such as heart disease, peripheral vascular disease, malnutrition, and anemia.

◗ Functions of the Skin

Although the skin has several functions, the three that are by far the most important are the following:

1. Protection of deeper tissues against drying and against invasion by pathogenic organisms or their toxins
2. Regulation of body temperature by dissipation of heat to the surrounding air
3. Receipt of information about the environment by means of the many nerve endings distributed throughout the skin. Much of this sensory infor-

Vitamin D

One function of the skin is the synthesis of vitamin D. This "vitamin" is actually a group of related fat-soluble compounds that aids in the absorption of calcium from foods. It is manufactured from a substance in the skin that is activated by the ultraviolet (UV) radiation in sunlight. Enzymes in the liver and kidneys then modify it into its most active form. Since it is manufactured in the body and carried in the blood to the tissues on which it acts, vitamin D is now considered to be a hormone.

A lack of vitamin D results in bone malformation. There is no dietary requirement for the vitamin because even moderate exposure to sunlight is enough to satisfy our needs. Exposure of the arms and face for only 1 hour per week is enough to avoid a deficiency. As a safeguard, all milk sold in the United States is fortified with vitamin D.

mation has a protective function. For example, it prompts one to withdraw from harmful stimuli, such as a hot stove.

The outermost layer of the skin, the stratum corneum, protects the body against invasions by pathogens and against dehydration. These dry, dead cells, composed of keratin, are found in a tight, interlocking pattern that is impervious to penetration by organisms and by water. The outermost cells are constantly being shed, causing the mechanical removal of pathogens. The function of epidermis as a water barrier is vital to provide the wet environment required by cells.

The regulation of body temperature, both the loss of excess heat and the protection from cold, is a very important function of the skin. Indeed, most of the blood that flows through the skin is concerned with temperature regulation. The skin forms a large surface for radiating body heat to the air. When the blood vessels dilate (enlarge), more blood is brought to the surface so that heat can be dissipated. The activation of sweat glands, and the evaporation of sweat from the surface of the body also helps cool the body. In cold conditions, the flow of small amounts of blood in veins deep in the subcutaneous tissue serves to heat the skin and protect

deeper tissues from excess heat loss. Special vessels that directly connect arteries and veins in the skin of the ears, nose, and other exposed locations provide the volume of blood flow needed to prevent freezing. As is the case with so many body functions, temperature regulation is complex and involves several parts of the body, including certain centers in the brain.

Another important function of the skin is obtaining information from the environment. Because of the many nerve endings and other special receptors for pain, touch, pressure, and temperature, which are located mostly in the dermis, the skin may be regarded as one of the chief sensory organs of the body. Many of the reflexes that make it possible for humans to adjust themselves to the environment begin as sensory impulses from the skin. Here, too, the skin works with the brain and the spinal cord to make these important functions possible.

The functions of absorption and excretion are minimal in the skin. Most medicated ointments used on the skin are for the treatment of local conditions only, and the injection of medication into the subcutaneous tissues is limited by the slow absorption that occurs here. The skin excretes a mixture of water and mineral salts in perspiration. Some nitrogen-containing wastes are also eliminated through the skin, but even in disease the amount of waste products excreted by the skin is small.

The human skin does not "breathe." The pores of the epidermis serve only as outlets for perspiration and oil from the sweat glands and sebaceous glands.

The skin has the ability to regenerate after injury. Skin that is injured can recover, even to the extent of growing a new epidermis from the cells of the hair follicle. During healing, a *scab,* or crust, made of dried pus and blood may form over the wound.

▶ Observation of the Skin

What can the skin tell you? What do its color, texture, and other attributes indicate? Is there any damage? Much can be learned by an astute observer. In fact, the first indication of a serious systemic disease (such as syphilis) may be a skin disorder.

The color of the skin depends on a number of factors, including the following:

1. The amount of pigment in the epidermis

2. The quantity of blood circulating in the surface blood vessels
3. The composition of the circulating blood
   a. Presence or absence of oxygen
   b. Concentration of hemoglobin
   c. Presence of bile, silver compounds, or other chemicals

Pigment

The main pigment of the skin, as we have noted, is called *melanin.* This pigment is also found in the hair, the middle coat of the eyeball, the iris of the eye, and certain tumors. Melanin is common to all races, but darker people have a much larger quantity in their tissues. The melanin in the skin helps to protect against damaging ultraviolet radiation from the sun. Thus, skin that is exposed to the sun shows a normal increase in this pigment, a response we call *tanning.*

Sometimes there are abnormal increases in the quantity of melanin, which may occur either in localized areas or over the entire body surface. For example, diffuse spots of pigmentation may be characteristic of some endocrine disorders.

Discoloration

A yellowish discoloration of the skin may be due to the presence of excessive quantities of bilirubin (bile pigment) in the blood. This condition, called *jaundice* (JAWN-dis), may be a symptom of a number of disorders, such as the following:

1. A tumor pressing on the common bile duct or a stone within the duct, either of which would obstruct the flow of bile into the small intestine
2. Inflammation of the liver (hepatitis)
3. Certain diseases of the blood in which red blood cells are rapidly destroyed.

Another cause of a yellowish discoloration of the skin is the excessive intake of carrots and other deeply colored vegetables. This condition is known as *carotenemia* (kar-o-te-NE-me-ah).

Chronic poisoning may cause gray or brown discoloration of the skin. A peculiar bronze cast is present in Addison's disease (malfunction of the adrenal gland). Many other disorders cause discoloration of the skin, but their discussion is beyond the scope of this chapter.

Lesions

A *lesion* (LE-zhun) is any wound or local damage to tissue. In examining the skin for lesions, it is important to make note of their type, arrangement, and location. Lesions may be flat or raised, or may extend below the surface of the skin (Fig. 6-3). A surface lesion is often called a *rash* or, if raised, an *eruption* (e-RUP-shun). Skin rashes may be localized, as in diaper rash, or generalized, as in measles and other systemic infections. Often these lesions are accompanied by *erythema* (er-eh-THE-mah), or redness of the skin. Some terms used to describe surface skin lesions are:

1. *Macules* (MAK-ules). These spots are neither raised nor depressed. They are typical of measles and descriptive of freckles.
2. *Papules* (PAP-ules). These are firm, raised areas, as in some stages of chickenpox and in the second stage of syphilis. Pimples are papules. A large firm papule is called a *nodule* (NOD-ule).
3. *Vesicles* (VES-ih-klz). These are blisters or small sacs that are full of fluid, such as may be found in some of the eruptions of chickenpox.
4. *Pustules* (PUS-tules). These are vesicles filled with pus. They may develop in chickenpox if the vesicles become infected.

A deeper lesion may develop from a surface lesion or may be caused by *trauma* (TRAW-mah), that is a wound or injury. Because such breaks may be followed by infection, wounds should be cared for to prevent the entrance of pathogens and toxins into deeper tissues and body fluids. Deeper injuries to the skin include:

1. *Excoriation* (eks-ko-re-A-shun), which is a scratch of the skin surface
2. *Laceration* (las-er-A-shun), which is a rough, jagged wound made by tearing of the skin
3. *Ulcer* (UL-ser), which is a sore associated with disintegration and death of tissue
4. *Fissure* (FISH-ure), which is a crack in the skin as seen in athlete's foot.

Effects of Aging on Skin, Hair, and Nails

As people age, wrinkles, or crow's feet, develop around the eyes and mouth owing to the loss of fat and collagen in the underlying tissues. The dermis becomes thinner, and the skin may become quite

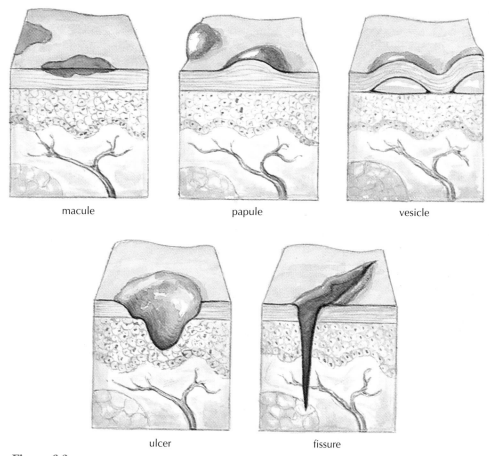

Figure 6-3

Some common skin lesions.

transparent and lose its elasticity, the effect of which is so-called "parchment skin." The hair does not replace itself as rapidly as before and thus becomes thinner. The formation of pigment also decreases with age, causing hair to become gray or white. However, there may be localized areas of extra pigmentation in the skin with the formation of brown spots, especially on areas exposed to the sun (*e.g.,* the hands). The sweat glands decrease in number, so there is less output of perspiration. The fingernails may flake, become brittle, or develop ridges. Toenails may become discolored or abnormally thickened.

Hair Testing for Genetic Identification

The living cells of the hair root can be used as sources of genetic material for the creation of a genetic fingerprint. Each person's genetic information (DNA) is easily separated into segments of varying lengths. The pattern of these segments is unique to each individual. Sorted, and then put in order by length, the resultant pattern looks much like a supermarket bar code. Comparisons of the codes of related individuals will show many matching bars. Pictures of the pattern can be stored and used later for identification (*e.g.,* of criminals or missing children).

▶ Care of the Skin

The most important factors in caring for the skin are those that ensure good general health. Proper nutrition and adequate circulation are vital to the maintenance of the skin. Regular cleansing removes dirt and dead skin debris and sustains the slightly

Burns

Burns of the skin may have serious consequences because they destroy the body's outer protective layer and may lead to infection, dehydration, and other complications.

Burns are classified according to the depth of the damage as burns of the first, second, and third degree. First-degree burns involve the outermost layer of the skin—the epidermis—which becomes red and painful. Most sunburns are first-degree burns. Pain and damage from a first-degree burn can be reduced by bathing the wound in cold water.

Second-degree burns penetrate into deeper layers and often cause the formation of blisters. Severe sunburns or scalding with hot water are examples of this type. Third-degree burns are the most serious and involve the full thickness of the skin, and often the underlying tissues such as muscle and connective tissue.

The percentage of the body's total skin area damaged by a burn is used as a measure of the seriousness of the injuries and the chances for the patient's survival. Skin grafts are used in the treatment of serious burns. Such grafts may be taken from the normal parts of the body or they may be cultured from a small sample of the skin obtained from the patient and grown in the laboratory.

acid environment that inhibits bacterial growth on the skin. Careful handwashing with soap and water, with attention to the undernail areas, is a simple measure that reduces the spread of disease.

The skin needs protection from continued exposure to sunlight to prevent premature aging and cancerous changes. Appropriate applications of sunscreens before and during time spent in the sun can prevent skin damage.

▶ Skin Diseases

Dermatosis and Dermatitis

Dermatosis (der-mah-TO-sis) is a general term referring to any skin disease. Inflammation of the skin is called *dermatitis* (der-mah-TI-tis). It may be due to many kinds of irritants, such as the oil of poison oak or poison ivy plants, detergents, and strong acids, alkalis, or other chemicals. Prompt removal of the irritant is the most effective method of prevention and treatment. A thorough soap-and-water bath taken as soon as possible after contact with plant oils may prevent the development of itching eruptions.

Atopic Dermatitis

Atopic dermatitis (ah-TOP-ik der-mah-TI-tis) or *eczema* (EK-ze-mah) is characterized by intense itching and skin inflammation. The affected areas show redness (erythema), blisters (vesicles), pimple-like lesions (papules), and scaling and crusting of the skin surface. Scratching of the skin can lead to a secondary bacterial infection. This disorder commonly first occurs in early childhood, with the recurrence of acute episodes throughout life. The skin may be excessively sensitive to many soaps, detergents, rough fabrics, or perspiration. The person with atopic dermatitis may also be subject to allergic disorders such as hay fever, asthma, and food allergies.

Sunburn

Sunlight may cause chemical and biologic changes in the skin. The skin first becomes reddened (erythematous) and then may become swollen and blistered. Some persons suffer from severe burns and become seriously ill. There is considerable evidence that continued excessive exposure to the sun is an important cause of skin cancer. Tanning requires the skin to protect itself by producing considerably more than usual amounts of melanin. This increase in pigmentation may have the effect of reducing the body's ability to profit from the desirable smaller amounts of sun available during some parts of the year.

Common Acne

Acne (AK-ne) is a disease of the sebaceous (oil) glands connected with the hair follicles. The common type, called *acne vulgaris* (vul-GA-ris), is found most often in individuals between the ages of 14 and 25. The infection of the oil glands takes the form of pimples, which generally surround blackheads. Acne is usually most severe at adolescence, when certain endocrine glands in the body that control the secretions of the sebaceous glands are particularly active.

Impetigo

Impetigo (im-peh-TI-go) is an acute contagious disease of staphylococcal or streptococcal origin that may be serious enough to cause death in newborn infants. It takes the form of blister-like lesions that become filled with pus and contain millions of virulent bacteria. It is found most frequently among poor and undernourished children. Affected persons may reinfect themselves or infect others.

Alopecia (Baldness)

Alopecia (al-o-PE-she-ah), or baldness, may be due to a number of factors. The most common type, known as male pattern baldness, is an expression of heredity and aging; it is influenced by male sex hormones. Topical applications of the drug minoxidil (used as an oral medication to control blood pressure) have produced growth of hair in this type of baldness. Alopecia may be the result of a systemic disease, such as uncontrolled diabetes, thyroid disease or malnutrition; in such cases, control of the disease will result in regrowth of hair. A growing list of drugs has been linked with baldness, including the chemotherapeutic drugs used in treating neoplasms.

Athlete's Foot

Fungi are the usual cause of athlete's foot, also known as *epidermophytosis* (ep-ih-der-mo-fi-TO-sis). The disease most commonly involves the toes and the soles but occasionally affects the fingers, the palms, and the groin region. In acute cases the lesions may include vesicles, fissures, and ulcers. Predisposition to fungal infection varies. Some individuals may be exposed to pathogenic fungi with no ill effects, whereas other persons develop severe skin infections with only mild exposure. Those who perspire a great deal are particularly susceptible to athlete's foot.

Other Disorders of the Skin

In addition to the disorders discussed above, other skin lesions include:

1. *Furuncle* (FU-rung-kl), or boil, which is a localized collection of pus in a cavity formed by the disintegration of tissue. They are caused by bacteria that enter hair follicles or sebaceous glands.

2. *Carbuncle* (KAR-bung-kl), which is a pus-producing lesion that results from the extension of an infectious process, such as a boil. They involve both the skin and subcutaneous tissues and have numerous drainage channels that extend to the skin surface.

3. *Psoriasis* (so-RI-ah-sis), which is characterized by sharply outlined, red, flat areas (plaques) covered with silvery scales. The cause of this chronic, recurrent skin disease is unknown.

4. *Herpes* (HER-peze) *simplex,* which is characterized by the formation of watery vesicles (cold sores, fever blisters) on the skin and mucous membranes, including the genital area (see Appendix 4 Table 3).

5. *Shingles* (herpes zoster), which is a viral disease seen in adults and is caused by the same virus that causes chickenpox. Infection follows nerve pathways, producing small lesions on the skin. Vesicular lesions may be noted along the course of a nerve. Pain, increased sensitivity, and itching are common symptoms that usually last longer than a year. Prompt treatment with antiviral drugs decreases the severity of this disease.

6. *Cancer* of the skin, which in the United States is most common among persons who have fair skin and who live in the Southwest, where exposure to the sun is consistent and may be intense. Early treatment in most cases means cure. Neglect can result in death.

7. *Urticaria* (ur-tih-KA-re-ah), or hives, which is an allergic reaction characterized by the temporary appearance of elevated red patches (wheals) often accompanied by severe *pruritus* (pru-RI-tus), or itching.

8. *Scleroderma* (skle-ro-DER-mah), which together with some of the metabolic diseases, such as certain forms of lupus erythematosus, causes thickening of the dermis.

9. *Decubitus* (de-KU-bih-tus) *ulcers* (bedsores, pressure ulcers) are ulcerous areas seen in bedridden, poorly nourished patients with decreased circulation. The immediate cause is impaired blood supply of an area of skin that is pressed between bone and the bed by the patient's weight. Prevention by frequent position change and adequate nutrition is far easier than treatment of an established ulcer.

SUMMARY

I. **Classification of the skin**
   A. Membrane (cutaneous)
   B. Organ
   C. System (integumentary)

II. **Structure of the integumentary system**
   A. Epidermis—surface layer of the skin
   B. Dermis—deeper layer of the skin
   C. Subcutaneous layer—under the skin

III. **Appendages of the skin**
   A. Sweat (sudoriferous) glands
   B. Sebaceous (oil) glands
   C. Hair
   D. Nails

IV. **Functions of the skin**
   A. Protection
   B. Regulation of body temperature
   C. Sensory perception

V. **Observation of the skin**
   A. Pigment
   B. Discoloration
   C. Lesions
      1. Surface lesions (rashes, eruptions)
      2. Deeper injuries
   D. Effects of aging
   E. Hair testing for genetic identification

VI. **Care of the skin**

VII. **Skin diseases**—dermatoses
   A. Dermatitis—inflammation
   B. Atopic dermatitis (eczema)—redness, blisters, lesions
   C. Sunburn—may lead to skin cancer
   D. Acne—disease of sebaceous glands related to increased endocrine secretions
   E. Impetigo—infectious disease of infants and children
   F. Alopecia—baldness
   G. Athlete's foot—fungal infection
   H. Other disorders

QUESTIONS FOR STUDY AND REVIEW

1. What characteristics of the skin classify it as a membrane? as an organ? as a system?
2. Of what type of cells is the epidermis composed?
3. Explain how the outermost cells of the epidermis are replaced.
4. Describe the structure of the dermis.
5. Describe the contents and functions of the subcutaneous layer.
6. Describe the location and function of the skin glands.
7. Explain the three most important functions of the skin.
8. What are the most important contributors to the color of the skin, normally?
9. Define melanin and explain its function.
10. Name and describe four types of surface skin lesions (rashes).
11. Name and describe four types of deeper skin lesions.
12. What changes may occur in the skin with age?
13. What is the difference between dermatosis and dermatitis?
14. What are some examples of irritants that frequently cause dermatitis?
15. What are the dangers of overexposure to the sun, and what precautions against it need to be considered?
16. Define *acne*. When is it usually most severe? Why?
17. What are the most common causes of baldness?
18. What are the best measures to take to prevent and control athlete's foot?
19. Define *furuncles, plaques, wheals,* and *pruritus.*
20. List the two best measures for preventing decubitus ulcers.
21. List some factors involved in proper care of the skin.

UNIT III

Movement and Support

This unit deals with the skeletal and muscular systems. It covers the functions of the skeletal system, going beyond support purposes and including those related to blood formation and to the storage and metabolism of certain mineral salts. The important muscles and their functions in various movements, as well as their ability to produce heat and to aid in the circulation of body fluids, are noted in Chapter 8, The Muscular System.

CHAPTER 7

The Skeleton: Bones and Joints

Selected Key Terms

The following terms are defined in the Glossary:

amphiarthrosis

arthritis

bursa

circumduction

diaphysis

diarthrosis

endosteum

epiphysis

fontanelle

joint

osteoblast

osteoclast

osteocyte

osteoporosis

periosteum

resorption

synarthrosis

synovial

Behavioral Objectives

After careful study of this chapter, you should be able to:

- Name the three different types of bone cells and describe the functions of each
- Differentiate between compact bone and spongy bone with respect to structure and location
- Describe the structure of a long bone
- Explain how a long bone grows
- Differentiate between red and yellow marrow with respect to function and location
- Name and describe various markings found on bones
- List the bones in the axial skeleton
- List the bones in the appendicular skeleton
- Describe five bone disorders
- Describe three abnormal curves of the spine
- Describe how the skeleton changes with age
- List and define six types of fractures
- Describe the three types of joints
- Describe the structure of a synovial joint and give six examples of synovial joints
- Define six types of movement that occur at synovial joints
- Describe four types of arthritis

Memmler, RL, Cohen, BJ, Wood, DL. *THE HUMAN BODY IN HEALTH AND DISEASE*, 8/e, © 1996 Lippincott-Raven Publishers

The bones are the framework of the body. They are a combination of several kinds of tissue and contain blood vessels and nerves. Bones are attached to each other at joints. The combination of bones and joints together with related connective tissues forms the skeletal system.

▶ The Bones

Bone Structure

The bones are composed chiefly of bone tissue, called *osseous* (OS-e-us) *tissue.* Bones are not lifeless. Even though the spaces between the cells of bone tissue are permeated with stony deposits of calcium salts, these cells themselves are very much alive. Bones are organs, with their own system of blood vessels, lymphatic vessels, and nerves.

In the embryo (the early developmental stage of a baby) most of the bones-to-be are composed of cartilage (Portions of the skull develop from fibrous connective tissue.) Bone formation begins during the second and third months of embryonic life. At this time, bone-building cells, called *osteoblasts* (OS-te-o-blasts), become very active. First, they manufacture a substance, the *intercellular* (in-ter-SEL-u-lar) *material,* which is located between the cells and contains large quantities of a protein called *collagen.* Then, with the help of enzymes, calcium compounds are deposited within the intercellular material. Once the intercellular material has hardened around these cells, they are called *osteocytes.* They are still living and continue to maintain the bone, but they do not produce new bone tissue. Other cells, called *osteoclasts* (OS-te-o-klasts), are responsible for the process of *resorption,* or the breakdown of bone. Enzymes also implement this process.

The complete bony framework of the body, known as the *skeleton* (Fig. 7-1), consists of bones of many different shapes. They may be flat (rib, skull), cube shaped (wrist, ankle), or irregular (vertebrae). However, typical bones, which make up most of the arms and legs, are described as *long bones.* This type of bone has a long shaft, called the *diaphysis* (di-AF-ih-sis), which has a central marrow cavity and two irregular ends, each called an *epiphysis* (eh-PIF-ih-sis) (Fig. 7-2).

There are two types of bone tissue. One type is *compact bone,* which is hard and dense. This makes up the main shaft of a long bone and the outer layer of other bones. The osteocytes in this type of bone are located in rings of bone tissue around a central *Haversian* (ha-VER-shan) *canal* containing nerves and blood vessels. The second type, called *spongy bone,* has more spaces than compact bone. It is made of a meshwork of small, bony plates filled with red marrow and is found at the ends of the long bones and at the center of other bones.

Bones contain two kinds of marrow: *red marrow,* found at the end of the long bones and at the center of other bones, which manufactures blood cells; and *yellow marrow* of the "soup bone" type, found chiefly in the central cavities of the long bones. Yellow marrow is largely fat.

Bones are covered on the outside (except at the joint region) by a membrane called the *periosteum* (per-e-OS-te-um). The inner layer of this membrane contains osteoblasts, which are essential in bone formation, not only during growth but also in the repair of fractures. Blood vessels and lymphatic vessels in the periosteum play an important role in the nourishment of bone tissue. Nerve fibers in the periosteum make their presence known when one suffers a fracture, or when one receives a blow, such as on the shinbone. A thinner membrane, the *endosteum* (en-DOS-te-um), lines the marrow cavity of a bone; it too contains cells that aid in the growth and repair of bone tissue.

Bone Growth and Repair

In a long bone, the transformation of cartilage into bone begins at the center of the shaft. Later, secondary bone-forming centers develop across the ends of the bones. The long bones continue to grow in length at these centers through childhood and into the late teens. Finally, by the late teens or early 20s, the bones stop growing in length. Each bone-forming region hardens and can be seen in x-ray films as a thin line across the end of the bone. Physicians can judge the future growth of a bone by the appearance of these lines on x-ray films.

As a bone grows in length, the shaft is remodeled so that it grows wider as the central marrow cavity increases in size. Thus, alterations in the shape of the bone are a result of the addition of bone tissue to some surfaces and its resorption from others.

The processes of bone formation and bone resorption continue throughout life, more rapidly at some times than at others. The bones of small children are relatively pliable because they contain

cranium

facial bones

mandible

clavicle

scapula

sternum

humerus

costal cartilage

ribs

vertebral column

ilium

radius

pelvis

ulna

sacrum

carpals

metacarpals

phalanges

femur

knee joint

patella

tibia

fibula

ankle

calcaneus

axial skeleton

appendicular skeleton

tarsals

metatarsals

phalanges

Figure 7-1

The skeleton.

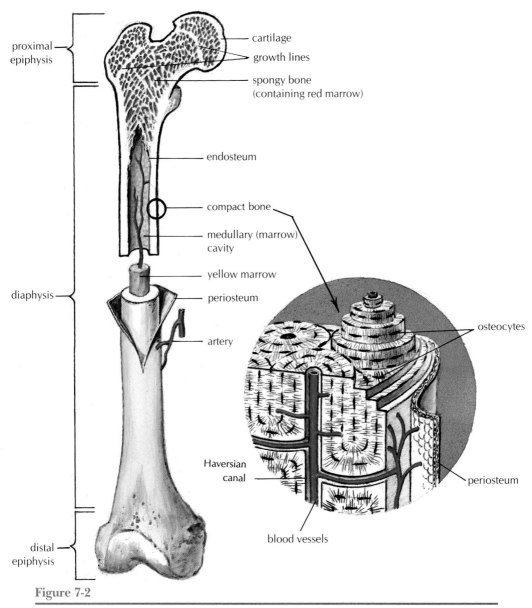

Figure 7-2

The structure of a long bone; the composition of compact bone.

a larger proportion of cartilage and are undergoing active bone formation. In the elderly, there is a slowing of the processes that continually renew bone tissue. As a result, the bones are weaker and more fragile. The elderly also have a decreased ability to form the protein framework on which calcium salts are deposited. Fractures in old people heal more slowly due to these decreases in bone metabolism.

Main Functions of Bones

Bones have a number of functions, many of which are not at all obvious. Some of these are listed below:

1. To serve as a firm framework for the entire body
2. To protect such delicate structures as the brain and the spinal cord
3. To serve as levers, working with attached muscles to produce movement

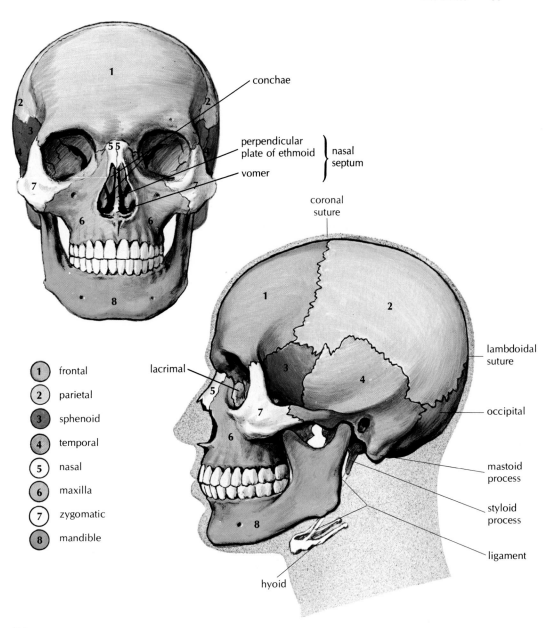

Figure 7-3

The skull, from the front and from the left.

1. frontal
2. parietal
3. sphenoid
4. temporal
5. nasal
6. maxilla
7. zygomatic
8. mandible

4. To serve as a storehouse for calcium salts, which may be resorbed into the blood if there is not enough calcium in the diet
5. To produce blood cells (in the red marrow).

Bone Markings

In addition to their general shape, bones have other distinguishing features called ***bone markings.***

These markings include raised areas and depressions that help to form joints or serve as points for muscle attachments, and various holes that allow the passage of nerves and blood vessels. Some of these identifying features are:

1. ***Projections***
   a. ***Head***, a rounded knob-like end separated from the rest of the bone by a slender region, the neck

maxilla
palatine } hard palate

zygomatic

vomer

sphenoid

styloid process

mastoid process

temporal

parietal

occipital

foramen
magnum

Figure 7-4

The skull from below, lower jaw removed.

b. ***Process***, a large projection of a bone, such as the upper part of the ulna in the forearm that creates the elbow

c. ***Crest***, a distinct border or ridge, often rough, such as over the top of the hip bone

d. ***Spine***, a sharp projection from the surface of a bone, such as the spine of the scapula (shoulder blade)

2. ***Depressions*** or ***holes***

a. ***Foramen*** (fo-RA-men), a hole that allows a vessel or a nerve to pass through or between bones. The plural is ***foramina*** (fo-RAM-ih-nah).

b. ***Sinus*** (SI-nus), an air space found in some skull bones

c. ***Fossa*** (FOS-sah), a depression on a bone surface. The plural is fossae (FOS-se).

d. ***Meatus*** (me-A-tus), a short channel or passageway, such as the channel in the temporal bone of the skull that leads to the inner ear.

Examples of these and other markings will be seen on the bones illustrated in this chapter.

Divisions of the Skeleton

The skeleton may be divided into two main groups of bones (see Fig. 7-1):

1. The ***axial*** (AK-se-al) ***skeleton***, which includes the bony framework of the head and the trunk

2. The ***appendicular*** (ap-en-DIK-u-lar) ***skeleton***, which forms the framework for the ***extremities*** (limbs) and for the shoulders and hips.

FRAMEWORK OF THE HEAD

The bony framework of the head, called the ***skull***, is subdivided into two parts: the cranium and the facial portion. Refer to Figures 7-3 through 7-6, which show different views of the skull, as you study the following descriptions. The numbers in brackets refer to the numbers in Figure 7-3.

A. The ***cranium*** is a rounded box that encloses the brain; it is composed of eight distinct cranial bones.

1. The ***frontal bone*** [1] forms the forehead, the front of the skull's roof, and part of the roof over the eyes and the nasal cavities. The ***frontal sinuses*** (air spaces) communicate with the

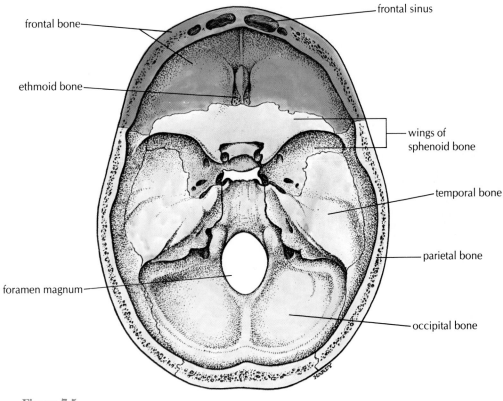

frontal sinus

frontal bone

ethmoid bone

wings of
sphenoid bone

temporal bone

parietal bone

foramen magnum

occipital bone

Figure 7-5

Base of the skull as seen from above, showing the internal surfaces of some of the cranial bones. (Chaffee EE, Lytle IM: *Basic Physiology and Anatomy*, 4th ed, p. 95. Philadelphia, JB Lippincott, 1980)

nasal cavities. These sinuses and others near the nose are described as *paranasal sinuses.*

2. The two *parietal* (pah-RI-eh-tal) bones [2] form most of the top and the side walls of the cranium.

3. The two *temporal bones* [4] form part of the sides and some of the base of the skull. Each one contains *mastoid sinuses* as well as the ear canal, the eardrum, and the entire middle and internal ears. The *mastoid process* of the temporal bone projects downward immediately behind the external part of the ear. It contains the mastoid air cells and serves as a place for muscle attachment.

4. The *ethmoid* (ETH-moyd) *bone* is a very light, fragile bone located between the eyes. It forms a part of the medial wall of the eye sockets, a small portion of the cranial floor, and most of the nasal cavity roof. It contains several air spaces—comprising some of the

paranasal sinuses. A thin, platelike, downward extension of this bone (the perpendicular plate) forms much of the nasal septum, a midline partition in the nose.

5. The *sphenoid* (SFE-noyd) *bone* [3], when seen from above, resembles a bat with its wings extended. It lies at the base of the skull in front of the temporal bones.

6. The *occipital* (ok-SIP-ih-tal) *bone* forms the back and a part of the base of the skull. The *foramen magnum*, located at the base of the occipital bone, is a large opening through which the spinal cord communicates with the brain (Figs. 7-4 and 7-5).

B. The *facial portion* of the skull is composed of 14 bones.

1. The *mandible* (MAN-dih-bl) [8], or lower jaw bone, is the only movable bone of the skull.

2. The two *maxillae* (mak-SIL-e) [6] fuse in the midline to form the upper jaw bone, includ-

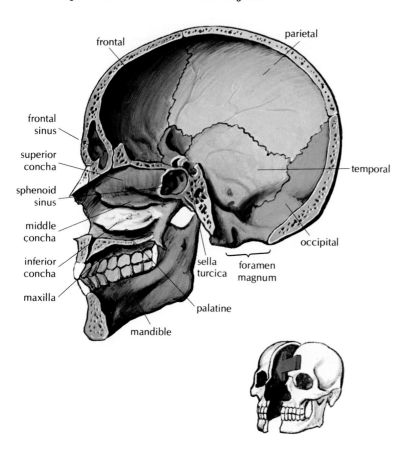

frontal
parietal
frontal
sinus
superior
concha
sphenoid
sinus
middle
concha
temporal
inferior
concha
occipital
maxilla
sella
turcica
foramen
magnum
palatine
mandible

Figure 7-6

The skull, internal view.

ing the front part of the hard palate (roof of the mouth). Each maxilla contains a large air space, called the *maxillary sinus,* that communicates with the nasal cavity.

3. The two *zygomatic* (zi-go-MAT-ik) *bones* [7], one on each side, form the prominences of the cheeks.
4. Two slender *nasal bones* [5] lie side by side, forming the bridge of the nose.
5. The two *lacrimal* (LAK-rih-mal) *bones,* each about the size of a fingernail, lie near the inside corner of the eye in the front part of the medial wall of the orbital cavity.
6. The *vomer* (VO-mer), shaped like the blade of a plow, forms the lower part of the nasal septum.
7. The paired *palatine bones* form the back part of the hard palate.
8. The two *inferior nasal conchae* (KON-ke) extend horizontally along the lateral wall (sides) of the nasal cavities. The paired superior and middle conchae are part of the ethmoid bone.

In addition to the bones of the cranium and the facial bones, there are three tiny bones, or *ossicles* (OS-sik-ls), in each middle ear (see Chap. 11) and a single horseshoe, or U-shaped, bone just below the skull proper, called the *hyoid* (HI-oyd) *bone,* to which the tongue and other muscles are attached.

Openings in the base of the skull provide spaces for the entrance and exit of many blood vessels, nerves, and other structures. Projections and slightly elevated portions of the bones provide for the attachment of muscles. Some portions contain delicate structures, such as the part of the temporal bone that encloses the middle and internal sections of the ear. The air sinuses provide lightness and serve as resonating chambers for the voice.

THE INFANT SKULL

The skull of the infant has areas in which the bone formation is incomplete, leaving so-called "soft spots." Although there are a number of these, the largest and best known is near the front at the junction of the two parietal bones with the frontal bone. It is called the *anterior fontanelle* (fon-tah-NEL), and

The Timetable of Bone Formation

The conversion of cartilage to bone, called *ossification*, begins in the fetus by the 3rd month. At birth, evidence of ossification should be noted in most of the bones. The long bones begin to ossify in the center of the shaft (diaphysis). At birth, secondary ossification centers are apparent at each end (epiphyses). The short bones usually have only one center of ossification.

During the first 12 years of life, bone formation continues throughout the entire skeleton. By 17 to 20 years of age the bones of the upper extremities, and the scapulae, but not the clavicles, are ossified. By age 18 to 23, the lower limbs, and the pelvis are converted to bone tissue. By age 23 to 25 the clavicles, the sternum, and the vertebrae are totally ossified, so that by 25 years of age bone formation is complete.

it usually does not close until the child is about 18 months old (Fig. 7-7).

FRAMEWORK OF THE TRUNK

The bones of the trunk include the *vertebral* (VER-teh-bral) *column* and the bones of the chest, or *thorax* (THO-raks). The vertebral column is made of a series of irregularly shaped bones. These number 33 or 34 in the child, but because of fusions that occur later in the lower part of the spine, there usually are just 26 separate bones in the adult spinal column (Figs. 7-8 and 7-9). Each of these *vertebrae* (VER-teh-bre), except the first two cervical vertebrae, has a drum-shaped body (centrum) located toward the front (anteriorly) that serves as the weight-bearing part; disks of cartilage between the vertebral bodies act as shock absorbers and provide flexibility.

In the center of each vertebra is a large hole, or foramen. When all the vertebrae are linked in series by strong connective tissue bands (ligaments), these spaces form the spinal canal, a bony cylinder that protects the spinal cord. Projecting backward from the bony arch that encircles the spinal cord is the *spinous process*, which usually can be felt just under the skin of the back. When viewed from the side the vertebral column can be seen to have a series of

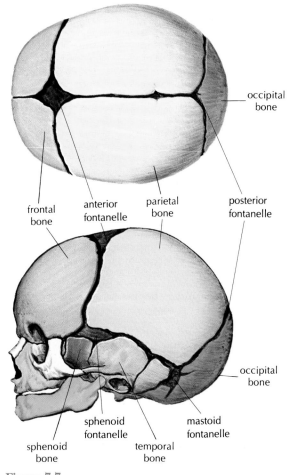

Figure 7-7

Infant skull showing fontanelles.

intervertebral foramina through which spinal nerves emerge as they leave the spinal cord.

The bones of the vertebral column are named and numbered from above downward, on the basis of location:

1. The *cervical* (SER-vih-kal) *vertebrae,* seven in number, are located in the neck. The first vertebra, called the *atlas,* supports the head; when one nods the head, the skull rocks on the atlas at the occipital bone. The second cervical vertebra, called the *axis,* serves as a pivot when the head is turned from side to side.
2. The *thoracic vertebrae,* 12 in number, are located in the thorax. The posterior ends of the 12 pairs of ribs are attached to these vertebrae.
3. The *lumbar vertebrae,* five in number, are located in the small of the back. They are larger and heav-

ier than the other vertebrae to support more weight.

4. The *sacral* (SA-kral) *vertebrae* are five separate bones in the child. However, they eventually fuse to form a single bone, called the *sacrum* (SA-krum), in the adult. Wedged between the two hip bones, the sacrum completes the posterior part of the bony pelvis.

5. The *coccyx* (KOK-siks), or tailbone, consists of four or five tiny bones in the child. These fuse to form a single bone in the adult.

When viewed from the side, the vertebral column can be seen to have four curves, corresponding to the four groups of vertebrae (see Fig. 7-8). In the newborn infant, the entire column is concave forward (curves away from a viewer facing the infant), as seen in Figure 7-10. This is the primary curve. When the infant begins to assume an erect posture, secondary curves, which are convex (curve toward the viewer) may be noted. For example, the cervical curve appears when the infant begins to hold up the head at about 3 months of age; the lumbar curve appears when he or she begins to walk. The thoracic and sacral curves remain the two primary curves. These curves of the vertebral column provide some of the resilience and spring so essential in balance and movement.

The bones of the *thorax* (Fig. 7-11) form a cone-shaped cage. Twelve pairs of *ribs* form the bars of this cage, completed by the *sternum* (STER-num), or breastbone, anteriorly.

The top portion of the sternum is the broadly T-shaped *manubrium* (mah-NU-bre-um) that joins at the top on each side with the clavicle (collarbone) and joins laterally with the first pair of ribs. The *body* of the sternum is long and bladelike. It joins along each side with ribs two through seven. Where the manubrium joins the body of the sternum, there is a slight elevation, the *sternal angle,* which easily can be felt as a surface landmark.

The lower end of the sternum consists of a small tip that is made of cartilage in youth but becomes bone in the adult. This is the *xiphoid* (ZIF-oyd) *process*. It is used as a landmark for CPR (cardio-pulmonary resuscitation) to locate the region for chest compression. The thorax protects the heart, lungs, and other organs.

All 24 of the ribs are attached to the vertebral column posteriorly. However, variations in the *anterior* attachment of these slender, curved bones have led to the following classification:

1. *True ribs,* the first seven pairs, are those that attach directly to the sternum by means of individual extensions called *costal* (KOS-tal) *cartilages.*

2. *False ribs* are the remaining five pairs. Of these, the 8th, 9th, and 10th pairs attach to the cartilage of the rib above. The last two pairs have no anterior attachment at all and are known as *floating ribs.*

The spaces between the ribs, called *intercostal spaces,* contain muscles, blood vessels, and nerves.

BONES OF THE APPENDICULAR SKELETON

The appendicular skeleton may be considered in two divisions: upper and lower. The upper division on each side includes the shoulder, the arm (between the shoulder and the elbow), the forearm (between the elbow and the wrist), the wrist, the hand, and the fingers. The lower division includes the hip (part of the pelvic girdle), the thigh (between the hip and the knee), the leg (between the knee and the ankle), the ankle, the foot, and the toes.

The bones of the upper division may be divided into two groups for ease of study:

A. The *shoulder girdle* (Fig. 7-12) consists of two bones:

1. The *clavicle* (KLAV-ih-kl), or collarbone, is a slender bone with two shallow curves. It joins the sternum anteriorly and the scapula laterally and helps to support the shoulder. Because it often receives the full force of falls on outstretched arms or of blows to the shoulder, it is the most frequently broken bone.

2. The *scapula* (SKAP-u-lah), or shoulder blade, is shown from a posterior view in Figure 7-13. The *spine* of the scapula is the raised ridge that can be felt behind the shoulder in the upper portion of the back. Muscles that move the arm attach to the fossae (depressions) above and below the scapular spine. The *acromion* (ah-KRO-me-on) is the process that joins the clavicle. This can be felt as the highest point of the shoulder. Below the acromion there is a shallow socket, the *glenoid cavity,* that forms a ball-and-socket joint with the arm bone (humerus). Medial to the glenoid cavity is the *coracoid process* to which muscles attach.

B. Each *upper extremity,* or upper limb, consists of the following bones:

1. The proximal bone, the *humerus* (HU-mer-us), or arm bone, forms a joint with the

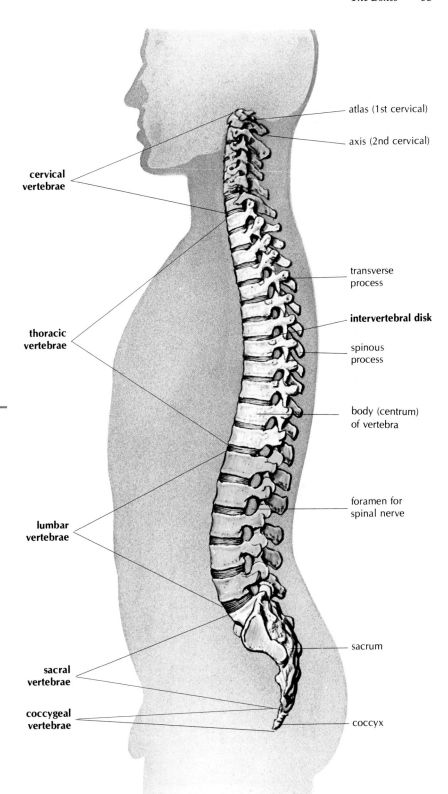

atlas (1st cervical)

axis (2nd cervical)

cervical vertebrae

transverse process

intervertebral disk

spinous process

thoracic vertebrae

body (centrum) of vertebra

Figure 7-8

Vertebral column from the side.

foramen for spinal nerve

lumbar vertebrae

sacrum

sacral vertebrae

coccygeal vertebrae

coccyx

front view of vertebral column   **vertebrae from above**

Figure 7-9

Front view of the vertebral column; vertebrae from above.

scapula above and with the two forearm bones at the elbow (Fig. 7-14).

2. The forearm bones are the ***ulna*** (UL-nah), which lies on the medial, or little finger, side, and the ***radius*** (RA-de-us), on the lateral, or

thumb, side. When the palm is up, or forward, the two bones are parallel; when the palm is turned down, or back, the lower end of the radius moves around the ulna so that the shafts of the two bones are crossed. In this position,

fetus   adult

Figure 7-10

Curves of the infant spine and adult spine.

a distal projection of the ulna (the styloid process) pops up at the outside of the wrist. The top of the ulna has the large *olecranon* (o-LEK-rah-non) that forms the point of the elbow. The pulley-like *trochlea* (TROK-le-ah) of the distal humerus fits into the deep *trochlear notch* of the ulna, allowing a hinge action at the elbow joint.

3. The wrist contains eight small *carpal* (KAR-pal) *bones* arranged in two rows of four each. The names of these eight different bones are given in Figure 7-15.
4. Five *metacarpal bones* are the framework for the palm of each hand. Their rounded distal ends form the knuckles.
5. There are 14 *phalanges* (fah-LAN-jeze), or finger bones, in each hand, two for the thumb and three for each finger. Each of these bones is called a *phalanx* (FA-lanx). They are identified as the first, or proximal, which is attached to a metacarpal; the second, or middle; and the third, or distal. Note that the thumb has only two phalanges, a proximal and a distal.

The bones of the lower division are grouped together in a similar fashion:

A. The bony pelvis supports the trunk and the organs in the lower abdomen, or pelvic cavity, including the urinary bladder, the internal reproductive organs, and parts of the intestine. The female pelvis is adapted for pregnancy and childbirth; it is broader and lighter than the male pelvis (Fig. 7-16).

 The *pelvic girdle* is a strong bony ring that forms the walls of the pelvis. It is composed of two hipbones, which form the front and the sides of the ring, and the sacrum, which joins with the hipbones to complete the ring at the back (posteriorly). Each hipbone, or *os coxae,* begins its development as three separate bones which later fuse:

1. The *ilium* (IL-e-um), which forms the upper, flared portion. The *iliac* (IL-e-ak) *crest* is the curved rim along the upper border of the ilium. It can be felt just below the waist. At either end of the crest are two bony projections. The most prominent of these is the *anterior superior iliac spine*, which is often used as a surface landmark in diagnosis and treatment.
2. The *ischium* (IS-ke-um), which is the lowest and strongest part. The *ischial* (IS-ke-al) *spine* at the back of the pelvic outlet is used as a point of reference during childbirth to indicate the progress of the presenting part (usually the baby's head) down the birth canal. Just below this spine is the large *ischial tuberosity*, which helps support the weight of the trunk when one sits down.
3. The *pubis* (PU-bis), which forms the anterior part. The joint formed by the union of the two hip bones anteriorly is called the *symphysis* (SIM-fih-sis) *pubis.*

 Portions of all three bones contribute to the formation of the *acetabulum* (as-eh-TAB-u-lum), the deep socket that holds the head of the femur (thigh bone) to form the hip joint. The largest foramina in the entire body are found in the pelvic girdle, near the front of each hipbone, one on each side of the symphysis pubis. Called the *obturator* (OB-tu-ra-tor) *foramina,* these are partially covered by a membrane.

(Text continues on page 100)

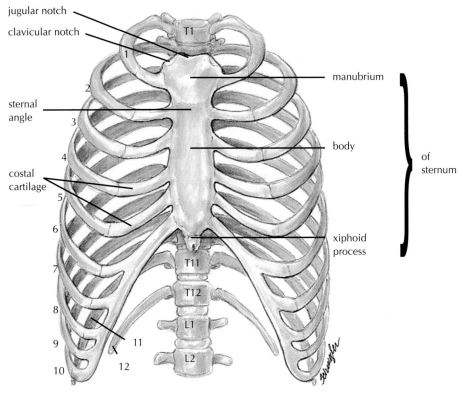

Figure 7-11

Bones of the thorax (anterior view). The first seven pairs of ribs are the true ribs; pairs 8 through 12 are the false ribs, of which the last two pairs are floating ribs.

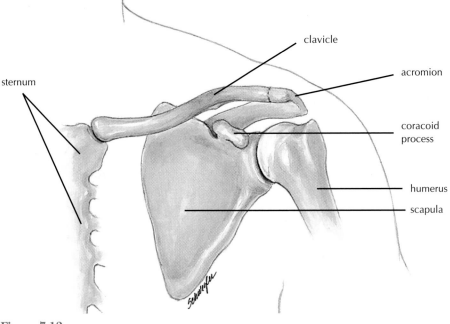

Figure 7-12

Bones of the shoulder girdle.

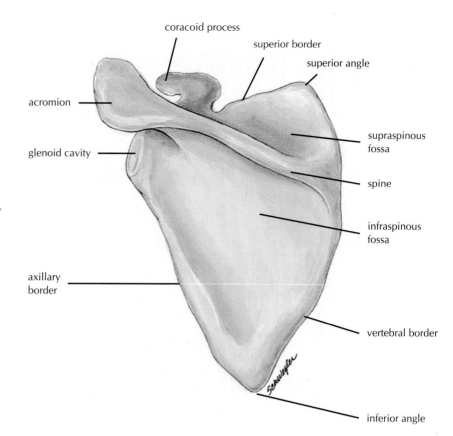

Figure 7-13

Left scapula (posterior view).

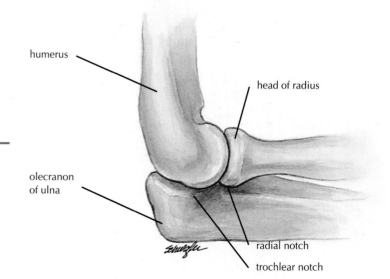

Figure 7-14

Lateral view of the right elbow.

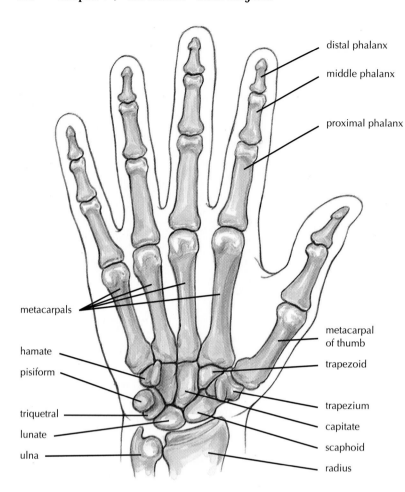

distal phalanx

middle phalanx

proximal phalanx

metacarpals

metacarpal
of thumb

hamate

trapezoid

pisiform

trapezium

triquetral

capitate

lunate

scaphoid

ulna

radius

Figure 7-15

Bones of the right hand, anterior view.

B. Each *lower extremity*, or lower limb, consists of the following bones:
1. The thigh bone, called the *femur* (FE-mer), is the longest and strongest bone in the body. Proximally it has a large ball-shaped head that joins the os coxae (Fig. 7-17). The large lateral projection at the top of the femur is the *greater trochanter* (tro-KAN-ter), used as a surface landmark. The *lesser trochanter*, a smaller elevation, is located on the medial side.
2. The *patella* (pah-TEL-lah), or kneecap, is embedded in the tendon of the large anterior thigh muscle, the quadriceps femoris, where it crosses the knee joint. It is an example of a *sesamoid* (SES-ah-moyd) *bone*, a type of bone that develops within a tendon or a joint capsule.
3. There are two bones in the leg. Medially (on the great toe side), the *tibia*, or shin bone, is the longer, weight-bearing bone. It has a sharp

anterior ridge that can be felt at the surface. Laterally, the slender *fibula* (FIB-u-lah) does not reach the knee joint; thus, it is not a weight-bearing bone. The *medial malleolus* (mal-LE-o-lus) is a downward projection at the lower end of the tibia; it forms the prominence on the inner aspect of the ankle (Fig. 7-18). The *lateral malleolus,* at the lower end of the fibula, forms the prominence on the outer aspect of the ankle.
4. The structure of the foot is similar to that of the hand. However, the foot supports the weight of the body, so it is stronger and less mobile than the hand. There are seven *tarsal bones* associated with the ankle and foot. These are named and illustrated in Figure 7-18. The largest of these is the *calcaneus* (kal-KA-ne-us), or heel bone. Five *metatarsal bones* form the framework of the instep, and the heads of these bones form the ball of the foot.

iliac crest

sacrum

ilium

anterior superior
iliac spine

ischial
spine

acetabulum
(socket for femur)

symphysis
pubis

ischium

pubis

pubic
arch

obturator
foramen

male pelvis from above

female pelvis from above

iliac crest

ilium

acetabulum

pubis

coccyx

ischium

side view of male pelvis

Figure 7-16

Pelvic girdle showing male pelvis and female pelvis.

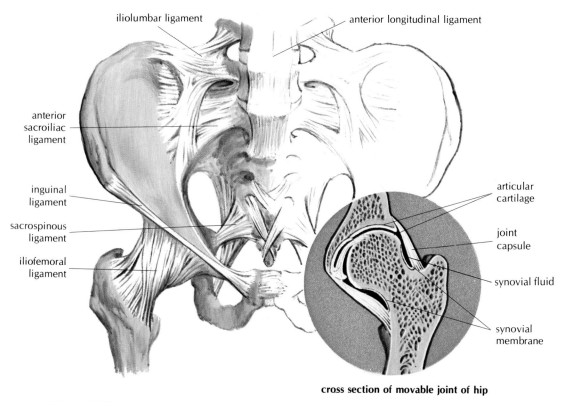

Figure 7-17

Hip joint and pelvis, showing ligaments.

5. The *phalanges* of the toes are counterparts of those in the fingers. There are three of these in each toe except for the great toe, which has only two.

Disorders of Bone

Cleft palate is a congenital deformity in which there is an opening in the roof of the mouth owing to faulty union of the maxillary bones. An infant born with this defect has difficulty nursing, because the mouth communicates with the nasal cavity above, and the baby therefore sucks in air rather than milk. Surgery is usually performed to correct the condition.

Osteoporosis (os-te-o-po-RO-sis) is a disorder of bone formation in which there is a lack of normal calcium salt deposits and a decrease in bone protein. There is an increased breakdown of bone tissue without increase in the deposit of new bone by osteoblasts. The bones thus become fragile and break easily. The condition is most common in postmeno-

pausal women as a result of a decrease in hormone levels; most often it involves the spine, pelvis, and long bones. Possible causes include poor nutrition, particularly a lack of protein and vitamins in the diet, as well as inactivity. Treatment has included administration of replacement hormones (estrogen, usually in combination with progesterone), an increase in exercise, and an increase in dietary calcium. It is recommended that all women take in 1000 mg to 1500 mg calcium daily throughout adult life.

Osteomyelitis (os-te-o-mi-eh-LI-tis) is an inflammation of bone caused by *pyogenic* (pi-o-JEN-ik) (pus-producing) bacteria. It may remain localized or it may spread through the bone to involve the marrow and the periosteum. The bacteria may reach the bone through the bloodstream or by way of an injury in which the skin has been broken. Before the advent of antibiotic drugs, bone infections were very resistant to treatment, and the prognosis for persons with such infections was poor. Now there are fewer cases because many bloodstream infections are prevented or treated early enough so they do not

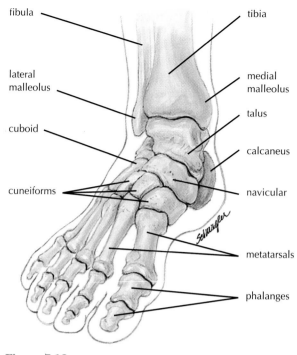

fibula

tibia

lateral malleolus

medial malleolus

cuboid

talus

cuneiforms

calcaneus

navicular

metatarsals

phalanges

Figure 7-18

Bones of the right foot.

progress to affect the bones. If those bone infections that do appear are treated promptly, the chance of a cure is usually excellent.

Tumors that develop in bone tissue may be benign, as is the case with certain cysts, or they may be malignant, as are osteosarcomas. In the latter case, the tumor originates most often in the bone tissue of a femur or a tibia, usually in a young person. In older persons, metastases from epithelial tumors or carcinomas of various organs may spread to many bones.

Abnormal calcium metabolism may cause various bone disorders. In one of these, called Paget's disease, or *osteitis deformans* (os-te-I-tis de-FOR-mans), the bones undergo periods of calcium loss followed by periods of excessive deposition of calcium salts. As a result the bones become deformed. Cause and cure are not known at the present time. The bones also can become decalcified owing to the effect of a tumor of the parathyroid gland (see Chap. 12).

Rickets is a rare childhood disease characterized by numerous bone deformities. Deficiency of the active form of vitamin D prevents the absorption of calcium and phosphorus salts through the intestine. These mineral salts are then not available for deposit

into bone. The bones remain soft and become distorted.

Abnormalities of the spinal curves, known as *curvatures of the spine* (Fig. 7-19), include an exaggeration of the thoracic curve, or *kyphosis* (ki-FO-sis) (hunchback), and an excessive lumbar curve, called *lordosis* (lor-DO-sis) (swayback). The most common of these disorders is *scoliosis* (sko-le-O-sis), a lateral curvature of the vertebral column. In extreme cases there may be compression of some of the internal organs. Scoliosis occurs in the rapid growth period of teens, more often in girls than in boys. Early discovery and treatment produce good results.

SKELETAL CHANGES IN THE AGING

The aging process includes significant changes in all connective tissues including bone. There is a loss of calcium salts and a decrease in the amount of protein formed in bone tissue. The reduction of collagen in bone and in tendons, ligaments, and skin contributes to the stiffness so often found in older people. Muscle tissue is also lost throughout adult life. Thus there is a tendency to a great decrease in the exercise that is so important to the maintenance of bone tissue.

Changes in the vertebral column with age lead to a loss in height. Approximately 1.2 cm (about 0.5 inches) are lost each 20 years beginning at age 40, owing primarily to a thinning of the intervertebral disks (between the bodies of the vertebrae). Even the vertebral bodies themselves may lose height in later years. The costal (rib) cartilages become calcified and less flexible, and the chest may decrease in diameter by 2 to 3 cm (about 1 inch), mostly in the lower part.

FRACTURES

Severe force is capable of causing a *fracture* in almost any bone (Fig. 7-20). The word *fracture* means "a break or rupture in a bone." Such injuries may be classified as follows:

1. *Closed fracture,* which is a simple fracture of the bone with no open wound
2. *Open fracture,* in which a broken bone protrudes through the skin or an external wound leads to a broken bone
3. *Greenstick fracture,* in which one side of the bone is broken and the other is bent. These are most common in children.

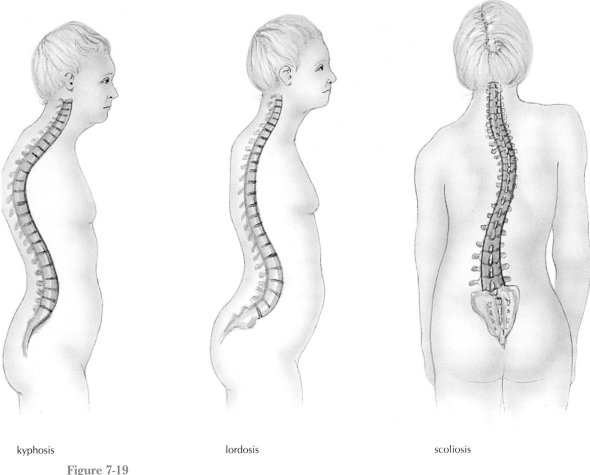

kyphosis   lordosis   scoliosis

Figure 7-19

Abnormalities of the spinal curves.

4. *Impacted fracture,* in which the broken ends of the bone are jammed into each other
5. *Comminuted* (KOM-ih-nu-ted) *fracture,* in which there is more than one fracture line and the bone is splintered or crushed
6. *Spiral fracture,* in which the bone has been twisted apart. These are relatively common in skiing accidents.

The most important step in first aid care of fractures is to prevent movement of the affected parts. Protection by simple splinting after careful evaluation of the situation, leaving as much as possible "as is," and a call for expert help are usually the safest measures. Persons who have back injuries may be spared serious spinal cord damage if they are carefully and correctly moved on a firm board or door. If trained paramedics or rescue personnel can reach the scene, a "hands off" rule for the untrained is strongly recommended. If there is no external bleeding, covering the victim with blankets may help combat shock. First aid should always be immediately directed toward the control of hemorrhage.

▶ The Joints

An *articulation,* or *joint,* is an area of junction or union between two or more bones.

Kinds of Joints
Joints are classified into three main types on the basis of the material between the bones (Fig. 7-21).

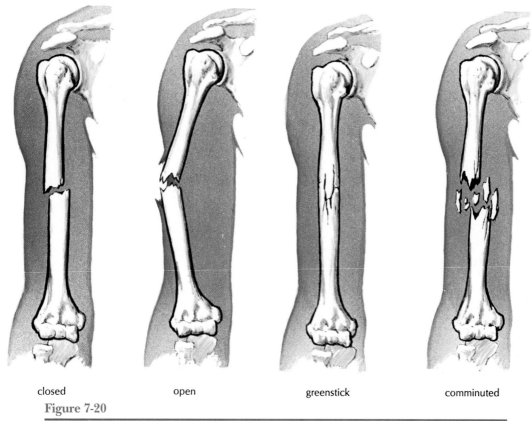

closed   open   greenstick   comminuted

Figure 7-20

Types of fractures.

They may also be classified according to the degree of movement permitted:

1. ***Fibrous joint.*** The bones in this type of joint are held together by fibrous connective tissue. An example is a ***suture*** (SU-chur) between bones of the skull. This type of joint is immovable and is termed a ***synarthrosis*** (sin-ar-THRO-sis).
2. ***Cartilaginous joint.*** The bones in this type of joint are connected by cartilage. Examples are the joint between the pubic bones of the pelvis and the joints between the bodies of the vertebrae. This type of joint is slightly movable and is termed an ***amphiarthrosis*** (am-fe-ar-THRO-sis).
3. ***Synovial*** (sin-O-ve-al) ***joint.*** The bones in this type of joint have a potential space between them. It is called the ***joint cavity,*** and it contains a small amount of thick colorless fluid that resembles uncooked egg white. This lubricant, ***synovial fluid,*** is secreted by the membrane that lines the joint cavity. This type of joint is freely movable and is termed a ***diarthrosis*** (di-ar-THRO-sis). Most joints are synovial joints; they are described in more detail below.

Table 7-1 gives a summary of joint types.

Structure of Synovial Joints

The bones in freely movable joints are held together by ***ligaments,*** bonds of fibrous connective tissue. Additional ligaments reinforce and help stabilize the joints at various points (see Fig. 7-17). Also, for strength and protection there is a ***joint capsule*** of connective tissue that encloses each joint and is continuous with the periosteum of the bones. The bone surfaces in freely movable joints are protected by a smooth layer of gristle called the ***articular*** (ar-TIK-u-lar) ***cartilage.***

Near some joints are small sacs called ***bursae*** (BER-se), which are filled with synovial fluid. These lie in areas subject to stress and help ease movement over and around the joints. Inflammation of a bursa, as a result of injury or irritation, is called ***bursitis.***

fibrous joint
(suture)

cartilaginous joint
(symphysis)

synovial
joint

Figure 7-21

Types of joints.

Types of Synovial Joints

Synovial joints may be classified according to the types of movement they allow as follows:

1. *Gliding* joint, in which the bone surfaces slide over one another. Examples are joints in the wrist and ankle.
2. *Hinge* joint, which allows movement in one direction, changing the angle of the bones at the joint. Examples are the elbow joint and the joints between the phalanges of the fingers and toes.
3. *Pivot* joint, which allows rotation around the length of the bone. Examples are the joint between the first and second cervical vertebrae and the joint at the proximal ends of the radius and ulna.
4. *Condyloid* joint, which allows movement in two directions. Examples are the joint between the wrist and the bones of the forearm; the joint between the occipital bone of the skull and the first cervical vertebra (the atlas).
5. *Saddle* joint, which is like a condyloid joint, but with deeper articulating surfaces. An example is the joint between the wrist and the metacarpal bone of the thumb.
6. *Ball-and-socket* joint, which allows movement in many directions around a central point. This type of joint gives the greatest freedom of movement. Examples are the shoulder and hip joints (see Fig. 7-17).

The knee is illustrated as an example of a synovial joint in Figure 7-22. Although basically a hinge

TABLE 7-1
Types of Joints

Name	Movement	Material Between the Bones	Examples
Fibrous	Immovable (Synarthrosis)	No joint cavity; fibrous connective tissue between bones	Sutures between bones of skull
Cartilaginous	Slightly movable (Amphiarthrosis)	No joint cavity; cartilage between bones	Pubic symphysis; joints between bodies of vertebrae
Synovial	Freely movable (Diarthrosis)	Joint cavity containing synovial fluid	Gliding, hinge, pivot, condyloid, saddle, ball-and-socket joints

joint, the knee is very complex and has some of the properties of a gliding joint as well. An x-ray picture of the knee is shown in Figure 7-23.

Function of Synovial Joints

The chief function of the freely movable joints is to allow for changes of position and so provide for motion. These movements are given names to describe the nature of the changes in the positions of the body parts (Fig. 7-24). For example, there are four kinds of angular movement, or movement that changes the angle between bones, as listed below:

1. *Flexion* (FLEK-shun) is a bending motion that decreases the angle between bones, as in bending the fingers to close the hand.
2. *Extension* is a straightening motion that increases the angle between bones, as in straightening the fingers to open the hand.

3. *Abduction* (ab-DUK-shun) is movement away from the midline of the body, as in moving the arms straight out to the sides.
4. *Adduction* is movement toward the midline of the body, as in bringing the arms back to their original position beside the body.

A combination of these angular movements enables one to execute a movement referred to as *circumduction* (ser-kum-DUK-shun). To perform this movement, stand with your arm outstretched and draw a large imaginary circle in the air. Note the smooth combination of flexion, abduction, extension, and adduction that makes circumduction possible. *Rotation* refers to a twisting or turning of a bone on its own axis, as in turning the head from side to side to say "no."

There are special movements that are characteristic of the forearm and the ankle:

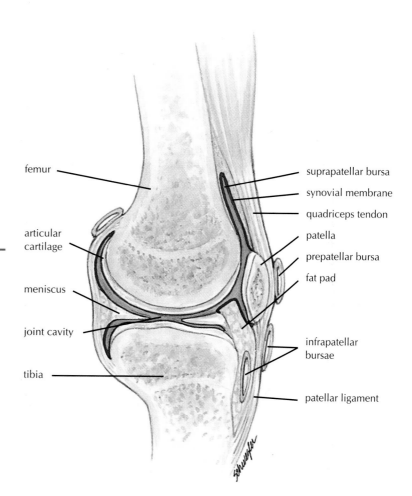

Figure 7-22

The knee joint (sagittal section).

femur

articular cartilage

meniscus

joint cavity

tibia

suprapatellar bursa

synovial membrane

quadriceps tendon

patella

prepatellar bursa

fat pad

infrapatellar bursae

patellar ligament

femur

patella

tibia

fibula

Figure 7-23

X-ray (radiograph) of the knee joint showing the articulating bones. (Greenspan A: Orthopedic Radiology, p. 6.2. New York: Gower 1988)

1. *Supination* (su-pin-A-shun) is the act of turning the palm up or forward; *pronation* (pro-NA-shun) turns the palm down or backward.
2. *Inversion* (in-VER-zhun) is the act of turning the sole inward, so that it faces the opposite foot; *eversion* (e-VER-zhun) turns the sole outward, away from the body.
3. In *dorsiflexion* (dor-sih-FLEK-shun) the foot is bent upward at the ankle, narrowing the angle between the leg and the top of the foot; in *plantar flexion* the toes point downward, as in toe dancing, flexing the arch of the foot.

Disorders of Joints

Joints are subject to certain disorders of a mechanical nature, examples of which are *dislocations* and *sprains.* A dislocation is a derangement of the parts of the joint. Ball-and-socket joints, which have the widest range of motion, also have the greatest tendency to dislocate. The shoulder joint is the most frequently dislocated joint in the body. A sprain is the wrenching of a joint with rupture or tearing of the ligaments.

ARTHRITIS

The most common type of joint disorder is termed *arthritis,* which means "inflammation of the joints." There are different kinds of arthritis, including the following:

1. *Osteoarthritis* (os-te-o-arth-RI-tis) is a degenerative joint disease (DJD) that usually occurs in elderly persons as a result of normal wear and tear. Although it appears to be a natural result of aging, such factors as obesity and repeated trauma can help bring it about. Osteoarthritis occurs mostly in joints used in weight bearing, such as the hips, knees, and spinal column. It involves degeneration of the joint cartilage, with growth of new bone at the edges of the joints. Degenerative changes include the formation of spurs at the edges of the articular surfaces, thickening of the synovial membrane, atrophy of the cartilage, and calcification of the ligaments.
2. *Rheumatoid arthritis* is a crippling condition characterized by swelling of the joints of the hands, the feet, and other parts of the body as a result of inflammation and overgrowth of the synovial

flexion/extension

abduction/adduction

circumduction

rotation

Figure 7-24

Movements at synovial joints.   *(continued)*

membranes and other joint tissues. The articular cartilage is gradually destroyed, and the joint cavity develops adhesions—that is, the surfaces tend to stick together—so that the joints stiffen and ultimately become useless. The exact cause of rheumatoid arthritis is uncertain. However, the disease shares many characteristics of disorders in which there is production of antibodies that attack normal body tissues. The role of inherited susceptibility is clear. The administration of steroids and gold salts may provide some relief.

3. ***Septic (infectious) arthritis*** arises when bacteria spread to involve joint tissue, usually by way of the blood stream. A variety of organisms is commonly involved, including *Streptococcus, Staphylococcus,* and *Neisseria.*

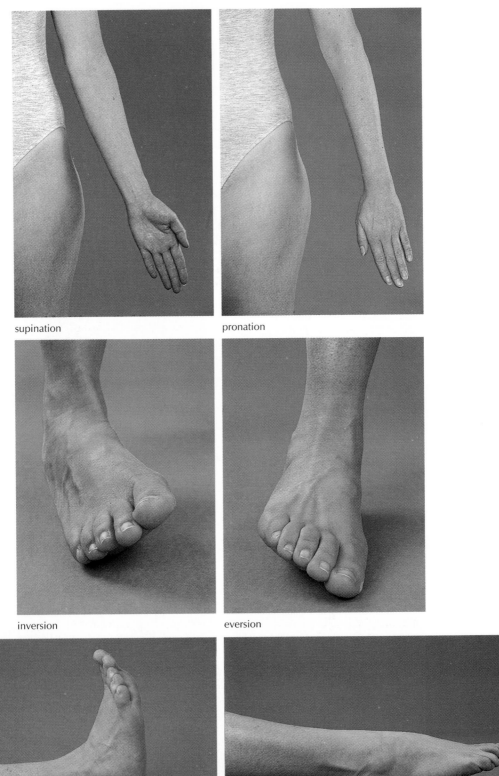

supination

pronation

inversion

eversion

dorsiflexion

plantar flexion

Figure 7-24 *(Continued)*

Joint Replacement

Since the first total hip replacement in the early 1960s, many thousands of joint replacements have been performed successfully. Joint replacements are needed when chronic arthritis or various injuries have caused serious damage. The hip (the head of the femur in the acetabulum) and the knee joint are the joints that have been most commonly and most effectively restored, but shoulder, elbow, and metacarpal joints have also been treated with desirable results.

Special materials are available for the artificial joints that are strong and long lasting and will not corrode under the effects of body fluids. Manufacturing time and cost have been reduced by using computers to design these joints.

The joints and the bones proper are subject to attack by the tuberculosis organism, and the result may be gradual destruction of parts of the bone near the joint. The organism is carried by the bloodstream, usually from a focus in the lungs or lymph nodes, and may cause considerable damage before it is discovered. The bodies of several vertebrae sometimes are affected, or one hip or other single joint may be diseased. Tuberculosis of the spine is called *Pott's disease*. The patient may complain only of difficulty in walking, and diagnosis is difficult unless an accompanying lung tuberculosis has been found. This disorder is most common in children.

4. *Gout* is a kind of arthritis caused by a disturbance of metabolism. One of the products of metabolism is uric acid, which normally is excreted in the urine. If there happens to be an overproduction of uric acid, or for some reason not enough is excreted, the accumulated uric acid forms crystals, which are deposited as masses about the joints and other parts of the body. As a result, the joints become inflamed and extremely painful. Any joint can be involved, but the one most commonly affected is the big toe. Most victims of gout are men past middle life.

BACKACHE

Backache is another common complaint. Some of its causes are listed below:

1. Diseases of the vertebrae, such as infections or tumors, and in older persons, osteoarthritis or atrophy (wasting away) of bone following long illnesses and lack of exercise
2. Disorders of the intervertebral disks, especially those in the lower lumbar region. Pain may be very severe, with muscle spasms and the extension of symptoms along the course of the sciatic nerve (back of the thigh to the toes)
3. Abnormalities of the lower vertebrae or of the ligaments and other supporting structures
4. Disorders involving abdominopelvic organs or those in the space behind the peritoneum (such as the kidney). Variations in the position of the uterus are seldom a cause.
5. Strains on the lumbosacral joint (where the lumbar region joins the sacrum) or strains on the sacroiliac joint (where the sacrum joins the ilium of the pelvis)

Backache can be prevented by attention to proper movement, good posture, and exercise. It is most important that the back itself not be used for lifting. Weight should be brought close to the body and the legs allowed to do the actual lifting. An adequate exercise program is also important.

SUMMARY

I. Bones
   A. Structure
      1. Cells
         a. Osteoblasts—bone-forming cells
         b. Osteocytes—mature bone cells that maintain bone
         c. Osteoclasts—bone cells that break down (resorb) bone
      2. Tissue
         a. Compact—shaft (diaphysis) of long bones; outside of other bones
         b. Spongy—end (epiphysis) of long bones; center of other bones
      3. Marrow
         a. Red—ends of long bones, center of other bones

      b. Yellow—center of long bones

    4. Membranes—contain bone-forming cells

      a. Periosteum—covers bone

      b. Endosteum—lines marrow cavity

  B. Long bone growth—begins in center of shaft and continues at ends of bone; growing area forms line across epiphysis

  C. Bone functions—serve as body framework; protect organs; serve as levers; store calcium salts; form blood cells

  D. Bone markings

    1. Projections—head, process, crest, spine

    2. Depressions and holes—foramen, sinus, fossa, meatus

II. Divisions of the skeleton

  A. Axial

    1. Head

      a. Cranial—frontal, parietal, temporal, ethmoid, sphenoid, occipital

      b. Facial—mandible, maxillae, zygomatic, nasal, lacrimal, vomer, palatine, inferior nasal conchae

      c. Other—ossicles (of ear), hyoid

      d. Infant skull—fontanelles (soft spots)

    2. Trunk

      a. Vertebral column—cervical, thoracic, lumbar, sacral, coccygeal

      b. Thorax

        (1) Sternum—manubrium, body, xiphoid process

        (2) Ribs

          (a) True—first seven pairs

          (b) False—remaining five pairs, including two floating ribs

  B. Appendicular

    1. Shoulder girdle—clavicle, scapula

    2. Upper extremity—humerus, ulna, radius, carpals, metacarpals, phalanges

    3. Pelvic girdle—os coxae: ilium, ischium, pubis

    4. Lower extremity—femur, patella, tibia, fibula, tarsals, metatarsals, phalanges

III. Disorders of bone

  A. Miscellaneous disorders—cleft palate, osteoporosis, osteomyelitis, tumors

  B. Metabolic disorders—osteitis deformans, rickets

  C. Curvatures of the spine–kyphosis (thoracic), lordosis (lumbar), scoliosis (lateral)

  D. Changes in aging—loss of calcium salts, decreased production of collagen, thinning of intervertebral disks

  E. Fractures—closed, open, greenstick, impacted, comminuted, spiral

IV. Joints

  A. Types

    1. Fibrous—immovable (synarthrosis)

    2. Cartilaginous—slightly movable (amphiarthrosis)

    3. Synovial—freely movable (diarthrosis)

  B. Structure of synovial joints

    1. Joint cavity—contains synovial fluid

    2. Ligaments—hold joint together

    3. Joint capsule—strengthens and protects joint

    4. Articular cartilage—covers ends of bones

    5. Bursae—fluid-filled sacs near joints; cushion and protect joints and surrounding tissue

  C. Types of synovial joints—gliding, hinge, pivot, condyloid, saddle, ball-and-socket

  D. Movement at synovial joints—flexion, extension, abduction, adduction, circumduction, rotation, supination, pronation, inversion, eversion, dorsiflexion, plantar flexion

V. Disorders of joints

  A. Dislocations and sprains

  B. Arthritis—osteoarthritis, rheumatoid arthritis, arthritis due to infection, gout

  C. Backache

QUESTIONS FOR STUDY AND REVIEW

1. Explain the difference between the terms in each of the following pairs:
   a. *osteoblast* and *osteoclast*
   b. *red marrow* and *yellow marrow*
   c. *periosteum* and *endosteum*
   d. *compact bone* and *spongy bone*
   e. *epiphysis* and *diaphysis*
2. Define *resorption* as it applies to bone.
3. Name five general functions of bones.
4. Name the main cranial and facial bones.
5. What is a sinus? Name several bones in which there are sinuses.
6. Define *fontanelle* and name the largest fontanelle.
7. What are the main divisions of the vertebral column? the ribs?
8. Name the bones of the upper and lower appendicular skeleton.
9. Name the bones on which you would find the following bone markings: mastoid process, acromion, olecranon, iliac crest, trochanter, medial malleolus, lateral malleolus, trochlea, ischial tuberosity, acetabulum, glenoid cavity, xiphoid process
10. What is a foramen? Give at least two examples of foramina.
11. Describe cleft palate, osteoporosis, osteomyelitis, and rickets.
12. Describe six types of fractures.
13. Name and describe three abnormal curves of the spine.
14. Name three effects of aging on the skeletal system.
15. List three types of joints. Describe the amount of movement that occurs at each type. Find an example of each type in the chapter illustrations.
16. Describe where each of the following is found in a synovial joint: joint capsule, articular cartilage, and synovial fluid.
17. Name six types of synovial joints and give one example of the location of each. Find one example of each in the chapter illustrations.
18. Differentiate between the terms in each of the following pairs:
    a. *flexion* and *extension*
    b. *abduction* and *adduction*
    c. *supination* and *pronation*
    d. *inversion* and *eversion*
    e. *circumduction* and *rotation*
    f. *dorsiflexion* and *plantar flexion*
19. Describe four joint diseases.

CHAPTER 8

The Muscular System

Selected Key Terms

The following terms are defined in the Glossary:

actin

antagonist

aponeurosis

bursitis

fascicle

glycogen

insertion

lactic acid

myalgia

myoglobin

myosin

neuromuscular junction

neurotransmitter

origin

oxygen debt

prime mover

tendinitis

tendon

tonus

Behavioral Objectives

After careful study of this chapter, you should be able to:

- List the characteristics of skeletal muscle
- Briefly describe how muscles contract
- List the substances needed in muscle contraction and describe the function of each
- Define the term *oxygen debt*
- List several compounds that delay the onset of oxygen debt
- Cite the effects of exercise on muscles
- Compare isotonic and isometric contractions
- Explain how muscles work in pairs to produce movement
- Compare the workings of muscles and bones to lever systems
- Explain how muscles are named
- Name some of the major muscles in each muscle group and describe the main function of each
- Describe how muscles change with age
- List the major muscular disorders

Memmler, RL., Cohen, BJ, Wood, DL. *THE HUMAN BODY IN HEALTH AND DISEASE*, 8/e,
© 1996 Lippincott-Raven Publishers

There are three kinds of muscle tissue: skeletal, smooth, and cardiac muscle. Smooth muscle makes up the walls of the hollow body organs as well as those of the blood vessels and respiratory passageways. It moves involuntarily, producing the wavelike motions of peristalsis that move substances through a system. Cardiac muscle, also involuntary, makes up the wall of the heart and creates the pulsing action of that organ. Skeletal muscle is attached to the bones and produces movement of the skeleton. This chapter is concerned with the skeletal muscles, which comprise the largest amount of muscle tissue of the body. There are more than 650 individual muscles in this muscular system; together they make up about 40% of the total body weight. Although each muscle is a distinct structure, muscles usually act in groups to execute body movements. Skeletal muscle is also known as *voluntary muscle*, because it is usually under conscious control.

▶ Characteristics of Skeletal Muscle

Structure

When seen through a microscope (Fig. 8-1), individual skeletal muscle cells are long and threadlike; therefore they are often called *muscle fibers.* An

muscle fiber nuclei (peripheral)   striations

Figure 8-1

Microscopic view of skeletal muscle fibers (cells) showing multiple nuclei and striations. (Cormack DH: Essential Histology. Philadelphia, JB Lippincott, 1993)

extraordinary feature of these cells is that each one has many nuclei and so is said to be *multinucleated*. Because of the striped or banded appearance of the cells, skeletal muscle is also described as **striated muscle**.

In forming whole muscles, these fibers are arranged in bundles, or **fascicles** (FAS-ih-kls), held together by connective tissue (Fig. 8-2). Deeper layers of connective tissue surround each individual fiber in the bundle. The entire muscle is then encased in a tough connective tissue sheath that is part of the **deep fascia.** All these supporting tissues merge to form the tendon that attaches the muscle to a bone. Each muscle also has a rich supply of blood vessels and nerves.

Muscles in Action

Like all cells, muscle fibers show **irritability**, which is the capacity to respond to a stimulus. Skeletal muscles usually are stimulated by the electrical energy of nerve impulses coming from the brain and the spinal cord (see Chap. 9). Nerve fibers carry impulses to the muscles, each fiber branching to supply from a few to more than 100 individual muscle cells. The point at which a nerve fiber contacts a muscle cell is called the **neuromuscular junction** (Fig. 8-3). It is here that chemicals called **neurotransmitters** are released from the neuron to stimulate the muscle fiber. The stimulus received by way of the nerve ending results in an electrical current, called an **action potential,** that is transmitted along the cell membrane.

Contractility is another important property of muscle tissue. This is the capacity of a muscle fiber to undergo shortening and to change its shape, becoming thicker. Studies of muscle chemistry and observation of cells under the powerful electron microscope have given a concept of how muscle cells work. These studies reveal that each skeletal muscle fiber contains many threads, or filaments, made of two kinds of proteins called **actin** (AK-tin) and **myosin** (MI-o-sin). Filaments made of actin are thin and light; those made of myosin are heavy and dark. These filaments are present in alternating bundles within the muscle cell (Fig. 8-4). It is the alternating bands of light actin and heavy myosin filaments that give skeletal muscle its striated appearance. They also give a view of what occurs when muscles contract.

In movement, the myosin filaments "latch on" to the actin filaments by means of many paddle-like

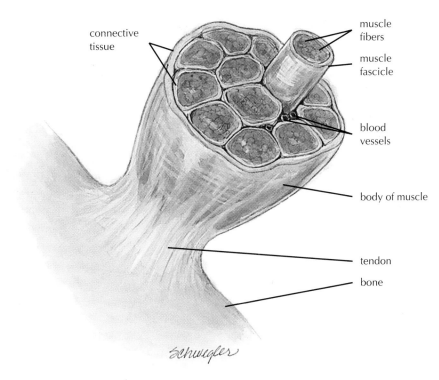

connective tissue

muscle fibers

muscle fascicle

blood vessels

body of muscle

tendon

bone

Schwegler

Figure 8-2

Structure of a skeletal muscle showing connective tissue coverings.

extensions. These attachments are described as *crossbridges* between the two types of filaments. Using the energy of ATP for repeated movements, these crossbridges, like the oars of a boat, push all the actin molecules closer together. As the overlapping filaments slide together, the muscle fiber contracts, becoming shorter and thicker (see Fig. 8-4).

It is well known that the mineral calcium is needed for muscle contraction. What calcium does is to uncover points on the actin where cross-bridges can form between the filaments so that the sliding action can begin. The calcium needed for contraction is stored within the endoplasmic reticulum (ER) of the muscle cell and is released into the cytoplasm when the cell is stimulated by a nerve fiber. Muscles relax when nerve stimulation stops and the calcium is pumped back into the ER.

Muscles and Energy

As noted earlier, all muscle contraction requires energy in the form of ATP. The source of this energy is the oxidation, or "burning," of nutrients within the cells.

To produce ATP, muscle cells must have an adequate supply of oxygen and glucose or other nu-

trients. These substances are constantly brought to the cells by the circulation, but muscle cells also store a small reserve supply of each. Additional oxygen is stored in the form of a compound similar to the blood's hemoglobin but located specifically in muscle cells, as indicated by the prefix *myo-* in its name, **myoglobin** (mi-o-GLO-bin). Additional glucose is stored as a compound that is built from glucose molecules and is called **glycogen** (GLI-kojen). Glycogen can be broken down into glucose when needed by the muscle cells, for example during vigorous exercise.

Oxygen plays an important role in preventing the accumulation of **lactic acid,** a waste product of metabolism that causes muscle fatigue. During strenuous activity, a person may not be able to breathe in oxygen rapidly enough to meet the needs of the hard-working muscles. The increased demands are met at first by the energy-rich compounds that are stored in the tissues. However, continual exercise depletes these stores. For a short time glucose may be used without the benefit of oxygen. In this case lactic acid is formed in the cells by an alternate pathway of metabolism. This anaerobic process (occurring without oxygen) permits greater magnitude of

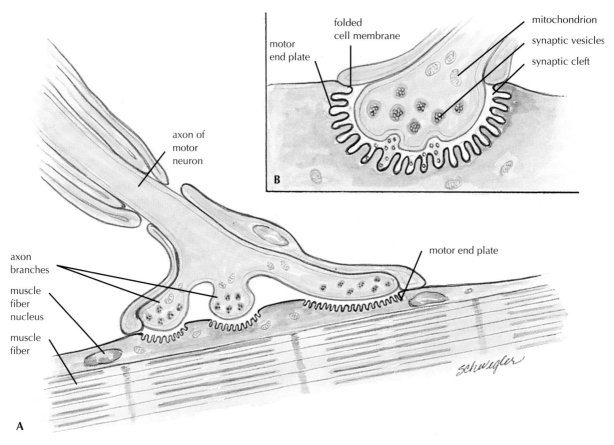

Figure 8-3

Neuromuscular junction. (**A**) The branched end of a motor neuron makes contact with the membrane of a muscle fiber (cell). (**B**) Enlarged view.

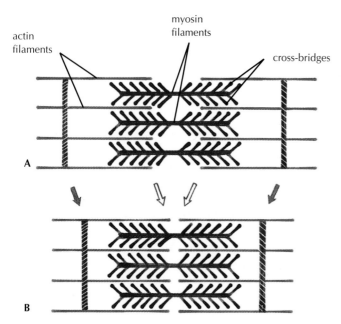

Figure 8-4

The sliding filament mechanism of skeletal muscle contraction (**A**) Muscle is in a relaxed state. (**B**) Muscle has contracted and the cross bridges have moved to a new site to prepare for another "push."

Rigor Mortis

 After death, muscles enter a stage of rigidity known as *rigor mortis*. The calcium that triggers contraction escapes from its storage areas and causes the muscle filaments to slide together. Because all metabolism has ceased, however, there is no ATP to power further movement of the cross-bridges, and the filaments remain in a contracted position. This stiffness lasts for about 24 hours, then gradually fades as body tissues begin to decay. The rate at which these changes occur varies with the individual and the environment.

activity than would otherwise be possible, as, for example, allowing sprinting instead of jogging.

Anaerobic metabolism can continue until buildup of lactic acid causes the muscle to fatigue. As lactic acid accumulates, the person is said to develop an *oxygen debt*. After stopping exercise, he or she must continue to take in more oxygen by continued rapid breathing (panting) until the debt is paid in full. That is, enough oxygen must be taken in to convert the lactic acid to other substances that can be used in metabolism. In addition, the energy-rich compounds that are stored in the cells must be replenished.

Effects of Exercise

Changes that occur as muscle metabolism increases during exercise cause *vasodilation* (vas-o-di-LA-shun)—an increase in the diameter of blood vessels—thereby allowing blood to flow more easily to these tissues. As muscles work, more blood is pumped back to the heart. The temporarily increased load on the heart acts to strengthen the heart muscle and to improve its circulation. The chambers of the heart gradually enlarge to accommodate more blood. Regular exercise also improves respiratory efficiency. Circulation in the capillaries surrounding the alveoli, or air sacs, is increased, and this brings about enhanced gas exchange and deeper breathing. Athletic training leads to more efficient distribution and use of oxygen so that the onset of oxygen debt is delayed. Even moderate regular exercise has the additional benefits of weight

control, strengthening of the bones, decreased blood pressure, and decreased risk of heart attacks. The effects of exercise on the body are studied in the fields of sports medicine and exercise physiology.

Types of Muscle Contractions

Muscle *tone* refers to a partially contracted state of the muscles that is normal even when the muscles are not in use. The maintenance of this tone, or *tonus* (TO-nus), is due to the action of the nervous system in keeping the muscles in a constant state of readiness for action. Muscles that are little used soon become flabby, weak, and lacking in tone.

In addition to the partial contractions that are responsible for muscle tone, there are two other types of contractions on which the body depends:

1. *Isotonic* (i-so-TON-ik) *contractions* are those in which the tone or tension within the muscle remains the same but the muscle as a whole shortens, producing movement. Lifting weights, walking, running, or any other activity in which the muscles become shorter and thicker (forming bulges) are isotonic contractions.
2. *Isometric* (i-so-MET-rik) *contractions* are those in which there is no change in muscle length but there is a great increase in muscle tension. For example, if you push against a brick wall, there is no movement, but you can feel the increased tension in your arm muscles.

Most muscles contract either isotonically or isometrically, but most movements of the body involve a combination of the two types of contraction.

▶ The Mechanics of Muscle Movement

Attachments of Skeletal Muscles

Most muscles have two or more attachments to the skeleton. The method of attachment varies. In some instances the connective tissue of the muscle ties directly to the periosteum of the bone. In other cases the connective tissue sheath and partitions within the muscle extend to form specialized structures that aid in attaching the muscle to the bone. Such an extension may take the form of a cord, called a *tendon* (Fig. 8-5); alternatively, a broad sheet called an *aponeurosis* (ap-o-nu-RO-sis) may attach muscles to bones or to other muscles, as in the

Anabolic Steroids

Anabolic steroids are hormones that promote metabolism and stimulate growth. They are sometimes used medically to promote muscle regeneration, for example, to prevent atrophy from disuse after surgery.

Anabolic steroids have been used illegally by athletes to boost the effects of training. These steroids, which function like the male sex hormone testosterone, are taken to increase muscle size and strength and to improve endurance. The doses needed to produce significant results are large enough to have serious side effects. These drugs are known to produce liver damage, heart disease, and abnormalities of the sex organs. They can cause infertility in both men and women. Mood swings and aggressive behavior are also observed.

abdomen (Fig. 8-6) or across the top of the skull (Fig. 8-7).

Whatever the nature of the muscle attachment, the principle remains the same: to furnish a means of harnessing the power of the muscle contractions. A muscle has two (or more) attachments, one of which is connected to a more freely movable part than the other. The less movable (more fixed) attachment is called the *origin;* the attachment to the part of the body that the muscle puts into action is called the *insertion.* When a muscle contracts, it pulls on both points of attachment, bringing the more movable insertion closer to the origin and thereby causing movement of the body part (see Fig. 8-5).

Muscles Work Together

Many of the skeletal muscles function in pairs. A movement is performed by muscle called the ***prime mover;*** the muscle that produces an opposite movement is known as the ***antagonist.*** Clearly, for any given movement, the antagonist must relax when

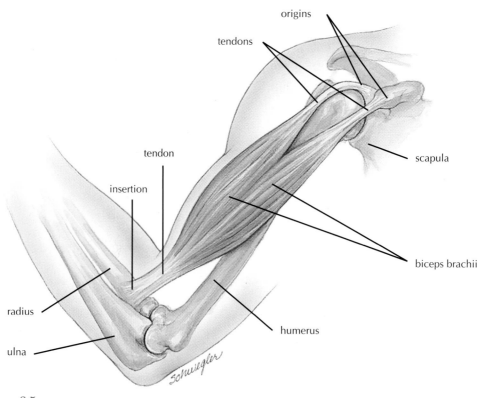

Figure 8-5

Diagram of a muscle showing three attachments to bones—two origins and one insertion.

temporalis

orbicularis oculi

orbicularis oris

masseter

sternocleidomastoid

trapezius

deltoid

pectoralis major

serratus anterior

biceps brachii

brachio-
radialis

flexor carpi

extensor carpi

external
oblique

intercostals

internal oblique

rectus abdominis

abdominal
aponeurosis

sartorius

adductor
longus

vastus lateralis

rectus
femoris

quadriceps femoris

peroneus
longus

tibialis
anterior

gastrocnemius

soleus

Figure 8-6

Superficial muscles, anterior (front) view.

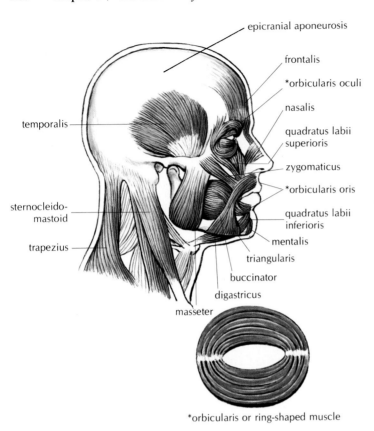

epicranial aponeurosis

frontalis

*orbicularis oculi

nasalis

quadratus labii superioris

zygomaticus

*orbicularis oris

quadratus labii inferioris

mentalis

triangularis

buccinator

digastricus

masseter

temporalis

sternocleido-mastoid

trapezius

*orbicularis or ring-shaped muscle

Figure 8-7

Muscles of the head.

the prime mover contracts. For example, when the biceps brachii at the front of the arm contracts to flex the arm, the triceps brachii at the back must relax; when the triceps brachii contracts to extend the arm, the biceps brachii must relax. In addition to prime movers and antagonists, there are also muscles that serve to steady body parts or to assist prime movers.

As the muscles work together, body movements are coordinated, and a large number of complicated movements can be carried out. At first, however, the nervous system must learn to coordinate any new, complicated movement. Think of a child learning to walk or to write, and consider the number of muscles she uses unnecessarily or forgets to use when the situation calls for them.

Levers and Body Mechanics

Proper body mechanics help conserve energy and ensure freedom from strain and fatigue; conversely, such ailments as lower back pain—a common complaint—can be traced to poor body mechanics. Body mechanics have special significance to

health workers, who are frequently called on to move patients and handle cumbersome equipment. Maintaining the body segments in correct relation to one another has a direct effect on the working capacity of the vital organs that are supported by the skeleton.

If you have had a course in physics, recall your study of levers. A lever is simply a rigid bar that moves about a fixed point, the fulcrum. There are three classes of levers, which differ only in the location of the fulcrum, the effort (force), and the resistance (weight). In a first-class lever, the fulcrum is located between the resistance and the effort; scissors, which you probably use every day, are an example of this class. The second-class lever has the resistance located between the fulcrum and the effort; a mattress lifted at one end is an illustration of this class. In the third-class lever, the effort is between the resistance and the fulcrum. A forceps or a tweezers is an example of this type of lever. The effort is applied in the center of the tool, between the fulcrum where the pieces join and the resistance at the tip.

The musculoskeletal system can be considered a system of levers, in which the bone is the lever, the joint is the fulcrum, and the force is applied by a muscle. Most lever systems in the body are of the third-class type. A muscle usually inserts over a joint and exerts force between the fulcrum and the resistance (see Fig. 8-5), in which the resistance is in the forearm and hand. By understanding and applying knowledge of levers to body mechanics, the health worker can improve his or her skill in carrying out numerous clinical maneuvers and procedures.

▶ Skeletal Muscle Groups

The study of muscles is made simpler by grouping them according to body regions. Knowing how muscles are named can also help in remembering them.

Naming of Muscles
A number of different characteristics are used in naming muscles. These include:

1. Location, near a bone, for example, lateral or medial, internal or external
2. Size, using terms such as maximus, major, minor, longus, brevis
3. Shape, such as circular (orbicularis), triangular (deltoid), trapezoid
4. Direction of fibers, including straight (rectus), angled (oblique)
5. Number of heads (attachment points) as indicated by the suffix -*ceps*, as in biceps, triceps, quadriceps
6. Action, as in flexor, extensor, adductor, abductor, levator.

Often, more than one feature is used in naming. Refer to Figures 8-6 and 8-8 as you study the locations and functions of some of the skeletal muscles and try to figure out why each has the name that it does. Although they are described in the singular, most of the muscles are present on both sides of the body. The information is summarized in Table 8-1.

Muscles of the Head
The principal muscles of the head are those of facial expression and of mastication (chewing) (see Fig. 8-7).

The muscles of facial expression include ring-shaped ones around the eyes and the lips, called the *orbicularis* (or-bik-u-LAH-ris) *muscles* because of their shape (think of "orbit"). The muscle sur-

rounding each eye is called the *orbicularis oculi* (OK-u-li), while the muscle of the lips is the *orbicularis oris*. These muscles, of course, all have antagonists. For example, the *levator palpebrae* (PAL-pe-bre) *superioris,* or lifter of the upper eyelid, is the antagonist for the orbicularis oculi.

One of the largest muscles of expression forms the fleshy part of the cheek and is called the *buccinator* (BUK-se-na-tor). Used in whistling or blowing, it is sometimes referred to as the *trumpeter's muscle.* You can readily think of other muscles of facial expression: for instance, the antagonists of the orbicularis oris can produce a smile, a sneer, or a grimace. There are a number of scalp muscles by means of which the eyebrows are lifted or drawn together into a frown.

There are four pairs of muscles of mastication, all of which insert on the mandible and move it. The largest are the *temporal* (TEM-po-ral), located above and near the ear, and the *masseter* (mas-SE-ter) at the angle of the jaw.

The tongue has two groups of muscles. The first group, called the *intrinsic muscles,* are located entirely within the tongue. The second group, the *extrinsic muscles,* originate outside the tongue. It is because of these many muscles that the tongue has such remarkable flexibility and can perform so many different functions. Consider the intricate tongue motions involved in speaking, chewing, and swallowing.

Muscles of the Neck
The neck muscles tend to be ribbon-like and extend up and down or obliquely in several layers and in a complex manner. The one you will hear of most frequently is the *sternocleidomastoid* (ster-no-kli-do-MAS-toyd), sometimes referred to simply as the *sternomastoid*. There are two of these strong muscles, which extend from the sternum upward, across either side of the neck, to the mastoid process. Working together, they bring the head forward on the chest (flexion). Working alone, each muscle tilts and rotates the head so as to orient the face toward the side opposite that muscle. If the head is abnormally fixed in this position, the person is said to have *torticollis* (tor-tih-KOL-is), or *wryneck;* this condition may be due to injury or spasm of the muscle. A portion of the trapezius muscle (described later) is located in the back of the neck where it helps hold the head up (extension). Other larger deep muscles are the chief extensors of the head and neck.

sternocleidomastoid

trapezius

teres minor

deltoid

teres major

triceps brachii

latissimus
dorsi

olecranon

lumbodorsal
fascia

gluteus maximus

iliotibial band

biceps femoris

semitendinosus

semimembranosus

gastrocnemius

peroneus longus

Achilles tendon

Figure 8-8

Superficial muscles, posterior (back) view.

TABLE 8-1
Review of Muscles

Name	Location	Function
Muscles of the Head and Neck		
Orbicularis oculi	Encircles eyelid	Closes eye
Levator palpebrae superioris	Back of orbit to upper eyelid	Opens eye
Orbicularis oris	Encircles mouth	Closes lips
Buccinator	Fleshy part of cheek	Flattens cheek; helps in eating, whistling, and blowing wind instruments
Temporal	Above and near ear	Closes jaw
Masseter	At angle of jaw	Closes jaw
Sternocleidomastoid	Along side of neck, to mastoid process	Flexes head; rotates head toward opposite side from muscle
Muscles of the Upper Extremities		
Trapezius	Back of neck and upper back, to clavicle scapula	Raises shoulder and pulls it back; extends head
Latissimus dorsi	Middle and lower back, to humerus	Extends and adducts arm behind back
Pectoralis major	Upper, anterior chest, to humerus	Flexes and adducts arm across chest; pulls shoulder forward and downward
Serratus anterior	Below axilla on side of chest to scapula	Moves scapula forward; aids in raising arm
Deltoid	Covers shoulder joint, to lateral humerus	Abducts arm
Biceps brachii	Anterior arm, to radius	Flexes forearm and supinates hand
Triceps brachii	Posterior arm, to ulna	Extends forearm
Flexor and extensor carpi groups	Anterior and posterior forearm, to hand	Flex and extend hand
Flexor and extensor digitorum groups	Anterior and posterior forearm, to fingers	Flex and extend fingers
Muscles of the Trunk		
Diaphragm	Dome-shaped partition between thoracic and abdominal cavities	Dome descends to enlarge thoracic cavity from top to bottom
Intercostals	Between ribs	Elevate ribs and enlarge thoracic cavity
External and internal oblique; transversus and rectus abdominis	Anterolateral abdominal wall	Compress abdominal cavity and expel substances from body; flex spinal column
Levator ani	Pelvic floor	Aids defecation
Sacrospinalis	Deep in back, vertical mass	Extends vertebral column to produce erect posture
Muscles of the Lower Extremities		
Gluteus maximus	Superficial buttock, to femur	Extends thigh
Gluteus medius	Deep buttock, to femur	Abducts thigh
Iliopsoas	Crosses front of hip joint, to femur	Flexes thigh
Adductor group	Medial thigh, to femur	Adduct thigh
Sartorius	Winds down thigh, ilium to tibia	Flexes thigh and leg (to sit cross-legged)
Quadriceps femoris	Anterior thigh, to tibia	Extends leg
Hamstring group	Posterior thigh, to tibia and fibula	Flexes leg
Gastrocnemius	Calf of leg, to calcaneus	Extends foot (as in tiptoeing)
Tibialis anterior	Anterior and lateral shin, to foot	Dorsiflexes foot (as in walking on heels); inverts foot (sole inward)
Peroneus longus	Lateral leg, to foot	Everts foot (sole outward)
Flexor and extensor digitorum groups	Posterior and anterior leg, to toes	Flex and extend toes

Muscles of the Upper Extremities
MOVEMENT OF THE SHOULDER AND ARM

The position of the shoulder depends to a large extent on the degree of contraction of the *trapezius* (trah-PE-ze-us), a triangular muscle that covers the back of the neck and extends across the back of the shoulder to insert on the clavicle and scapula. The trapezius muscles enable one to raise the shoulders and pull them back. The upper portion of each trapezius can also extend the head and turn it from side to side.

The *latissimus* (lah-TIS-ih-mus) *dorsi* originates from the vertebral spine in the middle and lower back and covers the lower half of the thoracic region. The fibers of each muscle converge to a tendon that inserts on the humerus. The latissimus dorsi powerfully extends the arm, bringing it down forcibly as, for example, in swimming.

A large *pectoralis* (pek-to-RAL-is) *major* is located on either side of the upper part of the chest at the front of the body. This muscle arises from the sternum, the upper ribs, and the clavicle and forms the anterior "wall" of the arm pit or axilla; it inserts on the upper part of the humerus. The pectoralis major flexes and adducts the arm, pulling it across the chest.

Below the axilla, on the side of the chest, is the *serratus* (ser-RA-tus) *anterior.* It originates on the upper eight or nine ribs on the side and the front of the thorax and inserts in the scapula on the side toward the vertebrae. The serratus anterior moves the scapula forward when, for example, one is pushing something. It also aids in raising the arm above the horizontal level.

The *deltoid* covers the shoulder joint and is responsible for the roundness of the upper part of the arm just below the shoulder. This area is often used as an injection site. Arising from the shoulder girdle (clavicle and scapula), the deltoid fibers converge to insert on the lateral side of the humerus. Contraction of this muscle abducts the arm, raising it laterally to the horizontal position.

The shoulder joint allows for a very wide range of movement. This freedom of movement is possible because of the shallow socket (glenoid cavity of the scapula), which requires the support of four deep muscles and their tendons. In certain activities, such as swinging a golf club, playing tennis, or pitching a baseball, these four muscles, called the *rotator cuff,* may be injured, even torn. Surgery is often required for repair of the rotator cuff.

MOVEMENT OF THE FOREARM AND HAND

The *biceps brachii* (BRA-ke-i), located on the front of the arm, is the muscle most often displayed when you "flex your muscles" to show your strength. It inserts on the radius and serves to flex the forearm. It is a supinator of the hand (see Fig. 8-5).

The *triceps brachii,* located on the back of the arm, inserts on the olecranon of the ulna. The triceps has been called the *boxer's muscle,* since it straightens the elbow when a blow is delivered. It is also important in pushing because it converts the arm and forearm into a sturdy rod.

Most of the muscles that move the hand and fingers originate from the radius and the ulna; some of them insert on the carpal bones of the wrist, whereas others have long tendons that cross the wrist and insert on bones of the hand and the fingers. The *flexor carpi* and the *extensor carpi muscles* are responsible for many movements of the hand. Muscles that produce finger movements are the several *flexor digitorum* (dij-e-TO-rum) and the *extensor digitorum muscles.* Special groups of muscles in the fleshy parts of the hand are responsible for the intricate movements that can be performed with the thumb and the fingers. The freedom of movement of the thumb has been one of the most useful endowments of humans.

Muscles of the Trunk
MUSCLES OF RESPIRATION

The most important muscle involved in the act of breathing is the *diaphragm.* This dome-shaped muscle forms the partition between the thoracic cavity above and the abdominal cavity below (Fig. 8-9). When the diaphragm contracts, the central dome-shaped portion is pulled downward, thus enlarging the thoracic cavity from top to bottom. The *intercostal muscles* are attached to and fill the spaces between the ribs. The external and internal intercostals run at angles in opposite directions. Contraction of the intercostal muscles serves to elevate the ribs, thus enlarging the thoracic cavity from side to side and from front to back. The mechanics of breathing are described in Chapter 18.

MUSCLES OF THE ABDOMEN AND PELVIS

The wall of the abdomen has three layers of muscle that extend from the back (dorsally) and around the sides (laterally) to the front (ventrally). They are the *external abdominal oblique* on the outside, the

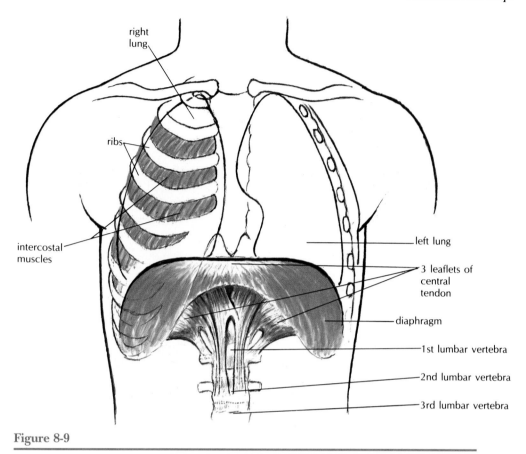

right
lung

ribs

intercostal
muscles

left lung

3 leaflets of
central
tendon

diaphragm

1st lumbar vertebra

2nd lumbar vertebra

3rd lumbar vertebra

Figure 8-9

The diaphragm forms the partition between the thoracic cavity and the abdominal cavity.

internal abdominal oblique in the middle, and the *transversus abdominis,* the innermost. The connective tissue from these muscles extends forward and encloses the vertical *rectus abdominis* of the anterior abdominal wall. The fibers of these muscles, as well as their connective tissue extensions (aponeuroses), run in different directions, resembling the layers in plywood and resulting in a very strong abdominal wall. The midline meeting of the aponeuroses forms a whitish area called the *linea alba* (LIN-e-ah Al-ba) which is an important landmark on the abdomen. It extends from the tip of the sternum to the pubic joint.

These four pairs of abdominal muscles act together to protect the internal organs and compress the abdominal cavity, as in coughing, emptying the bladder (urination) and bowel (defecation), sneezing, vomiting, and childbirth (labor). The two oblique muscles and the rectus abdominis help bend the trunk forward and sideways.

The pelvic floor, or *perineum* (per-ih-NE-um), has its own form of diaphragm, shaped somewhat like a shallow dish. One of the principal muscles of this pelvic diaphragm is the *levator ani* (le-VA-tor A-ni), which acts on the rectum and thus aids in defecation. The superficial and deep muscles of the female perineum are shown in Figure 8-10.

DEEP MUSCLES OF THE BACK

The deep muscles of the back, which act on the vertebral column itself, are thick vertical masses that lie under the trapezius and latissimus dorsi. The longest muscle is the *sacrospinalis* (sa-kro-spin-A-lis), which helps maintain the vertebral column in an erect posture.

Muscles of the Lower Extremities

The muscles in the lower extremities, among the longest and strongest muscles in the body, are specialized for locomotion and balance.

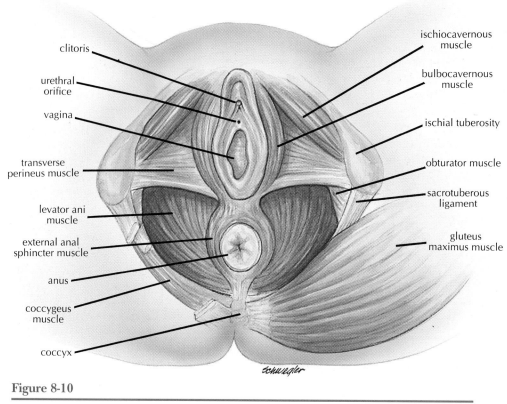

Figure 8-10

Muscles of the female perineum (pelvic floor).

In the figure, the following labels appear:

clitoris

urethral orifice

vagina

transverse perineus muscle

levator ani muscle

external anal sphincter muscle

anus

coccygeus muscle

coccyx

ischiocavernous muscle

bulbocavernous muscle

ischial tuberosity

obturator muscle

sacrotuberous ligament

gluteus maximus muscle

schwegler

MOVEMENT OF THE THIGH AND LEG

The *gluteus maximus* (GLU-te-us MAK-sim-us), which forms much of the fleshy part of the buttock, is relatively large in humans because of its support function when a person is standing in the erect position. This muscle extends the thigh and is very important in walking and running. The *gluteus medius,* which is partially covered by the gluteus maximus, serves to abduct the thigh, and is one of the sites used for intramuscular injections.

The *iliopsoas* (il-e-o-SO-as) arises from the ilium and the bodies of the lumbar vertebrae; it crosses the front of the hip joint to insert on the femur. It is a powerful flexor of the thigh and helps keep the trunk from falling backward when one is standing erect.

The *adductor muscles* are located on the medial part of the thigh. They arise from the pubis and ischium and insert on the femur. These strong muscles press the thighs together, as in grasping a saddle between the knees when riding a horse.

The *sartorius* (sar-TO-re-us) is a long, narrow muscle that begins at the iliac spine, winds downward and inward across the entire thigh, and ends on the upper medial surface of the tibia. It is called the *tailor's muscle* because it is used in crossing the legs in the manner of tailors, who in days gone by sat cross-legged on the floor.

The front and sides of the femur are covered by the *quadriceps femoris* (KWOD-re-seps FEM-or-is), a large muscle that has four heads of origin. The individual parts are: in the center, covering the anterior thigh, the *rectus femoris*; on either side the *vastus medialis* and *vastus lateralis*; deeper in the center, the *vastus intermedius.* One of these muscles (rectus femoris) originates from the ilium, and the other three are from the femur, but all four have a common tendon of insertion on the tibia. You may remember that this is the tendon that encloses the knee cap, or patella. This muscle extends the leg, as in kicking a ball. The vastus lateralis is also a site for intramuscular injections.

The *hamstring muscles* are located in the posterior part of the thigh. Their tendons can be felt behind the knee as they descend to insert on the tibia and fibula. The hamstrings flex the leg on the thigh, as in kneeling. Individually, moving from lateral to medial position, they are the *biceps femoris,* the *semimembranosus,* and the *semitendinosus.*

MOVEMENT OF THE FOOT

The *gastrocnemius* (gas-trok-NE-me-us) is the chief muscle of the calf of the leg. It has been called the *toe dancer's muscle* because it is used in standing on tiptoe. It ends near the heel in a prominent cord called the *Achilles tendon,* which attaches to the calcaneus (heel bone). The Achilles tendon is the largest tendon in the body.

Another leg muscle that acts on the foot is the *tibialis* (tib-e-A-lis) *anterior,* located on the front of the leg. This muscle performs the opposite function of the gastrocnemius. Walking on the heels will use the tibialis anterior to raise the rest of the foot off the ground (dorsiflexion). This muscle is also responsible for inversion of the foot. The muscle for eversion of the foot is the *peroneus* (per-o-NE-us) *longus,* located on the lateral side of the leg. The long tendon of this muscle crosses under the foot, forming a sling that supports the transverse (metatarsal) arch.

The toes, like the fingers, are provided with flexor and extensor muscles. The tendons of the extensor muscles are located in the top of the foot and insert on the superior surface of the toe bones (phalanges). The flexor digitorum tendons cross the sole of the foot and insert on the undersurface of the toe bones.

▶ Effects of Aging on Muscles

Beginning at about 40 years of age, there is a gradual loss of muscle cells with a resulting decrease in the size of each individual muscle. There is also a loss of power, notably in the extensor muscles, such as the large sacrospinalis near the vertebral column. This causes the "bent over" appearance of a hunchback (kyphosis), which in women is often referred to as the *dowager's hump.* Sometimes there is a tendency to bend (flex) the hips and knees. In addition to causing the previously noted changes in the vertebral column (see Chap. 7), these effects on the extensor muscles result in a further decrease in the elderly person's height. Activity and exercise throughout life delay and decrease these undesirable effects of aging.

▶ Muscular Disorders

A general disorder of the muscles called a *spasm* is a sudden and involuntary muscular contraction. A spasm is always painful. A spasm of the visceral muscles is called *colic,* a good example of which is the spasm of the intestinal muscles often referred to as *bellyache.* Spasms may occur also in the skeletal muscles; if the spasms occur in a series, the condition may be called a *seizure* or *convulsion. Atrophy* (AT-ro-fe) is a wasting or decrease in the size of a muscle when it cannot be used, such as when an extremity must be placed in a cast following a fracture.

Strains and *sprains* are typical injuries that often affect muscles. Severe and excessive exertion can cause detachment of muscles from bones or tearing of some of the muscle cells. Sprains can involve damage to other structures besides the ligaments, specifically blood vessels, nerves, and muscles. Much of the pain and swelling accompanying a sprain can be prevented by the immediate application of ice packs, which constrict some of the smaller blood vessels and reduce internal bleeding.

Muscular dystrophy (DIS-tro-fe) is really a group of disorders, most of which are hereditary. These disorders all progress at different rates. The most common type, which is found most frequently in male children, leads to complete helplessness and ends in death within 10 to 15 years of onset. The cure is not yet known for any muscular dystrophy. Progress toward treatment is possible now that the genetic defects that cause some forms of the disease have been identified.

Myasthenia gravis (mi-as-THE-ne-ah GRA-vis) is characterized by chronic muscular fatigue brought on by the slightest exertion. It affects adults and begins with the muscles of the head. Drooping of the eyelids (ptosis) is a common early symptom.

Myalgia (mi-AL-je-ah) means "muscular pain"; *myositis* (mi-o-SI-tis) is a term that indicates actual inflammation of muscle tissue. *Fibrositis* (fi-bro-SI-tis) means "inflammation of connective tissues" and refers particularly to those tissues connected with muscles and joints. Usually it appears as a combination disorder that is properly called *fibromyositis* but is commonly referred to as *rheumatism, lumbago,* or *charley horse.* The disorder may be acute, with severe pain on motion, or it may be chronic. Sometimes the application of heat, together with massage and rest, relieves the symptoms.

Bursitis is inflammation of a bursa, a fluid-filled sac that minimizes friction between tissues and bone. Some bursae communicate with joints; others are closely related to muscles. Sometimes bursae develop spontaneously in response to prolonged friction.

Bursitis can be very painful, with swelling and limitation of motion. Some examples of bursitis are listed below:

1. *Student's elbow,* in which the bursa over the point of the elbow (olecranon) is inflamed due to long hours of leaning on the elbow while studying
2. *Ischial bursitis,* which is said to be common among persons who must sit a great deal, such as taxicab drivers and truckers
3. *Housemaid's knee,* in which the bursa in front of the patella is inflamed. This form of bursitis is found in persons who must often be on their knees.
4. *Subdeltoid bursitis* in the shoulder region, a fairly common form.

In some cases a local anesthetic, corticosteroids, or both may be injected to relieve the pain of bursitis.

Bunions are enlargements commonly found at the base and medial side of the great toe. Usually, prolonged pressure has caused the development of a bursa, which has then become inflamed. Special shoes may be necessary if surgery is not performed.

Tendinitis (ten-din-I-tis), an inflammation of muscle tendons and their attachments, occurs most often in athletes who overexert themselves. It fre-

quently involves the shoulder, the hamstring muscle tendons at the knee, and the Achilles tendon near the heel. *Tenosynovitis* (ten-o-sin-o-VI-tis), which involves the synovial sheath that encloses tendons, is found most often in women in their 40s following an injury or surgery. It may involve swelling and severe pain with activity.

Carpal tunnel syndrome involves the tendons of the flexor muscles of the fingers as well as the nerves supplying the hand and fingers. Numbness and weakness of the hand is caused by pressure on the median nerve as it passes through a tunnel formed by the carpal bones of the wrist. It occurs in certain factory workers and others who use their hands and fingers strenuously, such as computer operators and musicians.

Muscle *cramps* include strong, painful contractions of leg and foot muscles during the night (recumbancy cramps). They are most likely to occur following unusually strenuous activity.

Flatfoot is a common disorder in which the arch of the foot, the normally raised portion of the sole, breaks down so that the entire sole rests on the ground. This condition may be congenital, in which case it usually gives little trouble. However, flatfoot can result from a progressive weakening of the muscles that support the arch, in which case it is usually accompanied by a great deal of pain. Incorrect use of the muscles that support the arch (such as toeing out when walking) or lack of exercise is thought to bring about flatfoot. Wearing properly fitted shoes and walking with the toes pointed straight forward may help prevent flatfoot and other painful foot disorders. Exercises performed under medical supervision can be helpful in strengthening the muscles that help maintain the foot arches.

SUMMARY

I. Structure of skeletal muscle tissue
A. Muscle fibers (cells): large, multinucleated, striated (banded)
B. Connective tissue
1. Around individual fibers
2. Around fascicles (bundles)
3. Around whole muscle

II. Characteristics of skeletal muscles
A. Irritability—capacity to respond to nerve impulses transmitted at neuromuscular junction
B. Contractility—ability to shorten
1. Contracting filaments—slide together to shorten muscle

  a. Actin—thin and light
  b. Myosin—heavy and dark
 2. ATP—supplies energy
 3. Myoglobin—stores energy
 4. Glycogen—stores glucose
 5. Oxygen debt—develops during strenuous exercise
  a. Anaerobic metabolism
  b. Yields lactic acid
C. Muscle contractions
 1. Tonus—partially contracted state
 2. Isotonic contractions—muscle shortens to produce movement
 3. Isometric contractions—tension increases but muscle does not shorten

III. Mechanics of movement
 A. Attachments of skeletal muscles
  1. Tendon—cord of connective tissue that attaches muscle to bone
   a. Origin—attached to more fixed part
   b. Insertion—attached to moving part
  2. Aponeurosis—broad band of connective tissue that attaches muscle to bone or other muscle
 B. Cooperation of muscles
  1. Prime mover—initiates movement
  2. Antagonist—produces opposite movement

  3. Others—steady body parts and assist prime mover
 C. Body mechanics—muscles function with skeleton as lever systems
  1. Lever—bone
  2. Fulcrum—joint
  3. Force—muscle contraction

IV. Skeletal muscle groups
 A. Naming of muscles—location, size, shape, direction of fibers, number of heads, action
 B. Muscle groups
  1. Muscles of the head and neck
  2. Muscles of the upper extremities
  3. Muscles of the trunk
  4. Muscles of the lower extremities

V. Effects of aging on muscles
 A. Decrease in size of muscles
 B. Weakening of muscles, especially extensors

VI. Muscular disorders
 A. Spasm—colic, seizure, convulsion
 B. Strains, sprains
 C. Muscular dystrophy—group of disorders
 D. Myasthenia gravis
 E. Bursitis, tendinitis
 F. Flatfoot

QUESTIONS FOR STUDY AND REVIEW

1. Describe the structure of muscles from fiber to deep fascia.

2. List the main characteristics of skeletal muscle cells.

3. Differentiate between the terms in each of the following pairs:
 a. *tendon* and *aponeurosis*
 b. *muscle origin* and *muscle insertion*
 c. *prime mover* and *antagonist*
 d. *isometric contraction* and *isotonic contraction*
 e. *bursitis* and *tendinitis*

4. Explain the role of each of the following in muscle contraction: actin and myosin, calcium, ATP, myoglobin, glycogen.

5. When does oxygen debt occur? What is the role of lactic acid in oxygen debt? How is oxygen debt eliminated?

6. What are some valuable effects of exercise?

7. What are levers and how do they work? The forceps is an example of which class of lever? When muscles and bones act as lever systems, what part of the body is the fulcrum?

8. List six characteristics according to which muscles are named.

9. Name and describe the functions of the principal muscles of the head and neck, upper extremities, trunk, and lower extremities.

10. Name three muscles used for intramuscular injections.

11. What effect does aging have on muscles?

12. Define *spasm* and give examples of spasms.

13. Define *atrophy* and give one cause.

14. What are muscular dystrophies, and what are some of their effects?

15. What is bursitis? Describe several forms.

UNIT IV

Coordination and Control

Of the four chapters in this unit, two describe the nervous system and some of its many parts and complex functions. The organs of special sense (eye, ear, taste buds, and so forth) are allotted an additional separate chapter. The fourth chapter in this unit discusses hormones and the organs that produce them. Working with the nervous system, these hormones play an important role in coordination and control.

The Nervous System: The Spinal Cord and Spinal Nerves

Selected Key Terms

The following terms are defined in the Glossary:

acetylcholine

action potential

afferent

autonomic nervous system

axon

dendrite

effector

efferent

epinephrine

ganglion

nerve

nerve impulse

neuritis

neurotransmitter

plexus

receptor

reflex

synapse

tract

Behavioral Objectives

After careful study of this chapter, you should be able to:

- Describe the organization of the nervous system according to structure and function
- Describe the structure of a neuron
- Briefly describe the transmission of a nerve impulse
- Explain the role of myelin in nerve conduction
- Briefly describe transmission at a synapse
- Define *neurotransmitter* and give several examples of neurotransmitters
- Name three types of nerves and explain how they differ from each other
- List the components of a reflex arc
- Describe the spinal cord and name several of its functions
- Define a reflex and give several examples
- Describe and name the spinal nerves and three of their main plexuses
- Compare the location and functions of the sympathetic and parasympathetic nervous systems
- Describe several disorders of the spinal cord and of the spinal nerves

Memmler, RL., Cohen, BJ, Wood, DL. *THE HUMAN BODY IN HEALTH AND DISEASE*, 8/e,
© 1996 Lippincott-Raven Publishers

N one of the body systems is capable of functioning alone. All are interdependent and work together as one unit so that normal conditions (homeostasis) within the body may prevail. The nervous system serves as the chief coordinating agency. Conditions both within and outside the body are constantly changing; one purpose of the nervous system is to respond to these internal and external changes (known as *stimuli*) and so cause the body to adapt itself to new conditions. It is through the instructions and directions sent to the various organs by the nervous system that a person's internal harmony and the balance between that person and the environment are maintained. The nervous system has been compared to a telephone exchange, in that the brain and the spinal cord act as switching centers and the nerve trunks act as cables for carrying messages to and from these centers.

▶ The Nervous System as a Whole

The parts of the nervous system may be grouped according to structure or function. The anatomic, or structural, divisions of the nervous system are as follows (Fig. 9-1):

1. The *central nervous system* (*CNS*) includes the brain and spinal cord.
2. The *peripheral* (per-IF-er-al) *nervous system* (*PNS*) is made up of all the nerves outside the CNS. It includes all the *cranial* and *spinal nerves*. Cranial nerves are those that carry impulses to and from the brain. Spinal nerves are those that carry messages to and from the spinal cord.

From the standpoint of structure, the CNS and PNS together include most of the nerve tissue in the body. However, certain peripheral nerves have a special function, and for this reason they are grouped together under the designation *autonomic* (aw-to-NOM-ik) *nervous system.* The reason for this separate classification is that the autonomic nervous system governs activities that go on automatically (involuntarily). This system carries impulses from the CNS to the glands, to the smooth muscles found in the walls of tubes and hollow organs, and to the cardiac muscle of the heart. Both cranial and spinal nerves carry autonomic nervous system impulses. The system is subdivided into the *sympathetic* and *parasympathetic*

nervous systems, both of which are explained later in this chapter.

The autonomic nervous system makes up a part of the *involuntary,* or *visceral, nervous system,* which controls smooth muscle, cardiac muscle, and glands. The *voluntary,* or *somatic* (so-MAH-tik), *nervous system* consists of all of the nerves that control the action of the skeletal muscles (described in Chap. 8), which are under conscious control.

▶ Neurons and Their Functions

The functional cells of the nervous system are highly specialized cells called *neurons* (Fig. 9-2). The cell body of each neuron contains the nucleus and other organelles. Extending from the cytoplasm of the cell body are threadlike fibers. Nerve cell fibers are of two kinds:

1. *Dendrites* (DEN-drites), which conduct impulses *to* the cell body
2. *Axons* (AK-sons), which conduct impulses *away from* the cell body.

Many dendrites have a highly branched, treelike appearance. In fact the name comes from a Greek word meaning "tree." The dendrites of sensory neurons, however, those which carry impulses toward the CNS, are different from those of other neurons: they may be long (as long as 1 meter) or short, but they are usually single instead of branched. Each of these dendrites functions as a *receptor,* where a stimulus is received and the sensory impulse begins. In the case of the special senses, such as hearing and sight (see Chap. 11), the dendrite is in contact with a modified cell that acts as the receptor.

Some axons are covered with a fatty material called *myelin* that insulates and protects the fiber (Fig. 9-3). In the peripheral nervous system this covering is produced by special myelin-containing cells, called *Schwann* (shwan) *cells,* that wrap around the axon like a jelly roll. When the sheath is complete, small spaces still remain between the cells. These tiny gaps, called *nodes,* are important in speeding the conduction of nerve impulses.

The outermost membranes of the Schwann cells form a thin coating known as the *neurilemma* (nu-rih-LEM-mah). This covering is a part of the mechanism by which some peripheral nerves repair themselves when injured. Under some circumstances, damaged nerve cell fibers may regenerate by

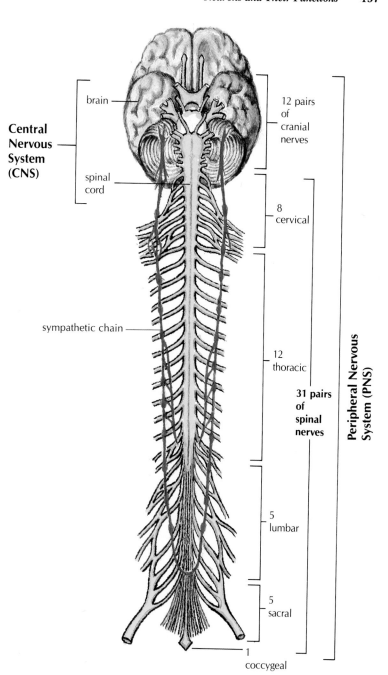

brain

Central Nervous System (CNS)

spinal cord

sympathetic chain

12 pairs of cranial nerves

8 cervical

12 thoracic

31 pairs of spinal nerves

5 lumbar

5 sacral

1 coccygeal

Peripheral Nervous System (PNS)

Figure 9-1

Anatomic divisions of the nervous system. (Cohen B: Medical Terminology: An Illustrated Guide, 2nd ed, p. 360. Philadelphia, JB Lippincott, 1994)

growing into the sleeve formed by the neurilemma. Cells of the brain and the spinal cord are myelinated, not by Schwann cells, but by neuroglia, the connective tissue cells of the CNS. As a result, they have no neurilemma. If they are injured, the injury is permanent. Even in the peripheral nerves, however, repair is a slow and uncertain process.

Axons covered with myelin are called *white fibers* and are found in the *white matter* of the brain and spinal cord as well as in the nerve trunks in all parts of the body. The fibers and cell bodies of the *gray matter* are not covered with myelin.

The job of neurons in the peripheral nervous system is to constantly relay information either to or from the central nervous system (CNS). Neurons that conduct impulses *to* the spinal cord and brain are described as *sensory,* or *afferent, neurons.* Those cells that carry impulses *from* the CNS out to muscles

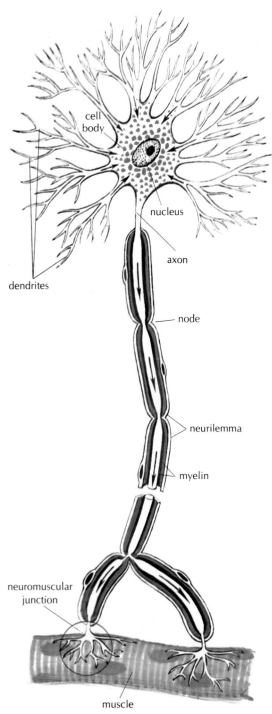

Figure 9-2

Diagram of a motor neuron. The break in the axon denotes length. The arrows show the direction of the nerve impulse.

and glands are *motor,* or *efferent, neurons.* Any organ, such as a muscle or gland, activated by a motor neuron is the *effector.*

The Nerve Impulse

The cell membrane of an unstimulated (resting) neuron carries an electric charge. Because of positive and negative ions concentrated on either side of the membrane, the inside of the membrane at rest is negative as compared with the outside. A *nerve impulse* is a local reversal in the charge on the nerve cell membrane that then spreads along the membrane like an electric current. This sudden electrical change in the membrane is called an *action potential.* A stimulus, then, is any force that can start an action potential. This electric change results from rapid shifts in sodium and potassium ions across the cell membrane. The reversal occurs very rapidly (in less than one thousandth of a second) and is followed by a rapid return of the membrane to its original state so that it can be stimulated again.

A myelinated nerve fiber conducts impulses more rapidly than an unmyelinated fiber of the same size because the electrical impulse "jumps" from node (space) to node in the myelin sheath instead of traveling continuously along the fiber.

The Synapse

Each neuron is a separate unit, and there is no anatomic connection between neurons. How then is it possible for neurons to communicate? In other words, how does the axon of one neuron make functional contact with the membrane of another neuron? This is accomplished by the *synapse* (SIN-aps), from a Greek word meaning "to clasp." Synapses are points of junction for the transmission of nerve impulses (Fig. 9-4).

Within the branched endings of the axon are small bubbles (vesicles) containing a type of chemical known as a *neurotransmitter.* When stimulated, the axon releases its neurotramsmitter into the narrow gap, the *synaptic cleft,* between the cells. The neurotransmitter then acts as a chemical signal to stimulate the next cell, described as the *postsynaptic cell.* On the receiving membrane, usually that of a dendrite, but sometimes another part of the cell, there are special sites, or *receptors*, ready to pick up and respond to specific neurotransmitters. Receptors in the cell membrane influence how or if that cell will respond to a given neurotransmitter.

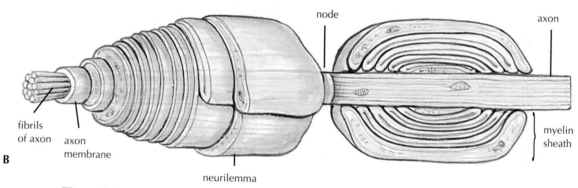

Figure 9-3

Formation of the myelin sheath. (**A**) Schwann cells wrap around the axon creating a myelin coating. (**B**) Outermost layer of the Schwann cell forms the neurilemma. Spaces between the cells are the nodes.

Although there are many known neurotransmitters, the main ones are *epinephprine* (ep-ih-NEF-rin), also called *adrenaline;* a related compound, *norepinephrine* (nor-ep-ih-NEF-rin), or *noradrenaline;* and *acetylcholine* (as-e-til-KO-lene). Acetylcholine (ACh) is the neurotransmitter released at the neuromuscular junction, the synapse between a neuron and a muscle cell. All three of the above neurotransmitters function in the autonomic nervous system. It is common to think of neurotransmitters as stimulating the cells they reach; in fact, they have been described as such in this discussion. Note, however, that some of these chemicals act to inhibit the postsynaptic cell and keep it from reacting.

The connections between neurons can be quite complex. One cell can branch to stimulate many receiving cells, or a single cell may be stimulated by a number of different axons (Fig. 9-5). The cell's response is based on the total effects of all the neurotransmitters it receives at any one time.

Nerves

A *nerve* is a bundle of nerve cell fibers located *outside* the CNS. Bundles of nerve cell fibers *within* the CNS are *tracts.* Tracts are located within the brain and also within the spinal cord to conduct impulses to and from the brain. A nerve or tract can be compared to an electric cable made up of many wires. As with muscles, the "wires," or nerve cell fibers in a nerve, are bound together with connective tissue.

A few of the cranial nerves have only sensory fibers for conducting impulses toward the brain. These are described as *sensory,* or *afferent, nerves.* A few of the cranial nerves contain only motor fibers

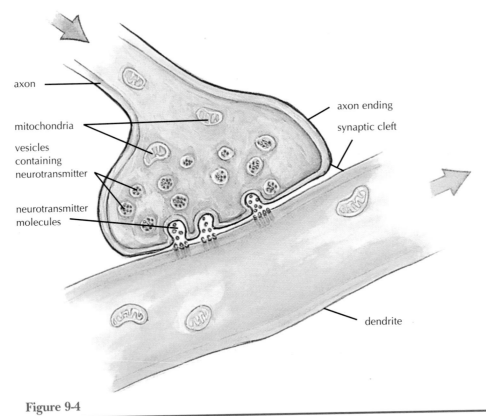

axon

mitochondria

vesicles
containing
neurotransmitter

neurotransmitter
molecules

axon ending

synaptic cleft

dendrite

Figure 9-4

Close-up view of a synapse. The axon ending has vesicles containing neurotransmitter, which
is released across the synaptic cleft to the membrane of the next cell.

for conducting impulses away from the brain and
are classified as *motor,* or *efferent, nerves.* However,
the remainder of the cranial nerves and *all* of the
spinal nerves contain both sensory *and* motor fibers
and are referred to as *mixed nerves.*

The Reflex Arc

As the nervous system functions, both external
and internal stimuli are received, interpreted, and
acted on. A complete pathway through the nervous
system from stimulus to response is termed a *reflex
arc* (Fig. 9-6). This is the basic functional pathway of
the nervous system. The parts of a typical reflex arc
are:

1. *Receptor*—the end of a dendrite or some specialized receptor cell, as in a special sense organ, that
detects a stimulus
2. *Sensory neuron,* or afferent neuron—a cell that
transmits impulses *toward* the CNS

3. *Central neuron*—a cell or cells within the CNS.
These neurons may carry impulses to and from
the brain, may function within the brain, or may
distribute impulses to different regions of the
spinal cord.
4. *Motor neuron,* or efferent neuron—a cell that
carries impulses *away from* the CNS
5. *Effector*—a muscle or a gland outside the CNS
that carries out a response.

At its simplest, a reflex arc can involve just two
neurons, one sensory and one motor, with a synapse
in the CNS. There are very few reflex arcs that require only this minimal number of neurons. The
knee jerk reflex described below is one of the few examples in humans. Most reflex arcs involve many
more, even hundreds, of connecting neurons within
the central nervous system. The many intricate patterns that make the nervous system so responsive
and adaptable also make it difficult to study, and

axon

cell body

dendrite

axon
endings

nerve
fiber

Figure 9-5

A single neuron is stimulated by axons of many other neurons.

investigation of the nervous system is one of the most active areas of research today.

▶ The Spinal Cord

Location of the Spinal Cord

In the embryo, the spinal cord occupies the entire spinal canal and so extends down into the tail portion of the vertebral column. However, the column of bone grows much more rapidly than the nerve tissue of the cord, so that eventually the end of the cord no longer reaches the lower part of the spinal canal. This disparity in growth continues to increase; in the adult the cord ends in the region just below the area to which the last rib attaches (between the first and second lumbar vertebrae).

Structure of the Spinal Cord

The spinal cord (see Fig. 9-6) has a small, irregularly shaped internal section that consists of gray matter (nerve cell bodies) and a larger area surrounding this gray part that consists of white matter (nerve cell fibers). The gray matter is so arranged that a column of cells extends up and down dorsally,

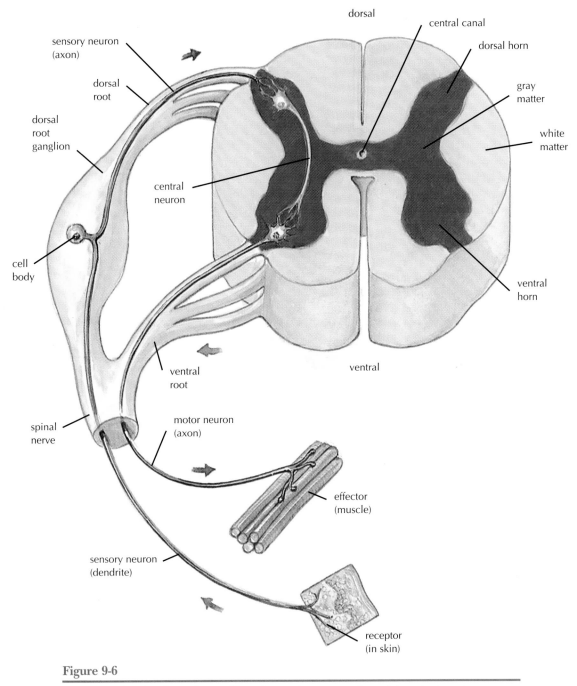

Figure 9-6

Reflex arc showing the pathway of impulses and a cross section of the spinal cord.

one on each side; another column is found in the ventral region on each side. These two pairs of columns, called the ***dorsal*** and ***ventral horns,*** give the gray matter an H-shaped appearance in cross section. In the center of the gray matter is a small channel, the ***central canal***, that contains cerebrospinal fluid, the liquid that circulates around the brain and spinal cord. The white matter consists of thousands of nerve cell fibers arranged in three areas external to the gray matter on each side.

Functions of the Spinal Cord

The spinal cord is the link between the spinal nerves and the brain. It is also a place where simple responses, known as *reflexes* can be coordinated even without involving the brain. The functions of the spinal cord may be divided into three categories:

1. *Conduction of sensory impulses* upward through ascending tracts to the brain
2. *Conduction of motor impulses* from the brain down through descending tracts to the efferent neurons that supply muscles or glands
3. *Reflex activities.* A reflex is a simple, rapid, and automatic response involving very few neurons.

When you fling out an arm or leg to catch your balance, withdraw from a painful stimulus, or blink to avoid an object approaching your eyes, you are experiencing reflex behavior. A reflex pathway that passes through the spinal cord alone and does not involve the brain is termed a *spinal reflex.* The *stretch reflex,* in which a muscle is stretched and responds by contracting, is one example. If you tap the tendon below the kneecap (the patellar tendon), the muscle of the anterior thigh (quadriceps femoris) contracts, eliciting the knee jerk. Such stretch reflexes may be evoked by appropriate tapping of most large muscles (such as the triceps brachii in the arm and the gastrocnemius in the calf of the leg). Because reflexes occur automatically, they are used in physical examinations to test the condition of the nervous system.

Medical Procedures Involving the Spinal Cord
LUMBAR PUNCTURE

Because the spinal cord is only about 18 inches long and ends some distance above the level of the hip line, a lumbar puncture or spinal tap is usually done between the third and fourth lumbar vertebrae, at about the level of the top of the hipbone. During this procedure, a small amount of the fluid that circulates around the brain and spinal cord may be removed from the space below the spinal cord. The fluid, called *cerebrospinal fluid (CSF),* can then be studied in the laboratory for evidence of disease or injury.

ADMINISTRATION OF DRUGS

Anesthetics or medications are sometimes injected into the space below the cord. The anesthetic agent temporarily blocks all sensation from the lower part of the body. This method of giving anesthesia has an advantage for certain types of procedures or surgery; the patient is awake but feels nothing in his or her lower body. This route can be used to administer pain medication effective for 24 hours.

Disorders Involving the Spinal Cord

An acute viral disease affecting both the spinal cord and the brain is *poliomyelitis* (po-le-o-mi-el-I-tis) ("polio"), which occurs most commonly in children. The polio virus enters the body through the nose and the throat; it multiplies in the gastrointestinal tract and then travels to the CNS, possibly by way of the blood. The virus may destroy the motor nerve cells in the spinal cord, in which case paralysis of one or more limbs results. The virus can also attack some of the cells of the brain and cause death. Prevention of poliomyelitis by means of the oral Sabin vaccine is one of many significant advances in preventive medicine.

Injuries to the spinal cord occur in instances in which bones of the spinal column are broken or dislocated, such as in swimming and diving accidents. Gunshot or shrapnel wounds may damage the cord to varying degrees. Because the nervous tissue of the brain and cord cannot repair itself, severing of the cord causes paralysis of all the muscles supplied by nerves below the level of the injury. Loss of sensation and motion in the lower part of the body is called *paraplegia* (par-ah-PLE-je-ah).

Other disorders of the spinal cord include tumors that grow from within the cord or that compress the cord from outside.

Multiple sclerosis (*sclerosis* means "hardening") involves the entire spinal cord as well as the brain. In this disease the myelin around axons disappears and the fibers themselves degenerate. Multiple sclerosis is an extremely disabling disease; however, it usually progresses very slowly, so that affected persons may have many years of relatively comfortable life.

Amyotrophic (ah-mi-o-TROF-ik) *lateral sclerosis* is a disorder of the nervous system in which motor neurons are destroyed. The progressive destruction causes muscle atrophy and loss of motor control until finally the affected person is unable to swallow or talk.

▶ Spinal Nerves

Location and Structure of the Spinal Nerves

There are 31 pairs of spinal nerves, each pair numbered according to the level of the spinal cord from which it arises. Each nerve is attached to the spinal cord by two roots, the **dorsal root** and the **ventral root** (see Fig. 9-6). On each dorsal root is a marked swelling of gray matter called the **dorsal root ganglion,** which contains the cell bodies of the sensory neurons. A **ganglion** (GANG-le-on) is any collection of nerve cell bodies located outside the central nervous system. Nerve cell fibers from sensory receptors throughout the body lead to these dorsal root ganglia.

Whereas sensory fibers form the dorsal roots, the ventral roots of the spinal nerves are a combination of motor (efferent) nerve fibers supplying voluntary muscles, involuntary muscles, and glands. The cell bodies of these neurons are located in the ventral gray matter (ventral horns) of the cord. Because the dorsal (sensory) and ventral (motor) roots are combined to form the spinal nerve, all spinal nerves are mixed nerves.

Branches of the Spinal Nerves

Each spinal nerve continues only a very short distance away from the spinal cord and then branches into small posterior divisions and rather large anterior divisions. The larger anterior branches interlace to form networks called **plexuses** (PLEK-sus-eze), which then distribute branches to the body parts. The three main plexuses are described as follows:

1. The **cervical plexus** supplies motor impulses to the muscles of the neck and receives sensory impulses from the neck and the back of the head. The phrenic nerve, which activates the diaphragm, arises from this plexus.
2. The **brachial** (BRA-ke-al) **plexus** sends numerous branches to the shoulder, arm, forearm, wrist, and hand. The radial nerve emerges from the brachial plexus.
3. The **lumbosacral** (lum-bo-SA-kral) **plexus** supplies nerves to the lower extremities. The largest of these branches is the **sciatic** (si-AT-ik) **nerve,** which leaves the dorsal part of the pelvis, passes beneath the gluteus maximus muscle, and extends down the back of the thigh. At its beginning it is nearly 1 inch thick, but it soon branches to the thigh muscles; near the knee it forms two subdivisions that supply the leg and the foot.

Disorders of the Spinal Nerves

Neuritis (nu-RI-tis) means "inflammation of a nerve." The term is also used to refer to degenerative and other disorders that may involve nerves. It may affect a single nerve or many nerves throughout the body, as a result of blows, bone fractures, or other mechanical injuries. Neuritis may be caused by diabetes, nutritional deficiency, or by various poisons such as alcohol, carbon monoxide, and barbitals.

Neuritis is fairly common in chronic alcoholics and is thought to be related to the severe malnourished state of these persons. Vitamin B deficiencies, especially of **thiamine** (THI-ah-min), are related to both the malnourished state and chronic alcoholism. Neuritis is really a symptom rather than a disease, so that thorough physical and laboratory studies may be needed to establish its cause.

Sciatica (si-AT-ih-kah) is a form of neuritis characterized by severe pain along the sciatic nerve and its branches. There are many causes of this disorder, but probably the most common are rupture of a disk between the lower lumbar vertebrae and arthritis of the lower part of the spinal column.

Herpes zoster, commonly known as *shingles,* is characterized by numerous blisters along the course of certain nerves, most commonly the intercostal nerves, which are branches of the thoracic spinal nerves in the waist area. The cause is the chicken pox virus, which attacks the sensory cell bodies inside the spinal ganglia. Recovery in a few days is usual, but neuralgic pains may persist for years and be very distressing. This infection may also involve the first branch of the fifth cranial nerve and cause pain in the eyeball and surrounding tissues.

▶ The Autonomic Nervous System

Parts of the Autonomic Nervous System

Although the internal organs such as the heart, lungs, and stomach contain nerve endings and nerve fibers for conducting sensory messages to the brain and cord, most of these impulses do not reach consciousness. These *afferent* impulses from the viscera are translated into reflex responses without

Dermatomes

Sensory neurons from all over the skin, except for the skin of the face and scalp, feed information into the spinal cord through the spinal nerves. The skin surface can be mapped into distinct regions that are supplied by a single spinal nerve. Each of these regions is called a *dermatome* (DER-mah-tome) (see illustration).

Stimulation of the skin within a given dermatome will be carried over the corresponding spinal nerve. This information can be used in diagnosis to identify the spinal nerve or spinal segment that is involved in an injury. In some areas, the dermatomes are not absolutely distinct. Some dermatomes may share a nerve supply with neighboring regions. For this reason, it is necessary to numb several adjacent dermatomes to achieve successful anesthesia.

(From Jones SA, Weigel A, White RD, McSwain Jr. NE, & Breiler M. Advanced Emergency Care for Paramedic Practice, Fig. 22-8, p. 545. Philadelphia, JB Lippincott, 1992)

reaching the higher centers of the brain; the sensory neurons from the organs are grouped with those that come from the skin and voluntary muscles. In contrast, the *efferent* neurons, which supply the glands and the involuntary muscles, are arranged very differently from those that supply the voluntary muscles. This variation in the location and arrange-ment of the *visceral efferent* neurons has led to their classification as part of a separate division called the **autonomic nervous system** (**ANS**) (Fig. 9-7). The ANS itself is comprised of **sympathetic** and **parasympathetic** divisions.

The autonomic nervous system has many ganglia that serve as relay stations. In these ganglia each

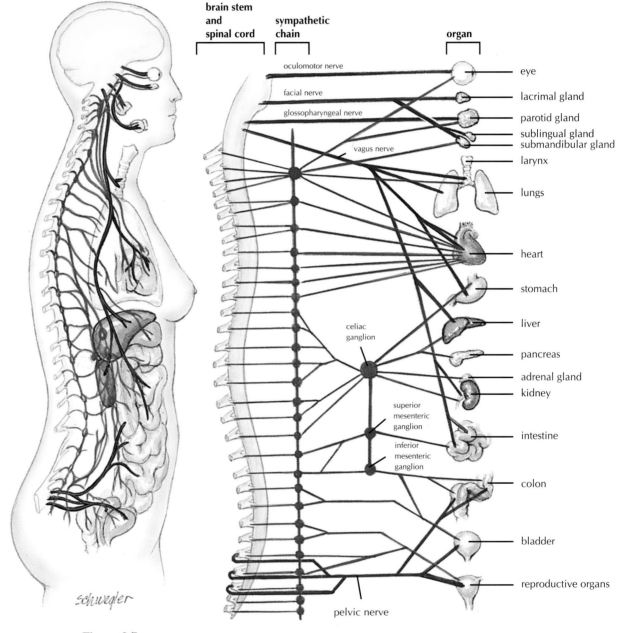

Figure 9-7

Autonomic nervous system (only one side is shown). The sympathetic system is shown in green; the parasympathetic system is shown in blue.

message is transferred at a synapse from the first neuron to a second one and from there to the muscle or gland cell. This differs from the voluntary (somatic) nervous system, in which each motor nerve fiber extends all the way from the spinal cord to the skeletal muscle with no intervening synapse. Some of the autonomic fibers are within the spinal nerves; some are within the cranial nerves (see Chap. 10). The distribution of the two divisions of the ANS is as follows:

1. The sympathetic pathways begin in the spinal cord with cell bodies in the thoracic and lumbar regions, the ***thoracolumbar*** (tho-rah-ko-LUM-bar) area. The sympathetic fibers arise from the spinal cord at the level of the first thoracic nerve down to the level of the second lumbar spinal nerve. From this part of the cord, nerve fibers extend to ganglia where they synapse with a second set of neurons, the fibers of which extend to the glands and involuntary muscle tissues. Many of the sympathetic ganglia form the ***sympathetic chains,*** two cordlike strands of ganglia that extend along either side of the spinal column from the lower neck to the upper abdominal region. The nerves that supply the organs of the abdominal and pelvic cavities synapse in three single ganglia farther from the spinal cord. The second neurons of the sympathetic nervous system act on the effectors by releasing the neurotransmitter epinephrine (adrenaline). This system is therefore described as ***adrenergic,*** which means "activated by adrenaline."

2. The parasympathetic pathways begin in the ***craniosacral*** (kra-ne-o-SAK-ral) areas, with fibers arising from cell bodies of the midbrain, medulla, and lower (sacral) part of the spinal cord. From these centers the first set of fibers extends to autonomic ganglia that are usually located near or within the walls of the effector organs. The pathways then continue along a second set of neurons that stimulate the involuntary tissues. These neurons release the neurotransmitter acetylcholine, leading to the description of this system as ***cholinergic*** (activated by acetylcholine).

Functions of the Autonomic Nervous System

The autonomic nervous system regulates the action of the glands, the smooth muscles of hollow organs and vessels, and the heart muscle. These actions are all carried on automatically; whenever

Cell Receptors— Getting the Message

The nervous system functions by means of neurotransmitters, chemicals that carry information from cell to cell across the synapse. Just as important as the neurotransmitters are the "docking sites" for these chemicals, the receptors on the receiving (postsynaptic) cell membranes. The same neurotransmitter may have different effects depending on the nature of the receptor.

Acetylcholine (ACh) stimulates muscle contraction at the neuromuscular junction, the point where an axon contacts a skeletal muscle. When functioning in the parasympathetic nervous system, however, ACh stimulates some tissues, such as the digestive organs, but inhibits others, for example, the heart. These differences depend on the receptors on the postsynaptic cell membrane.

Some drugs work by blocking receptors for specific neurotransmitters. For example, the beta-blocking agents used to regulate the heart in cardiac disease inhibit the receptors for adrenaline, a neurotransmitter that increases the rate and strength of heart contractions.

any changes occur that call for a regulatory adjustment, the adjustment is made without conscious awareness. The sympathetic part of the autonomic nervous system tends to act as an accelerator for those organs needed to meet a stressful situation. It promotes what is called the ***fight-or-flight response.*** If you think of what happens to a person who is frightened or angry, you can easily remember the effects of impulses from the sympathetic nervous system:

1. Stimulation of the central portion of the adrenal gland. This produces hormones, including epinephrine, that prepare the body to meet emergency situations in many ways (see Chap. 12). The sympathetic nerves and hormones from the adrenal gland reinforce each other.
2. Dilation of the pupil and decrease in focusing ability (for near objects)
3. Increase in the rate and force of heart contractions
4. Increase in blood pressure due partly to the more effective heartbeat and partly to constriction of small arteries in the skin and the internal organs

5. Dilation of blood vessels to skeletal muscles, bringing more blood to these tissues
6. Dilation of the bronchial tubes to allow more oxygen to enter
7. Increase in metabolism.

The sympathetic system also acts as a brake on those systems not directly involved in the response to stress, such as the urinary and digestive systems. If you try to eat while you are angry, you may note that your saliva is thick and so small in amount that you can swallow only with difficulty. Under these circumstances, when food does reach the stomach, it seems to stay there longer than usual.

The parasympathetic part of the autonomic nervous system normally acts as a balance for the sympathetic system once a crisis has passed. The parasympathetic system brings about constriction of the pupils, slowing of the heart rate, and constriction of the bronchial tubes. It also stimulates the formation and release of urine and activity of the digestive tract. Saliva, for example, flows more easily and profusely and its quantity and fluidity increase.

Most organs of the body receive both sympathetic and parasympathetic stimulation, the effects of the two systems on a given organ generally being opposite. Table 9-1 shows some of the actions of these two systems.

Disorders of the Autonomic Nervous System

Injuries due to wounds by penetrating objects or to tumors, hemorrhage, or spinal column dislocations or fractures may cause damage to the sympathetic chain. In addition to these rather obvious kinds of disorders, there are a great number of conditions in which symptoms such as heart palpitations, increased blood pressure, and stomachaches suggest autonomic malfunctions but in which the cause is not well understood. These disorders are related to psychologic factors involved in the functioning of the viscera and may arise from the close interrelations of the brain, brain stem, and spinal cord and the autonomic nervous system.

The autonomic nervous system, together with the endocrine system, regulates our responses to stress. This interrelation and the effects of prolonged stress are discussed in more detail in Chapter 12.

TABLE 9-1
Effects of the Sympathetic and Parasympathetic Systems on Selected Organs

Effector	Sympathetic System	Parasympathetic System
Pupils of eye	Dilation	Constriction
Sweat glands	Stimulation	None
Digestive glands	Inhibition	Stimulation
Heart	Increased rate and strength of beat	Decreased rate and strength of beat
Bronchi of lungs	Dilation	Constriction
Muscles of digestive system	Decreased contraction (peristalsis)	Increased contraction
Kidneys	Decreased activity	None
Urinary bladder	Relaxation	Contraction and emptying
Liver	Increased release of glucose	None
Penis	Ejaculation	Erection
Adrenal medulla	Stimulation	None
Blood vessels to		
Skeletal muscles	Dilation	Constriction
Skin	Constriction	None
Respiratory system	Dilation	Constriction
Digestive organs	Constriction	Dilation

SUMMARY

I. Organization of the nervous system
  A. Structural (anatomic) divisions
    1. Central nervous system (CNS)—brain and spinal cord
    2. Peripheral nervous system (PNS)—spinal and cranial nerves
  B. Functional (physiologic) divisions
    1. Somatic (voluntary) nervous system—supplies skeletal muscles
    2. Visceral (involuntary) nervous system—supplies smooth muscle, cardiac muscle, glands
      a. Autonomic nervous system—involuntary motor nerves

II. Function of the nervous system
  A. Neuron—nerve cell
    1. Cell body
    2. Cell fibers
      a. Dendrite—carries impulses to cell body
      b. Axon—carries impulses away from cell body
    3. Myelin sheath
      a. Covers some axons
      b. Protects axon and allows more rapid conduction
      c. Made by Schwann cells in PNS; neuroglia in CNS
      d. Neurilemma—outermost layer of Schwann cell; aids neuron repair
      e. White matter—myelinated tissue; gray matter—unmyelinated tissue
    4. Types of neurons
      a. Sensory (afferent)—carry impulses toward CNS
      b. Motor (efferent)—carry impulses away from CNS
  B. Nerve impulse (action potential)—electric current that spreads along nerve fiber
  C. Synapse—junction between neurons where a nerve impulse is transmitted from one neuron to the next
    1. Neurotransmitter—carries impulse across synapse
    2. Postsynaptic cell—picks up neurotransmitter at receptors in membrane

  D. Nerve—bundle of nerve cell fibers outside the CNS
    1. Sensory (afferent) nerve—contains only fibers that carry impulses toward the CNS (from the receptor)
    2. Motor (efferent) nerve—contains only fibers that carry impulses away from the CNS (to the effector)
    3. Mixed nerve—contains both sensory and motor fibers
  E. Reflex arc—pathway through the nervous system
    1. Receptor—detects stimulus
    2. Sensory neuron—receptor to CNS
    3. Central neuron(s)—in CNS
    4. Motor neuron—CNS to effector
    5. Effector—muscle or gland that responds

III. Spinal cord
  A. Structure—H-shaped area of gray matter surrounded by white matter
  B. Function
    1. Conduction of sensory impulses to brain—ascending tracts
    2. Conduction of motor impulses from brain—descending tracts
    3. Reflex activities
      a. Reflex—simple, rapid, automatic response using few neurons
      b. Examples—stretch reflex, eye blink, withdrawal reflex
      c. Spinal reflex—coordinated in spinal cord
  C. Disorders—poliomyelitis, multiple sclerosis, amyotrophic lateral sclerosis

IV. Spinal nerves—31 pairs
  A. Roots
    1. Sensory (dorsal)
    2. Motor (ventral)
  B. Plexuses—networks formed by anterior branches
    1. Cervical plexus
    2. Brachial plexus
    3. Lumbosacral plexus
  C. Disorders—neuritis, sciatica, herpes zoster

V. Autonomic nervous system—motor (efferent) division of the visceral (involuntary) nervous system
  A. Divisions
    1. Sympathetic system
      a. Thoracolumbar
      b. Adrenergic—uses adrenaline
      c. Stimulates response to stress
    2. Parasympathetic system
      a. Craniosacral
      b. Cholinergic—uses acetylcholine
      c. Reverses stress response
  B. Effectors—Involuntary muscles (smooth muscle and cardiac muscle) and glands

QUESTIONS FOR STUDY AND REVIEW

1. What is the main function of the nervous system?
2. Name the two structural divisions of the nervous system.
3. Name the organs or structures controlled by the somatic nervous system; by the visceral nervous system.
4. Define the following terms: *neuron, myelin, synapse, ganglion,* and *Schwann cell.*
5. Describe a nerve impulse. How does conduction along a myelinated fiber differ from conduction along an unmyelinated fiber?
6. What are neurotransmitters? Give several examples of neurotransmitters.
7. Differentiate between the terms in each of the following pairs:
   a. *axon* and *dendrite*
   b. *gray matter* and *white matter*
   c. *receptor* and *effector*
   d. *afferent* and *efferent*
   e. *sensory* and *motor*
   f. *nerve* and *tract*

8. Without the label, how would you know that the neuron in Figure 9-2 is a motor neuron? Is this neuron a part of the somatic or visceral nervous system?
9. What is a mixed nerve? Give several examples.
10. Name the components of a reflex arc.
11. Locate and describe the spinal cord. Name three of its functions.
12. Describe a reflex. Give several examples of reflexes.
13. Name and describe two spinal cord disorders.
14. Differentiate between the dorsal and ventral roots of a spinal nerve.
15. Define a *plexus.* Name the three main plexuses of the spinal nerves.
16. Name two disorders of the spinal nerves.
17. What are the functions of the sympathetic part of the autonomic nervous system, and how do these compare with those of the parasympathetic nervous system?

CHAPTER 10

The Nervous System: The Brain and Cranial Nerves

Selected Key Terms

The following terms are defined in the Glossary:

brain stem

cerebellum

cerebral cortex

cerebrospinal fluid

cerebrum

diencephalon

hypothalamus

medulla oblongata

meninges

midbrain

pons

thalamus

ventricle

Behavioral Objectives

After careful study of this chapter, you should be able to:

- Give the location and functions of the four main divisions of the brain
- Name and locate the three subdivisions of the brain stem
- Name and describe the three meninges
- Cite the function of cerebrospinal fluid and describe where and how this fluid is formed
- Cite one function of the cerebral cortex in each lobe of the cerebrum
- Cite the names and functions of the 12 cranial nerves
- List some disorders that involve the brain or the cranial nerves

Memmler, RL, Cohen, BJ, Wood, DL. *THE HUMAN BODY IN HEALTH AND DISEASE*, 8/e,
© 1996 Lippincott-Raven Publishers

▶ The Brain and its Protective Structures

Main Parts of the Brain

The brain occupies the cranial cavity and is covered by membranes, fluid, and the bones of the skull. Although the various regions of the brain are in communication and may function together, the brain may be divided into distinct areas for ease of study (Fig. 10-1):

1. The ***cerebrum*** (SER-e-brum) is the largest part of the brain. It is divided into right and left ***cerebral*** (SER-e-bral) ***hemispheres*** by a deep groove called the ***longitudinal fissure.***
2. The ***diencephalon*** (di-en-SEF-ah-lon) is the area between the cerebral hemispheres and the brain stem. It includes the thalamus and the hypothalamus.
3. The ***brain stem*** connects the cerebrum and diencephalon with the spinal cord. The upper portion of the brain stem is the ***midbrain.*** Below that, and plainly visible from the underview of the brain are the ***pons*** (ponz) and the ***medulla oblongata*** (meh-DUL-lah ob-long-GAH-tah). The pons connects the midbrain with the medulla, while the medulla connects the brain with the spinal cord through a large opening in the base of the skull (foramen magnum).
4. The ***cerebellum*** (ser-eh-BEL-um) is located immediately below the back part of the cerebral hemispheres and is connected with the cerebrum, brain stem, and spinal cord by means of the pons. The word *cerebellum* means "little brain."

Each of these divisions is described in greater detail later in this chapter.

Coverings of the Brain and Spinal Cord

The ***meninges*** (men-IN-jez) are three layers of connective tissue that surround both the brain

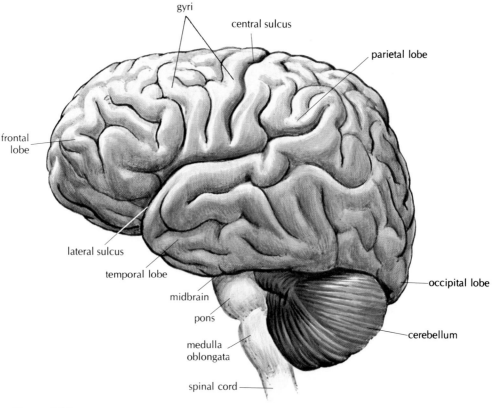

Figure 10-1

External surface of the brain showing the main parts and some of the lobes and sulci of the cerebrum.

and spinal cord to form a complete enclosure. The outermost of these membranes, the ***dura mater*** (DU-rah MA-ter), is the thickest and toughest of the meninges. Inside the skull, the dura mater splits in certain places to provide venous channels, called ***dural sinuses,*** for the drainage of blood coming from the brain tissue. The middle layer of the meninges is the ***arachnoid*** (ah-RAK-noyd). This membrane is loosely attached to the deepest of the meninges by weblike fibers allowing a space for the movement of cerebrospinal fluid between the two membranes. The innermost layer around the brain, the ***pia mater*** (PI-ah MA-ter), is attached to the nervous tissue of the brain and spinal cord and follows all the contours of these structures (Fig. 10-2). It is made of a delicate connective tissue in which there are many blood vessels. The blood supply to the brain is carried, to a large extent, by the pia mater.

INFLAMMATION OF THE MENINGES

Meningitis (men-in-JI-tis) is an inflammation of the brain and spinal cord coverings caused by pathogenic bacteria, notably a diplococcus called the ***meningococcus*** (me-ning-o-KO-us). If this organism attacks only the membranes around the spinal cord, the condition is called *spinal meningitis;* if it attacks the membranes around the brain as well, it is called *cerebrospinal meningitis*. Occasionally, other bacteria or viruses cause inflammation of the meninges. Sometimes the inflammatory process is so severe as to cause permanent brain damage or even death.

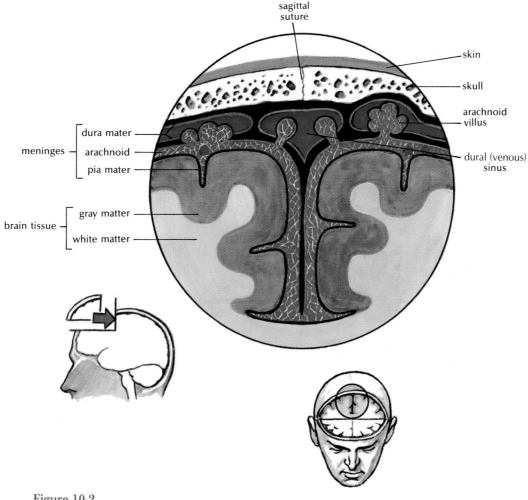

Figure 10-2

Frontal (coronal) section of the top of the head showing meninges and related parts.

TRAUMA TO THE MENINGES

Trauma to the head may cause bleeding between the skull and the brain. Arterial bleeding outside the dura causes an *epidural hematoma,* with rapidly progressing symptoms, including coma and dilated pupils. A tear in the wall of a dural sinus causes a *subdural hematoma,* with a slow leak and less dramatic symptoms. Frequent observation of level of consciousness, pupil response, and extremity reflexes is important in the patient with a head injury.

Cerebrospinal Fluid

Cerebrospinal (ser-e-bro-SPI-nal) *fluid* (CSF) is a clear liquid that circulates in and around the brain and spinal cord (Fig. 10-3). The function of CSF is to support neural tissue and to cushion shocks that would otherwise injure these delicate structures. This fluid also carries nutrients to the cells and transports waste products from the cells.

Cerebrospinal fluid flows freely through the brain and spinal cord and finally flows out into the subarachnoid space of the meninges. Much of the fluid is then returned to the blood through projections called *arachnoid villi* in the dural sinuses.

THE VENTRICLES

Cerebrospinal fluid is formed in four spaces within the brain called *ventricles* (VEN-trih-klz). A vascular network in each ventricle, the *choroid* (KOR-oyd) *plexus,* forms CSF by filtration of the blood and by cellular secretion.

Figure 10-3

Flow of CSF from choroid plexuses back to the blood in dural sinuses is shown by the black arrows; flow of blood is shown by the white arrows.

The four ventricles that produce CSF extend somewhat irregularly into the various parts of the brain (Fig. 10-4). The largest are the lateral ventricles in the two cerebral hemispheres. Their extensions into the lobes of the cerebrum are called *horns.* These paired ventricles communicate with a midline space, the third ventricle, by means of openings called *foramina* (fo-RAM-in-ah). The third ventricle is surrounded by the diencephalon. Continuing down from the third ventricle, a small canal, called the *cerebral aqueduct,* extends through the midbrain into the fourth ventricle. The latter is continuous with the central canal of the spinal cord. In the roof of the fourth ventricle are three openings that allow the escape of CSF to the area that surrounds the brain and spinal cord.

HYDROCEPHALUS

Any obstruction to the flow of CSF, for example in injury to the membranes around the three exit openings, may cause the condition called *hydrocephalus* (hi-dro-SEF-ah-lus). As the fluid accumulates, the mounting pressure can squeeze the brain against the skull and destroy brain tissue.

Hydrocephalus is more common in infants than in adults. Because the fontanelles of the skull have not closed in the infant, the cranium itself can become greatly enlarged; in contrast, in the adult, cranial enlargement cannot occur, so that even a slight increase in fluid results in symptoms of increased pressure within the skull and damage of brain tissue. A treatment for hydrocephalus involves the creation of a shunt to drain excess CSF from the brain.

Figure 10-4

Ventricles of the brain.

▶ Divisions of the Brain

The Cerebral Hemispheres

Each cerebral hemisphere is divided into four visible *lobes* named from the overlying cranial bones. These are the frontal, parietal, temporal, and occipital lobes (see Fig. 10-1). In addition, there is a small fifth lobe deep within each hemisphere that cannot be seen from the surface. Not much is known about this lobe, which is called the *insula.*

The outer nervous tissue of the cerebral hemispheres is gray matter that makes up the *cerebral cortex.* This thin layer of gray matter is the most highly evolved portion of the brain and is responsible for conscious thought, reasoning, and abstract mental functions. Specific functions are localized in the cortex of the different lobes, as described in greater detail below.

The cortex is arranged in folds forming elevated portions known as *gyri* (JI-ri), which are separated by shallow grooves called *sulci* (SUL-si) (see Fig. 10-1). Although there are many sulci, a few are especially important landmarks:

1. The *central sulcus,* which lies between the frontal and parietal lobes of each hemisphere at right angles to the longitudinal fissure
2. The *lateral sulcus,* which curves along the side of each hemisphere and separates the temporal lobe from the frontal and parietal lobes (see Fig. 10-1).

Internally, the cerebral hemispheres are made largely of white matter and a few islands of gray matter. The white matter consists of myelinated fibers that connect the cortical areas with each other and with other parts of the nervous system. An important band of white matter is the *corpus callosum* located at the bottom of the longitudinal fissure (see Fig. 10-3). This band acts as a bridge between the right and left hemispheres, permitting impulses to cross from one side of the brain to the other. The *internal capsule* is a crowded strip of white matter composed of many myelinated fibers (forming tracts). *Basal ganglia* are masses of gray matter located deep within each cerebral hemisphere. These groups of neurons help regulate body movement and facial expressions communicated from the cerebral cortex. The neurotransmitter *dopamine* (DO- pah-mene) is secreted by the neurons of the basal ganglia.

FUNCTIONS OF THE CEREBRAL CORTEX

It is within the cerebral cortex, the layer of gray matter that forms the surface of each cerebral hemisphere, that impulses are received and analyzed. These form the basis of knowledge. The brain "stores" information, much of which can be recalled on demand by means of the phenomenon called *memory.* It is in the cerebral cortex that thought processes such as association, judgment, and discrimination take place. It is also from the cerebral cortex that conscious deliberation and voluntary actions emanate.

Although the various areas of the brain act in coordination to produce behavior, particular functions are localized in the cortex of each lobe. Some of these are described below:

1. The *frontal lobe,* which is relatively larger in the human being than in any other organism, lies in front of the central sulcus. This lobe contains the motor area, which directs movement (Fig. 10-5). Because of the way in which motor fibers cross to opposite sides in the CNS, the left side of the brain governs the right side of the body, and the right side of the brain governs the left side of the body. The frontal lobe also contains two areas important in speech (the speech centers are discussed later).
2. The *parietal lobe* occupies the upper part of each hemisphere and lies just behind the central sulcus. This lobe contains the *sensory area,* in which impulses from the skin, such as touch, pain, and temperature, are interpreted. The estimation of distances, sizes, and shapes also take place here.
3. The *temporal lobe* lies below the lateral sulcus and folds under the hemisphere on each side. This lobe contains the *auditory area* for receiving and interpreting impulses from the ear. The *olfactory area,* concerned with the sense of smell, is located in the medial part of the temporal lobe; it is stimulated by impulses arising from receptors in the nose.
4. The *occipital lobe* lies behind the parietal lobe and extends over the cerebellum. This lobe contains the *visual area* for interpreting impulses arising from the retina of the eye.

COMMUNICATION AREAS

The ability to communicate by written and verbal means is an interesting example of the way in

Figure 10-5

Motor areas of the cerebral cortex (frontal lobe). The parts of the body are drawn in proportion to the area of control.

which areas of the cerebral cortex are interrelated (Fig. 10-6). The development and use of these areas are closely connected with the process of learning.

1. The *auditory areas* are located in the temporal lobe. In one of these areas sound impulses transmitted from the environment are detected, while

in the surrounding area (auditory speech center) the sounds are interpreted and understood. The beginnings of language are learned by auditory means, so the auditory area for understanding sounds is very near the auditory receiving area of the cortex. Babies often seem to understand what is being said long before they do any talking

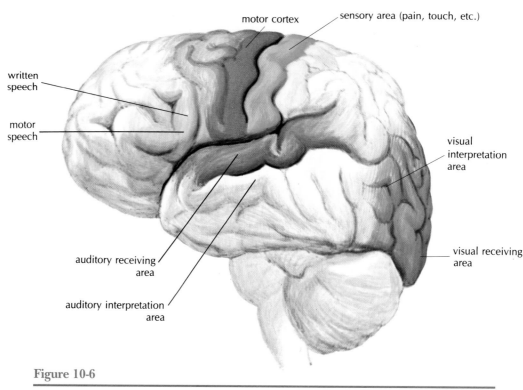

Figure 10-6

Functional areas of the cerebral cortex.

themselves. It is usually several years before children learn to read or write words.

2. The **motor areas** for communication by speech and writing are located in front of the lowest part of the motor cortex in the frontal lobe. Because the lower part of the motor cortex controls the muscles of the head and neck, it seems logical to think of the motor speech center as an extension forward in this area. Control of the muscles of speech (in the tongue, the soft palate, and the larynx) is carried out here. Similarly, the written speech center is located in front of the cortical area that controls the muscles of the arm and hand. The ability to write words is usually one of the last phases in the development of learning words and their meanings.

3. The **visual areas** of the cortex in the occipital lobe are also involved in communication. Here, visual images of language are received. These visual impulses are then interpreted as words in the visual area that lies in front of the receiving cortex. The ability to read with understanding is also developed in this area. You might *see* writing in the Japanese language, for example, but this would involve only the visual receiving area in the occipital lobe unless you could also *read* the words.

There is a functional relation among areas of the brain. Many neurons must work together to enable a person to receive, interpret, and respond to verbal and written messages as well as to touch (tactile stimulus) and other sensory stimuli.

MEMORY AND THE LEARNING PROCESS

Memory is the mental faculty for recalling ideas. In the initial stage of the memory process, sensory signals (*e.g.*, visual, auditory) are retained for a very short time, perhaps only fractions of a second. Nevertheless, they can be used for further processing. **Short-term memory** refers to the retention of bits of information for a few seconds or maybe a few minutes, after which the information is lost unless reinforced. **Long-term memory** refers to the storage of information that can be recalled at a later time. There is a tendency for a memory to become more fixed the more often a person repeats the remembered experience; thus, short-term memory signals can lead to long-term memories. Furthermore, the more often a memory is recalled, the more indelible it becomes; such a memory can be so deeply fixed in the brain that it can be recalled immediately.

Careful anatomic studies have shown that tiny extensions called *fibrils* form at the synapses in the cerebral cortex so impulses can travel more easily from one neuron to another. The number of these fibrils increases with age. Physiologic studies show that rehearsal (repetition) of the same information again and again accelerates and potentiates the degree of transfer of short-term memory into long-term memory. A person who is wide awake memorizes far better than a person who is in a state of mental fatigue. It has also been noted that the brain is able to organize information so that new ideas are stored in the same areas in which similar ones have been stored before.

The Diencephalon

The *diencephalon,* or interbrain, is located between the cerebral hemispheres and the brainstem. It can be seen by cutting into the central section of the brain. The *thalamus* (THAL-ah-mus) and the *hypothalamus* are included in the diencephalon (Fig. 10-7).

The two parts of the thalamus form the lateral walls of the third ventricle. Nearly all sensory impulses travel through the masses of gray matter that form the thalamus. The action of the thalamus is to sort out the impulses and direct them to particular areas of the cerebral cortex.

The hypothalamus is located in the midline area below the thalamus and forms the floor of the third

thalamus

hypothalamus

pituitary

Figure 10-7

Diagram showing the relationship among the thalamus, hypothalamus, and pituitary (hypophysis). (Chaffee EE, Lytle IM: Basic Physiology and Anatomy, 4th ed, p. 211. Philadelphia, JB Lippincott, 1980)

ventricle. It contains cells that help control body temperature, water balance, sleep, appetite, and some emotions, such as fear and pleasure. Both the sympathetic and parasympathetic divisions of the autonomic nervous system are under the control of the hypothalamus, as is the pituitary gland. The hypothalamus thus influences the heart beat, the contraction and relaxation of the walls of blood vessels, hormone secretion, and other vital body functions.

THE LIMBIC SYSTEM

Along the border between the cerebrum and the diencephalon is a region known as the *limbic system.* This system is involved in emotional states and behavior. It includes the *hippocampus* (shaped like a sea horse), located under the lateral ventricles, which functions in learning and the formation of long-term memory. It also includes regions that stimulate the *reticular formation,* a network that extends along the brain stem and governs wakefulness and sleep. The limbic system thus links the conscious functions

Sleep

Although everbody does it, there are still many mysteries about how, why, and when we sleep. Study of brain waves with the electroencephalograph has shown that sleep occurs in two phases. One of these, rapid eye movement (REM) sleep, is associated with dreaming.

As we descend from drowsiness into deep sleep, we progress through a non-rapid eye movement (NREM) phase. Although all the triggers for falling asleep are not known, a major factor is the amount of light received through the eyes. This signals the hypothalamus to regulate sleep–wake cycles. About 1 to 1½ hours after falling asleep, short periods of REM sleep with dreaming occur. These then alternate with NREM sleep at intervals of about 90 minutes. Extreme fatigue and sedatives decrease the amount of REM sleep in favor of deep NREM sleep. Also, infants have a much greater percentage of REM sleep; this declines with age.

Sleep rests the body, improves learning, and strengthens the immune system. There is also evidence that REM sleep helps the brain to sort memories.

of the cerebral cortex and the automatic functions of the brain stem.

The Brain Stem

The brain stem is composed of the midbrain, the pons, and the medulla oblongata. These structures connect the cerebrum and diencephalon with the spinal cord.

THE MIDBRAIN

The *midbrain,* located below the center of the cerebrum, forms the forward part of the brain stem. Four rounded masses of gray matter that are hidden by the cerebral hemispheres form the upper part of the midbrain; these four bodies act as relay centers for certain eye and ear reflexes. The white matter at the front of the midbrain conducts impulses between the higher centers of the cerebrum and the lower centers of the pons, medulla, cerebellum, and spinal cord. Cranial nerves III and IV originate from the midbrain.

THE PONS

The *pons* lies between the midbrain and the medulla, in front of the cerebellum. It is composed largely of myelinated nerve fibers, which serve to connect the two halves of the cerebellum with the brain stem as well as with the cerebrum above and the spinal cord below. The pons is an important connecting link between the cerebellum and the rest of the nervous system, and it contains nerve fibers that carry impulses to and from the centers located above and below it. Certain reflex (involuntary) actions, such as some of those regulating respiration, are integrated in the pons. Cranial nerves V through VIII originate from the pons.

THE MEDULLA OBLONGATA

The *medulla oblongata* of the brain is located between the pons and the spinal cord. It appears white externally because, like the pons, it contains many myelinated nerve fibers. Internally, it contains collections of cell bodies (gray matter) called *nuclei,* or *centers.* Among these are vital centers such as the following:

1. The *respiratory center* controls the muscles of respiration in response to chemical and other stimuli.
2. The *cardiac center* helps regulate the rate and force of the heart beat.

3. The *vasomotor* (vas-o-MO-tor) *center* regulates the contraction of smooth muscle in the blood vessel walls and thus controls blood flow and blood pressure.

The ascending sensory fibers that carry messages through the spinal cord up to the brain travel through the medulla, as do descending motor fibers. These groups of fibers form tracts (bundles) and are grouped together according to function. The motor fibers from the motor cortex of the cerebral hemispheres extend down through the medulla, and most of them cross from one side to the other (decussate) while going through this part of the brain. It is in the medulla that the shifting of nerve fibers occurs that causes the right cerebral hemisphere to control muscles in the left side of the body and the upper portion of the cortex to control muscles in the lower portions of the body. The medulla is an important reflex center; here, certain neurons end and impulses are relayed to other neurons. The last four pairs of cranial nerves (IX through XII) are connected with the medulla.

The Cerebellum

The *cerebellum* is made up of three parts: the middle portion (vermis) and two lateral hemispheres. Like the cerebral hemispheres, the cerebellum has an outer area of gray matter and an inner portion that is largely white matter. The functions of the cerebellum are:

1. To aid in the *coordination of voluntary muscles* so that they will function smoothly and in an orderly fashion. Disease of the cerebellum causes muscular jerkiness and tremors.
2. To aid in the *maintenance of balance* in standing, walking, and sitting, as well as during more strenuous activities. Messages from the internal ear and from sensory receptors in tendons and muscles aid the cerebellum.
3. To aid in the *maintenance of muscle tone* so that all muscle fibers are slightly tensed and ready to produce necessary changes in position as quickly as may be necessary.

▶ Brain Studies

Imaging the Brain

A major tool for clinical study of the brain is the *CT* (computed tomography) *scan,* which provides multiple x-ray pictures taken from different angles simultaneously. By means of a computer, the information is organized and displayed as photographs of the bone, soft tissue, and cavities of the brain (Fig. 10-8). Anatomic lesions such as tumors or scar tissue accumulations are readily seen.

MRI (magnetic resonance imaging) gives even clearer pictures of the brain without the use of dyes or x-rays. The method is based on computerized interpretation of the movements of atomic nuclei following their exposure to radio waves within a powerful magnetic field. Although the method is more expensive and takes longer than CT imaging, it gives more views of the brain and may reveal tumors, scar tissue, and hemorrhaging not shown by CT.

With *PET* (positron emission tomography) one can actually visualize the brain in action. With this method, a radioactively labeled substance, glucose for example, is followed as it moves through the brain. As tasks are performed, regions of the cortex that are involved become "hot."

The Electroencephalograph

The interactions of the billions of nerve cells in the brain give rise to measurable electric currents. These may be recorded by an instrument called the *electroencephalograph* (e-lek-tro-en-SEF-ah-lo-graf). The recorded tracings or brain waves produce an electroencephalogram (EEG).

▶ Disorders of the Brain

The scientific name for the brain is *encephalon;* therefore, inflammation of the brain is known as *encephalitis* (en-sef-ah-LI-tis). There are many causes of such disease, but the two chief pathogens are described below:

1. *Viruses* carried by insects cause some of the epidemic types of encephalitis sometimes found in the United States and other parts of the world.
2. Certain *protozoa,* called *trypanosomes* (tri-PAN-o-somes), cause the so-called *African sleeping sickness.* These protozoa, which are carried by a kind of fly (tsetse), are capable of invading the CSF of humans and infecting the surrounding tissue.

Stroke, or *cerebrovascular* (ser-e-bro-VAS-ku-lar) *accident* (CVA), is by far the most common kind of brain disorder. Rupture of a blood vessel, resulting in *cerebral hemorrhage,* or a blood clot in a cerebral vessel may cause destruction of brain tissue. Such disorders are most frequent in persons with artery wall disease, and hence are most common in persons over the age of 40. The effect of a stroke depends on the location of the artery and the extent of the involvement. A hemorrhage into the white matter of the internal capsule in the lower part of the cerebrum may cause extensive paralysis of the side opposite the affected area. Such a paralysis is called

Figure 10-8

CT scanner. (Photograph courtesy of Philips Medical Systems)

Magnetic Resonance Imaging

Since its introduction in the 1980s, magnetic resonance imaging (MRI) has greatly advanced the study of the nervous system and other tissues. It has been used to detect brain disorders, such as tumors and hemorrhages, to diagnose multiple sclerosis, and to locate herniated intervertebral disks. Unlike simple x-ray imaging, MRI gives pictures of soft tissues.

In producing an MRI scan, the area to be imaged is exposed to a carefully created magnetic field. This causes the common element hydrogen to align its magnetic field to the externally applied magnetic field. Radio waves in short pulses are introduced. These cause the hydrogen atoms to vibrate, or resonate. This extra energy is then released as radio waves. The time over which the waves are released gives information about the location of hydrogen and its freedom. Hydrogen can be free, as in water, loosely attached to a large protein, or tightly bound, as in bone.

hemiplegia (hem-e-PLE-je-ah), and the paralyzed person is described as *hemiplegic* (hem-e-PLE-jik).

Alzheimer's (ALZ-hi-merz) *disease* is a disorder of the brain resulting from an unexplained degeneration of the cerebral cortex and hippocampus. Although originally described as a premature senility in persons under 60 years, it is now more commonly diagnosed among the elderly. The disorder develops gradually and eventually causes severe intellectual impairment with mood changes and confusion. Memory loss, especially for recent events, is a common early symptom. At present there is no cure. Persons with Alzheimer's disease require help and understanding, as do those who must care for them.

Parkinson's disease appears most commonly in individuals over 60 years of age. It is characterized by tremors, slowness of movement, body rigidity, and inability to maintain posture. The disease originates in areas of the brain that make up the basal ganglia and is related to a deficiency of the neurotransmitter dopamine. The most successful treatment so far has been with a form of the neurotransmitter called *levodopa* (L-dopa). The outlook for the victim of Parkinson's disease is still poor.

Cerebral palsy (PAWL-ze) is a disorder caused by brain damage occurring before or during the birth process. It is characterized by diverse muscular disorders varying in degree: it may consist of only slight weakness of the lower extremity muscles or, at the other extreme, of paralysis of all four extremities as well as the speech muscles. With muscle and speech training and other therapeutic approaches, children with cerebral palsy can be helped.

Epilepsy is a chronic disorder involving an abnormality of the electric activity of the brain with or without apparent changes in the nervous tissues. One manifestation of epilepsy is seizure activity, which may be so mild as to be hardly noticeable or so severe as to result in loss of consciousness. In most cases the cause is not known. The study of brain waves obtained with the EEG usually shows abnormalities and is helpful in both diagnosis and treatment. Many persons with epilepsy can lead normal, active lives if they use appropriate medication as outlined by a physician.

Tumors of the brain may develop in persons of any age but are somewhat more common in young and middle-aged adults than in other groups. The majority of brain tumors originate from the neuroglia (connective tissue of the brain) and are called *gliomas.* The symptoms produced depend on the type of tumor, its location, its destructiveness, and the degree to which it compresses the brain tissue. Involvement of the frontal portion of the cerebrum often causes mental symptoms, such as changes in personality and in levels of consciousness. Early surgery, chemotherapy, and radiation therapy offer hope of cure in some cases.

Aphasia (ah-FA-ze-ah) is a term that refers to loss or defect in language communication. There may be a loss of the ability to speak or write (expressive aphasia) or a loss of understanding of written or spoken language (receptive aphasia). The type of aphasia present depends on what part of the brain is affected. The lesion that causes aphasia in the right-handed person is likely to be in the left cerebral hemisphere. Often much can be done for affected persons by patient retraining and much understanding. The brain is an organ that has a marvelous capacity for adapting itself to different conditions, and its resources are tremendous. Often some means of communication can be found even though speech areas are damaged.

) Cranial Nerves

Location of the Cranial Nerves

There are 12 pairs of cranial nerves (henceforth, when a cranial nerve is identified, a pair is meant). They are numbered according to their connection with the brain, beginning at the front and proceeding back (Fig. 10-9). The first 9 pairs and the 12th pair supply structures in the head.

General Functions of the Cranial Nerves

From a functional point of view, we may think of the kinds of messages the cranial nerves handle as belonging to one of four categories:

1. *Special sensory impulses,* such as those for smell, taste, vision, and hearing
2. *General sensory impulses,* such as those for pain, touch, temperature, deep muscle sense, pressure, and vibrations
3. *Somatic motor impulses* resulting in voluntary control of skeletal muscles
4. *Visceral motor impulses* producing involuntary control of glands and involuntary muscles (cardiac muscle and smooth muscle). These motor pathways are part of the autonomic nervous system, parasympathetic division.

Names and Functions of the Cranial Nerves

The 12 cranial nerves are always numbered according to the traditional Roman style. A few of the cranial nerves—I, II, and VIII—contain only sensory fibers; some—III, IV, VI, XI, and XII—contain all or mostly motor fibers. The remainder—V, VII, IX, and X—contain both sensory and motor fibers; they are known as *mixed nerves.* All 12 nerves are listed below:

I. The *olfactory nerve* carries smell impulses from receptors in the nasal mucosa to the brain.
II. The *optic nerve* carries visual impulses from the eye to the brain.
III. The *oculomotor nerve* is concerned with the contraction of most of the eye muscles.
IV. The *trochlear* (TROK-le-ar) *nerve* supplies one eyeball muscle.
V. The *trigeminal* (tri-JEM-in-al) *nerve* is the great sensory nerve of the face and head. It has three branches that transport general sense impulses (*e.g.,* pain, touch, temperature) from

the eye, the upper jaw, and the lower jaw. The third branch is joined by motor fibers to the muscles of mastication (chewing).

VI. The *abducens* (ab-DU-senz) *nerve* is another nerve sending controlling impulses to an eyeball muscle.
VII. The *facial nerve* is largely motor. The muscles of facial expression are all supplied by branches from the facial nerve. This nerve also includes special sensory fibers for taste (anterior two thirds of the tongue), and it contains secretory fibers to the smaller salivary glands (the submandibular and sublingual) and to the lacrimal (tear) gland.
VIII. The *vestibulocochlear* (ves-tib-u-lo-KOK-le-ar) *nerve* contains special sensory fibers for hearing as well as those for balance from the semicircular canals of the internal ear. This nerve is also called the *auditory* or *acoustic nerve.*
IX. The *glossopharyngeal* (glos-o-fah-RIN-ge-al) *nerve* contains general sensory fibers from the back of the tongue and the pharynx (throat). This nerve also contains sensory fibers for taste from the posterior third of the tongue, secretory fibers that supply the largest salivary gland (parotid), and motor nerve fibers to control the swallowing muscles in the pharynx.
X. The *vagus* (VA-gus) *nerve* is the longest cranial nerve. (Its name means "wanderer.") It supplies most of the organs in the thoracic and abdominal cavities. This nerve also contains motor fibers to the larynx (voice box) and pharynx, and to glands that produce digestive juices and other secretions.
XI. The *accessory nerve* (formerly called the *spinal accessory nerve*) is a motor nerve with two branches. One branch controls two muscles of the neck, the trapezius and sternocleidomastoid; the other supplies muscles of the larynx.
XII. The *hypoglossal nerve,* the last of the 12 cranial nerves, carries impulses controlling the muscles of the tongue.

Disorders Involving the Cranial Nerves

Destruction of optic nerve (II) fibers may result from increased pressure of the eye fluid on the nerves, as occurs in glaucoma, from the effect of poisons, and from some infections. Certain medications, when used in high doses for a long period of

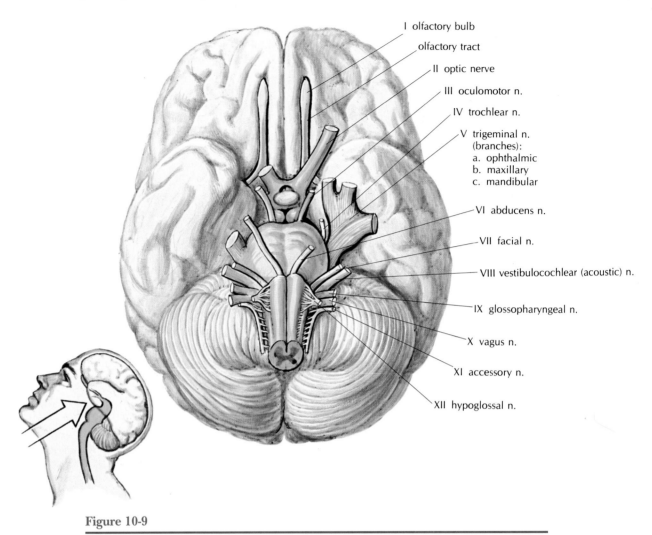

Figure 10-9

Base of the brain showing cranial nerves.

time, can damage the branch of the vestibulo-cochlear nerve responsible for hearing.

Injury to a nerve that contains motor fibers causes paralysis of the muscles supplied by these fibers. The oculomotor nerve (III) may be damaged by certain infections or various poisonous substances. Because this nerve supplies so many muscles connected with the eye, including the levator, which raises the eyelid, injury to it causes a paralysis that usually interferes with eye function. **Bell's palsy** is a facial paralysis caused by damage to the facial nerve

(VII), usually on one side of the face. This injury results in distortion of the face because of one-sided paralysis of the muscles of facial expression.

Neuralgia (nu-RAL-je-ah) in general means "nerve pain." A severe spasmodic pain affecting the fifth cranial nerve is known as *trigeminal neuralgia* or *tic douloureux* (tik du-lu-RU). At first the pain comes at relatively long intervals, but as time goes on, it is likely to appear at shorter intervals and to be of longer duration. Treatments include microsurgery and high-frequency current.

SUMMARY

I. **Protective structures of the nervous system**
   A. Meninges—coverings of brain and spinal cord
      1. Dura mater—tough outermost layer
      2. Arachnoid—weblike middle layer
      3. Pia mater—vascular innermost layer
   B. Cerebrospinal fluid (CSF)
      1. Cushions and protects
      2. Circulates around and within brain and spinal cord
      3. Ventricles—four spaces within brain where CSF is produced
      4. Choroid plexus—vascular network in ventricle that produces CSF

II. **Divisions of brain**
   A. Cerebrum—largest part of brain
      1. Right and left hemispheres
         a. Lobes—frontal, parietal, temporal, occipital, insula
         b. Cortex—outer layer of gray matter
   B. Diencephalon—area between cerebral hemispheres and brain stem
      1. Thalamus—directs sensory impulses to cortex
      2. Hypothalamus—maintains homeostasis, controls pituitary
      3. Limbic system
         a. Contains parts of cerebrum and diencephalon
         b. Controls emotion and behavior
   C. Brain stem
      1. Midbrain—involved in eye and ear reflexes

   2. Pons—connecting link for other divisions
   3. Medulla oblongata
      a. Connects with spinal cord
      b. Contains vital centers for respiration, heart rate, vasomotor activity
   D. Cerebellum—regulates coordination, balance, muscle tone

III. **Brain studies**
   A. Imaging methods
      1. CT—computed tomography
      2. MRI—magnetic resonance imaging
      3. PET—Positron emission tomography
   B. Electroencephalogram (EEG)

IV. **Disorders of brain and protective tissues**
   A. Meningitis, hydrocephalus
   B. Encephalitis
   C. Cerebrovascular accident (CVA)
   D. Alzheimer's disease, Parkinson's disease
   E. Cerebral palsy, epilepsy
   F. Tumors
   G. Aphasia

V. **Cranial nerves**—12 pairs
   A. Types
      1. Sensory (I, II, VIII)
      2. Motor (III, IV, VI, XI, XII)
      3. Mixed (V, VII, IX, X)
   B. Disorders—Bell's palsy, trigeminal neuralgia

QUESTIONS FOR STUDY AND REVIEW

1. Name and locate the main parts of the brain, and briefly describe the main functions of each.
2. Name the covering of the brain and the spinal cord. Name and describe its three layers. What is an infection of this covering called?
3. What is the purpose of the cerebrospinal fluid? Where and how is cerebrospinal fluid formed? Describe hydrocephalus.
4. Name the four surface lobes of the cerebral hemispheres and describe functions of the cortex in each.
5. Name and describe the speech centers.
6. What are the differences between short-term and long-term memory, and how is each attained?
7. Describe the thalamus; where is it located? What are its functions?
8. What activities does the hypothalamus regulate?
9. What is the limbic system and what are its functions?
10. Name and locate three divisions of the brain stem.
11. Name and describe six brain disorders.
12. Name four general functions of the cranial nerves.
13. Name and describe the functions of the 12 cranial nerves. What are some disorders of the cranial nerves?

CHAPTER 11

The Sensory System

Selected Key Terms

The following terms are defined in the Glossary:

accommodation

cataract

choroid

cochlea

conjunctiva

cornea

glaucoma

lacrimal

lens (crystalline lens)

organ of Corti

ossicle

proprioceptor

refraction

retina

sclera

semicircular canal

tympanic membrane

vestibule

vitreous body

Behavioral Objectives

After careful study of this chapter, you should be able to:

- Describe the function of the sensory system
- Differentiate between the special and general senses and give examples of each
- Describe the structure of the eye
- Define *refraction* and list the refractive media of the eye
- Differentiate between the rods and the cones of the eye
- List several disorders of the eye
- Describe the three divisions of the ear
- Describe the receptors for hearing and for equilibrium with respect to location and function
- List several disorders of the ear and hearing

Memmler, RL, Cohen, BJ, Wood, DL. *THE HUMAN BODY IN HEALTH AND DISEASE*, 8/e,
© 1996 Lippincott-Raven Publishers

▶ Senses and Sensory Mechanisms

The sensory system serves to protect the individual by detecting changes in the environment. An environmental change becomes a *stimulus* when it initiates a nerve impulse, which then travels to the CNS by way of a sensory (afferent) neuron. Many stimuli arrive from the external environment and are detected at or near the surface of the body. Others, such as the stimuli from the viscera, originate internally and help to maintain homeostasis. The part of the nervous system that detects a stimulus is the *receptor*. This may be:

1. The free dendrite of a sensory neuron, such as the receptors for pain
2. A modified ending, or *end-organ*, on the dendrite of an afferent neuron, such as those for touch and temperature
3. A special cell associated with an afferent neuron, such as the rods and cones of the retina of the eye and the receptors in the other special sense organs.

Regardless of the type of stimulus or the type of receptor involved, a stimulus becomes a sensation—something we experience—only when the nerve impulse it generates is interpreted by a specialized area of the cerebral cortex to which it travels.

One way of classifying the senses is according to the distribution of the receptors. A *special sense* is localized in a special sense organ; a *general sense* is widely distributed throughout the body.

1. *Special senses*
   a. *Vision* from receptors in the eye
   b. *Hearing* from receptors in the internal ear
   c. *Equilibrium* from receptors in the internal ear
   d. *Taste* from the tongue receptors
   e. *Smell* from receptors in the upper nasal cavities

2. *General senses*
   a. *Pressure*, *heat*, *cold*, *pain*, and *touch* from the skin and internal organs
   b. Sense of *position* from the muscles, tendons, and joints.

Mechanoreceptors

Sensory receptors can be classified according to the stimulus to which they respond. Thermoreceptors, such as those in the skin, detect change in temperature; photoreceptors in the retina of the eye respond to light; and chemoreceptors, such as those for taste and smell, detect chemicals.

Mechanoreceptors respond to movement, such as stretch, pressure, or vibration. The receptors that provide awareness of body position are examples, as are those that monitor stretching and tone of muscle fibers. Pressure and touch receptors in the skin are also mechanoreceptors.

In the ear, the receptors for hearing and equilibrium are mechanoreceptors. Both function by means of ciliated cells that respond to movement. In the vestibular apparatus, movements of the head trigger impulses that help to regulate balance. In the cochlea, sound waves cause movement of cilia and generate the impulses involved in hearing.

▶ The Eye

Protection of the Eyeball and Its Parts

In the embryo, the eye develops as an outpocketing of the brain. The eye is a delicate organ, and nature has carefully protected it by means of the following structures:

1. The skull bones form the walls of the eye orbit (cavity) and serve to protect more than half of the dorsal part of the eyeball.
2. The lids and eyelashes aid in protecting the eye anteriorly.
3. Tears lubricate the eye, wash away small foreign objects that enter the eye, and contain an enzyme that protects against infection.
4. A sac lined with an epithelial membrane separates the front of the eye from the eyeball proper and aids in the destruction of some of the pathogenic bacteria that may enter from the outside.

Coats of the Eyeball

The eyeball has three separate coats, or tunics (Fig. 11-1). The outermost tunic, called the *sclera*

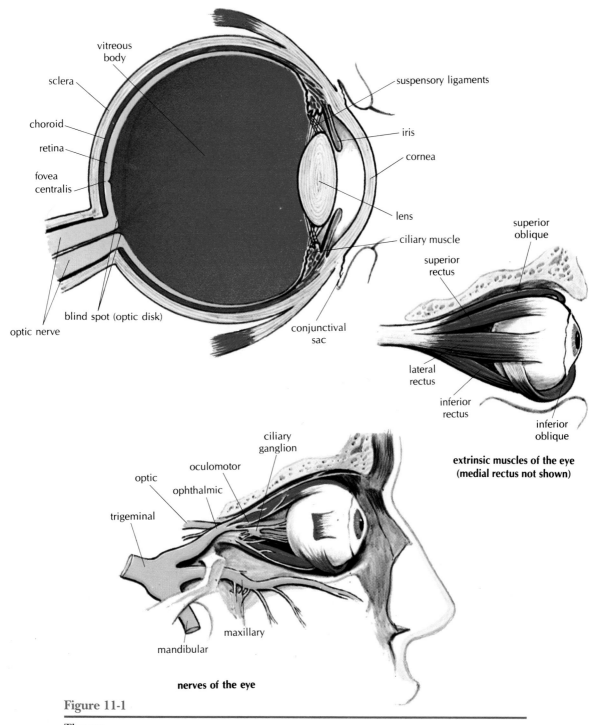

vitreous body

sclera

choroid

retina

fovea centralis

optic nerve

blind spot (optic disk)

suspensory ligaments

iris

cornea

lens

ciliary muscle

conjunctival sac

superior oblique

superior rectus

lateral rectus

inferior rectus

inferior oblique

extrinsic muscles of the eye (medial rectus not shown)

ciliary ganglion

oculomotor

optic

ophthalmic

trigeminal

maxillary

mandibular

nerves of the eye

Figure 11-1

The eye.

(SKLE-rah), is made of tough connective tissue. It is commonly referred to as the *white of the eye.* The second tunic of the eyeball is the **choroid** (KO-royd). Composed of a delicate network of connective tissue interlaced with many blood vessels, this coat contains much dark brown pigment. The choroid may be compared to the dull black lining of a camera in that it prevents incoming light rays from scattering and reflecting off the inner surface of the eye. The innermost tunic, called the **retina** (RET-ih-nah), includes ten layers of nerve cells, including the cells commonly called **rods** and **cones** (Fig. 11-2). These are the receptors for the sense of vision. The rods are highly sensitive to light and thus function in dim light but do not provide a very sharp image. The cones function in bright light and are sensitive to color. When you enter a darkened room, such as a movie theater, you cannot see for a short period of time. It is during this time that the rods are beginning to function, a change described as **dark adaptation.**

As far as is known, there are three types of cones, each sensitive to red, green, or blue light. Persons who completely lack cones are totally color blind; those who lack one type of cone are partially color blind. Color blindness is an inherited condition that occurs almost exclusively in males.

The rods and cones function by means of pigments that are sensitive to light. Manufacture of these pigments requires vitamin A, so a person who lacks vitamin A in the diet may have difficulty seeing in dim light, that is, may have **night blindness.**

Pathway of Light Rays

Light rays pass through a series of transparent, colorless eye parts. On the way they undergo a process of bending known as **refraction.** This refraction of the light rays makes it possible for light from a very large area to be focused on a very small surface, the retina, where the receptors are located. The following are, in order from outside in, the transparent refracting parts, or **media,** of the eye:

1. The **cornea** (KOR-ne-ah) is a forward continuation of the sclera, but it is transparent and colorless, whereas the rest of the sclera is opaque and white.
2. The **aqueous** (A-kwe-us) **humor,** a watery fluid that fills much of the eyeball in front of the lens, helps maintain the slight forward curve of the cornea.

3. The **lens,** technically called the *crystalline lens,* is a clear, circular structure made of a firm, elastic material.
4. The **vitreous** (VIT-re-us) **body** is a soft jellylike substance that fills the entire space behind the lens and keeps the eyeball in its spherical shape.

The cornea is referred to frequently as the *window* of the eye. It bulges forward slightly and is the most important refracting structure. Injuries caused by foreign objects or by infection may result in scar formation in the cornea, leaving an area of opacity through which light rays cannot pass. If such an injury involves the central area in front of the pupil (the hole in the center of the colored part of the eye), blindness may result. Eye banks store corneas obtained from donors, and corneal transplantation is a fairly common procedure.

The next light-bending medium is the aqueous humor, followed by the crystalline lens. The lens has two bulging surfaces, so it may be best described as biconvex. Because the lens is elastic, its thickness can be adjusted to focus light for near or distance vision.

The last of these transparent refracting parts of the eye is the vitreous body. Like the aqueous humor, it is important in maintaining the shape of the eyeball as well as in aiding in refraction.

Muscles of the Eye

The muscles inside the eyeball are **intrinsic** (in-TRIN-sik) **muscles;** those attached to bones of the eye orbit as well as to the sclera are **extrinsic** (eks-TRIN-sik) **muscles.**

The intrinsic muscles are found in two circular structures, as follows:

1. The **iris,** the colored or pigmented part of the eye, is composed of two types of muscles. The size of the central opening of the iris, called the **pupil,** is governed by the action of these two sets of muscles, one of which is arranged in a circular fashion, while the other extends in a radial manner like the spokes of a wheel.
2. The **ciliary body** is shaped somewhat like a flattened ring with a hole the size of the outer edge of the iris. This muscle alters the shape of the lens during the process of accommodation (described below).

The purpose of the iris is to regulate the amount of light entering the eye. If a strong light is flashed

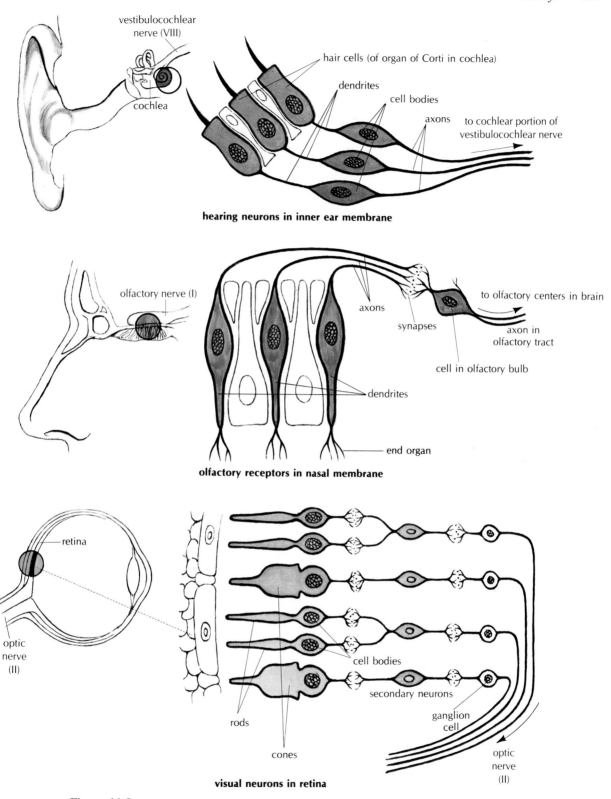

hearing neurons in inner ear membrane

olfactory receptors in nasal membrane

visual neurons in retina

Figure 11-2

Diagram of neurons for receiving impulses from the special sense organs.

in the eye, the circular muscle fibers of the iris, which form a sphincter, contract and thus reduce the size of the pupil. In contrast, if the light is very dim, the radial involuntary iris muscles, which are attached at the outer edge, contract; the opening is thereby pulled outward and thus enlarged. This pupillary enlargement is known as *dilation* (di-LA-shun).

The pupil changes size, too, according to whether one is looking at a near object or a distant one. Viewing a near object causes the pupil to become smaller; viewing a distant object causes it to enlarge.

The muscle of the ciliary body is similar in direction and method of action to the radial muscle of the iris. When the ciliary muscle contracts, it draws forward and removes the tension on the *suspensory ligaments,* which hold the lens in place (see Fig. 11-1). The elastic lens then recoils and becomes thicker in much the same way that a rubber band thickens when the pull on it is released. When the ciliary body relaxes, the lens becomes flattened. These actions change the refractive ability of the lens.

The process of *accommodation* involves coordinated eye changes to enable one to focus on near objects. The ciliary body contracts, thereby thickening the lens, and the circular muscle fibers of the iris contract to decrease the size of the pupillary opening. In young persons the lens is elastic, and therefore its thickness can be readily adjusted according to the need for near or distance vision. With aging, the lens loses its elasticity and therefore its ability to adjust to near vision by thickening, making it difficult to focus clearly on close objects. This condition is called *presbyopia* (pres-be-O-pe-ah), which literally means "old eye".

The six extrinsic muscles connected with each eye are ribbonlike and extend forward from the apex of the orbit behind the eyeball (see Fig. 11-1). One end of each muscle is attached to a bone of the skull, while the other end is attached to the sclera. These muscles pull on the eyeball in a coordinated fashion that causes the two eyes to move together to center on one visual field. Another muscle located within the orbit is attached to the upper eyelid. When this muscle contracts, it keeps the eye open; as the muscle becomes weaker with age, the eyelids may droop (a condition called *ptosis*) and interfere with vision.

Nerve Supply to the Eye

Two sensory nerves supply the eye (see Fig. 11-1):

1. The *optic nerve* (cranial nerve II) carries visual impulses from the retinal rods and cones to the brain.
2. The *ophthalmic* (of-THAL-mik) *branch of the trigeminal nerve* (cranial nerve V) carries impulses of pain, touch, and temperature from the eye and surrounding parts to the brain.

The optic nerve arises from the retina a little toward the medial or nasal side of the eye. Visual impulses are transmitted from the retina, ultimately to the cortex of the occipital lobe. There are no rods and cones in the retina near the area of the optic nerve fibers, so this circular white area is a blind spot, known as the *optic disk.* Near the optic disk there is a tiny depressed area in the retina, called the *fovea centralis* (FO-ve-ah sen-TRA-lis), which contains a high concentration of cones and is the point of most acute vision.

There are three nerves that carry motor impulses to the muscles of the eyeball. The largest is the oculomotor nerve (cranial nerve III), which supplies voluntary and involuntary motor impulses to all the muscles but two. The other two nerves, the trochlear (cranial nerve IV) and the abducens (cranial nerve VI), each supply one voluntary muscle.

The Conjunctiva and the Lacrimal Apparatus

The *conjunctiva* (kon-junk-TI-vah) is a membrane that lines the eyelid and covers the anterior part of the sclera. As the conjunctiva extends from the eyelid to the front of the eyeball, recesses, or sacs, are formed. Tears, produced by the *lacrimal* (LAK-rih-mal) *gland,* serve to keep the conjunctiva moist. As tears flow across the eye from the lacrimal gland, located in the upper lateral part of the orbit, the fluid carries away small particles that have entered the conjunctival sacs. The tears are then carried into ducts near the nasal corner of the eye where they drain into the nose by way of the *nasolacrimal* (na-zo-LAK-rih-mal) *duct* (Fig. 11-3). Any excess of tears causes a "runny nose"; a greater overproduction of them results in the spilling of tears onto the cheeks.

With age there is often a thinning and drying of the conjunctiva, resulting in inflammation and en-

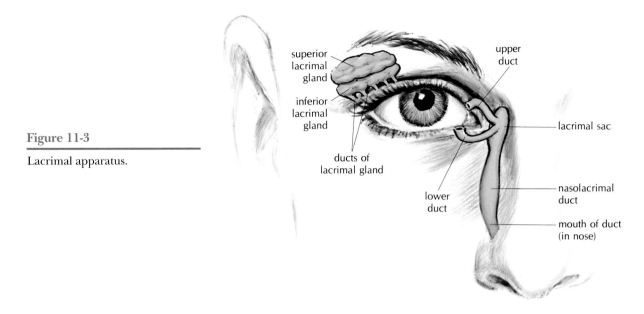

Figure 11-3

Lacrimal apparatus.

larged blood vessels. The lacrimal gland produces less secretion, but those tears that are produced may overflow to the cheek because of plugging of the nasolacrimal ducts, which would normally carry the tears away from the eye.

Eye Disorders
INFECTIONS

Inflammation of the membrane that lines the eyelid and covers the front of the eyeball is called *conjunctivitis* (kon-junk-tih-VI-tis). It may be acute or chronic, and may be caused by a variety of irritants and pathogens. "Pinkeye" is a highly contagious acute conjunctivitis that is usually caused by cocci or bacilli. Sometimes irritants such as wind and excessive glare cause an inflammation, which may in turn lead to susceptibility to bacterial infection.

Inclusion conjunctivitis is an acute eye infection caused by *Chlamydia trachomatis* (klah-MID-e-ah trah-KO-mah-tis). This same organism causes a sexually transmitted disease of the reproductive tract. In underdeveloped countries where this infection appears in chronic form it is known as *trachoma.* If not treated, scarring of the conjunctiva and cornea can cause blindness. The use of antibiotics as well as hygiene to prevent reinfection have reduced the incidence of blindness from this disorder in many parts of the world.

An acute infection of the eye of the newborn, caused by organisms acquired during passage through the birth canal, is called **ophthalmia neonatorum** (of-THAL-me-ah ne-o-na-TO-rum). The usual cause is gonococcus, chlamydia, or some other sexually transmitted organism. Preventive doses of an appropriate antiseptic or antibiotic solution are routinely administered into the conjunctiva of newborns just after birth.

The iris, choroid coat, ciliary body, and other parts of the eyeball may become infected by various organisms. Such disorders are likely to be very serious; fortunately, they are not very common. Syphilis spirochetes, tubercle bacilli, and a variety of cocci may cause these painful infections. They may follow a sinus infection, tonsillitis, conjunctivitis, or other disorder in which the infecting agent can spread from nearby structures. The care of these conditions usually should be in the hands of an *ophthalmologist* (of-thal-MOL-o-jist), a physician who specializes in the diagnosis and treatment of disorders of the eye.

INJURIES

The most common eye injury is a laceration or scratch of the cornea caused by a foreign body that becomes trapped under the eyelid. Severe traumatic injuries to the eye, such as penetration of deeper structures may not be subject to surgical repair. In such cases, an operation to remove the eyeball, a procedure called *enucleation* (e-nu-kle-A-shun) is performed.

It is very important to prevent infection in cases of injury to the eye. Even a tiny scratch can become so seriously infected that blindness may result. Injuries by pieces of glass and other sharp objects are a frequent cause of eye damage. The incidence of accidents involving the eye has been greatly reduced by the use of protective goggles.

VISUAL DEFECTS

Hyperopia (hi-per-O-pe-ah), or farsightedness, is usually due to an abnormally short eyeball (Fig. 11-4). In this situation, the focal point is behind the retina because light rays cannot bend sharply enough to focus on the retina. Objects must be moved away from the eye to be seen clearly. Farsightedness is nor-

mal in the infant but usually corrects itself by the time the child uses the eyes more for near vision. To a certain extent, the ciliary muscle can thicken the lens (accommodate) and thereby enable a person to focus objects on the too-near retina. Glasses with convex lenses that increase the refraction of light rays can correct for the visual disturbance.

Myopia (mi-O-pe-ah), or nearsightedness, is another defect of the eye also related to development. In this case the eyeball is too long or the bending of the light rays too sharp, so that the focal point is in front of the retina (see Fig. 11-4). Distant objects appear blurred and may appear clear only if brought very near the eye. A concave lens is used to correct for myopia. This alters the angle of refraction so

hyperopia

myopia

convergent strabismus   divergent strabismus

Figure 11-4

Visual defects.

that the focal point is moved backward. Nearsightedness in a young person becomes worse each year until the person reaches the 20s.

Another rather common visual defect, *astigmatism* (ah-STIG-mah-tism), is due to irregularity in the curvature of the cornea or the lens. As a result, light rays are incorrectly bent, causing blurred vision. Astigmatism is often found in combination with hyperopia or myopia, so a careful eye examination is needed to obtain properly fitted glasses.

Strabismus (strah-BIZ-mus) means that the muscles of the eyeballs do not coordinate, so that the two eyes do not work together (see Fig. 11-4). There are several types of strabismus. One common type is *convergent strabismus,* in which the eye deviates toward the nasal side, or medially. This disorder gives an appearance of cross-eyedness. A second type of strabismus is *divergent strabismus,* in which the affected eye deviates laterally. If these disorders are not corrected early, the transmission and interpretation of visual impulses from the affected eye to the brain will be decreased. The brain will not develop ways to "see" images from the eye. Care by an ophthalmologist as soon as the condition is detected may result in restoration of muscle balance. In some cases, glasses and patching of one eye correct the defect, while in others surgery may be required.

CATARACTS

A *cataract* is an opacity of the lens or its capsule. Sometimes the areas of opacity can be seen through the pupil, which may become greatly enlarged to increase the amount of light that reaches the retina. In other cases there is very gradual loss of vision, and frequent changes in glasses may aid in maintaining useful vision for some time. Removal of the lens may restore some vision, but the addition of eyeglasses or a contact lens is required to achieve satisfactory visual acuity (sharpness). Surgical techniques are now available to implant an artificial lens, and this procedure has been successful in restoring normal vision. A cataract, if untreated, will lead to complete vision loss. Although the cause of cataracts is not known, there is some evidence that excess exposure to ultraviolet rays is a factor. Eye doctors recommend the wearing of good sunglasses that filter out ultraviolet rays when the eyes are exposed to the sun.

GLAUCOMA

Glaucoma is a condition characterized by excess pressure of the aqueous humor. This fluid is produced constantly from the blood, and after circulation it is reabsorbed into the bloodstream. Interference with the normal reentry of this fluid to the bloodstream leads to an increase in pressure inside the eyeball. The most common type of glaucoma usually progresses rather slowly, with vague visual disturbances. Halos around lights, headaches, and decreased field of vision are symptoms that should be investigated by an ophthalmologist.

Measurement of the pressure in the eye is part of a routine eye examination for persons over 35 or those with a family history of glaucoma. The high pressure of the aqueous humor will eventually cause destruction of the optic nerve fibers. Continuous treatment with medications to reduce pressure frequently results in the preservation of vision.

DISORDERS INVOLVING THE RETINA

Diabetes as a cause of blindness is increasing in the United States. In *diabetic retinopathy* (ret-in-OP-ah-the), the retina is damaged by blood vessel hemorrhages and growth of new vessels. Other disorders of the eye directly related to diabetes include optic atrophy, in which the optic nerve fibers die; cataracts, which occur earlier and with greater frequency among diabetics than among nondiabetics; and retinal detachment.

In cases of *retinal detachment* the retina becomes separated from the underlying layer of the eye as a result of trauma or an accumulation of fluid or tissue between the layers. This disorder may develop slowly or may occur suddenly. If it is left untreated, complete detachment can occur, resulting in blindness. Surgical treatment includes a sort of "spot welding" with an electric current or a weak laser beam. A series of pinpoint scars (connective tissue) develop to reattach the retina.

Macular degeneration is another leading cause of blindness. This name refers to the macula, the yellow area of the retina that contains the fovea centralis. This disorder is due to age-related changes, including hardening and obstruction of the blood vessels in the retina. When blood vessels form to replace obstructed ones, they form on the surface of the retina, blocking light to the cones, especially in this area of acute vision.

▶ The Ear

The ear is a sense organ for both hearing and equilibrium (Fig. 11-5). It may be divided into three main sections:

1. The ***external ear*** includes the outer projection and a canal.
2. The ***middle ear*** is an air space containing three small bones.
3. The ***internal ear*** is the most complex because it contains the sensory receptors for hearing and equilibrium.

The External Ear

The projecting part of the ear is known as the ***pinna*** (PIN-nah), or the ***auricle*** (AW-rih-kl). Its purpose is to direct sound waves into the ear, but it is probably of little importance in the human. Then follows an opening, the ***external auditory canal,*** or ***meatus*** (me-A-tus), which extends medially for about

Figure 11-5

The ear, showing the external, middle, and internal subdivisions. (Adapted from Chaffee EE, Lytle IM: Basic Physiology and Anatomy, 4th ed, p. 227. Philadelphia, JB Lippincott, 1980)

2.5 cm or more, depending on which wall of the canal is measured. The skin lining this tube is very thin, and in the first part of the canal contains many *ceruminous* (seh-RU-mih-nus) *glands.* The *cerumen* (seh-RU-men), or wax, may become dried and impacted in the canal and must then be removed. The same kinds of disorders that involve the skin elsewhere—atopic dermatitis, boils, and other infections—may also affect the skin of the external auditory canal.

At the end of the auditory canal is the *tympanic* (tim-PAN-ik) *membrane,* or eardrum, which serves as a boundary between the external auditory canal and the middle ear cavity. It vibrates freely as sound waves enter the ear. The *eustachian* (u-STA-shun) *tube* (auditory tube) connects the middle ear cavity and the throat, or *pharynx* (FAR-inks). This tube opens to allow pressure to equalize on the two sides of the tympanic membrane.

The Middle Ear

The middle ear cavity is a small, flattened space that contains air and three small bones, or *ossicles* (OS-ih-klz). Air enters the cavity from the pharynx through the eustachian tube. The mucous membrane of the pharynx is continuous through the eustachian tube into the middle ear cavity, and infection may travel along the membrane, causing middle ear disease. At the back of the middle ear cavity is an opening into the mastoid air cells, which are spaces inside the mastoid process of the temporal bone.

The three ossicles are joined in such a way that they amplify the sound waves received by the tympanic membrane and then transmit the sounds to the fluid in the internal ear. The handle-like part of the first bone, or *malleus* (MAL-e-us), is attached to the tympanic membrane, while the head-like portion is connected with the second bone, called the *incus* (ING-kus). The innermost of the ossicles is shaped somewhat like a stirrup and is called the *stapes* (STA-peze). It is connected with the membrane of the *oval window,* which in turn vibrates and transmits these waves to the fluid of the internal ear.

The Internal Ear

The most complicated and important part of the ear is the internal portion, which consists of three separate spaces hollowed out inside the temporal bone (Fig. 11-6). This part of the ear, called the *bony* *labyrinth* (LAB-ih-rinth), consists of three divisions. One is the *vestibule,* next to the oval window. The second and third divisions are the *cochlea* (KOK-le-ah) and the *semicircular canals.* All three divisions contain a fluid called *perilymph* (PER-e-limf). The cochlea is a bony tube, shaped like a snail shell and located toward the front; the semicircular canals are three projecting bony tubes toward the back. In the fluid of the bony semicircular canals are the *membranous* (MEM-brah-nus) *canals,* which contain another fluid called *endolymph* (EN-do-limf). A *membranous cochlea,* situated in the perilymph of the bony cochlea, also is filled with endolymph.

The organ of hearing, called the *organ of Corti,* consists of receptors connected with nerve fibers in the *cochlear nerve* (a part of the eighth cranial nerve). This organ is located inside the membranous cochlea, or *cochlear duct.* The sound waves enter the external auditory canal and cause the tympanic membrane to vibrate. These vibrations are amplified by the ossicles and transmitted by them to the perilymph. They then are conducted by the perilymph through the membrane to the endolymph. The waves of the endolymph stimulate the tiny, hairlike cilia on the receptor cells, setting up nerve impulses that are then conducted to the brain.

Other sensory receptors in the internal ear include those related to equilibrium, some of which are located at the base of the semicircular canals (see Fig. 11-6). These canals are connected with two small sacs in the vestibule, one of which contains sensory cells for obtaining information on the position of the head. Receptors for the sense of equilibrium are also ciliated cells. As the head moves, a shift in the position of the cilia generates a nerve impulse. Nerve fibers from these sacs and from the canals form the *vestibular* (ves-TIB-u-lar) *nerve,* which joins the cochlear nerve to form the vestibulocochlear nerve, the eighth cranial nerve (formerly called the *auditory* or *acoustic nerve).*

Disorders of the Ear

Sudden great changes in the pressure on either side of the tympanic membrane may cause excessive stretching and inflammation of the membrane. There may even be perforation of the tympanic membrane to relieve the pressure.

Infection of the middle ear cavity, called *otitis media* (o-TI-tis ME-de-ah), is rather common. A variety of bacteria as well as viruses may cause otitis media. It is also a frequent complication of measles,

Figure 11-6

The internal ear, including a section showing the crista ampullaris, where sensory receptors for balance are located.

influenza, and other infections, especially those of the pharynx. Transmission of pathogens from the pharynx to the middle ear happens most often in children, partly because the eustachian tube is relatively short and horizontal in the child; in the adult the tube is longer and tends to slant down toward the pharynx. Antibiotic drugs have reduced complications and have caused a marked reduction in the amount of surgery done to drain middle ear in-

fections. However, in some cases pressure from pus or exudate in the middle ear can be relieved only by cutting the tympanic membrane, a procedure called a *myringotomy* (mir-in-GOT-o-me). Placement of a **tympanostomy** (tim-pan-OS-to-me) **tube** in the ear drum will allow pressure to equalize and prevent further damage to the ear drum.

Another disorder of the ear is hearing loss, which may be partial or complete. When the loss is complete, the condition is called *deafness.* The two main types of hearing loss are *conductive hearing loss* and *sensorineural hearing loss.* Conductive hearing loss is due to interference with the passage of sound waves from the outside to the inner ear. In this condition there may be obstruction of the external canal by wax or a foreign body. Blockage of the eustachian tube prevents the equalization of air pressure on both sides of the tympanic membrane, thereby decreasing the ability of the membrane to vibrate. Another cause of conductive hearing loss is damage to the tympanic membrane and ossicles resulting from chronic otitis media or from *otosclerosis* (o-to-skle-RO-sis), a hereditary disease that causes bone changes in the stapes that prevent its normal vibration. Surgical removal of the diseased stapes and its replacement with an artificial device allow conduction of sound from the ossicles to the oval window and the cochlea. Sensorineural hearing loss may involve the cochlea, the vestibulocochlear nerve, or the brain areas concerned with hearing. It may result from prolonged exposure to loud noises, to the use of certain drugs for long periods, or to exposure to various infections and toxins.

Presbycusis (pres-be-KU-sis) is a slowly progressive hearing loss that often accompanies aging. The condition involves gradual atrophy of the sensory receptors and the nerve fibers in the cochlear nerves. As a result, the affected person may experience a sense of isolation and depression, so that psychologic help may be desirable. Since the ability to hear high-pitched sounds is usually lost first, it is important to address elderly persons in clear, low-pitched tones.

▶ Other Special Sense Organs

Sense of Taste

The sense of taste involves receptors in the tongue and two different nerves that carry taste impulses to the brain (Fig. 11-7). The taste receptors, known as *taste buds,* are located along the edges of small depressed areas called *fissures.* Taste buds are stimulated only if the substance to be tasted is in solution. Receptors for four basic tastes are localized in different regions, forming a "taste map" of the tongue:

1. *Sweet* tastes are most acutely experienced at the tip of the tongue (hence the popularity of lollipops and ice cream cones).

2. *Sour* tastes are most effectively detected by the taste buds located at the sides of the tongue.
3. *Salty* tastes are most acute at the anterior sides of the tongue.
4. *Bitter* tastes are detected at the back part of the tongue.

The nerves of taste include the facial and the glossopharyngeal cranial nerves (VII and IX). The interpretation of taste impulses probably is accomplished by the lower front portion of the brain, although there may be no sharply separate taste, or *gustatory* (GUS-tah-to-re), *center.*

Sense of Smell

The receptors for smell are located in the *olfactory* (ol-FAK-to-re) *epithelium* of the upper part of the nasal cavity. Again, the chemicals detected must be in solution in the fluids that line the nose. Because these receptors are high in the nasal cavity, one must "sniff" to bring odors upward in the nose. The impulses from the receptors for smell are carried by the olfactory nerve (I), which leads directly to the olfactory center in the brain. The interpretation of smell is closely related to the sense of taste, but a greater variety of dissolved chemicals can be detected by smell than by taste. The smell of foods is just as important in stimulating appetite and the flow of digestive juices as is the sense of taste.

The olfactory receptors deteriorate with age, with the result that food becomes less appealing. It is important when presenting food to elderly persons that the food look inviting so as to stimulate their appetites.

▶ General Senses

Unlike the *special* sensory receptors, which are limited to a relatively small area, the *general* sensory receptors are scattered throughout the body. These include receptors for pressure, heat, cold, pain, touch, position, and balance (Fig. 11-8).

Sense of Pressure

It has been found that even when the skin is anesthetized, there is still consciousness of pressure. These sensory end-organs for deep sensibility are located in the subcutaneous and deeper tissues. They are sometimes referred to as *receptors for deep touch.*

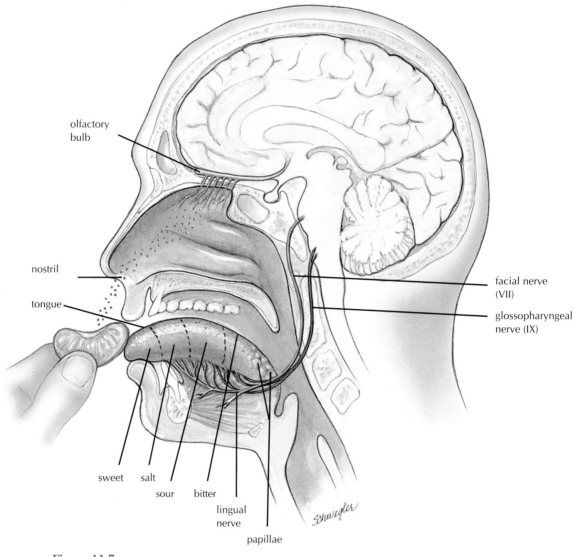

Figure 11-7

Organs of taste and smell.

Sense of Temperature

There are separate receptors for heat and cold and they vary in distribution. A warm object stimulates only the heat receptors, and a cool object affects only the cold receptors. As in the case of other sensory receptors, continued stimulation results in *adaptation;* that is, the receptors adjust themselves in such a way that the sensation becomes less acute if the original stimulus is continued. For example, if you immerse your hand in very warm water, it may be uncomfortable; however, if you leave your hand there, very soon the water will feel less hot (even if it has not cooled appreciably). Receptors for

warmth, cold, and light pressure adapt rapidly. In contrast, those for pain adapt slowly; this is nature's way of being certain that the warnings of the pain sense are heeded.

Sense of Touch

The touch receptors are small, rounded bodies called *tactile* (TAK-til) *corpuscles.* They are found mostly in the dermis and are especially numerous and close together in the tips of the fingers and the toes. The lips and the tip of the tongue also contain many of these receptors and are very sensitive to

Figure 11-8

Diagram showing the superficial receptors and the deeper cell bodies and synapses, suggesting the continuity of sensory pathways into the CNS.

touch. Other areas, such as the back of the hand and the back of the neck, have fewer receptors and are less sensitive to touch.

Sense of Pain

Pain is the most important protective sense. The receptors for pain are very widely distributed. They are found in the skin, muscles, and joints, and to a lesser extent in most internal organs (including the blood vessels and viscera). Pain receptors are not enclosed in capsules as are the sensory end-organs; rather, they are merely branchings of the nerve fiber, called *free nerve endings.*

There are two pathways for the transmission of pain to the CNS, one for acute, sharp pain and

Referred Pain

Referred pain is a term used to describe pain that seems to be in an outer part of the body—particularly the skin—but actually originates in an internal organ located near that particular area of skin. Liver and gallbladder disease often cause referred pain in the skin over the right shoulder. Spasm of the coronary arteries that supply the heart may cause pain in the left shoulder and the left arm.

The reason for this referral of pain is that some neurons have the twofold duty of conducting impulses from visceral pain receptors and from pain receptors in neighboring areas of the skin. The brain cannot differentiate between these two possible sources, but because most pain sensations originate in the skin, the brain automatically assigns the pain to this more likely place of origin.

the other for slow, chronic pain. Thus, a single strong stimulus produces the immediate sharp pain followed in a second or so by the slow, diffuse, burning pain that increases in severity with the passage of time.

Sometimes the cause of pain cannot be remedied quickly, and occasionally it cannot be remedied at all. In the latter case it is desirable to relieve the pain. Some pain relief methods that have been found to be effective include the following:

1. *Analgesic drugs.* An analgesic (an-al-JE-zik) is a drug that relieves pain. There are two main categories of such agents:
   a. *Non-narcotic analgesics* act locally to reduce inflammation and are effective for mild to moderate pain. Most of these drugs are commonly known as nonsteroidal anti-inflammatory drugs (NSAIDs). Examples are **acetaminophen** (ah-set-ah-MIN-o-fen) and **aspirin**.
   b. *Narcotics* act on the central nervous system to alter the perception and response to pain. Effective for severe pain, narcotics are administered by several methods including orally and by intramuscular injection. They are also

effectively administered into the space surrounding the spinal cord. An example of a narcotic drug is **morphine**.

2. *Anesthetics.* Although most commonly used to prevent pain during surgery, anesthetic injections are also used to relieve certain types of chronic pain.

3. *Endorphins* (en-DOR-fins) are released naturally from certain regions of the brain and are associated with the control of pain. Massage, acupressure, and electric stimulation are among the techniques that are thought to activate this system of natural pain relief.

4. Applications of **heat** or **cold** can be a simple but effective means of pain relief, either alone or in combination with medications. Care must be taken to avoid injury caused by excessive heat or cold.

5. **Relaxation** or **distraction techniques** include several methods that reduce perception of pain in the CNS. Relaxation techniques counteract the fight-or-flight response to pain and complement other pain-control methods.

Sense of Position

Receptors located in muscles, tendons, and joints relay impulses that aid in judging one's position and changes in the locations of body parts in relation to each other. They also inform the brain of the amount of muscle contraction and tendon tension. These rather widespread end-organs, known as *proprioceptors* (pro-pre-o-SEP-tors), are aided in this function by the equilibrium receptors of the internal ear. Information received by these receptors is needed for the coordination of muscles and is important in such activities as walking, running, and many more complicated skills, such as playing a musical instrument. These muscle receptors also play an important part in maintaining muscle tone and good posture, as well as allowing for the adjustment of the muscles for the particular kind of work to be done. The nerve fibers that carry impulses from these receptors enter the spinal cord and ascend to the brain in the posterior part of the cord. The cerebellum is a main coordinating center for these impulses.

SUMMARY

I. **Sensory system**—protects by detecting changes (stimuli) in the environment
   A. Special senses—vision, hearing, equilibrium, taste, smell
   B. General senses—pain, touch, temperature, pressure, position

II. **Eye**
   A. Protective structures—eyelids, bony orbit, eyelashes, tears (lacrimal apparatus), conjunctiva
   B. Coats (tunics)
      1. Sclera—white of the eye
         a. Cornea—anterior
      2. Choroid—pigmented; contains blood vessels
      3. Retina—nervous tissue
         a. Rods—detect white and black; function in dim light
         b. Cones—detect color; function in bright light
   C. Refraction
      1. Bending of light rays
      2. Refracting parts (media)—cornea, aqueous humor, crystalline lens, vitreous body
   D. Muscles
      1. Extrinsic muscles—six move each eyeball
      2. Intrinsic muscles
         a. Iris—colored ring around pupil; regulates the amount of light entering the eye
         b. Ciliary body—regulates the thickness of the lens for accommodation
   E. Nerve supply
      1. Optic nerve (II)—carries impulses from retina to brain
      2. Ophthalmic branch of trigeminal (V)—sensory nerve
      3. Oculomotor (III), trochlear (IV), abducens (VI) nerves—move eyeball
   F. Disorders
      1. Infections—conjunctivitis, ophthalmia neonatorum
      2. Injuries
      3. Visual defects—hyperopia (farsightedness), myopia (nearsightedness), astigmatism, strabismus
      4. Cataracts
      5. Glaucoma
      6. Retinal disorders

III. **Ear**
   A. Divisions
      1. External ear—pinna, auditory canal (meatus), tympanic membrane (eardrum)
      2. Middle ear—ossicles (malleus, incus, stapes)
         a. Eustachian tube—connects middle ear with pharynx to equalize pressure
      3. Internal ear—bony labyrinth
         a. Cochlea—contains receptors for hearing (organ of Corti)
         b. Vestibule—contains receptors for equilibrium
         c. Semicircular canals—contain receptors for equilibrium
   B. Nerve—vestibulocochlear (auditory) nerve (VIII)
   C. Disorders—otitis media, hearing loss, otosclerosis

IV. **Other special senses**
   A. Taste—gustatory sense
      1. Receptors—taste buds
      2. Tastes—sweet, sour, salty, bitter
      3. Nerves—facial (VII) and glossopharyngeal (IX)
   B. Smell—olfactory sense
      1. Receptors—in upper part of nasal cavity
      2. Nerve—olfactory nerve (I)

V. **General senses**
   A. Pressure
   B. Temperature
   C. Touch
   D. Pain—receptors are free nerve endings
      1. Relief—analgesic drugs, anesthetics, endorphins, heat, cold, relaxation and distraction techniques
   E. Position—receptors are proprioceptors in muscles, tendons, joints

QUESTIONS FOR STUDY AND REVIEW

1. What is the function of the sensory system?
2. Define *sensory receptor* and give several examples.
3. List the five special senses.
4. List six general senses.
5. Name the main parts of the eye.
6. Trace the path of a light ray from the outside of the eye to the retina.
7. List the extrinsic muscles of the eye. What is their function?
8. List the intrinsic muscles of the eye. What is their function?
9. What is near accommodation? What structures of the eye are necessary for near accommodation?
10. List five protective devices for the eye.
11. List five cranial nerves associated with the eye and give the function of each.
12. Describe three changes that occur in the eye with age.
13. Describe three eye infections and four visual defects. What are some possible causes of blindness?
14. Describe the structures that sound waves pass through in traveling through the ear to the receptors for hearing.
15. What cranial nerve carries impulses from the ear? Name the two branches.
16. Name and describe two ear disorders and list some causes of hearing loss.
17. Name the four kinds of taste. Where are the taste receptors?
18. Describe the olfactory apparatus.
19. What does *adaptation* mean with respect to the senses? How does sensory adaptation differ among the senses?
20. Name several types of pain-relieving drugs. Describe several methods for relieving pain that do not involve drugs.
21. Where are the receptors for the senses of position and balance (equilibrium) located?
22. Differentiate between the terms in each of the following pairs:
    a. *special sense* and *general sense*
    b. *hyperopia* and *myopia*
    c. *presbyopia* and *presbycusis*
    d. *rods* and *cones*
    e. *endolymph* and *perilymph*
    f. *gustatory* and *olfactory*

CHAPTER 12

The Endocrine System: Glands and Hormones

Selected Key Terms

The following terms are defined in the Glossary:

endocrine

hormone

hypothalamus

pituitary (hypophysis)

prostaglandin

receptor

steroid

target tissue

Behavioral Objectives

After careful study of this chapter, you should be able to:

- Compare the effects of the nervous system and the endocrine system in controlling the body
- Describe the functions of hormones
- Explain how hormones are regulated
- Identify the glands of the endocrine system on a diagram
- List the hormones produced by each endocrine gland and describe the effects of each on the body
- Describe the effects of hyposecretion and hypersecretion of the various hormones
- Describe how the hypothalamus controls the anterior and posterior pituitary
- List tissues other than the endocrine glands that produce hormones
- Explain how the endocrine system responds to stress

Memmler, RL., Cohen, BJ, Wood, DL. *THE HUMAN BODY IN HEALTH AND DISEASE*, 8/e,
© 1996 Lippincott-Raven Publishers

The nervous system and the endocrine system are the two main controlling and coordinating systems of the body. The nervous system controls such rapid activity as muscle movement and intestinal activity by means of electric and chemical stimuli. The effects of the endocrine system occur more slowly and over a longer period. The glands of the endocrine system produce chemical messengers, or *hormones*, which have widespread effects. The two systems, however, are closely related. The activity of the pituitary gland, which in turn regulates other glands, is controlled by the nervous system. The connections between the nervous system and the endocrine system enable endocrine function to adjust to the demands of a changing environment.

▶ Functions of Hormones

Hormones are the chemical messengers that have specific regulatory effects on certain other cells or organs in the body. Originally, the word *hormone* applied to the secretions of the endocrine glands only. The term now includes various substances produced in the body that have regulatory actions, either locally or at a distance from where they are produced. Many body tissues produce substances that have strong effects in regulating the local environment.

Hormones are released into tissue fluid and the bloodstream and are carried to all parts of the body. They regulate growth, metabolism, reproduction, and other body processes. Some affect many tissues, for example, thyroid hormone and insulin. Others affect only specific tissues. One pituitary hormone, thyroid-stimulating hormone (TSH), acts only on the thyroid gland; another (ACTH) stimulates only the outer portion of the adrenal gland. The specific tissue acted upon by each hormone is the *target tissue.* The cells that make up these tissues have *receptors* in the cell membrane or within the cytoplasm to which the hormone attaches. Once attached to the cell, the hormone affects metabolism by regulating the manufacture of proteins.

Some hormones are effective for relatively long periods (thyroid hormones act for about 2 weeks). Others act rapidly and are removed rapidly. The level of antidiuretic hormone (ADH) in the circulation can vary from low to high in 15 minutes.

Chemically, hormones fall into several categories. The principal ones are:

1. *Amino acid compounds.* Most of the hormones are proteins or related compounds made of amino acids. All hormones except those of the adrenal cortex and the sex glands fall into this category.
2. *Steroids* are hormones derived from lipids. They are produced by the adrenal cortex and the sex glands.

Substances known for other chemical effects act as hormones under certain circumstances—for example, neurotransmitters.

Regulation of Hormones

The amount of each hormone that is secreted is normally kept within a specific range. Negative feedback is the method used to regulate these levels. The endocrine gland tends to oversecrete its hormone, exerting more effect on the target tissue. When the target tissue becomes too active, there is a negative effect on the endocrine gland which decreases its secretory action. The release of hormones may fall into a rhythmic pattern. Hormones of the adrenal cortex follow a 24-hour cycle related to a person's sleeping pattern, with the level of secretion greatest just before arising and least at bedtime. Hormones of the female menstrual cycle follow a monthly pattern.

The remainder of this chapter deals with the main hormone-secreting organs. Refer to Figure 12-1 to locate each of the organs as you study it.

▶ The Endocrine Glands and Their Hormones

The Pituitary

The *pituitary* (pih-TU-ih-tar-e), or *hypophysis* (hi-POF-ih-sis), is a small gland about the size of a cherry. It is located in a saddle-like depression of the sphenoid bone just behind the point at which the optic nerves cross. It is surrounded by bone except where it connects with the brain by a stalk called the *infundibulum* (in-fun-DIB-u-lum). The gland is divided into two parts, the *anterior lobe* and the *posterior lobe.* The hormones released from each lobe are shown in Figure 12-2.

The pituitary is often called the *master gland* because it releases hormones that affect the working of other glands, such as the thyroid, gonads, and

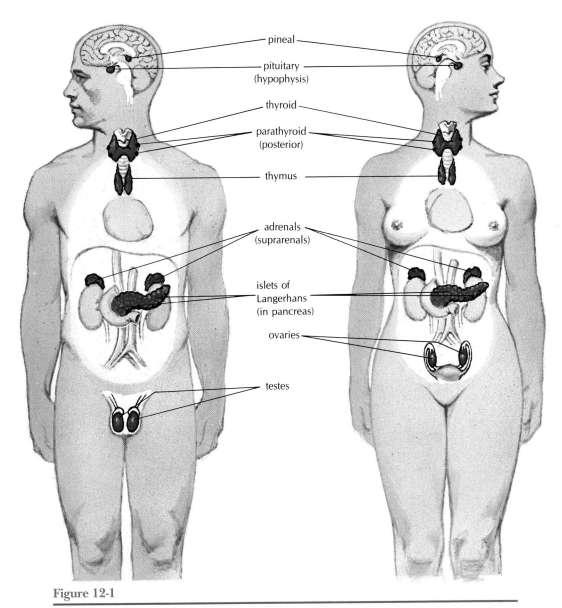

Figure 12-1

The main hormone-secreting organs.

adrenal glands. (Hormones that stimulate other glands may be recognized by the ending *-tropin,* as in *thyrotropin,* which means "acting on the thyroid gland.") However, the pituitary itself is controlled by the **hypothalamus** of the brain to which it is connected by the infundibulum or stalk (see Fig. 12-2).

The hormones produced in the anterior pituitary are not released from the gland until chemical messengers called **releasing hormones** arrive from the hypothalamus. These releasing hormones are sent to the anterior pituitary by way of a special type of circulatory pathway called a **portal system.** By this circulatory "detour," some of the blood that leaves the hypothalamus travels to capillaries in the anterior pituitary before returning to the heart. As the blood circulates through the capillaries, it delivers the hormones that stimulate the release of anterior pituitary secretions.

The two hormones of the posterior pituitary (ADH and oxytocin) are actually produced in the hypothalamus and stored in the posterior pituitary. Their release is controlled by nerve impulses that

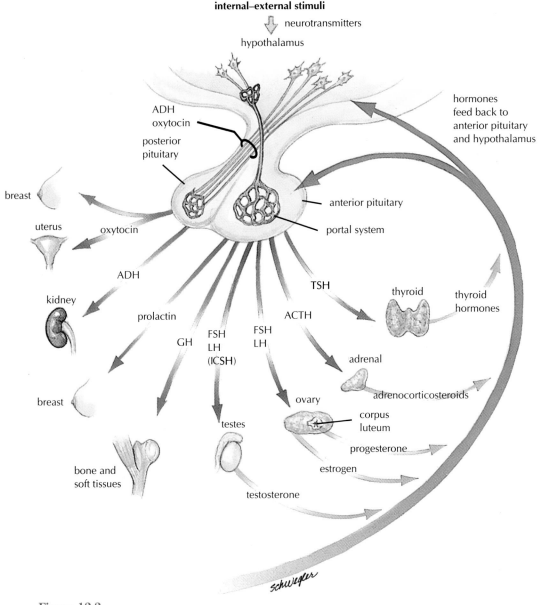

Figure 12-2

Pituitary gland and its relations with the brain and target tissues. Hypothalamic releasing hormones influence the anterior pituitary through a portal system. Tropic hormones from the anterior pituitary affect the working of various other glands. The hypothalamus communicates with the posterior pituitary through tracts.

travel over pathways (tracts) between the hypothalamus and the posterior pituitary.

HORMONES OF THE ANTERIOR LOBE

1. ***Growth hormone (GH),*** or ***somatotropin*** (so-mah-to-TRO-pin), acts directly on most body tissues, promoting protein deposits that are essential for growth. Produced throughout life, GH causes in-

crease in size and height to occur in youth, before the closure of the epiphyses of long bones. A young person with a deficiency of GH will remain small, though well-proportioned, unless treated with adequate hormone.

2. ***Thyroid-stimulating hormone (TSH),*** or ***thyrotropin*** (thi-ro-TRO-pin) stimulates the thyroid gland to produce thyroid hormones.

Growth Hormone

Growth hormone, also called *somatotropin*, is produced by the anterior pituitary gland. Its release is regulated by growth hormone–releasing hormone (GH–RH) and growth hormone–inhibiting hormone (GH–IH), also called *somatostatin*, from the hypothalamus.

Although growth hormone mainly affects the development of bones and muscles, it has a general effect on most other tissues as well, stimulating an increase in protein synthesis. It also stimulates the liver to release fatty acids for energy when the level of blood glucose drops. In this capacity, it is released during times of stress. The production of growth hormone is controlled by its concentration in the blood, by the level of other hormones, including those from the adrenal, thyroid, and sex glands, and by the level of nutrients in the blood.

Insufficient growth hormone results in pituitary dwarfism. This condition can be prevented by administering growth hormone to children who lack it. Growth hormone is now produced by genetic engineering. The gene for the hormone is introduced into harmless bacteria, which then produce large quantities in laboratory cultures.

3. *Adrenocorticotropic* (ad-re-no-kor-te-ko-TRO-pik) *hormone (ACTH)* stimulates the cortex of the adrenal glands.
4. *Prolactin* (pro-LAK-tin) *(PRL)* stimulates the production of milk in the female.

Gonadotropins (gon-ah-do-TRO-pinz), acting on the gonads, regulate the growth, development, and functioning of the reproductive systems in both males and females. The two gonadotropins are the following:

5. *Follicle-stimulating hormone (FSH)* stimulates the development of eggs in the ovaries and sperm cells in the testes.
6. *Luteinizing* (LU-te-in-i-zing) *hormone (LH)* causes ovulation in females and sex hormone secretion in both males and females; in males the hormone is called *interstitial cell–stimulating hormone (ICSH)*.

HORMONES OF THE POSTERIOR LOBE

1. *Antidiuretic* (an-ti-di-u-RET-ik) *hormone (ADH)* promotes the reabsorption of water from the kidney tubules and thus decreases the excretion of water. Large amounts of this hormone cause contraction of the smooth muscle of blood vessels and raise blood pressure. Inadequate amounts of ADH cause excessive loss of water and result in a disorder called *diabetes insipidus*. This type of diabetes should not be confused with diabetes mellitus, which is due to inadequate amounts of insulin.
2. *Oxytocin* (ok-se-TO-sin) causes contraction of the muscle of the uterus and milk ejection from the breasts. Under certain circumstances, commercial preparations of this hormone are administered during or after childbirth to cause the uterus to contract.

TUMORS OF THE PITUITARY

The effects of pituitary tumors depend on the types of cells the new or excess tissue contains. Some of these tumors contain an excessive number of the cells that produce growth hormone. A person who develops such a tumor in childhood will grow to an abnormally tall stature, a condition called *gigantism* (ji-GAN-tizm). Although persons with this condition are large, they are usually very weak.

If the GH-producing cells become overactive in the adult, a disorder known as *acromegaly* (ak-ro-MEG-ah-le) develops. In acromegaly the bones of the face, hands, and feet widen. The fingers resemble a spatula, and the face takes on a grotesque appearance: the nose widens, the lower jaw protrudes, and the forehead bones may bulge. Often these pituitary tumors involve the optic nerves and cause blindness.

Tumors or disease may destroy the secreting tissues of the gland so that signs of underactivity develop. Patients with this condition often become obese and sluggish and may exhibit signs of underactivity of other endocrine glands such as the ovaries, testes, or thyroid. Evidence of tumor formation in the pituitary gland may be obtained by x-ray examinations of the skull; the saddle-like space for the pituitary is distorted by the pressure of the tumor. CT and MRI scans are also used to diagnose pituitary abnormalities.

The Thyroid Gland

The largest of the endocrine glands is the *thyroid*, which is located in the neck (Fig. 12-3). It has two oval parts called the *lateral lobes*, one on either side

thyroid gland
(right lobe)

isthmus of
thyroid gland

epiglottis
hyoid bone
thyrohyoid
membrane
larynx
(thyroid cartilage)

trachea

Figure 12-3

Thyroid gland (anterior view) in relation to the larynx and trachea.

of the larynx (voice box). A narrow band called the *isthmus* (IS-mus) connects these two lobes. The entire gland is enclosed by a connective tissue capsule. The principal hormone produced by this gland is *thyroxine* (thi-ROK-sin). The hormone *calcitonin* (kal-sih-TO-nin), produced in the thyroid and active in calcium metabolism, will be discussed later in connection with the parathyroid gland.

The main function of thyroid hormones is to increase the rate of metabolism of most body cells. In particular, these hormones increase energy metabolism and protein metabolism. Thyroid hormones are necessary along with growth hormones for normal growth to occur. For the thyroid gland to produce hormones, there must be an adequate supply of iodine in the blood. Iodine deficiency is rare now due to widespread availability of this mineral in vegetables, seafood, dairy products, and processed foods.

A *goiter* is an enlargement of the thyroid gland, which may or may not be associated with overproduction of hormone. A *simple goiter* is the uniform overgrowth of the thyroid gland, with a smooth appearance. An *adenomatous* (ad-eh-NO-mah-tus), or *nodular, goiter* is an irregular-appearing goiter accompanied by tumor formation.

For various reasons, the thyroid gland may become either underactive or overactive. Underactivity of the thyroid, known as *hypothyroidism* (hi-po-THI-royd-izm), shows up as two characteristic states:

1. *Cretinism* (KRE-tin-izm), a condition resulting from hypothyroidism in infants and children. The usual cause is a failure of the thyroid gland to form during fetal development. The infant suffers lack of physical growth and lack of mental development. Early and continuous treatment with replacement hormone can alter the outlook of this disease. Most states require newborns to have a blood test to detect this disorder.

2. *Myxedema* (mik-seh-DE-mah), the result of atrophy (wasting) of the thyroid in the adult. The patient becomes sluggish both mentally and physically. The skin and the hair become dry, and a peculiar swelling of the tissues of the face develops. Since thyroid hormone can be administered by mouth, the patient with myxedema can be restored to health very easily, though treatment must be maintained throughout the remainder of his or her life.

Hyperthyroidism is the opposite of hypothyroidism, that is, overactivity of the thyroid gland with excessive secretion of hormone. A common form of hyperthyroidism is *exophthalmic* (ek-sof-THAL-mik) *goiter*, or *Graves' disease*, which is characterized by a goiter, bulging of the eyes, a strained appearance of the face, intense nervousness, weight loss, a rapid pulse, sweating, tremors, and an abnormally quick metabolism. Treatment of hyperthyroidism may take the form of:

1. Suppression of hormone production with medication

2. Destruction of thyroid tissue with radioactive iodine

3. Surgical removal of part of the thyroid gland.

An exaggerated form of hyperthyroidism with a sudden onset is called a *thyroid storm*. Untreated, it is usually fatal, but with appropriate care, most affected persons can be saved.

TESTS FOR THYROID FUNCTION

The most frequently used tests for thyroid function are blood tests in which the uptake of radioactive iodine added to the blood sample is measured. These very sensitive tests are used to detect abnormal thyroid function and to monitor response to drug therapy. A test for the level of thyroid-stimulating hormone (from the pituitary) is frequently done at the same time. Further testing involves giving the person radioactive iodine orally. The amount and distribution of the rays emitted from the radioactive iodine accumulated by the thyroid gland are then measured.

The Parathyroid Glands

Behind the thyroid gland, and embedded in its capsule, are four tiny epithelial bodies called the *parathyroid glands* (Fig. 12-4). The secretion of these glands, *parathyroid hormone (PTH),* is one of three substances that regulate calcium metabolism. The other two are *calcitonin* and *hydroxycholecalciferol* (hi-drok -se-ko-le-kal-SIF-eh-rol). PTH promotes the release of calcium from bone tissue, thus increasing the amount of calcium circulating in the bloodstream. Calcitonin, which is produced in the thyroid gland, acts to lower the amount of calcium circulating in the blood. Hydroxycholecalciferol is produced from vitamin D after first being modified by the liver and then the kidney. It increases absorption of calcium by the intestine to raise blood calcium levels.

If the parathyroid glands are removed, there follows a series of muscle contractions involving particularly the hand and the face muscles. These spasms are due to a low concentration of blood calcium, and the condition is called *tetany* (TET-ah-ne), which should not be confused with the infection called *tetanus* (lockjaw). In contrast, if there is excess production of PTH, as may happen in tumors of the parathyroids, calcium is removed from its normal storage place in the bones and released

into the bloodstream. Because this calcium is finally excreted by the kidneys, the formation of kidney stones is common in such cases. The loss of calcium from the bones leads to fragile bones that fracture easily.

The Adrenal Glands

The *adrenals,* or *suprarenals,* are two small glands located above the kidneys. Each adrenal gland has two parts that act as separate glands. The inner area is called the *medulla,* and the outer portion is called the *cortex.*

HORMONES FROM THE ADRENAL MEDULLA

The hormones of the adrenal medulla are released in response to stimulation by the sympathetic nervous system. The principal hormone produced by the medulla is one we already have learned about because it acts as a neurotransmitter. It is *epinephrine,* also called *adrenaline.* Another hormone, *norepinephrine,* is closely related chemically and is similar in its actions. These are referred to as the *fight-or-flight hormones* because of their effects during emergency situations. Some of these effects are as follows:

1. Stimulation of the involuntary muscle in the walls of the arterioles, causing these muscles to contract and blood pressure to rise accordingly
2. Conversion of the glycogen stored in the liver into glucose. The glucose is poured into the blood and brought to the voluntary muscles, permitting them to do an extraordinary amount of work.
3. Increase in the heart rate
4. Increase in the metabolic rate of body cells
5. Dilation of the bronchioles, through relaxation of the smooth muscle of their walls.

Figure 12-4

Posterior view of the thyroid gland showing the parathyroid glands embedded in its surface.

HORMONES FROM THE ADRENAL CORTEX

There are three main groups of hormones secreted by the adrenal cortex:

1. *Glucocorticoids* (glu-ko-KOR-tih-koyds) maintain the carbohydrate reserve of the body by controlling the conversion of amino acids into sugar instead of protein. These hormones are produced in larger than normal amounts in times of stress and so aid the body in responding to unfavorable conditions. They have the ability to suppress the inflammatory response and are often administered as medication for this purpose. A major hormone of this group is *cortisol,* which is also called *hydrocortisone.*
2. *Mineralocorticoids* (min-er-al-o-KOR-tih-koyds) are important in the regulation of electrolyte balance. They control the reabsorption of sodium and the secretion of potassium by the kidney tubules. The major hormone of this group is *aldosterone* (al-DOS-ter-one).
3. *Sex hormones* are normally secreted, but in small amounts; their effects in the body are slight.

Hypofunction of the adrenal cortex gives rise to a condition known as *Addison's disease,* a disease characterized chiefly by muscle atrophy (loss of tissue), weakness, skin pigmentation, and disturbances in salt and water balance. Hyperfunction of the adrenal cortex results in a condition known as *Cushing's syndrome,* the symptoms of which include obesity with a round face, thin skin that bruises easily, muscle weakness, bone loss, and elevated blood sugar. Use of steroid drugs also may produce these symptoms.

Adrenal gland tumors give rise to a wide range of symptoms resulting from an excess or a deficiency of the hormones secreted.

The Pancreas, Insulin, and Diabetes

Scattered throughout the *pancreas* are small groups of specialized cells called *islets* (I-lets), also known as the *islets of Langerhans* (LAHNG-er-hanz). A microscopic view of the pancreas showing the islet cells is presented in Figure 12-5. These cells make up the endocrine portion of the pancreas and function independently from the exocrine part of the pancreas, which produces digestive juices and releases them through ducts. The most important hormone secreted by these islets is *insulin.*

Figure 12-5

Microscopic view of pancreatic cells. Light staining islet cells are seen among the cell clusters (acini) that produce digestive juices.   (Courtesy of Dana Morse Bittus and B. J. Cohen)

Insulin is active in the transport of glucose across the cell membrane. Once inside the cell, glucose can be used for energy. Insulin also increases the transport of amino acids into the cells and improves their use in the manufacture of proteins. When blood sugar and insulin decrease, most body cells switch to using fat for basic energy needs. Brain cells are an exception, because they use only glucose for energy and do not require insulin for glucose transport and metabolism. Insulin increases the rate at which the liver changes excess sugar into fatty acids, which can then be stored in fat cells. Through its actions, insulin has the effect of lowering the level of sugar in the blood.

A second hormone produced by the islet cells is *glucagon* (GLU-kah-gon), which works with insulin to regulate blood sugar levels. Glucagon causes the liver to release stored glucose into the bloodstream. Glucagon also increases the rate at which glucose is made from proteins in the liver. In these two ways, glucagon increases blood sugar.

If for some reason the pancreatic islets fail to produce enough insulin, sugar is not oxidized ("burned") in the tissues for transformation into energy. Instead, the sugar remains in the blood and then must be excreted in the urine. This condition, called *diabetes mellitus* (di-ah-BE-teze mel-LI-tus), is the most common endocrine disorder.

Diabetes is divided into two types: insulin-dependent and non-insulin–dependent.

The less common but more severe insulin-dependent diabetes mellitus (IDDM) is also known as *type I diabetes* (formerly known as *juvenile diabetes).*

This disease usually appears by age 30 and is characterized by an autoimmune destruction of the insulin-producing cells in the islets. People with IDDM need close monitoring of blood sugar levels and injections of insulin.

Non-insulin–dependent diabetes mellitus (NIDDM), or *type II diabetes,* characteristically occurs in overweight adults. These people retain the ability to secrete varying amounts of insulin, depending on the severity of the disease. However, the ability of their body cells to respond to the hormone is diminished. This disease can be controlled with diet, oral medication to improve insulin production and increase its effectiveness, and, for the obese patient, weight reduction. Treatment with injectable insulin may be necessary during illness or other stress.

Diabetic persons learn to give themselves their injections, adjust their diets, and exercise to maintain the proper balance between intake of insulin and sugar needs. A diabetic person should carry a special identification card to let people know that he or she is a diabetic taking insulin and that a dazed condition might indicate a need for some sugar.

Special pumps that provide a round-the-clock supply of insulin are now available. The insulin is placed in a device that then injects it into the subcutaneous tissues of the abdomen. The consistent blood sugar level thus achieved results in a more nearly normal metabolism.

Insulin has been and still is obtained from animal pancreases, but human insulin produced by bacteria through genetic engineering is also available.

Long-term complications of diabetes are many. Resistance to infection is lessened. The arteries, including those of the retina, kidney, and heart, may be seriously damaged. The peripheral nerves are affected also, with accompanying pain and loss of sensation. Recent studies document the advantages of tight control of the blood sugar of diabetics in reducing the severity of these long-term complications. Tight control is achieved by testing the blood sugar 4 times a day and injecting insulin twice a day.

The Sex Glands

The sex glands, the ovaries of the female and the testes of the male, not only produce the sex cells but also are important endocrine organs. The hormones produced by these organs are needed in the development of the sexual characteristics, which usually appear in the early teens, and for the maintenance of the reproductive organs once full development has been attained. Those features that typify a male or female other than the structures directly concerned with reproduction are termed *secondary sex characteristics.* They include a deep voice and facial and body hair in males; and wider hips and a greater ratio of fat to muscle in females.

The main male sex hormone or androgen (AN-dro-jen) produced by the male sex glands is *testosterone* (tes-TOS-ter-one).

In the female, the hormones that most nearly parallel testosterone in their actions are the *estrogens* (ES-tro-jens). Estrogens contribute to the development of the female secondary sex characteristics and stimulate the development of the mammary glands, the onset of menstruation, and the development and functioning of the reproductive organs.

The other hormone produced by the female sex glands, called *progesterone* (pro-JES-ter-one), assists in the normal development of pregnancy. All the sex hormones are discussed in more detail in Chapter 23. Table 12-1 gives a summary of the major endocrine glands and the hormones they secrete.

▶ Other Hormone-Producing Tissues

The *thymus gland* is a mass of lymphoid tissue that lies in the upper part of the chest above the heart. This gland is important in the development of immunity. Its hormone, *thymosin* (THI-mo-sin), assists in the maturation of certain white blood cells called *T-lymphocytes* (T cells) after they have left the thymus gland and taken up residence in lymph nodes throughout the body.

The *pineal* (PIN-e-al) *body* is a small, flattened, cone-shaped structure located posterior to the midbrain and connected to the roof of the third ventricle. The pineal produces the hormone, *melatonin* (mel-ah-TO-nin), during the dark period of each day. Little hormone is produced during daylight hours. This pattern of hormone secretion influences the regulation of sleep/wake cycles. Melatonin also seems to delay the onset of puberty.

Cells in the lining of the stomach and small intestine secrete hormones that stimulate the production of digestive juices and help regulate the process of digestion.

In response to a decreased supply of oxygen, the kidneys produce a hormone called *erythropoietin*

TABLE 12-1
The Major Endocrine Glands and Their Hormones

Gland	Hormone	Principal Functions
Anterior pituitary	GH (growth hormone)	Promotes growth of all body tissues
	TSH (thyroid-stimulating hormone)	Stimulates thyroid gland to produce thyroid hormones
	ACTH (adrenocorticotropic hormone)	Stimulates adrenal cortex to produce cortical hormones; aids in protecting body in stress situations (injury, pain)
	PRL (prolactin)	Stimulates secretion of milk by mammary glands
	FSH (follicle-stimulating hormone)	Stimulates growth and hormone activity of ovarian follicles; stimulates growth of testes; promotes development of sperm cells
	LH (luteinizing hormone); ICSH (interstitial cell–stimulating hormone) in males	Causes development of corpus luteum at site of ruptured ovarian follicle in female; stimulates secretion of testosterone in male
Posterior pituitary	ADH (antidiuretic hormone; vasopressin)	Promotes reabsorption of water in kidney tubules; stimulates smooth muscle tissue of blood vessels to constrict
	Oxytocin	Causes contraction of muscle of uterus; causes ejection of milk from mammary glands
Thyroid	Thyroid hormone (thyroxine and triiodothyronine)	Increases metabolic rate, influencing both physical and mental activities; required for normal growth
	Calcitonin	Decreases calcium level in blood
Parathyroids	Parathyroid hormone	Regulates exchange of calcium between blood and bones; increases calcium level in blood
Adrenal medulla	Epinephrine and norephinephrine	Increases blood pressure and heart rate; activates cells influenced by sympathetic nervous system plus many not affected by sympathetic nerves
Adrenal cortex	Cortisol (95% of glucocorticoids)	Aids in metabolism of carbohydrates, proteins, and fats; active during stress
	Aldosterone (95% of mineralocorticoids)	Aids in regulating electrolytes and water balance
	Sex hormones	May influence secondary sexual characteristics
Pancreatic islets	Insulin	Aids transport of glucose into cells; required for cellular metabolism of foods, especially glucose; decreases blood sugar levels
	Glucagon	Stimulates liver to release glucose, thereby increasing blood sugar levels
Testes	Testosterone	Stimulates growth and development of sexual organs (testes, penis, others) plus development of secondary sexual characteristics such as hair growth on body and face and deepening of voice; stimulates maturation of sperm cells
Ovaries	Estrogens (*e.g.*, estradiol)	Stimulates growth of primary sexual organs (uterus, tubes, etc.) and development of secondary sexual organs such as breasts, plus changes in pelvis to ovoid, broader shape
	Progesterone	Stimulates development of secretory parts of mammary glands; prepares uterine lining for implantation of fertilized ovum; aids in maintaining pregnancy

(e-rith-ro-POY-eh-tin), which stimulates red blood cell production in the bone marrow.

The atria of the heart produce a substance called *atrial natriuretic* (na-tre-u-RET-ik) *peptide (ANP)* in response to increased filling of the atria. ANP increases loss of salt by the kidneys and lowers blood pressure.

The *placenta* (plah-SEN-tah) produces several hormones during pregnancy. These cause changes in the uterine lining and later in pregnancy help prepare the breasts for lactation. Pregnancy tests are based on the presence of placental hormones.

Prostaglandins

Prostaglandins (pros-tah-GLAN-dins) are a group of local hormones made by most body tissues. They are produced, act, and are rapidly inactivated in or close to the sites of origin. A bewildering array of functions has been ascribed to these substances. Some prostaglandins cause constriction of blood vessels, of bronchial tubes, and of the intestine, whereas others cause dilation of these same structures. Prostaglandins are active in promoting inflammation; certain anti-inflammatory drugs, such as aspirin, act by blocking the production of prostaglandins. Some of the prostaglandins have been used to induce labor or abortion and have been recommended as possible contraceptive agents. Overproduction of prostaglandins by the uterine lining (endometrium) can cause painful cramps of the muscle of the uterus. Treatment with drugs that are prostaglandin inhibitors has been successful in some cases. Much has been written about these substances, and extensive research on them continues.

▶ Hormones and Treatment

Many hormones may be extracted from animal tissues for use as medication. However, some hormone and hormone-like substances are available in synthetic form, meaning that they are manufactured in commercial laboratories. Still others are produced by the genetic engineering technique of recombinant DNA. In this method, a gene for the cellular manufacture of a given product is introduced in the laboratory into the common bacterium, *E. coli*. The organisms are then grown in quantity and the desired substance harvested and purified.

A few examples of the use of natural and synthetic hormones in treatment are noted here:

1. *Growth hormone* is used for the treatment of children with a deficiency of this hormone. Adequate supplies are available from recombinant DNA techniques.
2. *Insulin* is used in the treatment of diabetes mellitus. Types of insulin available include those obtained from animal pancreases and "human" insulin produced by recombinant DNA methods.
3. *Adrenal steroids,* primarily the glucocorticoids, are used for the relief of inflammation in such diseases as rheumatoid arthritis, lupus erythematosus, asthma, and cerebral edema; for immunosuppression after organ transplants; and for the relief of the stress symptoms of shock.
4. *Epinephrine* (adrenaline) has many uses, including stimulation of the heart muscle when rapid response is required; treatment of asthmatic attacks by relaxation of the muscles of the small bronchial tubes; and treatment of the acute allergic reaction called *anaphylaxis* (an-ah-fi-LAK-sis).
5. *Thyroid hormones,* either extracted from animal thyroids or manufactured synthetically, are used in the treatment of hypothyroid conditions (cretinism and myxedema) and as replacement ther-

Seasonal Affective Disorder

We all sense that long dark days make us blue and sap our motivation. Are these learned responses or is there a physical basis for them? Studies have shown that the amount of light in the environment truly does have a physical effect on behavior.

Evidence that light alters mood comes from people who are intensely affected by the dark days of winter—people who suffer from *seasonal affective disorder*, aptly abbreviated SAD. When days shorten, these people feel sleepy, depressed, and anxious. They tend to overeat, especially carbohydrates. Daily exposure to bright lights has been found to improve the mood of most people with SAD.

Light exerts its effects on the pineal gland of the brain. As it strikes the retina of the eye, it sets up impulses that decrease the amount of melatonin produced by the pineal. Because melatonin is known to depress mood, the final effect of light is to act as a mood elevator.

apy following surgical removal of the thyroid gland.

6. *Oxytocin* is used to contract the uterine muscle.

7. *Androgens,* including testosterone and androsterone, are used in severe chronic illness to aid tissue building and promote healing.

8. *Oral contraceptives* (birth control pills; "the pill") contain estrogen and progesterone. They are highly effective in preventing pregnancy. Occasionally they give rise to unpleasant side effects, such as nausea. More rarely, they cause serious complications such as thrombophlebitis, embolism, and hypertension. Any woman taking birth control pills should have a medical examination every 6 months.

▶ Hormones and Stress

Stress in the form of physical injury, disease, emotional anxiety, and even pleasure calls forth a specific response from the body that involves both the nervous system and the endocrine system. The nervous system response, the "fight-or-flight" response, is mediated by parts of the brain, especially the hypothalamus, and by the autonomic nervous system. The hypothalamus also triggers the release of ACTH from the anterior pituitary. The hormones released from the adrenal cortex as a result of stimulation by ACTH raise the blood sugar level, inhibit inflammation, decrease the immune response, and limit the release of histamine. Growth hormone, thyroid hormones, sex hormones, and insulin may also be released. These substances help the body meet stressful situations, but unchecked they are harmful to the body and may lead to such stress-related disorders as high blood pressure, heart disease, ulcers, back pain, and headaches.

Although no one would enjoy a life totally free of stress in the form of stimulation and challenge, unmanaged stress, or "distress," has negative effects on the body. For this reason, techniques such as biofeedback and meditation to control stress are useful. Simply setting priorities, getting adequate periods of relaxation, and getting regular physical exercise are important in maintaining total health.

SUMMARY

I. **Hormones**
   A. Functions
      1. Affect certain other cells or organs—target tissue
      2. Carried by bloodstream
      3. Widespread effects on growth, metabolism, reproduction
      4. Local effects
   B. Main types
      1. Amino acid compounds—proteins and related compounds
      2. Steroids
         a. Derived from lipids
         b. Produced by adrenal cortex and sex glands
   C. Regulation—negative feedback

II. **Endocrine glands**
   A. Pituitary
      1. Regulated by hypothalamus
         a. Anterior pituitary—releasing hormones sent through portal system
         b. Posterior pituitary—stores hormones; released by nervous stimulation
      2. Anterior lobe hormones
         a. Growth hormone (GH)
         b. Thyroid-stimulating hormone (TSH)
         c. Adrenocorticotropic hormone (ACTH)
         d. Prolactin (PRL)
         e. Follicle-stimulating hormone (FSH)
         f. Luteinizing hormone (LH)
      3. Posterior lobe hormones
         a. Antidiuretic hormone (ADH)
         b. Oxytocin
      4. Pituitary tumors—may cause underactivity or overactivity of pituitary
   B. Thyroid
      1. Hormones
         a. Thyroxine—influences cell metabolism
         b. Calcitonin—decreases blood calcium levels
      2. Abnormalities
         a. Goiter—enlarged thyroid
         b. Hypothyroidism—causes cretinism or myxedema

    c. Hyperthyroidism—Graves' disease (exophthalmic goiter), a common example

C. Parathyroids—secrete parathyroid hormone (PTH), which increases blood calcium levels

D. Adrenals

  1. Hormones of adrenal medulla (inner region)

    a. Epinephrine and norepinephrine—act as neurotransmitters

  2. Hormones of adrenal cortex (outer region)

    a. Glucocorticoids—released during stress; example, cortisol

    b. Mineralocorticoids—regulate water and electrolyte balance; example, aldosterone

    c. Sex hormones—produced in small amounts

E. Islet cells of pancreas—secrete hormones

  1. Insulin

    a. Lowers blood glucose

    b. Lack causes diabetes mellitus

  2. Glucagon

    a. Raises blood glucose

F. Sex glands—needed for reproduction and development of secondary sex characteristics

  1. Testes—secrete testosterone

  2. Ovaries—secrete estrogen and progesterone

III. Other hormone-producing tissues

A. Thymus—secretes thymosin, which aids in development of T-lymphocytes

B. Pineal—secretes melatonin

  1. Regulates sexual development and sleep/wake cycles

  2. Controlled by environmental light

C. Stomach and small intestine—secrete hormones that regulate digestion

D. Kidneys—secrete erythropoietin, which increases production of red blood cells

E. Atria of heart—ANP causes loss of sodium by kidney and lowers blood pressure

F. Placenta—secretes hormones that maintain pregnancy and prepare breasts for lactation

G. Other—cells throughout body produce prostaglandins, which have varied effects

IV. Hormones and treatment

A. Growth hormone—treatment of deficiency in children

B. Insulin—treatment of diabetes mellitus

C. Steroids—reduction of inflammation, suppression of immunity

D. Epinephrine—treatment of asthma, anaphylaxis, shock

E. Thyroid hormone—treatment of hypothyroidism

F. Oxytocin—contraction of uterine muscle

G. Androgens—promote healing

H. Estrogen and progesterone—contraception

QUESTIONS FOR STUDY AND REVIEW

1. Compare the actions of the nervous system and the endocrine system.
2. Define *hormone*. Describe some general functions of hormones.
3. Give three examples of amino acid hormones; of steroid hormones.
4. Define *target tissue.*
5. Name the two divisions of the pituitary gland. Name the hormones released from each division and describe the effects of each.
6. What type of system connects the anterior pituitary with the hypothalamus? What is carried to the pituitary by this system?
7. Where is the thyroid gland located? What is its main hormone, and what does it do?
8. Describe the effects of hypothyroidism and hyperthyroidism.
9. How is thyroid function measured?
10. What is the main purpose of PTH? What are the effects of removal of the parathyroid glands? of excess secretion?
11. Name the two divisions of the adrenal glands and describe the effects of the hormones from each.

12. What are the results of hypofunction of the adrenal cortex?

13. What are the results of hyperfunction of the adrenal cortex? What common therapy produces similar symptoms?

14. What are the main purposes of insulin in the body? Name and describe the condition characterized by insufficient production of insulin.

15. Name the male and female sex hormones and briefly describe what each does.

16. Name the hormone produced by the thymus gland; by the pineal body. What are the effects of each?

17. Name five organs other than the endocrine glands that secrete hormones.

18. What are some of the various functions that have been ascribed to prostaglandins?

19. List several hormones that are released during stress.

UNIT V

Circulation and Body Defense

The chapters of this unit will discuss the blood, the heart, the blood vessels, and the lymphatic system as well as body defenses, immunity, and vaccines. It is the purpose of this unit to emphasize the importance of transportation and immune systems in maintaining normal body functions.

CHAPTER 13

The Blood

Selected Key Terms

The following terms are defined in the Glossary:

agglutination

anemia

antiserum

centrifuge

coagulation

erythrocyte

fibrin

hematocrit

hemoglobin

hemolysis

hemorrhage

hemostasis

leukemia

leukocyte

plasma

platelet (thrombocyte)

serum

thrombocytopenia

Behavioral Objectives

After careful study of this chapter, you should be able to:

- List the functions of the blood
- List the main ingredients in plasma
- Name the three types of formed elements in the blood and give the function of each
- Describe five types of leukocytes
- Describe the formation of blood cells
- Cite three steps in hemostasis
- Briefly describe the steps in blood clotting
- Define *blood type* and explain the relation between blood type and transfusions
- List the possible reasons for transfusions of whole blood and blood components
- Define *anemia* and list the causes of anemia
- Define *leukemia* and name the two types of leukemia
- Describe several forms of clotting disorders
- Describe the tests used to study blood

Memmler, RL, Cohen, BJ, Wood, DL. *THE HUMAN BODY IN HEALTH AND DISEASE*, 8/e,
© 1996 Lippincott-Raven Publishers

Blood is classified as a connective tissue because nearly half of it is made up of cells. Blood cells share many characteristics of origination and development with other connective tissues. However, blood differs from other connective tissues in that its cells are not fixed in position; instead, they move freely in the liquid portion of the blood, the *plasma.*

Blood is a viscous (thick) fluid that varies in color from bright scarlet to dark red, depending on how much oxygen it is carrying. The quantity of circulating blood differs with the size of the person; the average adult male, weighing 70 kg (154 pounds), has about 5 liters (5.2 quarts) of blood. This volume accounts for about 8% of total body weight.

The blood is carried through a closed system of vessels pumped by the heart. The circulating blood is of fundamental importance in maintaining the internal environment in a constant state (homeostasis).

▶ Functions of the Blood

The circulating blood serves the body in three ways: transportation, regulation, and protection.

Transportation
1. Oxygen from inhaled air diffuses into the blood through the thin lung membranes and is carried to all the tissues of the body. Carbon dioxide, a waste product of cell metabolism, is carried from the tissues to the lungs, where it is breathed out.
2. The blood transports nutrients and other needed substances, such as electrolytes (salts) and vitamins, to the cells. These materials may enter the blood from the digestive system or may be released into the blood from body stores.
3. The blood transports the waste products from the cells to the sites from which they are released. The kidney removes excess water, electrolytes, and urea (from protein metabolism) and maintains the acid–base balance of the blood. The liver removes bile pigments and drugs.
4. The blood carries hormones from their sites of origin to the organs they affect.

Regulation
1. Buffers in the blood help keep the pH of body fluids at about 7.4.

2. The blood serves to regulate the amount of fluid in the tissues by means of substances (mainly proteins) that maintain the proper osmotic pressure.
3. The blood transports heat that is generated in the muscles to other parts of the body, thus aiding in the regulation of body temperature.

Protection
1. The blood carries the cells that are among the body's defenders against pathogens. It also contains substances (antibodies) that are concerned with immunity to disease.
2. The blood contains factors that protect against blood loss from the site of an injury.

▶ Blood Constituents

The blood is composed of two prime elements: as already mentioned, the liquid element is called *plasma;* the cells and fragments of cells are called *formed elements* or *corpuscles* (KOR-pus-ls) (Fig. 13-1). The formed elements are classified as follows:

1. *Erythrocytes* (eh-RITH-ro-sites), from *erythro,* meaning "red," are the red blood cells, which transport oxygen.
2. *Leukocytes* (LU-ko-sites), from *leuko,* meaning "white," are the several types of white blood cells, which protect against infection.
3. *Platelets,* also called *thrombocytes* (THROM-bo-sites), are cell fragments that participate in blood clotting.

Blood Plasma
More than half of the total volume of blood is plasma. The plasma itself is 90% water. Many different substances, dissolved or suspended in the water, make up the other 10%. The plasma content varies somewhat, because substances are removed and added as the blood circulates to and from the tissues. However, the body tends to maintain a fairly constant level of these substances. For example, the level of glucose, a simple sugar, is maintained at a remarkably constant level of about one tenth of one percent (0.1%) in solution.

After water, the next largest percentage of material in the plasma is *protein.* Proteins are the princineutrophil eosinophil basophil

blood smear

red blood cells
and platelets

lymphocyte monocyte

Figure 13-1

Normal blood smear and close-up view of individual blood cells.

pal constituents of cytoplasm and are essential to the growth and the rebuilding of body tissues. The plasma proteins include the following:

1. ***Albumin*** (al-BU-min), the most abundant protein in plasma, is important for maintaining the osmotic pressure of the blood. This protein is manufactured in the liver.
2. Blood ***clotting factors*** are also manufactured in the liver.

3. ***Antibodies*** combat infection.
4. A system of enzymes made of several proteins, collectively known as ***complement,*** helps antibodies in their fight against pathogens (see Chap. 17).

Nutrients are also found in the plasma. The principal carbohydrate found in the plasma is ***glucose,*** which is absorbed by the capillaries of the intestine following digestion. Glucose is stored mainly in the liver as glycogen and is released as needed to supply energy.

Amino acids, the products of protein digestion, are also found in the plasma. These are also absorbed into the blood through the intestinal capillaries.

Lipids constitute a small percentage of blood plasma. Lipids include fats. They may be stored as fat for reserve energy or carried to the cells as a source of energy.

The *electrolytes* in the plasma appear primarily as chloride, carbonate, or phosphate salts of sodium, potassium, calcium, and magnesium. These salts have a variety of functions, including the formation of bone (calcium and phosphorus), the production of hormones by certain glands (iodine for the production of thyroid hormone), the transportation of oxygen (iron), and the maintenance of the acid–base balance (sodium and potassium carbonates and phosphates). Small amounts of other elements also help maintain homeostasis.

Many other materials, such as vitamins, hormones, waste products and drugs, are transported in the plasma.

Hemoglobin

The hemoglobin molecule is a protein made of four chains of amino acids (the "globin" part of the molecule), each of which holds an iron-containing "heme" unit. It is the heme portion that binds oxygen.

Hemoglobin allows the blood to carry much more oxygen than it could carry simply dissolved in the plasma. Hemoglobin picks up oxygen in the lungs and releases it in the body tissues. When cells are active, they need more oxygen. At the same time, they generate heat and acidity. These changing conditions promote the release of oxygen from hemoglobin.

Hemoglobin is produced by red cells in the red bone marrow. It is constantly broken down as red blood cells die and disintegrate. Some of its components are recycled, but we still need protein, iron, and vitamin B_{12} in our diet to maintain supplies.

The Formed Elements
ERYTHROCYTES

Erythrocytes, the red cells, are tiny, disk-shaped bodies with a central area that is thinner than the edges (see Fig. 13-1). They are different from other cells in that the mature form found in the circulating blood is lacking a nucleus and most of the other organelles commonly found in cells. As red cells mature, these components are lost to provide more space for the cells to carry oxygen. This vital gas is bound in the red cells to *hemoglobin* (he-mo-GLO-bin), a protein that contains iron. Hemoglobin combined with oxygen gives the blood its characteristic red color. The more oxygen carried by the hemoglobin, the brighter is the red color of the blood. Therefore, the blood that goes from the lungs to the tissues is a bright red because it carries a great supply of oxygen; in contrast, the blood that returns to the lungs is a much darker red, because it has given up much of its oxygen to the tissues. Hemoglobin that has given up its oxygen is able to carry hydrogen ions; in this way, hemoglobin acts as a buffer and plays an important role in acid–base balance (see Chap. 21). The red cells also carry a small amount of carbon dioxide from the tissues to the lungs for elimination by exhalation.

The ability of hemoglobin to carry oxygen can be blocked by carbon monoxide. This harmful gas combines with hemoglobin to form a stable compound that can severely restrict the ability of the erythrocytes to carry oxygen. Carbon monoxide is a by-product of the incomplete burning of fuels such as gasoline and other petroleum products, coal, wood, and other carbon-containing materials. It also occurs in cigarette smoke and auto exhaust.

The erythrocytes are by far the most numerous of the corpuscles, averaging from 4.5 to 5 million per cubic millimeter (mm^3) of blood. (A cubic millimeter is the same as a microliter [µL], one millionth of a liter, another way of expressing the concentration of blood cells). Because mature red cells have no nucleus and cannot divide, they must be replaced constantly. The production of red cells is stimulated by the hormone *erythropoietin* (eh-rith-ro-POY-eh-tin). This hormone is released from the kidney in response to a decrease in oxygen supply.

LEUKOCYTES

The *leukocytes,* or white blood cells, are very different from the erythrocytes in appearance, quantity, and function (see Fig. 13-1). The cells themselves are round, but they contain nuclei of varying shapes and sizes. Occurring at a concentration of 5,000 to 10,000 per cubic millimeter of blood, leukocytes are outnumbered by red cells by about 700 to 1. Whereas the red cells have a definite color, the leukocytes tend to be colorless.

The different types of white cells are identified by their size, the shape of the nucleus, and the appearance of granules in the cytoplasm when the cells are stained, usually with Wright's blood stain. *Granulocytes* (GRAN-u-lo-sites) include *neutrophils* (NU-tro-fils), which show lavender granules; *eosinophils* (e-o-SIN-o-fils), which have beadlike, bright pink granules; and *basophils* (BA-so-fils), which have large, dark blue granules that often obscure the nucleus. The neutrophils, which are active in fighting infections, are the most numerous of the white cells, constituting up to 60% of all leukocytes. The eosinophils and basophils make up a very small percentage of the white blood cells but increase in number during allergic reactions.

Because the nuclei of the neutrophils are of various shapes, they are also called *polymorphs* (meaning "many forms") or simply *polys.* Other nicknames are *PMNs,* an abbreviation of *p*olymorphonuclear *n*eutrophils, and *segs.* Before reaching full maturity and becoming segmented, the nucleus of the neutrophil looks like a thick, curved band. An increase in the number of these *band cells* (also called *stab* or *staff cells)* is a sign of infection and active production of neutrophils.

The *agranulocytes,* so named because they lack easily visible granules, are the *lymphocytes* and *monocytes.* The ratio of the different types of leukocytes is often a valuable clue in arriving at a diagnosis. The relative percentages and the functions of the various leukocytes are given in Table 13-1.

The most important function of leukocytes is to destroy pathogens. Whenever pathogens enter the tissues, as through a wound, certain white blood cells (neutrophils and monocytes) are attracted to that area. They leave the blood vessels and proceed by *ameboid* (ah-ME-boyd) or ameba-like motion to the area of infection. There they engulf the invaders by the process of *phagocytosis* (fag-o-si-TO-sis) (Fig. 13-2). If the pathogens are extremely virulent (strong) or numerous, they may destroy the leukocytes. A collection of dead and living bacteria, together with dead and living leukocytes, forms *pus.* A collection of pus localized in one area is known as an *abscess.*

Leukocytes may enter the tissues, undergo transformation, and become active in many different ways. Monocytes mature to become *macrophages* (MAK-ro-faj-ez) or fuse together to become giant cells, highly active in disposing of invaders or foreign material. Some lymphocytes become *plasma cells,*

TABLE 13-1
Leukocytes (White Blood Cells)

Type of Cell	Relative Percentage (Adult)	Function
Granulocytes		
Neutrophils (NU-tro-fils)	54%–62%	Phagocytosis
Eosinophils (e-o-SIN-o-fils)	1%–3%	Allergic reactions; defense against parasites
Basophils (BA-so-fils)	less than 1%	Allergic reactions; inflammatory reactions
Agranulocytes		
Lymphocytes (LIM-fo-sites)	25%–38%	Immunity (T cells and B cells)
Monocytes (MON-o-sites)	3%–7%	Phagocytosis

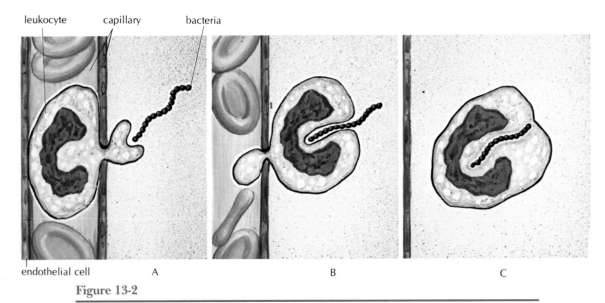

leukocyte capillary bacteria

endothelial cell A B C

Figure 13-2

Phagocytosis. (**A**) White blood cell squeezes through a capillary wall in the region of an infection. (**B,C**) White cells engulfs the bacteria.

active in the production of circulating antibodies. The activities of the various white cells are further discussed in Chapter 17.

PLATELETS

Of all the formed elements, the blood *platelets* (thrombocytes) are the smallest (see Fig. 13-1). These tiny structures are not cells in themselves, but fragments of cells. The number of platelets in the circulating blood has been estimated to range from 150,000 to 450,000 per cubic millimeter. Platelets are essential to blood *coagulation* (clotting). When, as a result of injury, blood comes in contact with any tissue other than the lining of the blood vessels, the platelets stick together and form a plug that seals the wound. They then release chemicals that take part in a series of reactions that eventually results in the formation of a clot. The last step in these reactions is the conversion of a plasma protein called *fibrinogen* (fi-BRIN-o-jen) into solid threads of *fibrin,* which form the clot.

Origin of the Formed Elements

The red blood cells (erythrocytes), the platelets (thrombocytes), and most of the white blood cells (leukocytes) are formed in the red marrow, which is located in the ends of long bones and in the inner mass of all other bones.

The ancestors of blood cells are called *stem cells.* Each kind of stem cell develops into one of the blood cell types found within the red marrow. One group of white cells, the lymphocytes, develops to maturity in lymphoid tissue and can multiply in this tissue as well (see Chap. 16).

As already mentioned, when an invader enters the tissues, the leukocytes in the blood are attracted to the area. In addition, the bone marrow goes into emergency production, with the result that the number of leukocytes in the blood is enormously increased. Detection in a blood examination of an abnormally large number of white cells may thus be an indication of an infection.

The platelets are believed to originate in the red marrow as fragments of certain giant cells called *megakaryocytes* (meg-ah-KAR-e-o-sites), which are formed in the red marrow. Platelets do not have nuclei or DNA, but they do contain active enzymes and mitochondria.

The life spans of the different types of blood cells vary considerably. For example, after leaving the bone marrow, erythrocytes circulate in the bloodstream for approximately 120 days. Leukocytes may appear in the circulating blood for only 6 to 8 hours. However, they may then enter the tissues, where they survive for longer periods—days, months, or even years. Blood platelets have a life span of about 10 days. In comparison with other tissue cells, most of those in the blood are very short lived. Thus, the need for constant replacement of blood cells means that normal activity of the red bone marrow is absolutely essential to life.

▶ Hemostasis

Hemostasis (he-mo-STA-sis) is the process that prevents the loss of blood from the circulation when a blood vessel is ruptured by an injury. Events in hemostasis include the following:

1. *Contraction* of the smooth muscles in the wall of the blood vessel. This reduces the flow of blood and loss from the defect in the vessel wall. The term for this reduction in the diameter of a vessel is *vasoconstriction.*
2. Formation of a *platelet plug.* Activated platelets become sticky and adhere to the defect to form a temporary plug.
3. Formation of a *blood clot.*

Blood Clotting

The many substances necessary for blood clotting, or coagulation, are normally inactive in the bloodstream. A balance is maintained between compounds that promote clotting, known as *procoagulants,* and those that prevent clotting, known as *anticoagulants.* In addition, there are chemicals in the circulation that act to dissolve clots. Under normal conditions the substances that prevent clotting prevail. However, when an injury occurs, the procoagulants are activated and a clot is formed.

Basically, the clotting process consists of the following essential steps (Fig. 13-3):

1. The injured tissues release *thromboplastin* (throm-bo-PLAS-tin), a substance that triggers the clotting mechanism.
2. Thromboplastin reacts with certain protein factors and calcium ions to form *prothrombin* activator, which in turn reacts with calcium ions to convert the prothrombin to *thrombin.*
3. Thrombin, in turn, converts soluble fibrinogen into insoluble fibrin. *Fibrin* forms a network of threads that entraps red blood cells and platelets to form a clot.

When blood is removed from the body it will automatically clot as it comes into contact with some

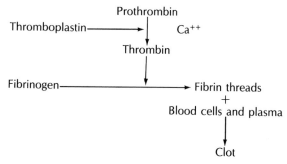

Figure 13-3

Formation of a blood clot.

surface other than the lining of a blood vessel. The fluid that remains after clotting has occurred is called *serum* (pl. *sera*). Serum contains all the components of blood plasma *except* the clotting factors.

Several methods used to measure the body's ability to coagulate blood are described later in this chapter.

▶ Blood Typing and Transfusions

Blood Groups

If for some reason the amount of blood in the body is severely reduced, through *hemorrhage* (HEM-eh-rij) (excessive bleeding) or disease, the body cells suffer from lack of oxygen and nutrients. The obvious measure to take in such an emergency is to administer blood from another person into the veins of the patient, a procedure called *transfusion.*

Care must be taken in transferring blood from one person to another, however, because the patient's plasma may contain substances, called *antibodies,* or *agglutinins,* that can cause the red cells of the donor's blood to rupture and release their hemoglobin. Such cells are said to be *hemolyzed* (HE-mo-lized), and the resulting condition can be very dangerous.

These reactions are determined by certain proteins, called *antigens* (AN-ti-gens), or *agglutinogens,* on the red cell membranes. There are many types of these proteins, but only two groups are particularly likely to cause a transfusion reaction, the so-called *A and B antigens* and the *Rh factor.* Four blood types involving the A and B antigens have been recognized: A, B, AB, and O. These letters indicate the type of antigen present on the red cells. If only the A antigen is present on the red cells, the person is type A; if only the B antigen is present, he or she is type B. Type AB red cells have both antigens, and type O have neither. It is these antigens on the donor's red cells that react with the antibodies in the patient's plasma and cause a transfusion reaction.

Blood sera containing antibodies to the A and B antigens are used to test for blood type. Blood serum containing antibodies that can agglutinate and destroy red cells with A antigen on the surface is called *anti-A serum;* blood serum containing antibodies that can destroy red cells with B antigen on the surface is called *anti-B serum.* When combined in the laboratory, each antiserum will cause the corresponding red cells to clump together in a process known as *agglutination* (ah-glu-tih-NA-shun). The blood's pattern of agglutination, when mixed *separately* with these two sera, reveals its blood type (Fig. 13-4). Type A reacts with A antiserum only; type B reacts with B antiserum only. Type AB agglutinates with both, and type O agglutinates with neither A nor B.

Of course no one has antibodies to his or her own blood type antigens or their own cells would be destroyed by their plasma. Each person does, however, develop antibodies that react with the AB antigens he or she is lacking.

An individual's blood type is determined by heredity, and the percentage of persons with each of the different blood types varies in different populations. About 45% of the white population of the United States is type O. In an emergency, type O blood can be given to any ABO type because the cells lack both A and B antigens and will not react with either A or B antibodies. Conversely, the blood of type AB individuals contains no antibodies to agglutinate red cells and they can therefore receive blood from any type of ABO donor (Table 13-2).

Usually a person can safely give blood to any person with the same blood type. However, because of other factors that may be present in the blood, determination of blood type must be accompanied by additional tests (cross matching) for compatibility before a transfusion is given.

The Rh Factor

More than 85% of the population of the United States has another red cell antigen group called the *Rh factor,* named for *Rh*esus monkeys in which it was first found. Rh is also known as the *D antigen.* Such individuals are said to be *Rh positive.* Those persons who lack this protein are said to be *Rh negative.* If Rh-positive blood is given to an Rh-negative person, he or she may produce antibodies to the "foreign" Rh antigens. The blood of this "Rh-sensitized" person will then destroy any Rh-positive cells received in a later transfusion.

Rh incompatibility is a potential problem in certain pregnancies. A mother who is Rh negative may develop antibodies to the Rh protein of an Rh-positive fetus (the fetus having inherited this factor from the father). The response is evoked by red cells from the fetus that enter the mother's circulation during childbirth. In a subsequent pregnancy with an Rh-positive fetus, some of the anti-Rh antibodies may pass from the mother's blood into the blood of her fetus and cause destruction of the fetus's red cells. This condition is called *erythroblastosis fetalis* (eh-rith-ro-blas-TO-sis fe-TA-lis), or *hemolytic disease of the newborn.* Should this occur, the newborn can sometimes be saved by a transfusion during which much of the blood is replaced.

Erythroblastosis fetalis is now prevented by administration of immune globulin $Rh_o(D)$, or Rho-GAM, to the mother during pregnancy and shortly after delivery. These preformed antibodies clear the mother's circulation of Rh antigens and prevent the activation of her immune system.

Blood Banks

Blood can be packaged and kept in blood banks for emergencies. To keep the blood from clotting, a solution such as citrate–phosphate–dextrose–adenine (CPDA-1) is added. The blood may then be stored for up to 35 days. The supplies of blood in the bank are dated with an expiration date so that blood in which red cells may have disintegrated is not used. Blood banks usually have all types of blood available. However, it is important that there be a particularly large supply of type O, Rh-negative blood, since in an emergency this type can be used for any patient. As already mentioned, a patient's blood must be tested for compatibility with the donor's blood before any transfusion is begun.

▶ Uses of Blood and Blood Components

The transfer of whole human blood from a healthy person to a patient is often a lifesaving process. Blood transfusions may be used for any condition in

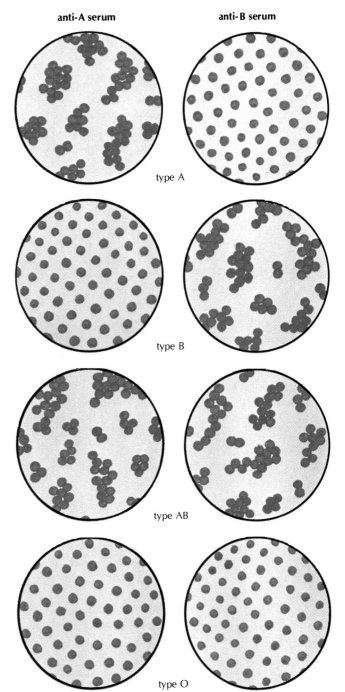

anti-A serum   **anti-B serum**

type A

type B

type AB

type O

Figure 13-4

Blood typing. Red cells in type A blood are agglutinated (clumped) by anti-A serum; those in type B
blood are agglutinated by anti-B serum. Type AB
blood cells are agglutinated by both sera, and type
O blood is not agglutinated by either serum.

which there is loss of a large volume of blood, for
example:

1. In the treatment of massive hemorrhage from
serious mechanical injuries
2. For blood loss during internal bleeding, as from
bleeding ulcers
3. For blood replacement in the treatment of
erythroblastosis fetalis
4. During or after an operation that causes considerable blood loss.

Caution and careful evaluation of the need for
blood or blood component transfusion is the rule

TABLE 13-2
The ABO Blood Group System

Blood Type	Red Blood Cell Antigen	Reacts with Antiserum	Plasma Antibodies	Can Take From	Can Donate to
A	A	Anti-A	Anti-B	A, O	A, AB
B	B	Anti-B	Anti-A	B, O	B, AB
AB	A, B	Anti-A, anti-B	None	AB, A, B, O	AB
O	None	None	Anti-A, anti-B	O	O, A, B, AB

due to the risk of transmission of viral diseases, particularly hepatitis.

Most often, when some ingredient of the blood is needed, it is not whole blood but a blood component that is given. Blood can be broken down into its various components, and the substances derived from it may be used for a number of purposes. One of the more common of these processes is separation of the blood plasma from the formed elements. This is accomplished by means of a *centrifuge* (SEN-trih-fuje), a machine that spins in a circle at high speed to separate components of a mixture according to density. When a container of blood is spun rapidly, all the formed elements of the blood are pulled into a clump at the bottom of the container. They are thus separated from the plasma, which is less dense. The formed elements may be further separated and used for specific purposes. Blood losses to the donor can be minimized by removal of the blood, separation of the desired components, and return of the remainder to the donor. The general term for this procedure is *hemapheresis* (hem-ah-fer-E-sis). If the plasma is removed and the formed elements returned to the donor, the procedure is called *plasmapheresis* (plas-mah-fer-E-sis).

Blood plasma is a very useful substance; it may be given as an emergency measure to combat shock and replace blood volume. Plasma is especially useful in situations in which blood typing and the use of whole blood are not possible, such as in natural disasters or in emergency rescues. Because the red cells have been removed from the plasma, there are no incompatibility problems; plasma can be given to anyone. Plasma separated from the cellular elements is usually further separated by chemical means into various components such as plasma protein fraction, serum albumin, immune serum, and clotting factors.

The packaged plasma that is currently available is actually plasma protein fraction. Further separation yields serum albumin which is available in solutions of 5% or 25% concentration. In addition to use in treatment of shock, these solutions are given in cases when plasma proteins are deficient; they increase the osmotic pressure of the blood and thus draw fluids back into circulation. There is increasing use of plasma proteins and serum albumin, because these blood components can be treated with heat to prevent transmission of viral diseases.

Fresh plasma may be frozen and saved. When frozen plasma is thawed, a white precipitate called *cryoprecipitate* (kri-o-pre-SIP-ih-tate) forms in the bottom of the container. Plasma frozen when it is less than 6 hours old and cryoprecipitate contain most of the factors needed for clotting and may be given when there is a special need for these factors.

Gamma globulin is the fraction of the plasma that contains the antibodies produced by lymphocytes when they come in contact with foreign agents such as bacteria and viruses. Antibodies play an important role in the immune system (see Chap. 17). Commercially prepared immune serums are available for administration to those in immediate need of antibodies, such as infants born to mothers with active hepatitis.

▶ Blood Disorders

Abnormalities involving the blood may be divided into three groups:

1. *Anemia* (ah-NE-me-ah), a disorder in which there is an abnormally low level of hemoglobin or red cells in the blood and thus impaired delivery of oxygen to the tissues. Anemia may result from the following:

Reticulocytes

As they mature, red blood cells go through a series of stages in which they lose their nucleus and most other organelles. In one of the last stages of development, ribosomes that are still remaining in the cell appear as a network when stained. These cells are called *reticulocytes* (reh-TIK-u-lo-sites). A small percentage of reticulocytes normally appears in the blood. Changes in these numbers can be used in diagnosis.

When red cells are lost or destroyed, as from bleeding or some form of hemolytic anemia, production of red cells is "stepped up" to compensate for the loss. Greater numbers of reticulocytes are then released into the blood before reaching full maturity, and counts will increase above the average. A decrease in the number of circulating reticulocytes suggests that there is a problem with the manufacture of red cells, as in cases of deficiency anemias or suppression of bone marrow activity.

a. *Excessive loss or destruction of red blood cells.* This may occur with hemorrhage or with conditions that cause hemolysis, the rupture of red cells.
b. *Impaired production of red cells or hemoglobin.* Both nutritional deficiencies and suppression of bone marrow may cause anemia.
2. *Neoplastic disease* of the blood-forming organs. A neoplasm is a new growth of abnormal cells or tissues. Neoplastic diseases, which may be cancerous, include *leukemia* (lu-KE-me-ah), a disorder characterized by an increase in the number of white blood cells.
3. *Clotting disorders.* These disorders are characterized by an abnormal tendency to bleed due to a breakdown in the body's clotting mechanism.

Anemia
ANEMIA DUE TO EXCESSIVE LOSS OR DESTRUCTION OF RED CELLS

Excessive loss of red cells occurs with hemorrhage, which may be sudden and acute or gradual and chronic.

The average adult has about 5 liters of blood. If a person loses as much as 2 liters suddenly, death

usually results. However, if the loss is gradual, over a period of weeks, or months, the body can compensate and withstand the loss of as much as 4 or 5 liters. If the cause of the chronic blood loss, such as bleeding ulcers, excessive menstrual flow, and bleeding hemorrhoids (piles), can be corrected, the body is usually able to restore the blood to normal. However, this process can take as long as 6 months, and until the blood returns to normal, the affected person may have anemia.

Anemia caused by the excessive destruction of red cells is called *hemolytic* (he-mo-LIH-tik) *anemia.* The spleen, along with the liver, normally destroys old red blood cells. Occasionally, an overactive spleen destroys the cells too rapidly, causing anemia. Infections may also cause the loss of red blood cells. For example, the malarial parasite multiplies in red cells and destroys them, and certain bacteria, particularly streptococci, produce a toxin that causes hemolysis.

Certain inherited diseases that involve the production of abnormal hemoglobin may also result in hemolytic anemia. The hemoglobin in normal adult cells is of the A type and is designated *HbA.* In the inherited disease *sickle cell anemia,* the hemoglobin in many of the red cells is abnormal. When these cells give up their oxygen to the tissues, they are transformed from the normal disk shape into a sickle shape (Fig. 13-5). These sickle cells are fragile and tend to break easily. Because of their odd shape, they also tend to become tangled in masses that can block smaller blood vessels. When obstruction occurs, there may be severe joint swelling and pain, especially in the fingers and toes, as well as abdominal pain. This development is referred to as *sickle cell crisis.*

Sickle cell anemia is seen almost exclusively in black people. About 8% of African Americans have one of the genes for the abnormal hemoglobin and are said to have the *sickle cell trait.* It is only when the involved gene is transmitted from both parents that the clinical disease appears. About 1% of African Americans have two of these genes and thus have *sickle cell disease.*

ANEMIA DUE TO IMPAIRED PRODUCTION OF RED CELLS OR HEMOGLOBIN

Many factors can interfere with normal red cell production. Anemia that results from a deficiency of some nutrient is referred to as *nutritional anemia.* These conditions may arise from a deficiency of the

normal blood cells

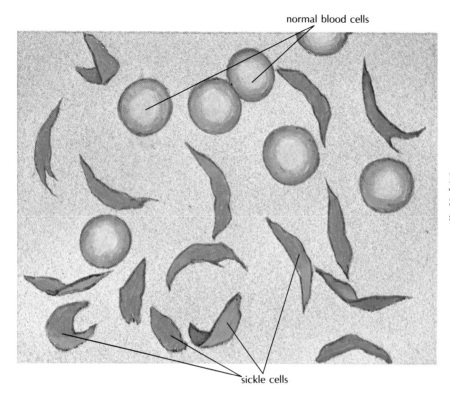

Figure 13-5

Sickling of red blood cells in sickle cell anemia.

sickle cells

specific nutrient in the diet, from an inability to absorb the nutrient, or from drugs that interfere with the body's use of the nutrient.

DEFICIENCY ANEMIA

The most common nutritional anemia is *iron deficiency anemia*. Iron is an essential constituent of hemoglobin. The average diet usually provides enough iron to meet the needs of the adult male, but this diet often is inadequate to meet the needs of growing children and women of childbearing age.

A diet deficient in proteins or vitamins can also result in anemia. Folic acid, one of the B complex vitamins, is necessary for the production of blood cells. Folic acid deficiency anemia occurs in alcoholics, in elderly persons on poor diets, and in infants or others suffering from intestinal disorders that interfere with the absorption of this water-soluble vitamin.

Pernicious (per-NISH-us) *anemia* is characterized by a deficiency of vitamin B_{12}, a substance essential for the proper formation of red blood cells. The cause is a permanent deficiency of *intrinsic factor,* a secretion in the gastric juice that is responsible for the absorption of vitamin B_{12} from the intestine. Neglected pernicious anemia can bring about deterio-

ration in the nervous system, causing difficulty in walking, weakness and stiffness in the extremities, mental changes, and permanent damage to the spinal cord. Early treatment, including the intramuscular injection of vitamin B_{12} and attention to a prescribed diet, ensures an excellent outlook. This treatment must be kept up for the rest of the patient's life if good health is to be maintained.

BONE MARROW SUPPRESSION

The decreased production of red blood cells may also be brought about by *bone marrow suppression* or failure. One type of bone marrow failure, *aplastic* (a-PLAS-tik) *anemia,* may be caused by a variety of physical and chemical agents. Chemical substances that injure the bone marrow include benzene, arsenic, nitrogen mustard, gold compounds, and, in some persons, certain prescribed drugs. Physical agents that may injure the marrow include x-rays, atomic radiation, radium, and radioactive phosphorus. The damaged bone marrow fails to produce either red or white cells, so that the anemia is accompanied by *leukopenia* (lu-ko-PE-ne-ah), a drop in the number of white cells. Removal of the toxic agent, followed by blood transfusions until the marrow is able to resume its activity, may result in recovery. Bone marrow transplants have also been successful.

A less severe depression of the bone marrow may develop in persons with certain chronic diseases, such as cancer, kidney or liver disorders, and rheumatoid arthritis.

Neoplastic Blood Disease

Neoplastic blood diseases are characterized by an enormous increase in the number of white cells due to a cancer of the tissues that produce these cells. As noted earlier, the white cells have two main sources: red marrow, also called *myeloid tissue,* and lymphoid tissue. If this wild proliferation of white cells stems from a tumor of the bone marrow, the condition is called *myelogenous* (mi-eh-LOJ-en-us) *leukemia.* When the cancer arises in the lymphoid tissue, so that the majority of abnormal cells are lymphocytes, the condition is called *lymphocytic* (lim-fo-SIT-ik) *leukemia.*

At present the cause of leukemia is unknown. Both inborn factors and various environmental agents have been implicated. Among the latter are chemicals (such as benzene), x-rays, radioactive substances, and viruses.

Patients with leukemia exhibit the general symptoms of anemia. In addition, they have a tendency to bleed easily due to a lack of platelets. The spleen is greatly enlarged, and several other organs may be increased in size because of the accumulation of white cells within them. Treatment consists of x-ray therapy along with drugs, but the disease is malignant and thus may be fatal. With new methods of chemotherapy, the outlook is improving, and many patients survive for years.

Clotting Disorders

A characteristic common to all clotting disorders is a disruption of the coagulation process, which brings about abnormal bleeding. *Hemophilia* (he-mo-FIL-e-ah) is a rare hereditary bleeding disorder that influenced history by its occurrence in some of the royal families of Russia and Western Europe. All forms of hemophilia are characterized by a deficiency of a specific clotting factor, so that any cut or bruise may cause serious abnormal bleeding. The needed clotting factors are now available in concentrated form for treatment in cases of injury, preparation for surgery, or painful bleeding into the joints, a frequent occurrence in hemophilia.

The most common clotting disorder is a deficient number of circulating platelets (thrombocytes). The condition, called *thrombocytopenia* (throm-bo-si-to-PE-ne-ah), results in hemorrhages in the skin or mucous membranes. The decrease in the number of platelets may be due to decreased production or to increased destruction of the platelets. A number of causes have been suggested, including diseases of the red bone marrow, liver disorders, and various drug toxicities. When a drug is the cause of the disorder, its withdrawal leads to immediate recovery.

A serious disorder of clotting involving excessive coagulation is *disseminated intravascular coagulation (DIC).* This disease occurs in cases of tissue damage due to massive burns, trauma, certain acute infections, cancer, and some disorders of childbirth. During the progress of DIC, platelets and various clotting factors are used up faster than they can be produced, and serious hemorrhaging may result.

▶ Blood Studies

Many kinds of studies may be made of the blood, and some of these have become a standard part of a routine physical examination. Machines that are able to perform several tests at the same time have largely replaced manual procedures, particularly in large institutions. Standard blood tests are listed in Appendix 3, Tables 2 and 3.

The Hematocrit

The *hematocrit* (he-MAT-o-krit), the volume percentage of red blood cells in whole blood, is determined by spinning of a blood sample in a high-speed centrifuge for 3 to 5 minutes. In this way the cellular elements are separated out from the plasma.

The hematocrit is expressed as the volume of packed red cells per unit volume (100 mL) of whole blood. For example, if a laboratory report states "hematocrit, 38%," that means that there are 38 mL red cells per 100 mL whole blood. In other words, 38% of the whole blood is red cells. For adult men, the normal range is 42 mL to 54 mL per 100 mL blood, whereas for adult women the range is slightly lower, 36 mL to 46 mL per 100 mL blood. These normal ranges, like all normal ranges for humans, may vary depending on the method used and the interpretation of the results by an individual laboratory. Hematocrit values much below or much above these figures point to an abnormality requiring further study.

Hemoglobin Tests

A sufficient amount of hemoglobin in red cells is required for adequate delivery of oxygen to the tissues. To measure its level, the hemoglobin is released from the red cells and the color of the blood is compared with a known color scale. Hemoglobin is expressed in grams per 100 mL (dL) whole blood. Normal hemoglobin concentrations for adult males range from 14 to 17 g per 100 mL blood. Values for adult females are in a somewhat lower range, at 12 to 15 g per 100 mL blood. A decrease in hemoglobin indicates anemia.

Normal and abnormal types of hemoglobin can be separated and measured by the process of *electrophoresis* (e-lek-tro-fo-RE-sis). In this procedure an electric current is passed through the liquid that contains the hemoglobin. This test is useful in the diagnosis of sickle cell anemia and other disorders caused by abnormal types of hemoglobin.

Blood Cell Counts

Most laboratories use automated methods for obtaining the data for blood counts. However, visual counts using a *hemocytometer* (he-mo-si-TOM-eh-ter) are still done for reference and quality control in all automated laboratories (Fig. 13-6).

The normal count for red blood cells varies from 4.5 to 5.5 million cells per cubic millimeter (μL) of blood. The leukocyte count varies from 5000 to 10,000 cells per cubic millimeter of blood.

An increase in the red blood cell count is called *polycythemia* (pol-e-si-THE-me-ah). This condition may be found in persons who live at high altitudes and in persons with the disease *polycythemia* (pol-e-si-THE-me-ah) *vera.*

In *leukopenia* the white count is below 5000 cells per cubic millimeter. This condition is indicative of depressed bone marrow or a neoplasm of the bone marrow. In *leukocytosis* (lu-ko-si-TO-sis) the white cell count is in excess of 10,000 cells per cubic millimeter. This condition is characteristic of most bacterial infections. It may also occur after hemorrhage and in gout and uremia, a result of kidney disease.

It is difficult to count platelets directly because they are so small. More accurate counts can be obtained with automated methods. These counts are necessary for the evaluation of platelet loss (throm-bocytopenia) such as occurs following radiation therapy or cancer chemotherapy. The normal platelet count ranges from 150,000 to 450,000 per cubic millimeter of blood, but counts may fall to 100,000 or less without causing serious bleeding problems. If a count is very low, a platelet transfusion may be given.

The Blood Slide (Smear)

In addition to the above tests, the *complete blood count* (*CBC*) includes the examination of a stained blood slide. In this procedure a drop of blood is spread very thinly and evenly over a glass slide and a special stain (Wright's) is applied to differentiate the otherwise colorless white cells. The slide is then studied under the microscope. The red cells are examined for abnormalities in size, color, or shape and for variations in the percentage of immature forms (reticulocytes). The number of platelets is estimated. Parasites such as the malarial organism and others may be found. In addition, a *differential white count* is done. This is an estimation of the percentage of each type of white blood cell in the smear. Because each type of white blood cell has a specific function, changes in their proportions can be a valuable diagnostic aid.

Blood Chemistry Tests

Batteries of tests on blood serum are often done by machine. One machine, the Sequential Multiple Analyzer (SMA), provides for the running of some 20 tests per minute. Tests for electrolytes, such as sodium, potassium, chloride, and bicarbonate, as well as for blood glucose, blood urea nitrogen (BUN), and *creatinine* (kre-AT-in-in), another nitrogen waste product, may be performed at the same time.

Other tests check for enzymes. Increased levels of *CPK* (creatine phosphokinase), *LDH* (lactic dehydrogenase), and other enzymes indicate tissue damage, such as that which may occur in heart disease. An excess of *alkaline phosphatase* (FOS-fah-tase) could indicate a liver disorder or metastatic cancer involving bone (See Appendix 3, Table 3).

Blood can be tested for amounts of lipids such as cholesterol and triglycerides (fats) or for amounts of plasma proteins. Many of these tests help in evaluating disorders that may involve various vital organs. For example, the presence of more than the normal amount of glucose (sugar) dissolved in the blood, a

Figure 13-6

Parts of a hemocytometer.

condition called *hyperglycemia* (hi-per-gli-SE-me-ah), is found most frequently in persons with unregulated diabetes. Sometimes several evaluations of sugar content are done following the administration of a known amount of glucose. This procedure is called the *glucose tolerance test* and usually is given along with another test that determines the amount of sugar in the urine. This combination of tests can indicate faulty cell metabolism. The list of blood chemistry tests is extensive and is constantly increasing. We may now obtain values for various hormones, vitamins, antibodies, and toxic or therapeutic drug levels.

Coagulation Studies

Nature prevents the excessive loss of blood from small vessels by the formation of a clot. Preceding surgery and under some other circumstances, it is important to know that the time required for coagulation to take place is not long. Because clotting is a rather complex process involving many reactants, a delay may be due to a number of different causes, including lack of certain hormone-like substances, calcium salts, or vitamin K.

Each of the various clotting factors has been designated by a Roman numeral, ranging from I to XIII. Factor I is fibrinogen, factor II is prothrombin, factor III is thromboplastin, and factor IV is assigned to

calcium ions. The amounts of all these factors may be determined and evaluated on a percentage basis, aiding in the diagnosis and treatment of some bleeding disorders.

Additional tests for coagulation include tests for bleeding time, clotting time, capillary strength, and platelet function.

Bone Marrow Biopsy

A special needle is used to obtain a small sample of red marrow from the sternum, sacrum, or iliac crest in a procedure called a ***bone marrow biopsy***. If marrow is taken from the sternum, the procedure may be referred to as a ***sternal puncture***. Examination of the cells by a trained person gives valuable information that can aid in the diagnosis of bone marrow disorders, including leukemia and certain kinds of anemia.

SUMMARY

I. Functions of blood
  A. Transportation—of oxygen, carbon dioxide, nutrients, minerals, vitamins, waste, hormones
  B. Regulation—of pH, fluid balance, body temperature
  C. Protection—against foreign organisms, blood loss

II. Blood constituents
  A. Plasma—liquid component
    1. Water—main ingredient
    2. Proteins—albumin, clotting factors, antibodies, complement
    3. Nutrients—carbohydrates, lipids, amino acids
    4. Minerals
    5. Waste products
    6. Hormones and other materials
  B. Formed elements
    1. Erythrocytes—red blood cells
    2. Leukocytes—white blood cells
      a. Granulocytes—neutrophils (polymorphs, segs), eosinophils, basophils
      b. Agranulocytes—lymphocytes, monocytes
    3. Platelets (thrombocytes)
  C. Origin of formed elements—produced in red bone marrow from stem cells

III. Hemostasis—prevention of blood loss
  A. Contraction of blood vessels
  B. Formation of platelet plug
  C. Formation of blood clot
    1. Factors
      a. Procoagulants—promote clotting
      b. Anticoagulants—prevent clotting
    2. Main steps in blood clotting
      a. Thromboplastin converts prothrombin to thrombin
      b. Thrombin converts fibrinogen to solid threads of fibrin
      c. Threads form clot
    3. Serum—fluid that remains after blood has clotted

IV. Blood typing
  A. AB antigens—result in A, B, AB, and O blood types
  B. Rh antigens—positive or negative
  C. Tested by mixing with antisera to different antigens
  D. Incompatible transfusions will cause destruction of donor red cells

V. Uses of blood and blood components
  A. Whole blood—used only to replace large blood losses
  B. Formed elements—separated by centrifugation

C. Plasma
    1. Protein fractions
    2. Cryoprecipitate—obtained by freezing; contains clotting factors
    3. Gamma globulin—contains antibodies
VI. Blood disorders
  A. Anemia—lack of hemoglobin
    1. Loss of cells
    2. Destruction of cells
    3. Impaired production of cells
      a. Deficiency anemia
      b. Pernicious anemia
      c. Bone marrow suppression
  B. Leukemia—excess production of white cells
    1. Myelogenous leukemia—cancer of bone marrow
    2. Lymphocytic leukemia—cancer of lymphoid tissue

C. Clotting disorders
    1. Hemophilia—lack of clotting factors
    2. Thrombocytopenia—lack of platelets
    3. Disseminated intravascular coagulation (DIC)
VII. Blood studies
  A. Hematocrit—measures percentage of packed red cells in whole blood
  B. Hemoglobin tests—color test, electrophoresis
  C. Blood cell counts
  D. Blood slide (smear)
  E. Blood chemistry tests—electrolytes, waste products, enzymes, glucose, hormones
  F. Coagulation studies—clotting factor assays, bleeding time, clotting time, capillary strength, platelet function
  G. Bone marrow biopsy

QUESTIONS FOR STUDY AND REVIEW

1. How does the color of blood vary with the amount of oxygenation?

2. Name the three main purposes of blood.

3. Name the two main fractions of blood.

4. Name four main ingredients of blood plasma. What are their purposes?

5. Name and describe the three main types of formed elements in blood.

6. Name and give the functions of the five types of leukocytes.

7. What is the main function of erythrocytes? leukocytes? platelets? Where do they originate?

8. Name the four blood types in the ABO system. What determines the different types?

9. What is the Rh factor? What proportion of the population of the United States possesses this factor? In what situations is this factor of medical importance? Why?

10. Define *hemostasis*. Describe the three main steps in hemostasis.

11. Describe the three basic steps involved in the clotting process.

12. What are the advantages of blood banks? Are there any disadvantages? If so, what are they, and how might they be counteracted?

13. What are some of the conditions for which blood transfusions are useful?

14. What precautions should always be taken before a transfusion is given?

15. What blood components may be useful in the treatment of the sick, and in what way is each of these blood derivatives used?

16. Name the three general categories of blood disorders.

17. Differentiate among the types of anemia and give an example of each.

18. Name two kinds of leukemia. What are the symptoms of leukemia?

19. Name the main characteristics of clotting disorders and give several examples of clotting disorders.

20. What does the hematocrit measure? Cite normal hematocrit ranges for males and females.

21. Differentiate between leukopenia and leukocytosis. What are some of the conditions that result in abnormal white counts?

22. What can be learned by studying the blood smear? What is determined by a differential white count?

23. What are some evaluations made by blood chemistry tests?

24. What can be determined by a bone marrow biopsy?

CHAPTER 14

The Heart and Heart Disease

Selected Key Terms

The following terms are defined in the Glossary:

arrhythmia

atherosclerosis

atrium

coronary

diastole

echocardiograph

electrocardiograph

endocardium

epicardium

infarct

ischemia

murmur

myocardium

pacemaker

pericardium

septum

stenosis

systole

thrombus

valve

ventricle

Behavioral Objectives

After careful study of this chapter, you should be able to:

- Describe the three layers of the heart
- Compare the functions of the right heart and left heart
- Name the four chambers of the heart
- Name the valves at the entrance and exit of each ventricle
- Briefly describe blood circulation through the myo-cardium
- Explain the effects of the autonomic nervous system on the heart rate
- Briefly describe the cardiac cycle
- Name the components of the heart's conduction system
- List and define several terms that describe different heart rates
- Explain what produces the two main heart sounds
- Describe several common types of heart disease
- List six areas that might contribute to the prevention of heart disease
- Briefly describe five instruments used in the study of the heart
- Describe three approaches to the treatment of heart disease

Memmler, RL., Cohen, BJ, Wood, DL. *THE HUMAN BODY IN HEALTH AND DISEASE, 8/e,*
© 1996 Lippincott-Raven Publishers

Circulation and the Heart

The next two chapters investigate the manner in which the blood delivers oxygen and nutrients to the cells and carries away the waste products of cell metabolism. This continuous one-way movement of the blood is known as its *circulation.* The prime mover that propels blood throughout the body is the *heart.* We shall have a look at the heart before going into the subject of the blood vessels in detail.

The heart is a muscular pump that drives the blood through the blood vessels. Slightly bigger than a fist, this organ is located between the lungs in the center and a bit to the left of the midline of the body. The strokes (contractions) of this pump average about 72 per minute and are carried on unceasingly for the whole of a lifetime.

The importance of the heart has been recognized for centuries. The fact that its rate of beating is affected by the emotions may be responsible for the very frequent references to the heart in song and poetry. However, the vital functions of the heart and its disorders are of more practical importance to us.

Structure of the Heart

Layers of the Heart Wall

The heart is a hollow organ, the walls of which are formed of three different layers. Just as a warm coat might have a smooth lining, a thick and bulky interlining, and an outer layer of a third fabric, so the heart wall has three tissue layers (Fig. 14-1).

1. The *endocardium* (en-do-KAR-de-um) is a very thin smooth layer of cells that resembles squamous epithelium. This membrane lines the interior of the heart. The valves of the heart are formed by reinforced folds of this material.
2. The *myocardium* (mi-o-KAR-de-um), the muscle of the heart, is the thickest layer and is responsible for pumping blood through the vessels. The unique structure of cardiac muscle is described in more detail below.
3. The *epicardium* (ep-ih-KAR-de-um) forms the thin, outermost layer of the heart wall and is continuous with the serous lining of the fibrous sac that encloses the heart. These membranes together make up the *pericardium* (per-ih-KAR-de-um). The serous lining of the pericardial sac is separated from the epicardium on the heart surface by a thin film of fluid.

SPECIAL FEATURES OF THE MYOCARDIUM

Cardiac muscle cells are lightly striated (striped) and have specialized partitions between the cells that appear faintly under the light microscope (Fig. 14-2). These *intercalated* (in-TER-cah-la-ted) *disks* are actually modified cell membranes that allow for rapid transfer of electric impulses between the cells. The adjective *intercalated* means "inserted between."

Another feature of cardiac muscle tissue is the branching of the muscle fibers (cells). These fibers are interwoven so that the stimulation that causes the contraction of one fiber results in the contraction of a whole group. These structural features play an important role in the working of the heart muscle.

Two Hearts and a Partition

Health professionals often refer to the *right heart* and the *left heart.* This is because the human heart is really a double pump (Fig. 14-3). The right side pumps blood low in oxygen to the lungs through the *pulmonary circuit.* The left side pumps oxygenated blood to the remainder of the body through the *systemic circuit.* The two sides are completely separated from each other by a partition called the *septum.* The upper part of this partition is called the *interatrial* (in-ter-A-tre-al) *septum,* while the larger, lower portion is called the *interventricular* (in-ter-ven-TRIK-u-lar) *septum.* The septum, like the heart wall, consists largely of myocardium.

Four Chambers

On either side of the heart are two chambers, one a receiving chamber (atrium) and the other a pumping chamber (ventricle):

1. The *right atrium* is a thin-walled chamber that receives the blood returning from the body tissues. This blood, which is low in oxygen, is carried in *veins,* the blood vessels leading *to* the heart from the body tissues.
2. The *right ventricle* pumps the venous blood received from the right atrium into the lungs.
3. The *left atrium* receives blood high in oxygen content as it returns from the lungs.
4. The *left ventricle,* which is the chamber with the thickest walls, pumps oxygenated blood to all

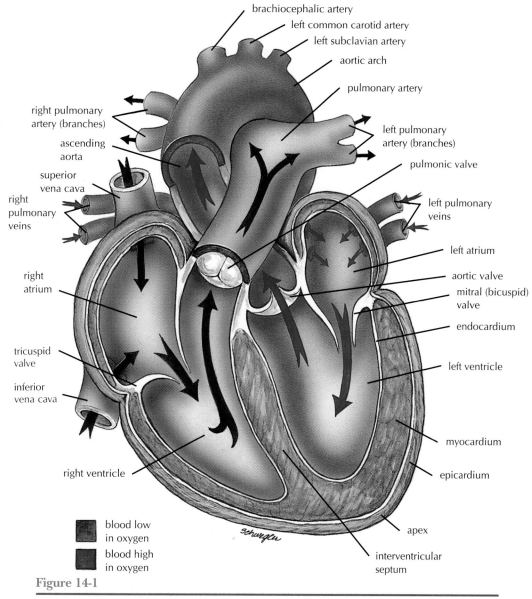

brachiocephalic artery
left common carotid artery
left subclavian artery
aortic arch
pulmonary artery
right pulmonary artery (branches)
ascending aorta
superior vena cava
right pulmonary veins
right atrium
tricuspid valve
inferior vena cava
right ventricle
left pulmonary artery (branches)
pulmonic valve
left pulmonary veins
left atrium
aortic valve
mitral (bicuspid) valve
endocardium
left ventricle
myocardium
epicardium
apex
interventricular septum

blood low in oxygen
blood high in oxygen

Figure 14-1

Heart and great vessels.

parts of the body. This blood goes through the **arteries,** the vessels that take blood *from* the heart to the tissues.

Four Valves

One-way valves that direct the flow of blood through the heart are located at the entrance and the exit of each ventricle. The entrance valves are the **atrioventricular** (a-tre-o-ven-TRIK-u-lar) **valves;** the exit valves are the **semilunar** (sem-e-LU-nar) **valves.** (So named because each flap of these valves resembles a half-moon.) Each valve has a specific name, as follows:

1. The **right atrioventricular (AV) valve** is also known as the **tricuspid** (tri-KUS-pid) **valve,** because it has three cusps, or flaps, that open and close. When this valve is open, blood flows freely from the right atrium into the right ventricle. However, when the right ventricle begins to contract, the valve closes so that blood cannot return to the right atrium; this ensures forward flow into the pulmonary artery.

Figure 14-2

Cardiac muscle tissue. Note the central nuclei, branching of fibers, and intercalated disks. (Courtesy of Nancy C. Maguire, Thomas Jefferson University)

2. The *left atrioventricular (AV) valve* is the bicuspid valve, but it is usually referred to as the *mitral* (MI-tral) *valve.* It has two rather heavy cusps that permit blood to flow freely from the left atrium into the left ventricle. However, the cusps close when the left ventricle begins to contract; this prevents blood from returning to the left atrium and ensures the forward flow of blood into the *aorta* (a-OR-tah). Both the tricuspid and mitral valves are attached by means of thin fibrous threads to muscles in the walls of the ventricles (see Fig. 14-7). The function of these threads, called the *chordae tendineae* (KOR-de ten-DIN-e-e), is to stabilize the valve flaps when the ventricles contract so that the force of the blood will not push them up into the atria. In this manner they help to prevent a backflow of blood when the heart beats.

3. The *pulmonic* (pul-MON-ik) *valve*, also called the *pulmonary valve*, is a semilunar valve located between the right ventricle and the pulmonary artery that leads to the lungs. As soon as the right ventricle has finished emptying itself, the valve closes to prevent blood on its way to the lungs from returning to the ventricle.

4. The *aortic* (a-OR-tik) *valve* is a semilunar valve located between the left ventricle and the aorta. Following contraction of the left ventricle, the aortic valve closes to prevent the flow of blood back from the aorta to the ventricle.

The appearance of the heart valves in the closed position is illustrated in Figure 14-4.

Blood Supply to the Myocardium

Only the endocardium comes into contact with the blood that flows through the heart chambers.

Therefore, the myocardium must have its own blood vessels to provide oxygen and nourishment and to remove waste products. The arteries that supply blood to the muscle of the heart are called the *right* and *left coronary arteries* (Fig. 14-5). These arteries, which are the first branches of the aorta, arise just above the aortic semilunar valve (see Fig. 14-4). They receive blood when the heart relaxes and branch to all regions of the heart muscle. After passing through capillaries in the myocardium, blood drains into the cardiac veins and finally into the *coronary sinus* for return to the right atrium.

❯ Function of the Heart

The Work of the Heart

Although the right and left sides of the heart are separated from each other, they work together. The blood is squeezed through the chambers by a contraction of heart muscle beginning in the thin-walled upper chambers, the atria, followed by a contraction of the thick muscle of the lower chambers, the ventricles. This active phase is called *systole* (SIS-to-le), and in each case it is followed by a resting period known as *diastole* (di-AS-to-le). The contraction of the walls of the atria is completed at the time the contraction of the ventricles begins. Thus, the resting phase (diastole) begins in the atria at the same time as the contraction (systole) begins in the ventricles. After the ventricles have emptied, both chambers are relaxed for a short period as they fill with blood. Then another beat begins with contraction of the atria followed by contraction of the ventricles. This sequence of heart relaxation and

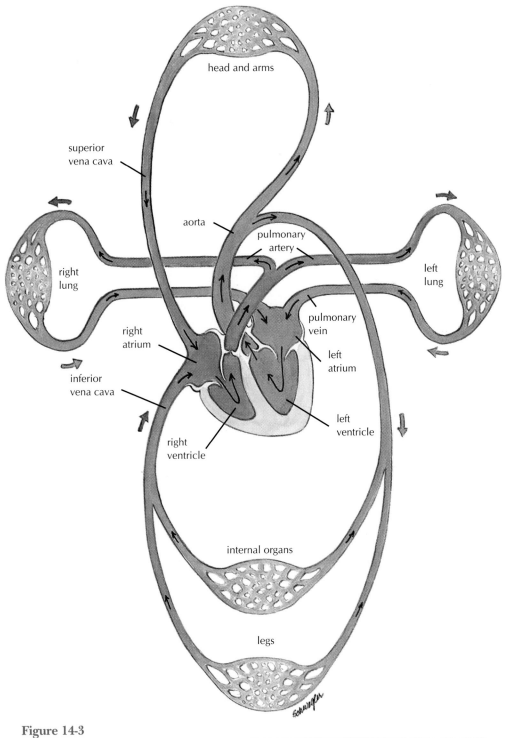

Figure 14-3

The heart is a double pump. The pulmonary circuit carries blood to the lungs to be oxy-genated; the systemic circuit carries blood to all other parts of the body.

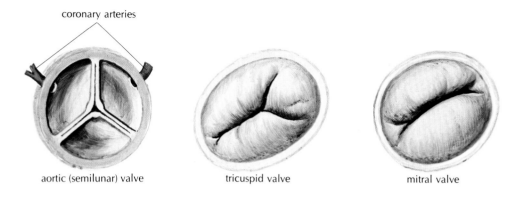

coronary arteries

aortic (semilunar) valve   tricuspid valve   mitral valve

Valves of the heart, seen from above, in the closed position.

contraction is called the *cardiac cycle.* Each cycle takes an average of 0.8 seconds (Fig. 14-6).

A unique property of heart muscle is its ability to adjust the strength of contraction to the amount of blood received. When the heart chamber is filled and the wall stretched (within limits), the contraction is strong. As less blood enters the heart, the contraction becomes less forceful. Thus, as more blood enters the heart, as occurs during exercise, the muscle contracts with greater strength to push the larger volume of blood out into the blood vessels.

The volume of blood pumped by each ventricle in 1 minute is termed the *cardiac output.* It is the product of the *stroke volume*—the volume of blood ejected from the ventricle with each beat, and the *heart rate*—the number of times the heart beats per minute.

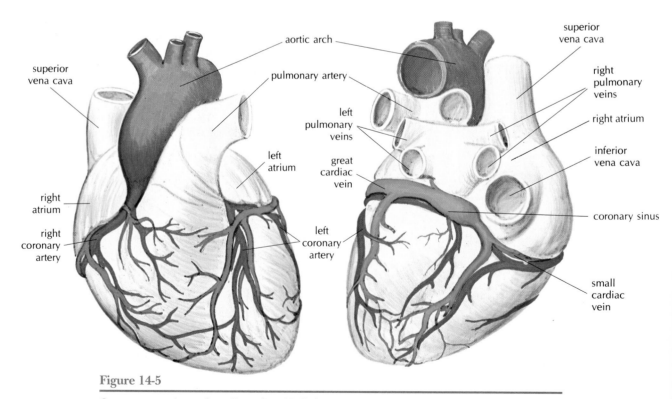

Figure 14-5

Coronary arteries and cardiac veins. (*Left*) Anterior view. (*Right*) Posterior view.

Diastole
Atria fill with blood which begins to flow into ventricles as soon as their walls relax.

Atrial Systole
Contraction of atria pumps blood into the ventricles.

Ventricular Systole
Contraction of ventricles pumps blood into aorta and pulmonary arteries.

Figure 14-6

Pumping cycle of the heart.

The Conduction System of the Heart

The cardiac cycle is regulated by specialized areas in the heart wall that form the conduction system of the heart (Fig. 14-7). Two of these areas are tissue masses called *nodes;* the third is a group of fibers called the *atrioventricular bundle.* The *sinoatrial (SA) node,* which is located in the upper wall of the right atrium, initiates the heartbeat and for this reason is called the *pacemaker.* The second node, located in the interatrial septum at the bottom of the right atrium, is called the *atrioventricular (AV) node.* The *atrioventricular bundle,* also known as the *bundle of His,* is located at the top of the interventricular septum; it has branches that extend to all parts of the ventricular walls. Fibers travel first down both sides of the interventricular septum in groups called the *right* and *left bundle branches.* Smaller *Purkinje* (pur-KIN-je) *fibers* then travel in a branching network throughout the myocardium of the ventricles. The special membranes between the cells (intercalated disks) allow the rapid flow of impulses throughout the heart muscle. The order in which these impulses travel is as follows:

1. The sinoatrial node generates the electric impulse that begins the heartbeat.
2. The excitation wave travels throughout the muscle of each atrium, causing it to contract.
3. The atrioventricular node is stimulated. The relatively slower conduction through this node allows time for the atria to contract and complete the filling of the ventricles.
4. The excitation wave travels rapidly through the bundle of His and then throughout the ventric-

 Cardiac Reserve

Like many other organs, the heart has great reserves of strength. The *cardiac reserve* is a measure of how many times more than average the heart can produce when needed. Based on a heart rate of 75 beats/minute and a stroke volume of 70 mL/beat, the average cardiac output for an adult at rest is about 5L/minute. This means that *at rest* the heart pumps the equivalent of the total blood volume each minute.

During mild exercise this volume might double and even double again during strenuous exercise. For most people the cardiac reserve is 4 to 5 times the resting output. In athletes exercising vigorously the ratio may reach 6 to 7 times. In contrast, those with heart disease may have little or no cardiac reserve. They may be fine at rest but quickly become short of breath or fatigued when exercising.

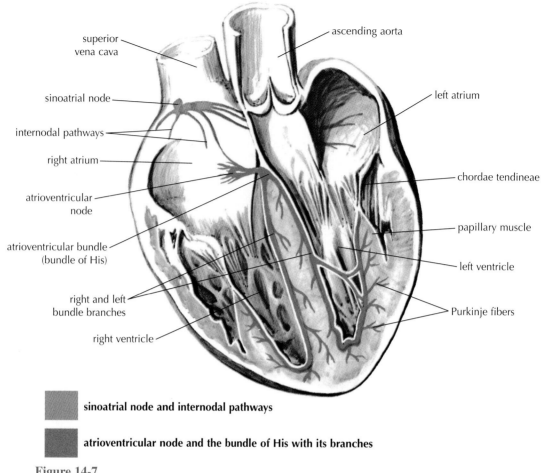

superior vena cava

ascending aorta

sinoatrial node

left atrium

internodal pathways

right atrium

chordae tendineae

atrioventricular node

papillary muscle

atrioventricular bundle (bundle of His)

left ventricle

right and left bundle branches

Purkinje fibers

right ventricle

■ sinoatrial node and internodal pathways

■ atrioventricular node and the bundle of His with its branches

Figure 14-7

Conduction system of the heart.

ular walls by means of the bundle branches and Purkinje fibers. The entire musculature of the ventricles contracts practically at once.

As a safety measure, a region of the conduction system other than the sinoatrial node can generate a heartbeat if the sinoatrial node fails, but it does so at a slower rate. A normal heart rhythm originating at the SA node is termed a *sinus rhythm.*

Control of the Heart Rate

Although the fundamental beat of the heart originates within the heart itself, the heart rate can be influenced by the nervous system and by other factors in the internal environment. The autonomic nervous system (ANS) plays a major role in modifying the heart rate according to need (Fig. 14-8). Stimulation from the sympathetic nervous system increases the heart rate. During a fight-or-flight re-

sponse, the sympathetic nerves can boost the cardiac output two to three times the resting value by increasing the rate and force of heart contractions. Stimulation from the parasympathetic nervous system decreases the heart rate to restore homeostasis. The parasympathetic nerve that supplies the heart is the vagus nerve (cranial nerve X). It slows the heart rate by acting on the SA and AV nodes. These influences allow the heart to meet changing needs rapidly. The heart rate is also affected by such factors as hormones, ions, and drugs in the blood.

Heart Rates

1. *Bradycardia* (brad-e-KAR-de-ah) is a relatively slow heart rate of less than 60 beats/minute. During rest and sleep, the heart may beat less than 60 beats/minute but usually does not fall below 50 beats/minute.

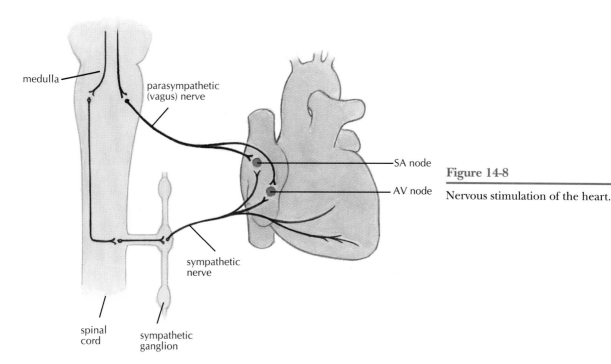

Figure 14-8

Nervous stimulation of the heart.

2. *Tachycardia* (tak-e-KAR-de-ah) refers to a heart rate over 100 beats/minute.
3. *Sinus arrhythmia* (ah-RITH-me-ah) is a regular variation in heart rate due to changes in the rate and depth of breathing. It is a normal phenomenon.
4. *Premature beat,* also called *extrasystole,* is a beat that comes before the expected normal beat. These may occur in normal persons initiated by caffeine, nicotine, or psychologic stresses. They are also common in persons with heart disease.

Heart Sounds and Murmurs

The normal heart sounds are usually described by the syllables "lubb" and "dupp." The first is a longer, lower-pitched sound that occurs at the start of ventricular systole. It is probably caused by a combination of things, mainly closure of the atrioventricular valves. The second, or "dupp," sound is shorter and sharper. It occurs at the beginning of ventricular relaxation and is due in large part to sudden closure of the semilunar valves. An abnormal sound is called a *murmur* and is usually due to faulty action of a valve. For example, if a valve fails to close tightly and blood leaks back, a murmur is heard. Another condition giving rise to an abnormal sound is the narrowing (stenosis) of a valve opening. The many conditions that can cause abnormal heart sounds include congenital defects, disease, and physiologic variations. A murmur due to rapid filling of the ventricles is called a *functional* or *flow murmur;* such a murmur is not abnormal. To differentiate, an abnormal sound caused by any structural change in the heart or the vessels connected with the heart is called an *organic murmur.*

▶ Heart Disease

Classification of Heart Disease

There are many ways of classifying heart disease. The anatomy of the heart forms the basis for one grouping of heart pathology:

1. *Endocarditis* (en-do-kar-DI-tis) means "inflammation of the lining of the heart cavities," but it most commonly refers to valvular disease.
2. *Myocarditis* (mi-o-kar-DI-tis) is inflammation of heart muscle.
3. *Pericarditis* (per-ih-kar-DI-tis) refers to disease of the serous membrane on the heart surface, as well as that lining the pericardial sac.

These inflammatory diseases are often due to infection. They may also occur secondary to respiratory or other systemic diseases.

Another classification of heart disease is based on causative factors:

1. *Congenital* heart disease is present at birth.
2. *Rheumatic* heart disease begins with an attack of rheumatic fever in childhood or in youth.
3. *Coronary* (KOR-o-na-re) heart disease involves the walls of the blood vessels that supply the muscle of the heart.
4. *Heart failure* is due to deterioration of the heart tissues and is frequently the result of disorders of long duration such as high blood pressure.

Congenital Heart Disease

Congenital heart diseases are usually due to defects in the development of the fetus. The most common single defect is a hole in the septum between the two ventricles. Two other congenital disorders represent the abnormal persistence of structures that are part of the normal fetal circulation. The circulation of the fetus differs from that of the child after birth, due to the fact that the lungs are not used until the child is born. Before birth the non-functional lungs are bypassed by a blood vessel (ductus arteriosus) connecting the pulmonary artery and the aorta. This vessel normally closes of its own accord once the lungs are in use. Also present in the fetus is a small hole (foramen ovale) in the septum between the atria that allows some blood to flow directly from the right atrium into the left atrium, bypassing the lungs. If it causes no difficulty, the persistence of this hole is often not diagnosed until an adult is being examined for other cardiac problems.

In each of the three defects described above, part of the output of the left side of the heart goes back to the lungs instead of out to the body. This greatly increases the work of the left ventricle, and may lead to heart failure.

Other congenital defects involve restriction of blood flow out of the heart. *Coarctation* (ko-ark-TA-shun) *of the aorta* is a localized narrowing of the arch of the aorta. Another example is obstruction or narrowing of the pulmonary artery that prevents blood from passing in sufficient quantity from the right ventricle to the lungs.

In many cases, several heart defects occur together. The most common combination is that of four specific defects known as the *tetralogy of Fallot.* The so-called "blue baby" is often a victim of this disorder. The blueness, or *cyanosis* (si-ah-NO-sis), of the skin and mucous membranes is caused by a relative lack of oxygen. (See Chap. 18 for other causes.)

In recent years it has become possible to remedy many congenital defects by heart surgery, one of the more spectacular advances in modern medicine. A persistent ductus arteriosus may also respond to drug treatment. During fetal life, prostaglandins act to keep the ductus arteriosus open. Drugs that inhibit prostaglandins can promote closing of the duct after birth.

Rheumatic Fever and the Heart

Streptococcal infections are indirectly responsible for rheumatic fever and rheumatic heart disease. The toxin produced by the streptococci causes an immune reaction that may be followed some 2 to 4 weeks later by rheumatic fever with marked swelling of the joints. The antibodies formed to combat the toxin may then attack the heart valves, producing a condition known as *rheumatic endocarditis.* The heart valves, particularly the mitral valve, become inflamed, and the normally flexible valve cusps thicken and harden so they do not open sufficiently (mitral stenosis) or close effectively (mitral regurgitation). Either condition interferes with the flow of blood from the left atrium into the left ventricle, causing pulmonary congestion, an important characteristic of mitral heart disease. Although antibiotic treatment of the prior streptococcal infection is available, children may not receive adequate treatment and fall victim to rheumatic heart disease.

Coronary Heart Disease

The heart muscle receives its own blood supply through the coronary arteries. A common cause of sudden death, or at least disability, from heart disease is *coronary occlusion* (ok-LU-shun), that is, closure of one or more branches of the coronary arteries. The result is *ischemia* (is-KE-me-ah), a lack of blood supply to the areas fed by those arteries. Such interference with the blood supply results in damage to the myocardium. The degree of injury sustained depends on a number of factors, including the size of the involved artery and whether the occlusion is gradual or sudden.

Arteries of the heart and of the rest of the body can undergo degenerative changes. It may happen that the lumen (space) inside the vessel narrows gradually due to progressive thickening and hardening of the arteries, a process called *atherosclerosis* (ath-er-o-skleh-RO-sis) (Fig. 14-9). As a consequence, the volume of blood supplied to the heart muscle is reduced and the action of the heart weakened.

Degenerative changes of the artery wall may cause the inside surface to become roughened as well. This roughened artery wall is highly conductive to the formation of a ***thrombus*** (THROM-bus), or blood clot (Fig. 14-10). The thrombus may cause sudden closure of the vessel with complete obstruction of the blood flow, a life-threatening condition known as ***coronary thrombosis.*** The area that has been cut off from its blood supply is called an ***infarct*** (IN-farkt). The outcome of a myocardial infarction depends on the extent of the damage and whether other branches of the artery can supply enough

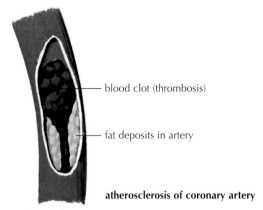

atherosclerosis of coronary artery

Figure 14-10

Development of thrombosis.

blood to maintain the heart's action. Death may occur swiftly if a large area of heart muscle is suddenly deprived of blood supply. If a smaller area is involved, the heart may continue to function. However, complete and prolonged lack of blood to any part of the myocardium results in death of tissue and weakening of the heart wall. In some cases the weakened area ruptures, but in most cases a scar forms in the area of the infarct.

Many effective techniques are now used for the treatment of heart attacks. Patients are admitted to coronary care units for monitoring and care. Chest pain is treated with IV (intravenous) morphine and drugs are used to maintain a steady heart rhythm. Rest and administration of oxygen reduce the workload of the heart. Special tubes (catheters) are inserted to monitor the ability of the heart to pump blood. Coronary occlusion may be treated with ***thrombolytic*** (throm-bo-LIT-ik) ***drugs,*** which act to dissolve the clots blocking coronary arteries. Therapy must be initiated promptly to be effective in preventing permanent damage to heart muscle.

Although myocardial infarction is one of the leading causes of death in the United States, prompt and effective treatment enables many individuals to survive a "heart attack" or "coronary." Despite heart damage, a person can usually lead a fairly normal life so long as he or she exercises moderately, gets sufficient rest, follows a prescribed diet, and tries to minimize stressful situations.

Inadequate blood flow to the heart muscle results in a characteristic discomfort, felt in the region of the heart and in the left arm and the shoulder, called ***angina pectoris*** (an-JI-nah PEK-to-ris). Severe

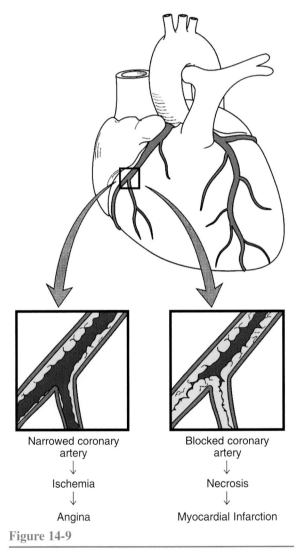

Figure 14-9

Coronary atherosclerosis. (Jones SA, Weigel A, White RD, McSwain Jr NE, & Breiter M: Advanced Emergency Care for Paramedic Practice, p. 419. Philadelphia. JB Lippincott, 1992)

What Are All Those DLs?

 Although cholesterol has received a lot of bad press in recent years, it is a necessary substance in the body. It is found in bile salts needed for digestion of fats, in hormones, and in the cell membrane. However, high levels of cholesterol in the blood have been associated with heart disease.

It now appears that the total amount of blood cholesterol is not as important as the form in which it occurs. Cholesterol is transported in the blood in combination with other lipids and with protein, forming compounds called *lipoproteins*. These compounds are distinguished by their relative density. High-density lipoprotein (HDL) is about one-half protein, whereas low-density lipoprotein (LDL) has a higher proportion of cholesterol and less protein. VLDLs, or very-low-density lipoproteins, are substances that are converted to LDLs.

LDLs carry cholesterol from the liver to the tissues. HDLs remove cholesterol from the tissues, such as the walls of the arteries, and carry it back to the liver for reuse or disposal. Thus, high levels of HDLs indicate efficient removal of arterial plaques, whereas high levels of LDLs suggest that arteries will become clogged.

angina pectoris may be accompanied by a feeling of suffocation and a general sensation of forthcoming doom. Coronary artery disease is a common cause of angina pectoris, although the condition has other causes as well.

ABNORMALITIES OF HEART RHYTHM

Myocardial infarction commonly results in an abnormality in the rhythm of the heartbeat, or *arrhythmia* (ah-RITH-me-ah). Extremely rapid but coordinated contractions, numbering up to 300 per minute, are described as *flutter*. Rapid, wild, and uncoordinated contractions of the heart muscle are called *fibrillation*. Fibrillation begins as small, local contractions of a few cells followed by multiple areas of cellular contraction within the muscle wall. It usually affects only the atria but may involve the ventricles. Ventricular fibrillation is a very serious cardiac disorder requiring first aid procedures, including cardiopulmonary resuscitation (CPR) and transfer by a paramedical unit to a hospital.

An interruption of electric impulses in the conduction system of the heart is called *heart block.* The seriousness of this condition depends on how completely the impulses are blocked. It may result in independent beating of the chambers if the ventricles respond to a second pacemaker.

Heart Failure

During a person's lifetime many toxins, infections, and other kinds of injuries may cause weakening of the heart muscle. Over a period of years high blood pressure, known as *hypertension* (hi-per-TEN-shun), may cause an enlargement of the heart and finally heart failure. Malnutrition, chronic infections, and severe anemias may cause degeneration of the heart muscle. Hyperthyroidism, with its tendency to cause overactivity of all parts of the body, including the heart, is another cause of heart failure. Although heredity may be responsible for undue susceptibility to degenerative heart disease, poor health habits and various diseases contribute to heart damage.

The Heart in the Elderly

As a result of accumulated damage, the heart may become less efficient with age. On the average, by the age of 70 there has been a decrease in cardiac output of about 35%. Because of a decrease in the reserve strength of the heart, elderly persons are often limited in their ability to respond to emergencies, infections, blood loss, or stress. A disorder known as *Adams–Stokes syndrome,* which is characterized by periodic loss of consciousness because of heart block, is most common in the elderly.

▶ Prevention of Heart Ailments

Although there may not be complete agreement on any set of rules for preventing or delaying the onset of heart disease, many authorities concur on the following:

1. Proper nutrition, including all the basic food elements, will help maintain all tissues in optimal condition. Amounts of fats, especially animal fats, and sugar in the diet should be limited. Foods high in fiber, such as whole grains, fruits, and vegetables, should be included in the diet. Total food intake should be reduced if necessary to prevent obesity.
2. Infections should be avoided as much as possible. Mild infections should be treated to prevent se-

rious complications from developing. Dental care should include the treatment of abscesses and the cleaning and filling of decayed teeth.

3. Temperate habits and adequate rest are desirable. Persons who wish to avoid heart disease need to avoid the use of tobacco, as well as excesses of alcohol and food. On the basis of numerous studies, it appears that smokers are 10 times more likely to die of coronary heart disease than are nonsmokers. Emotional upsets and psychologic upheavals may not be conducive to maintaining a healthy heart. Playing too hard is just as damaging as working too hard.

4. Regular physical examinations are important, particularly for elderly persons and persons who have had symptoms that suggest the presence of disease.

5. About 20% of adults in the United States suffer from hypertension. One of the serious problems of hypertension is that it usually produces no symptoms until damage to vital structures has already occurred. Blood pressure should be checked regularly, and treatment should be given if the readings are abnormal. If modification of diet, weight control, and exercise programs are unsuccessful, treatment with medication may be needed.

6. Appropriate exercise programs are an important part of prevention and treatment of heart disease. Regular exercise programs, with gradual increases in the length and difficulty of activity, have been found to be very helpful. To begin with, simple walking may be most effective, followed by such activities as swimming, bicycling, and jogging. Even a minimal amount of regular exercise, such as walking 30 minutes 3 times a week, has been shown to be of major benefit.

▶ Instruments Used in Heart Studies

The *stethoscope* (STETH-o-skope) is a relatively simple instrument used for conveying sounds from within the patient's body to the ear of the examiner. Experienced listeners can gain much information using this device.

The *electrocardiograph* (*EKG* or *ECG*) is used for making records of the changes in electric currents produced by the contracting heart muscle. It may thus reveal certain myocardial injuries. Electric activity is picked up by electrodes placed on the sur-

face of the skin and appears as *waves* on the EKG tracing. The P wave represents the activity of the atria; the QRS and T waves represent the activity of the ventricles (Fig. 14-11). Changes in the waves and the intervals between them are used to diagnose heart damage and arrhythmias.

The *fluoroscope* (flu-OR-o-scope), an instrument for examining deep structures with x-rays, may be used to reveal heart action and to establish the sizes and relations of some of the thoracic organs. It may be used in conjunction with *catheterization* (kath-eh-ter-i-ZA-shun) of the heart. In right heart catheterization, an extremely thin tube (catheter) is passed through the veins of the right arm or right groin and then into the right side of heart. During the procedure, the fluoroscope is used for observation of the route taken by the catheter. The tube is passed all the way through the pulmonic valve into the large lung arteries. Samples of blood are obtained along the way and removed for testing, pressure readings being taken meanwhile.

In left heart catheterization, a catheter is passed through an artery in the left groin or arm to the heart. Dye can then be injected into the coronary arteries. The tube may also be passed through the aortic valve into the left ventricle for further studies.

Ultrasound consists of sound waves generated at a frequency above the range of sensitivity of the human ear. In *echocardiography* (ek-o-kar-de-OG-rah-fe), also known as *ultrasound cardiography,* high-frequency sound waves are sent to the heart from a small instrument on the surface of the chest. The ultrasound waves bounce off the heart and are recorded as they return, showing the heart in action. Movement of the echoes is traced on an electronic instrument called an *oscilloscope* and recorded on film. (The same principle is employed by submarines to detect ships.) The method is safe and painless, and it does not use x-rays. It provides information on the sizes and shapes of heart structures, on cardiac function, and on possible heart defects.

▶ Treatment of Heart Disease

Medications

One of the oldest and still the most important drug for many heart patients is *digitalis* (dij-ih-TAL-is). This agent, which slows and strengthens contractions of the heart muscle, is obtained from the leaf of the foxglove, a plant originally found growing wild in many parts of Europe. Foxglove is now

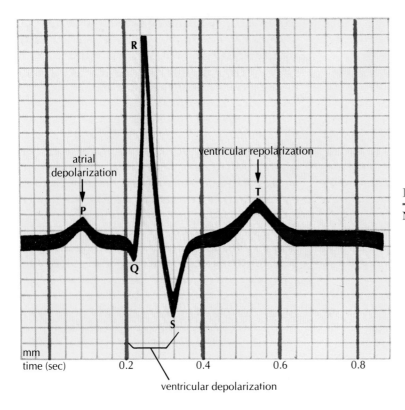

atrial
depolarization

ventricular repolarization

P

R

Q

S

T

ventricular depolarization

mm

time (sec)   0.2   0.4   0.6   0.8

Figure 14-11

Normal EKG showing one cardiac cycle.

cultivated to ensure a steady supply of digitalis for medical purposes.

Several forms of *nitroglycerin* are used to relieve angina pectoris. This drug dilates (widens) the vessels in the coronary circulation and improves the blood supply to the heart.

Beta-adrenergic blocking agents ("beta blockers") control sympathetic stimulation of the heart. They reduce the rate and strength of heart contractions, thus reducing the heart's oxygen demand. Propanolol is one example.

Antiarrhythmic agents (*e.g.,* quinidine) are used to regulate the rate and rhythm of the heartbeat.

Slow channel calcium blockers aid in the treatment of coronary heart disease and hypertension by several mechanisms. They may dilate vessels, control the force of heart contractions, or regulate conduction through the atrioventricular node. Their actions are based on the fact that calcium ions must enter muscle cells before contraction can occur.

Anticoagulants (an-ti-ko-AG-u-lants) are valuable drugs for heart patients. They may be used to prevent clot formation in persons with damage to heart valves or blood vessels or in persons who have had a myocardial infarction.

Artificial Pacemakers

Electric, battery-operated pacemakers that supply impulses to regulate the heartbeat have been implanted under the skin of many thousands of individuals. The site of implantation is usually in the left chest area. Electrode catheters attached to the pacemaker are then passed into the heart and anchored to the chest wall. The frequency of battery pack replacement varies with the type of pacemaker. Many people whose hearts cannot beat effectively alone have been saved by this rather simple device. In an emergency, a similar stimulus can be supplied to the heart muscle through electrodes placed externally on the chest wall.

Heart Surgery

The heart–lung machine has made possible many operations on the heart and other thoracic organs that otherwise could not have been done. There are several types of machines in use, all of which serve as temporary substitutes for the patient's heart and lungs.

The machine siphons off the blood from the large vessels entering the heart on the right side so that no blood passes through the heart and lungs. While passing through the machine, the blood is oxygenated by means of an oxygen inlet, and carbon dioxide is removed by various chemical means. These are the processes that normally take place between the blood and the air in the lung tissue. While in the machine, the blood is also "defoamed," or rid of air bubbles, since such bubbles could be fatal to the patient by obstructing blood vessels. An electric motor in the machine serves as a pump during the surgical procedure to return the processed blood to the general circulation by way of a large artery.

Coronary bypass surgery to relieve obstruction in the coronary arteries is a common and often successful treatment. While the damaged coronary arteries remain in place, healthy segments of blood vessels from other parts of the patient's body are inserted in such a way that blood bypasses whatever obstruction exists in the coronary arteries. Usually parts of the saphenous vein (a superficial vein in the leg) are used. Sometimes as many as six or seven segments are required to provide an adequate blood supply if the coronary vessels are seriously damaged. The mortality associated with this operation is low. Most patients are able to return to a nearly normal lifestyle following recovery from the surgery. The effectiveness of this procedure diminishes over a period of years, however, due to the blockage of the replacement vessels.

Less invasive surgical procedures include the technique of *angioplasty* (AN-je-o-plas-te), which is used to open restricted arteries in the heart and other areas of the body. A catheter with a balloon is guided by fluoroscope to the affected area. There it is inflated to break up the blockage in the coronary artery, thus restoring effective circulation to the heart muscle.

Diseased valves may become so deformed and scarred from endocarditis that they are ineffective and often obstructive. In most cases there is so much damage that *valve replacement* is the best treatment. Substitute valves made of plastic materials have proved to be a lifesaving measure for many patients. Very thin butterfly valves made of Dacron or other synthetic material have been used successfully, as have animal valves, such as pig valves.

The news media have given considerable attention to the *surgical transplantation* of human hearts and sometimes of lungs and hearts together. This surgery is done in specialized centers and is available to some patients with degenerative heart disease who are otherwise in good health. Tissues of the recently deceased donor and of the recipient must be as closely matched as possible to avoid rejection. Efforts to replace a damaged heart with a completely artificial heart have not been successful so far. There are devices available, however, to assist the pumping of a damaged heart during recovery from heart attack or while awaiting a donor heart.

SUMMARY

I. Structure of the heart
  A. Layers
    1. Endocardium—thin inner layer
    2. Myocardium—thick muscle layer
      a. Lightly striated, intercalated disks, branching of fibers
    3. Epicardium—thin outer layer
  B. Pericardium—membrane-lined sac that encloses the heart
  C. Chambers
    1. Atria—left and right receiving chambers
    2. Ventricles—left and right pumping chambers
    3. Septa—partitions between chambers
  D. Valves—prevent backflow of blood
    1. Tricuspid—right atrioventricular valve

  2. Mitral (bicuspid)—left atrioventricular valve
  3. Pulmonic (semilunar) valve—at entrance to pulmonary artery
  4. Aortic (semilunar) valve—at entrance to aorta
  E. Blood supply to myocardium
    1. Coronary arteries—first branches of aorta; fill when heart relaxes
    2. Coronary sinus—collects venous blood from heart and empties into right atrium

II. Function of the heart
  A. Cardiac cycle
    1. Diastole—relaxation phase
    2. Systole—contraction phase

B. Cardiac output—volume pumped by each ventricle per minute
1. Stroke volume—amount pumped with each beat
2. Heart rate—number of beats per minute

C. Conduction system
1. Sinoatrial node (pacemaker)—at top of right atrium
2. Atrioventricular node—between atria and ventricles
3. Atrioventricular bundle (bundle of His)—at top of interventricular septum
   a. Bundle branches—right and left, on either side of septum
   b. Purkinje fibers—branch through myocardium of ventricles

D. Control of heart rate
1. Autonomic nervous system
   a. Sympathetic system—speeds heart rate
   b. Parsympathetic system—slows heart rate via vagus nerve
2. Others—hormones, ions, drugs

E. Heart rates
1. Bradycardia—slower rate than normal; less than 60 beats/minute
2. Tachycardia—faster rate than normal; more than 100 beats/minute
3. Extrasystole—premature beat

F. Heart sounds
1. Normal
   a. "Lubb"—occurs at closing of atrioventricular valves
   b. "Dupp"—occurs at closing of semilunar valves
2. Abnormal—murmur

III. Heart disease

A. Anatomic classification
1. Endocarditis
2. Myocarditis
3. Pericarditis

B. Causal classification
1. Congenital heart diseases—present at birth
   a. Holes in septum
   b. Failure of fetal lung bypasses to close
   c. Coarctation of aorta
   d. Pulmonary artery obstruction
2. Rheumatic heart disease

   a. Mitral stenosis—valve cusps do not open
   b. Mitral regurgitation—valve cusps do not close
3. Coronary heart disease
   a. Coronary occlusion—closure of coronary arteries, as by a thrombus
   b. Ischemia—lack of blood to area fed by occluded arteries
   c. Infarct—area of damaged tissue
   d. Arrhythmia—abnormal rhythm
      (1) Flutter—rapid, coordinated beats
      (2) Fibrillation—rapid, uncoordinated contractions of heart muscle
      (3) Heart block—interruption of electric conduction
   e. Angina pectoris—pain caused by lack of blood to heart muscle
4. Heart failure—due to hypertension, disease, malnutrition, anemia, age

IV. Prevention of heart ailments—proper diet, avoidance of infections, temperate habits, physical examinations, control of hypertension, exercise

V. Instruments used in heart studies

A. Stethoscope—used to listen to heart sounds
B. Electrocardiograph (EKG, ECG)—records electric activity as waves
C. Fluoroscope—examines deep tissue with x-rays
D. Catheter—thin tube inserted into heart for blood samples, pressure readings, etc.
E. Echocardiograph—uses ultrasound to record picture of heart in action

VI. Treatment of heart disease

A. Medications—digitalis, nitroglycerin, beta-adrenergic blocking agents, antiarrhythmic agents, slow channel calcium blockers, anticoagulants
B. Artificial pacemakers—electronic devices implanted under skin to regulate heartbeat
C. Surgery
1. Bypass—other vessels inserted to augment blood supply
2. Angioplasty
3. Valve replacement
4. Heart transplants
5. Artificial hearts—experimental

QUESTIONS FOR STUDY AND REVIEW

1. What are the three layers of the heart wall?
2. Describe the characteristics of heart muscle tissue.
3. Name the sac around the heart.
4. What is a partition in the heart called? Name two.
5. Name the chambers of the heart and tell what each does.
6. Name the valves of the heart and explain the purpose of each valve.
7. Why does the myocardium need its own blood supply? Name the arteries that supply blood to the heart.
8. Explain the contraction pattern of the cardiac cycle.
9. How does the heart's ability to contract differ from that of other muscles? What is required to maintain an effective rate of heartbeat?
10. Define *cardiac output*. What determines cardiac output?
11. What are the parts of the heart's conduction system called and where are these structures located? Outline the order in which the excitation waves travel.
12. Compare the effects of the sympathetic and parasympathetic nervous systems on the working of the heart.
13. What two syllables are used to indicate normal heart sounds, and at what time in the heart cycle can they be heard?
14. Inflammation of the heart structures is the basis for a classification of heart pathology. List the three terms in this classification and explain each.
15. What is meant by *congenital heart disease*? Give an example.
16. What part does infection play in rheumatic heart disease?
17. What type of heart disease is the most frequent cause of sudden death?
18. What is the effect of a thrombus in an artery supplying the heart wall? What are some of the precautions that need to be taken by a person recovering from thrombosis?
19. What are some of the factors that contribute to heart failure?
20. What are some rules that may delay or avoid the onset of heart ailments?
21. What is a fluoroscope and what is its purpose in heart studies?
22. Of what value is heart catheterization and how is it carried out?
23. What is an electrocardiograph and what is its purpose?
24. How does echocardiography differ from electrocardiography?
25. How does digitalis help a person who has heart muscle damage?
26. Why are artificial pacemakers used?
27. What is the purpose of coronary bypass surgery?
28. What is coronary angioplasty and how does it benefit the heart?
29. What are artificial heart valves made of and how valuable are they?
30. Differentiate between the terms in each of the following pairs:
    a. *Pulmonary* and *systemic circuit*
    b. *Interatrial* and *interventricular*
    c. *Systole* and *diastole*
    d. *Stroke volume* and *heart rate*
    e. *Tachycardia* and *bradycardia*
    f. *Functional murmur* and *organic murmur*
    g. *Flutter* and *fibrillation*

CHAPTER 15

Blood Vessels and Blood Circulation

Selected Key Terms

The following terms are defined in the Glossary:

anastomosis

aorta

arteriole

artery

atherosclerosis

capillary

embolus

endothelium

hemorrhage

hypertension

hypotension

ischemia

phlebitis

pulse

shock

sinusoid

sphygmomanometer

vasoconstriction

vasodilation

vein

venule

Behavioral Objectives

After careful study of this chapter, you should be able to:

- Differentiate among the three main types of vessels in the body with regard to structure and function
- Compare the locations and functions of the pulmonary and systemic circuits
- Name the four sections of the aorta
- Name the main branches of the aorta
- Name the main vessels that drain into the superior and inferior venae cavae
- Define *venous sinus* and give four examples
- Describe the structure and function of the hepatic portal system
- Explain how materials are exchanged across the capillary wall
- Describe the factors that regulate blood flow
- Define *pulse* and list factors that affect the pulse rate
- List several factors that affect blood pressure
- Explain how blood pressure is commonly measured
- List some disorders that involve the blood vessels
- List four types of shock

Memmler, RL., Cohen, BJ, Wood, DL. *THE HUMAN BODY IN HEALTH AND DISEASE*, 8/e,
© 1996 Lippincott-Raven Publishers

The vascular system will be easier to understand if you refer to the appropriate illustrations in this chapter as the vessels are described. If this information is added to what you already know about the blood and the heart, a picture of the circulatory system as a whole will emerge.

❯ Blood Vessels

Functional Classification

The blood vessels, together with the four chambers of the heart, form a closed system for the flow of blood; only if there is an injury to some part of the wall of this system does any blood escape. On the basis of function, blood vessels may be divided into three groups:

1. *Arteries* carry blood from the ventricles (pumping chambers) of the heart out to the capillaries in the tissues. The smallest arteries are called *arterioles* (ar-TE-re-olz).
2. *Veins* drain capillaries in the tissues and return the blood to the heart. The smallest veins are the *venules* (VEN-ulz).
3. *Capillaries* allow for exchanges between the blood and body cells, or between the blood and air in the lung tissues. The capillaries connect the arterioles and venules.

All the vessels together may be subdivided into two groups or circuits: pulmonary and systemic. The vessels in these two circuits are shown in Figure 15-1; the anatomic relation of the circuits to the heart is shown in Figure 14-3.

1. The *pulmonary circuit* eliminates carbon dioxide from the blood and replenishes its supply of oxygen. The pulmonary vessels that carry blood to and from the lungs include:
   a. the pulmonary artery and its branches that carry blood from the right ventricle to the lungs
   b. the capillaries in the lungs through which gases are exchanged
   c. the pulmonary veins that carry blood back to the left atrium

The pulmonary vessels differ from those in the systemic circuit in that the pulmonary arteries carry blood that is *low* in oxygen and the pulmonary veins carry blood that is *high* in oxygen.

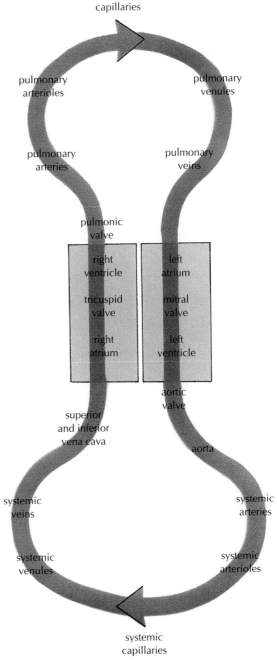

Figure 15-1

Blood vessels form a closed system for the flow of blood. Blood high in oxygen (oxygenated) is shown in red; blood low in oxygen (deoxygenated) is shown in blue. Changes in oxygen content occur as blood flows through capillaries.

2. The *systemic* (sis-TEM-ik) *circuit* serves the rest of the body. These vessels supply nutrients and oxygen to all the tissues and carry waste materials away from the tissues for disposal.

Structure
ARTERY WALLS

The arteries have thick walls because they must be strong enough to receive blood pumped under pressure from the ventricles of the heart (Fig. 15-2). The three coats (tunics) of the arteries resemble the three tissue layers of the heart:

1. The innermost membrane of simple, flat epithelial cells makes up the *endothelium* (en-do-THE-le-um), forming a smooth surface over which the blood may easily move.
2. The second and thickest layer is made of smooth (involuntary) muscle combined with elastic connective tissue. This tissue allows the arteries to stretch when receiving blood and then return to their original size.
3. An outer tunic is made of a supporting connective tissue.

Blood Vessels in the Fetus

 The developing fetus has several adaptations in the cardiovascular system that change at birth. These adaptations serve to bypass the lungs, which are not functional because fetal blood is oxygenated by the placenta. Blood comes from the placenta to the fetus in the umbilical vein, which is contained within the umbilical cord. Most of this blood joins the inferior vena cava and is carried to the heart. A portion of the blood entering the right atrium flows directly into the left atrium through a hole in the septum called the *foramen ovale*, thus bypassing the pulmonary circuit. Blood that does enter the right ventricle gets pumped into the pulmonary artery. However, most of this blood shunts directly into the systemic circulation through a small vessel, the *ductus arteriosus*, that connects the pulmonary artery to the aorta. Blood goes back to the placenta through two umbilical arteries. All these adaptations close at birth when the lungs begin to function.

The largest artery, the *aorta* (a-OR-tah), is about 2.5 cm (1 inch) in diameter and has the thickest wall because it receives blood under the highest pressure from the left ventricle. The smallest subdivisions of arteries, the arterioles, have thinner walls in which there is very little elastic connective tissue but relatively more smooth muscle.

CAPILLARY WALLS

The microscopic branches of these tiny connecting vessels have the thinnest walls of any vessels: one cell layer. The capillary walls are transparent and are made of smooth, platelike cells that continue from the lining of the arteries. Because of the thinness of these walls, exchanges between the blood and the body cells are possible. The capillary boundaries are the most important center of activity for the entire circulatory system. Their function is explained later in this chapter.

VEIN WALLS

The smallest veins, the *venules,* are formed by the union of capillaries, and their walls are only slightly thicker than those of the capillaries. As veins become larger, their walls become thicker. However, the wall of a vein is much thinner than the wall of an artery of comparable size and has less elastic tissue. As a result, the blood within the veins is carried under much lower pressure. Although there are three layers of tissue in the walls of the larger veins, as in the artery walls, the middle tunic is relatively thin in the veins. Therefore, veins are easily collapsed, and only slight pressure on a vein by a tumor or other mass may interfere with return blood flow. Most veins are equipped with one-way valves that permit blood to flow in only one direction: toward the heart (see Fig. 15-2). Such valves are most numerous in the veins of the extremities.

Figure 15-3 is a cross section of an artery and a vein as seen through a microscope.

▶ Names of Systemic Arteries

The Aorta and Its Parts
The aorta extends upward and to the right from the left ventricle. Then it curves backward and to the left. It continues down behind the heart just in front of the vertebral column, through the diaphragm,

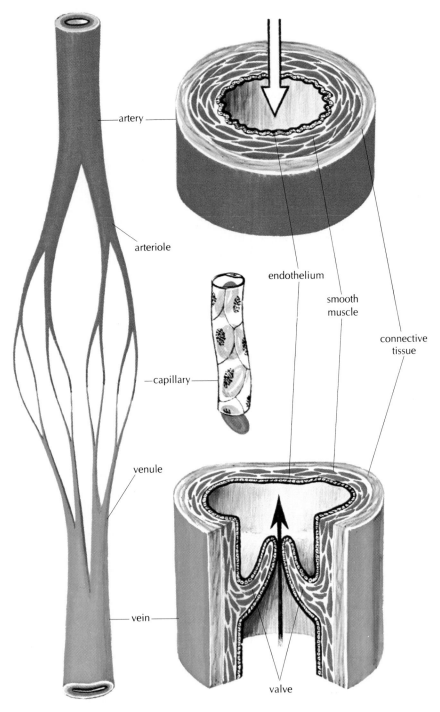

artery

arteriole

endothelium

smooth
muscle

connective
tissue

capillary

venule

vein

valve

Figure 15-2

Sections of small blood vessels
showing the thick arterial wall,
the thin wall of a vein, and the
single-layered wall of a capillary. A venous valve also is
shown. The arrows indicate the
direction of blood flow.

and into the abdomen (Figs. 15-4 and 15-5). The
aorta is one continuous artery, but it may be divided
into sections:

1. The *ascending aorta* is near the heart and inside
   the pericardial sac.

2. The *aortic arch* curves from the right to the left
   and also extends backward.

3. The *thoracic aorta* lies just in front of the vertebral
   column behind the heart and in the space behind
   the pleura.

smooth muscle

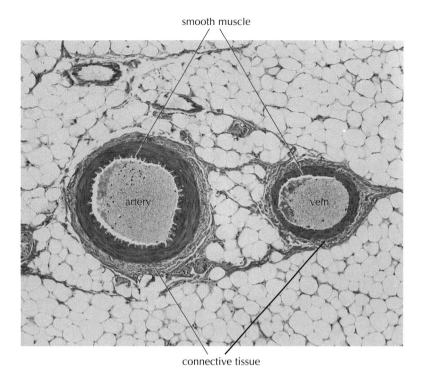

artery

vein

connective tissue

Figure 15-3

Cross section of an artery and a vein as seen through a microscope. (Cormack DH: Essential Histology, plate 11–1. Philadelphia, JB Lippincott, 1993)

4. The *abdominal aorta* is the longest section of the aorta, spanning the abdominal cavity.

The thoracic and abdominal aorta together make up the descending aorta.

Branches of the Ascending Aorta

The first, or ascending, part of the aorta has two branches near the heart, called the *left* and *right coronary arteries,* that supply the heart muscle. These form a crown around the base of the heart and give off branches to all parts of the myocardium.

Branches of the Aortic Arch

The arch of the aorta, located immediately beyond the ascending aorta, gives off three large branches.

1. The *brachiocephalic* (brak-e-o-seh-FAL-ik) *trunk* is a short artery formerly called the *innominate.* Its name means that it supplies the head and the arm. After extending upward somewhat less than 5 cm (2 inches), it divides into the *right subclavian* (sub-KLA-ve-an) *artery,* which supplies the right upper extremity (arm), and the *right common carotid* (kah-ROT-id) *artery,* which supplies the right side of the head and the neck.

2. The *left common carotid artery* extends upward from the highest part of the aortic arch. It supplies the left side of the neck and the head.
3. The *left subclavian artery* extends under the left collar bone (clavicle) and supplies the left upper extremity. This is the last branch of the aortic arch.

Branches of the Thoracic Aorta

The thoracic aorta supplies branches to the chest wall, to the *esophagus* (e-SOF-ah-gus), and to the bronchi (the subdivisions of the trachea) and their treelike subdivisions in the lungs. There are usually 9 to 10 pairs of *intercostal* (in-ter-KOS-tal) *arteries* that extend between the ribs, sending branches to the muscles and other structures of the chest wall.

Branches of the Abdominal Aorta

As in the case of the thoracic aorta, there are unpaired branches extending forward and paired arteries extending toward the side. The unpaired vessels are large arteries that supply the abdominal viscera. The most important of these visceral branches are listed below:

1. The *celiac* (SE-le-ak) *trunk* is a short artery about 1.25 cm (1/2 inch) long that subdivides into

(Text continues on page 244)

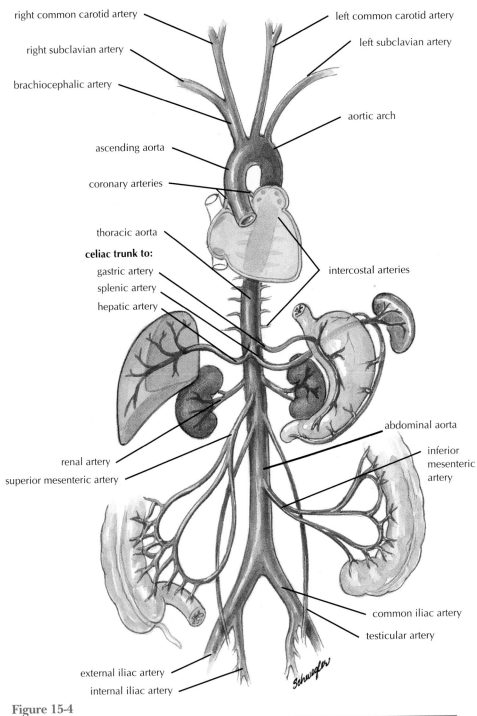

right common carotid artery

right subclavian artery

brachiocephalic artery

ascending aorta

coronary arteries

thoracic aorta

celiac trunk to:

gastric artery

splenic artery

hepatic artery

renal artery

superior mesenteric artery

left common carotid artery

left subclavian artery

aortic arch

intercostal arteries

abdominal aorta

inferior mesenteric artery

common iliac artery

testicular artery

external iliac artery

internal iliac artery

Figure 15-4

Aorta and its branches.

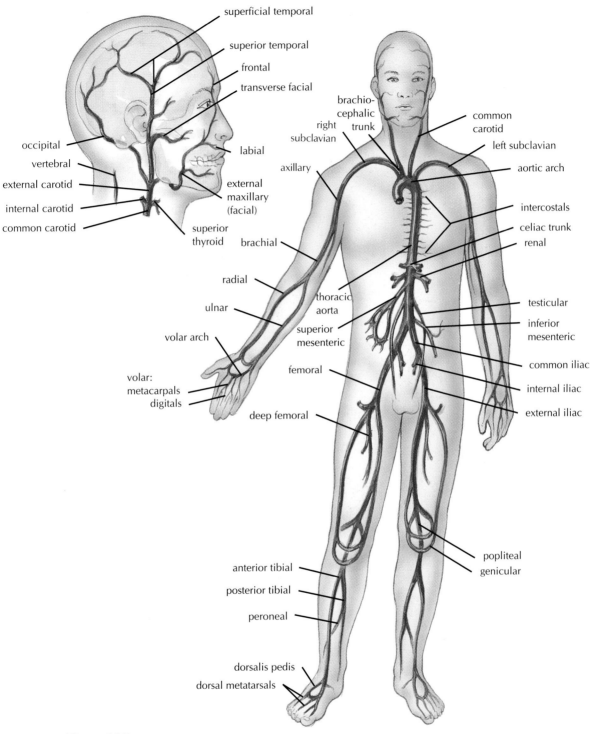

Figure 15-5

Principal systemic arteries.

three branches: the *left gastric artery* goes to the stomach, the *splenic* (SPLEN-ik) *artery* goes to the spleen, and the *hepatic* (heh-PAT-ik) *artery* carries oxygenated blood to the liver.

2. The *superior mesenteric* (mes-en-TER-ik) *artery,* the largest of these branches, carries blood to most of the small intestine and to the first half of the large intestine.

3. The much smaller *inferior mesenteric artery,* located below the superior mesenteric and near the end of the abdominal aorta, supplies the second half of the large intestine.

The paired lateral branches of the abdominal aorta include the following right and left vessels:

1. The *phrenic* (FREN-ik) *arteries* supply the diaphragm.
2. The *suprarenal* (su-prah-RE-nal) *arteries* supply the adrenal (suprarenal) glands.
3. The *renal* (RE-nal) *arteries,* the largest in this group, carry blood to the kidneys.
4. The *ovarian arteries* in the female and *testicular* (tes-TIK-u-lar) *arteries* in the male (formerly called the *spermatic arteries)*, supply the sex glands.
5. Four pairs of *lumbar* (LUM-bar) *arteries* extend into the musculature of the abdominal wall.

Iliac Arteries and Their Subdivisions

The abdominal aorta finally divides into two *common iliac* (IL-e-ak) *arteries.* Both of these vessels, about 5 cm (2 inches) long, extend into the pelvis, where each one subdivides into an *internal* and an *external iliac artery.* The internal iliac vessels then send branches to the pelvic organs, including the urinary bladder, the rectum, and some of the reproductive organs. Each external iliac artery continues into the thigh as the *femoral* (FEM-or-al) *artery.* This vessel gives off branches in the thigh and then becomes the *popliteal* (pop-LIT-e-al) *artery,* which subdivides below the knee. The subdivisions include the *tibial artery* and the *dorsalis pedis* (dor-SA-lis PE-dis), which supply the leg and the foot.

Other Subdivisions of Systemic Arteries

Just as the larger branches of a tree give off limbs of varying sizes, so the arterial tree has a multitude of subdivisions. Hundreds of names might be included, but we shall mention only a few. For example, each common carotid artery gives off branches to the thyroid gland and other structures in the neck

before dividing into the *external* and *internal carotid arteries,* which supply parts of the head. The hand receives blood from the subclavian artery, which becomes the *axillary* (AK-sil-ar-e) in the axilla (armpit). The longest part of this vessel, the *brachial* (BRA-ke-al) *artery,* is in the arm proper. It subdivides into two branches near the elbow: the *radial artery,* which continues down the thumb side of the forearm and wrist, and the *ulnar artery,* which extends along the medial or little finger side into the hand.

Anastomoses

A communication between two vessels is called an *anastomosis* (ah-nas-to-MO-sis). By means of arterial anastomoses, blood reaches vital organs by more than one route. Some examples of such unions of end arteries are described below:

1. The *circle of Willis* (Fig. 15-6) receives blood from the two internal carotid arteries and from the *basilar* (BAS-il-ar) *artery,* which is formed by the union of two vertebral arteries. This arterial circle lies just under the center of the brain and sends branches to the cerebrum and other parts of the brain.
2. The *volar* (VO-lar) *arch* is formed by the union of the radial and ulnar arteries in the hand. It sends branches to the hand and the fingers.
3. The *mesenteric arches* are made of communications between branches of the vessels that supply blood to the intestinal tract.
4. *Arterial arches* are formed by the union of branches of the tibial arteries in the foot. Similar anastomoses are found in other parts of the body.

Arteriovenous anastomoses are blood shunts found in a few areas, including the external ears, the hands, and the feet. Vessels with muscular walls connect arteries directly with veins and thus bypass the capillaries (Fig. 15-7). This provides a more rapid flow and a greater volume of blood to these areas than elsewhere, thus protecting these exposed parts from freezing in cold weather.

▌ Names of Systemic Veins

Superficial Veins

Whereas most arteries are located in protected and rather deep areas of the body, many veins are found near the surface (Fig. 15-8). The most important of these superficial veins are in the extremities. These include the following:

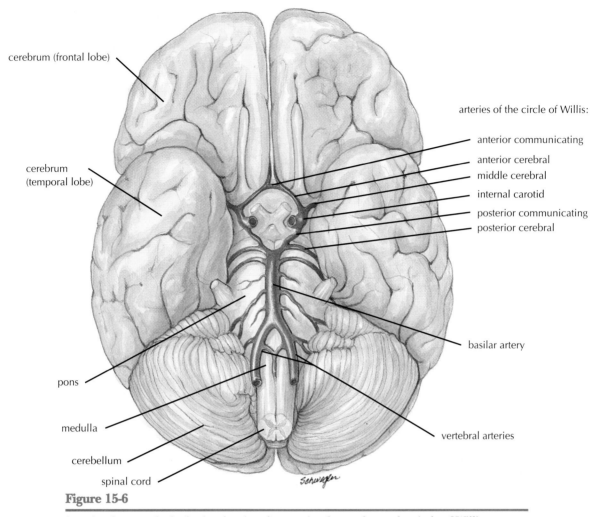

cerebrum (frontal lobe)

arteries of the circle of Willis:

anterior communicating

anterior cerebral

middle cerebral

cerebrum
(temporal lobe)

internal carotid

posterior communicating

posterior cerebral

basilar artery

pons

medulla

vertebral arteries

cerebellum

spinal cord

Figure 15-6

Arteries that supply the brain, showing the arteries that make up the circle of Willis.

1. The veins on the back of the hand and at the front of the elbow. Those at the elbow are often used for removing blood samples for test purposes, as well as for intravenous injections. The largest of this group of veins are the ***cephalic*** (seh-FAL-ik), the ***basilic*** (bah-SIL-ik), and the ***median cubital*** (KU-bih-tal) ***veins.***
2. The ***saphenous*** (sah-FE-nus) ***veins*** of the lower extremities, which are the longest veins of the body. The great saphenous vein begins in the foot and extends up the medial side of the leg, the knee, and the thigh. It finally empties into the femoral vein near the groin.

Deep Veins

The deep veins tend to parallel arteries and usually have the same names as the corresponding ar-

teries. Examples of these include the ***femoral*** and the ***iliac*** vessels of the lower part of the body and the ***brachial, axillary,*** and ***subclavian*** vessels of the upper extremities. However, exceptions are found in the veins of the head and the neck. The ***jugular*** (JUG-u-lar) ***veins*** drain the areas supplied by the carotid arteries. Two ***brachiocephalic*** (innominate) ***veins*** are formed, one on each side, by the union of the subclavian and the jugular veins. (Remember, there is only *one* brachiocephalic artery.)

Superior Vena Cava

The veins of the head, neck, upper extremities, and chest all drain into the ***superior vena cava*** (VE-nah KA-vah), which goes to the heart. It is formed by the union of the right and left brachiocephalic veins, which drain the head, neck, and upper ex-

Figure 15-7

Capillary network showing an arteriovenous shunt (anastomosis).

tremities. The **azygos** (AZ-ih-gos) **vein** drains the veins of the chest wall and empties into the superior vena cava just before the latter empties into the heart (see Fig. 15-8).

Inferior Vena Cava

The **inferior vena cava,** which is much longer than the superior vena cava, returns the blood from the parts of the body below the diaphragm. It begins in the lower abdomen with the union of the two common iliac veins. It then ascends along the back wall of the abdomen, through a groove in the posterior part of the liver, through the diaphragm, and finally through the lower thorax to empty into the right atrium of the heart.

Drainage into the inferior vena cava is more complicated than drainage into the superior vena cava. The large veins below the diaphragm may be divided into two groups:

1. The right and left veins that drain paired parts and organs. They include the *iliac veins* from near the groin, four pairs of *lumbar veins* from the dorsal part of the trunk and from the spinal cord, the *testicular veins* from the testes of the male and the *ovarian veins* from the ovaries of the female, the *renal* and *suprarenal veins* from the kidneys

and adrenal glands near the kidneys, and finally the large *hepatic veins* from the liver. For the most part, these vessels empty directly into the inferior vena cava. The left testicular in the male and the left ovarian in the female empty into the left renal vein, which then takes this blood to the inferior vena cava; these veins thus constitute exceptions to the rule that the paired veins empty directly into the vena cava.

2. Unpaired veins that come from the spleen and from parts of the digestive tract (stomach and intestine) empty into a vein called the *hepatic portal vein.* Unlike other lower veins, which empty into the inferior vena cava, the hepatic portal vein is part of a special system that enables blood to circulate through the liver before returning to the heart.

Venous Sinuses

The word *sinus* means "space" or "hollow." A **venous sinus** is a large channel that drains deoxygenated blood but does not have the usual tubular structure of the veins. An important example of a venous sinus is the **coronary sinus,** which receives most of the blood from the veins of the heart wall (see Fig. 14-5). It lies between the left atrium and left

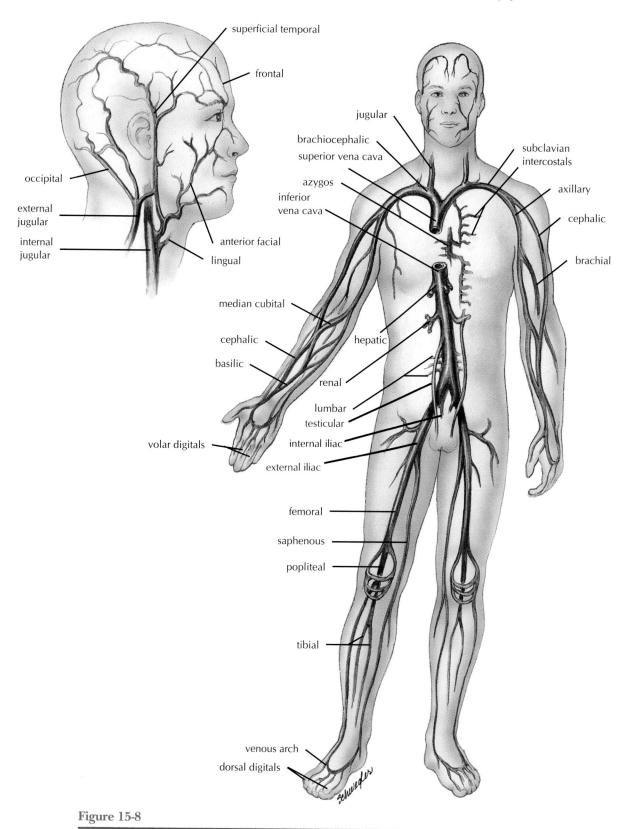

superficial temporal

frontal

jugular

brachiocephalic

superior vena cava

azygos

inferior vena cava

occipital

external jugular

internal jugular

anterior facial

lingual

subclavian

intercostals

axillary

cephalic

brachial

median cubital

cephalic

basilic

hepatic

renal

lumbar

testicular

internal iliac

external iliac

volar digitals

femoral

saphenous

popliteal

tibial

venous arch

dorsal digitals

Figure 15-8

Principal systemic veins.

ventricle on the posterior surface of the heart, and it empties directly into the right atrium along with the two venae cavae.

Other important venous sinuses are the *cranial venous sinuses,* which are located inside the skull and drain the veins that come from all over the brain (Fig. 15-9). The largest of the cranial venous sinuses are described below.

1. The two *cavernous sinuses,* situated behind the eyeballs, serve to drain the *ophthalmic* (of-THAL-mik) *veins* of the eyes.
2. The *superior sagittal* (SAJ-ih-tal) *sinus* is a single long space located in the midline above the brain and in the fissure between the two hemispheres of the cerebrum. It ends in an enlargement called the *confluence* (KON-flu-ens) *of sinuses.*
3. The two *transverse sinuses,* also called the *lateral sinuses,* are large spaces between the layers of the dura mater (the outermost membrane around the brain). They begin posteriorly, in the region of the confluence of sinuses, and then extend toward either side. As each sinus extends around the inside of the skull, it receives blood draining those parts not already drained by the superior sagittal and other sinuses that join the back portions of the transverse sinuses. This means that nearly all the blood that comes from the veins of the brain eventually empties into one or the other of the transverse sinuses. On either side the sinus extends far enough forward to empty into an internal jugular vein, which then passes through a hole in the skull to continue downward in the neck.

Hepatic Portal System

Almost always, when blood leaves a capillary bed it flows directly back to the heart. In a portal system, however, blood circulates through a second capillary bed, usually in a second organ, before it returns to the heart. A portal system is a kind of detour in the pathway of venous return that can transport materials directly from one organ to another. The largest portal system in the body is the *hepatic portal system,* which carries blood from the abdominal organs to the liver (Fig. 15-10). In a similar fashion, a small, local portal system is located in the brain to carry substances from the hypothalamus to the pituitary (see Chap. 12). The hepatic portal system includes the veins that drain blood from capillaries in the spleen, stomach, pancreas, and intestine. Instead of

emptying their blood directly into the inferior vena cava, they deliver it by way of the hepatic portal vein to the liver. The largest tributary of the portal vein is the *superior mesenteric vein.* It is joined by the *splenic vein* just under the liver. Other tributaries of the portal circulation are the *gastric, pancreatic,* and *inferior mesenteric veins.*

On entering the liver, the portal vein divides and subdivides into ever smaller branches. Eventually, the portal blood flows into a vast network of sinus-like vessels called *sinusoids* (SI-nus-oyds). These enlarged capillary channels allow liver cells close contact with the blood coming from the abdominal organs. (Similar blood channels are found in the spleen and endocrine glands, including the thyroid and adrenals.) After leaving the sinusoids, blood is finally collected by the hepatic veins, which empty into the inferior vena cava.

The purpose of the hepatic portal system of veins is to transport blood from the digestive organs and the spleen to the liver sinusoids so the liver cells can carry out their functions. For example, when food is digested, most of the end products are absorbed from the small intestine into the bloodstream and transported to the liver by the portal system. In the liver, these nutrients are processed, stored, and released as needed into the general circulation.

▶ The Physiology of Circulation

How Capillaries Work

In a general way, the circulating blood might be compared to a train that travels around the country, loading and unloading freight in each of the cities it serves. For example, as blood flows through capillaries surrounding the air sacs in the lungs, it picks up oxygen and unloads carbon dioxide. Later, when this oxygenated blood is pumped to capillaries in other parts of the body, it unloads the oxygen and picks up carbon dioxide (Fig. 15-11) and other substances resulting from cellular activities. The microscopic capillaries are of fundamental importance in these activities. It is only through and between the cells of these thin-walled vessels that the aforementioned exchanges can take place.

All living cells are immersed in a slightly salty liquid called *tissue fluid.* Looking again at Figure 15-11, one can see how this fluid serves as "middleman" between the capillary membrane and the neighboring cells. As water, oxygen, and other materials necessary

superior sagittal sinus

inferior longitudinal sinus

straight sinus

cavernous sinus

ophthalmic vein

confluence

transverse sinus

petrosal sinuses

jugular vein

Figure 15-9

Cranial venous sinuses. The paired transverse sinuses, which carry blood from the brain to the jugular veins, are shown in the inset.

for cellular activity pass through the capillary walls, they enter the tissue fluid. Then these substances make their way by diffusion to the cells. At the same time, carbon dioxide and other end products of cell metabolism come from the cells and move in the opposite direction. These substances enter the capil-lary and are carried away in the bloodstream, to reach other organs or to be eliminated from the body.

Diffusion is the main process by which sub-stances move between the cells and the capillary blood. A secondary force is the pressure of the blood

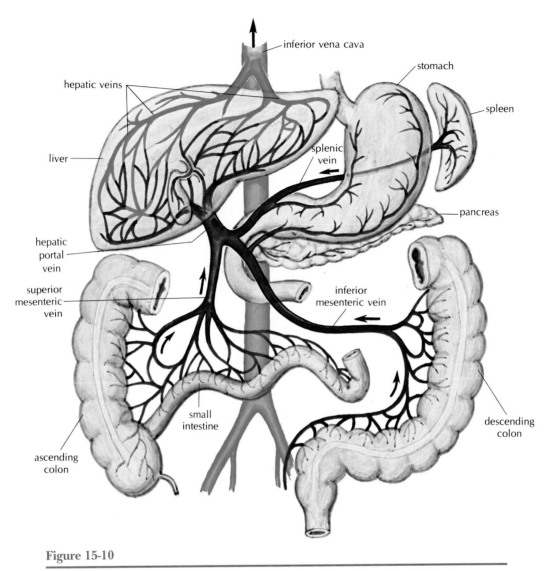

Figure 15-10

Hepatic portal circulation.

as it flows through the capillaries. This force acts to filter, or "push," water and dissolved materials out of the capillary into the tissue fluid. Fluid is drawn back into the capillary by osmotic pressure, the "pulling force" of substances dissolved and suspended in the blood. Osmotic pressure is maintained by plasma proteins (mainly albumin), which are too large to go through the capillary wall. These processes result in the constant exchange of fluids across the capillary wall.

The movement of blood through the capillaries is relatively slow due to the much larger size of the cross-sectional area of the capillaries compared with that of the larger vessels from which capillaries branch. This slow progress through the capillaries allows time for exchanges to occur.

Dynamics of Blood Flow

The flow of blood is carefully regulated to supply the needs of the tissues without unnecessary burden on the heart. Some organs, such as the brain, liver, and kidneys, require large quantities of blood even at rest. The requirements of some tissues, such as those of skeletal muscles and digestive organs, increase greatly during periods of activity. (The blood

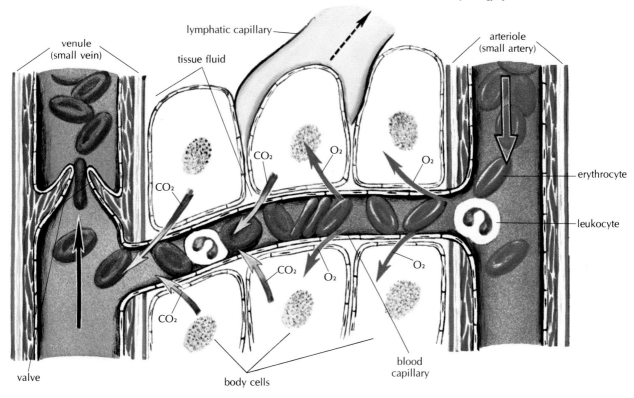

Figure 15-11

Diagram showing the connection between the small blood vessels through capillaries. Note the lymphatic capillary, which aids in tissue drainage.

flow in muscle can increase 25 times during exercise.) The volume of blood flowing to a particular organ can be regulated by changing the size of the blood vessels supplying that organ.

VASODILATION AND VASOCONSTRICTION

An increase in the diameter of a blood vessel is called *vasodilation.* This change allows for the delivery of more blood to an area. *Vasoconstriction* is a decrease in the diameter of a blood vessel, causing a decrease in blood flow. These *vasomotor activities* result from the contraction or relaxation of smooth muscle in the walls of the blood vessels, mainly the arterioles. A *vasomotor center* in the medulla of the brain stem regulates these activities, sending its messages through the autonomic nervous system.

The flow of blood into an individual capillary is regulated by a smooth muscle fiber that encircles the entrance to the capillary (see Fig. 15-7). The *precapillary sphincter* widens to allow more blood to enter when tissues need oxygen.

RETURN OF BLOOD TO THE HEART

By the time blood arrives in the veins, little force remains from the pumping action of the heart. Also, because the veins tend to expand under pressure, considerable amounts of blood are stored in the venous system. Blood from the extremities is pushed toward the heart by the contraction of skeletal muscles, which compresses the veins and squeezes the blood forward. The valves in the veins ensure that the blood flows toward the heart. Changes in pressures in the abdominal and thoracic cavities during breathing also promote return of blood in the venous system. During inhalation, the diaphragm flattens and puts pressure on the large abdominal veins. At the same time, expansion of the chest causes pressure to drop in the thorax. Together, these actions serve to push and pull blood through these cavities and return it to the heart. As evidence of these effects, if a person stands completely motionless, especially on a hot day when the vessels dilate, enough blood can accumulate in the lower extremities to cause fainting from insufficient oxygen to the brain.

‣ Pulse and Blood Pressure

Meaning of the Pulse

The ventricles pump blood into the arteries regularly about 70 to 80 times a minute. The force of ventricular contraction starts a wave of increased pressure that begins at the heart and travels along the arteries. This wave, called the *pulse,* can be felt in any artery that is relatively close to the surface, particularly if the vessel can be pressed down against a bone. At the wrist the radial artery passes over the bone on the thumb side of the forearm, and the pulse is most commonly obtained here. Other vessels sometimes used for obtaining the pulse are the carotid artery in the neck and the dorsalis pedis on the top of the foot.

Normally, the pulse rate is the same as the heart rate. Only if a heartbeat is abnormally weak, or if the artery is obstructed, may the beat not be detected as a pulse. In checking the pulse of another person, it is important to use your second or third finger. If you use your thumb, you may find that you are getting your own pulse. When taking a pulse, it is important to gauge the strength as well as the regularity and the rate.

Various factors may influence the pulse rate. We will enumerate just a few:

1. The pulse is somewhat faster in small persons than in large persons and usually is slightly faster in women than in men.
2. In a newborn infant the rate may be from 120 to 140 beats/minute. As the child grows, the rate tends to become slower.
3. Muscular activity influences the pulse rate. During sleep the pulse may slow down to 60 beats/minute, while during strenuous exercise the rate may go up to well over 100 beats/minute. In a person in good condition, the pulse does not remain rapid despite continued exercise.
4. Emotional disturbances may increase the pulse rate.
5. In many infections, the pulse rate increases with the increase in temperature.
6. An excessive amount of secretion from the thyroid gland may cause a rapid pulse.

Blood Pressure and Its Determination

Blood pressure is the force exerted by the blood against the walls of the vessels. Blood pressure is the product of the output of the heart and the resistance in the vessels. The output of the heart is influenced by:

1. Strength of the contraction of the heart
2. Total blood volume, which controls the volume each beat pushes out.

The resistance in the vessels is affected by:

1. Vasomotor changes. Vasoconstriction increases resistance to flow; vasodilation lowers resistance.
2. Elasticity of blood vessels. Blood vessels lose elasticity and often become hard as a part of aging, thus increasing resistance.
3. Thickness of the blood, called *viscosity.* Increased numbers of red blood cells increase viscosity.

The measurement and careful interpretation of blood pressure may prove a valuable guide in the care and evaluation of a person's health. Because blood pressure decreases as the blood flows from arteries into capillaries and finally into veins, measurements ordinarily are made of arterial pressure only. The instrument used is called a *sphygmomanometer* (sfig-mo-mah-NOM-eh-ter), and two variables are measured:

1. *Systolic pressure,* which occurs during heart muscle contraction, averages around 120 and is expressed in millimeters of mercury (mm Hg).
2. *Diastolic pressure,* which occurs during relaxation of the heart muscle, averages around 80 mm Hg.

The sphygmomanometer is essentially a graduated column of mercury connected to an inflatable cuff. The cuff is wrapped around the patient's upper arm and inflated with air until the brachial artery is compressed and the blood flow cut off. Then, listening with a stethoscope, the doctor or nurse slowly lets air out of the cuff until the first pulsations are heard. At this point the pressure in the cuff is equal to the systolic pressure, and this pressure is read off the mercury column. Then, more air is let out until a characteristic muffled sound indicates the point at which the diastolic pressure is to be read. Considerable practice is required to ensure an accurate reading. The blood pressure is reported as a fraction, with the systolic pressure above and the diastolic pressure below, such as 120/80.

Special Methods for Measuring Blood Pressure

 Measurement of blood pressure with a simple inflatable cuff cannot reflect the pressure in all parts of the cardiovascular system. In specialized hospital units, more specific readings can be obtained by using a catheter (thin tube) inserted directly into the heart and large vessels. One type commonly used is the Swan-Ganz catheter, which has an inflatable balloon at the tip. This device is threaded through a nearby vein into the right side of the heart, where atrial and ventricular pressures can be recorded. As the catheter continues into the pulmonary artery, pressure in this vessel can be read. When the balloon is inflated, the catheter becomes wedged in a branch of the pulmonary artery, blocking blood flow. The reading obtained is called the *pulmonary capillary wedge (PCW) pressure*. It gives information on pressure in the left side of the heart and on resistance in the lungs. Combined with other tests it can be used to diagnose cardiac and pulmonary disorders.

Abnormal Blood Pressure

Lower-than-normal blood pressure is called *hypotension* (hi-po-TEN-shun). However, because of individual variations in normal pressure levels, what would be a low pressure for one person might be a normal or even a high pressure for someone else. For this reason, hypotension is best evaluated in terms of how well the body tissues are being supplied with blood. A person whose systolic blood pressure drops to below his normal range may experience episodes of fainting because of inadequate blood flow to the brain. The sudden lowering of blood pressure below a person's normal level is one symptom of shock; it may also occur in certain chronic diseases and in heart block.

Hypertension, or high blood pressure, has received a great deal of attention. Often it occurs temporarily as a result of excitement or exertion. It may be persistent in a number of conditions, including the following:

1. Kidney disease and uremia or other toxic conditions

2. Endocrine disorders such as hyperthyroidism and acromegaly
3. Artery disease, including hardening of the arteries (atherosclerosis), which reduces elasticity
4. Tumors of the central portion (medulla) of the adrenal gland.

Hypertension that has no apparent medical cause is called *essential hypertension.* This condition, which is fairly common, may cause strokes, heart failure, or kidney damage. Treatment should be begun while the patient is young if the diastolic pressure is over 90 mm Hg. Excess of an enzyme produced in the kidney, called *renin* (RE-nin), seems to play a role in the severity of this kind of hypertension. Renin acts to constrict the blood vessels and to promote the retention of salt and water by the kidneys. Drugs may be given to block excessive renin production and to prevent fluid retention. General health measures, such as weight control, avoidance of excessive alcohol and cigarette consumption, proper diet, and adequate exercise, are all beneficial.

Although stress has been placed on the systolic blood pressure, in many cases the diastolic pressure is even more important, since the condition of small arteries may have more of an effect on diastolic pressure. At any rate, the determination of what really constitutes hypertension depends on each person's normal range. As previously noted, a pressure that is normal for one individual may be abnormal for another.

Disorders Involving the Blood Vessels

Aneurysm of the Aorta

An *aneurysm* (AN-u-rizm) is a bulging sac in the wall of a blood vessel due to a localized weakness in that part of the vessel. The aorta is the vessel most commonly involved. The damage to the wall may be congenital, due to infections, or due to degenerative changes referred to as *hardening of the arteries*. Whatever the cause, the aneurysm may continue to grow in size. As it swells it may cause some derangement of other structures, in which case definite symptoms are manifested. Eventually, however, the walls of the weakened area yield to the pressure, and the aneurysm bursts like a balloon, usually causing immediate death. Surgical replacement of the damaged segment with a synthetic graft may be lifesaving.

Arterial Degeneration

Changes in the walls of arteries frequently lead to loss of elasticity, which is accompanied by irregular thickening of the artery wall at the expense of the lumen (space inside the vessel). Areas of yellow, fat-like material, called *plaque* (PLAK), may accumulate in the vessels and separate the muscle and elastic connective tissue, a disorder called *atherosclerosis* (ath-er-o-skle-RO-sis) (Fig. 15-12). Sometimes the lining of the artery is damaged, and a blood clot (thrombus) may form at this point. Such a thrombus may more or less completely obstruct the vessel, as it sometimes does in coronary thrombosis. In other cases calcium salts and scar tissue (fibrous connective tissue) may cause hardening of the arteries, known as *arteriosclerosis* (ar-te-re-o-skle-RO-sis). A diet high in fats, particularly saturated fats, is known to contribute to atherosclerosis. Cigarette smoking increases the extent and severity of arteriosclerosis.

Artery damage may be present for years without causing any noticeable symptoms. As the thickening of the wall continues and the diameter of the passage for blood flow is decreased, a variety of symptoms appears. The nature of these disturbances will vary with the parts of the body affected and with the extent of the changes in the artery walls. Some examples are listed below:

1. Leg cramps, pain, and sudden lameness while walking may be due to insufficient blood supply to the lower extremities as a result of artery wall damage.

2. Headaches, dizziness, and mental disorders may be the result of cerebral artery sclerosis.

3. Hypertension may be due to a decrease in size of the lumina within many arteries all over the body. Although hypertension may be present in many young persons with no apparent artery damage, and arteriosclerosis may be present without causing hypertension, the two are often found together in elderly people.

4. Palpitations, dyspnea (difficulty in breathing), paleness, weakness, and other symptoms may be the result of arteriosclerosis of the coronary arteries. The severe pain of angina pectoris may follow the lack of oxygen and myocardial damage associated with sclerosis of the vessels that supply the heart.

5. An increase in the amount of urine with the appearance of albumin. This is a normal plasma protein usually found in the urine only if there is kidney damage. Other symptoms referable to the kidneys may be due to damage to the renal arteries.

Gradual narrowing of the interior of the arteries, with a consequent reduction in the volume of blood that passes through them, gives rise to *ischemia* (is-KE-me-ah), a lack of blood. Those parts that are supplied by the damaged artery suffer from an inadequate blood supply, the result being that vital cells in these organs gradually die. The death of cells, for whatever cause, is called *necrosis* (neh-KRO-sis).

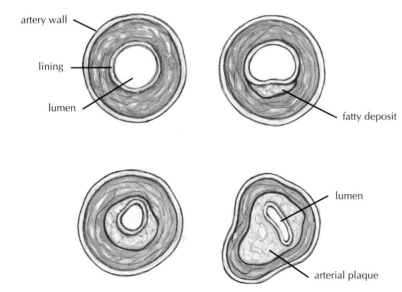

artery wall

lining

lumen

fatty deposit

lumen

arterial plaque

Figure 15-12

Stages in atherosclerosis.

Once these vital cells die, the organ loses its effectiveness. One example of necrosis due to ischemia is the death of certain cells of the brain, with mental disorders as a possible result. Another example is the chain of complications resulting from gradual closure of the arteries of the leg or (rarely) of the arm. The circulation of blood in the toes or the fingers, never too brisk even at the best of times, may cease altogether. Necrosis occurs; the dead tissue is invaded by bacteria, and putrefaction sets in. This condition, called *gangrene* (GANG-grene), can result from a number of disorders that may injure the arteries, such as diabetes. Diabetic gangrene is a fairly common occurrence in elderly diabetic patients.

Hemorrhage and First Aid

A profuse escape of blood from the vessels is known as *hemorrhage,* a word that means "a bursting forth of blood." Such bleeding may be external or internal, may be from vessels of any size, and may involve any part of the body. Capillary oozing usually is stopped by the normal process of clot formation. Flow from larger vessels can be stopped by appropriate first aid measures carried out at the scene. In most cases, pressure with a clean bandage directly on the wound stops the bleeding effectively.

The loss of blood from a cut artery may be rapid. Often it is quickly fatal, but immediate appropriate action can be lifesaving. The Red Cross and other organizations that give instructions in first aid agree that excessive loss of blood can and should be prevented in all circumstances. Because hemorrhage is the most frequent problem in case of an accident, everyone should know that certain arteries can be pressed against a bone to stop hemorrhage. The most important of these "pressure points" are as follows:

1. The *facial artery,* which may be pressed against the lower jaw as the vessel extends along the side of the face, for hemorrhage around the nose, mouth, and cheek
2. The *temporal artery,* which may be pressed against the side of the skull just in front of the ear to stop hemorrhage on the side of the face and around the ear
3. The *common carotid artery* in the neck, which may be pressed back against the spinal column for bleeding in the neck and the head. Avoid prolonged compression, which can result in lack of oxygen to the brain.

4. The *subclavian artery,* which may be pressed against the first rib by a downward push with the thumb to stop bleeding from the shoulder or arm
5. The *brachial artery,* which may be pressed against the humerus (arm bone) by a push inward along the natural groove between the two large muscles of the arm. This stops hand, wrist, and forearm hemorrhage.
6. The *femoral artery* (in the groin), which may be pressed to avoid serious hemorrhage of the lower extremity.

It is important not to leave the pressure on too long, because it may cause damage to tissues past the pressure point.

Shock

The word *shock* has a number of meanings. However, in terms of the circulating blood, it refers to a life-threatening condition in which there is inadequate blood flow to the tissues of the body. The factor common to all cases of shock is inadequate output by the heart. Shock can be instigated by a wide range of conditions that reduce the effective circulation. The exact cause of shock is often not known; however, a widely used classification is based on causative factors, the most important of which include the following:

1. *Cardiogenic* (kar-de-o-JEN-ik) *shock,* sometimes called *pump failure,* is often a complication of heart muscle damage as is found in myocardial infarction. It is the leading cause of shock death.
2. *Septic shock* is second only to cardiogenic shock as a cause of shock death. It is usually due to an overwhelming bacterial infection.
3. *Hypovolemic* (hi-po-vo-LE-mik) *shock* is due to a decrease in the volume of circulating blood and may follow severe hemorrhage or burns.
4. *Anaphylactic* (an-ah-fih-LAK-tik) *shock* is a severe allergic reaction to foreign substances to which the person has been sensitized.

When the cause is not known, shock is classified according to its severity.

In *mild shock* regulatory mechanisms act to relieve the circulatory deficit. Symptoms are often subtle changes in heart rate and blood pressure. Constriction of small blood vessels and the detouring of blood away from certain organs increase the effective circulation. Mild shock may develop into a severe, life-threatening circulatory failure.

Severe shock is characterized by poor circulation, which causes further damage and deepening of the shock. Symptoms of late shock include clammy skin, anxiety, very low blood pressure, rapid pulse, and rapid, shallow breathing. Contractions of the heart are weakened owing to the decrease in blood supply to the heart muscle. The muscles in the blood vessel walls are also weakened so that they dilate. The capillaries become more permeable and lose fluid owing to the accumulation of metabolic wastes.

The victim of shock should first be placed in a horizontal position and covered with a blanket. If there is bleeding, it should be stopped. The patient's head should be kept turned to the side to prevent aspiration (breathing in) of vomitus, an important cause of death in shock cases. Further treatment of shock depends largely on treatment of the causative factors. For example, shock due to fluid loss such as hemorrhage or burns is best treated with blood products or plasma expanders (intravenous fluids). Shock due to heart failure should be treated with drugs that improve the contractions of the heart muscle. In any case, all measures are aimed at supporting the circulation and improving the output of the heart. Oxygen is frequently administered to improve the delivery of oxygen to the tissues.

Varicose Veins

Varicose veins are a condition in which superficial veins have become swollen, tortuous, and ineffective. It may be a problem in the esophagus or in the rectum, but the veins most commonly involved are the saphenous veins of the lower extremities (Fig. 15-13). This condition is found frequently in persons who spend a great deal of time standing, such as salespeople. Pregnancy, with its accompanying pressure on the veins in the pelvis, may also be a predisposing cause. Varicose veins in the rectum are called *hemorrhoids* (HEM-o-royds), or *piles*. The general term for varicose veins is *varices* (VAR-ih-seze), the singular form being *varix* (VAR-iks).

Phlebitis

Inflammation of a vein, called *phlebitis* (fleh-BI-tis), causes marked pain and often considerable swelling. The entire vein wall is involved. A blood clot may form, causing a dangerous condition called *thrombophlebitis* (throm-bo-fleh-BI-tis), in which a piece of the clot may become loose and float in the blood as an *embolus* (EM-bo-lus). An embolus is car-

Figure 15-13

Varicose veins. (Bates B: Guide to Physical Examination and History Taking, 6th ed, p. 439. Philadelphia, JB Lippincott, 1995)

ried through the circulatory system until it lodges in a vessel. If it reaches the lungs, sudden death from *pulmonary embolism* (EM-bo-lizm) may result. Prevention of infection, early activity to ensure circulation following an injury or an operation, and the use of anticoagulant drugs when appropriate have greatly reduced the incidence of this condition.

Stasis Dermatitis and Ulcers

Stasis means a standing still or stoppage in normal blood flow. Often the valves in the veins of the lower extremity become incompetent and blood is not returned from the legs. The skin becomes inflamed and undergoes scaling; fissures (cracks) form, followed by ulcers. Treatment of the causes and attention to the local lesions should be in the hands of a physician, and patient cooperation is most important.

Arterial Obstruction

When the lumen of an artery is completely blocked by arteriosclerosis, the condition is known as *arteriosclerosis obliterans* (o-BLIT-er-ans). The complete obstruction to blood flow causes intense pain;

if the disease is allowed to progress and is not treated, the affected part or parts may become ulcerated or gangrenous.

Small areas of blockage can be relieved by direct surgical removal of the obstruction. In other cases, the affected vessel can be dilated by angioplasty with a balloon catheter. When the obstruction cannot be relieved by these methods, a bypass graft is placed surgically. The most successful of these grafts uses the saphenous vein from the patient's own body.

SUMMARY

I. Blood vessels
 A. Functional classification
  1. Arteries—carry blood away from heart
   a. Arterioles—small arteries
  2. Veins—carry blood toward heart
   a. Venules—small veins
  3. Capillaries—allow for exchanges between blood and tissues, or blood and air in lungs; connect arterioles and venules
 B. Circuits
  1. Pulmonary circuit—carries blood to and from lungs
  2. Systemic circuit—carries blood to and from rest of body
 C. Structure
  1. Tissue layers
   a. Innermost—single layer of flat epithelial cells (endothelium)
   b. Middle—thicker layer of smooth muscle and elastic connective tissue
   c. Outer—connective tissue
  2. Arteries—all three layers; highly elastic
  3. Arterioles—thinner walls, less elastic tissue, more smooth muscle
  4. Capillaries—only endothelium; single layer of cells
  5. Veins—all three layers; thinner walls than arteries, less elastic tissue

II. Systemic arteries
 A. Aorta—largest artery
  1. Divisions
   a. Ascending aorta
    (1) Left and right coronary arteries
   b. Aortic arch
    (1) Brachiocephalic trunk
    (2) Left common carotid artery
    (3) Left subclavian artery
   c. Descending aorta
    (1) Thoracic aorta
    (2) Abdominal aorta

 B. Iliac arteries—final division of aorta; branch to pelvis and legs
 C. Anastomoses—communications between vessels

III. Systemic veins
 A. Location
  1. Superficial—near surface
  2. Deep—usually parallel to arteries with same names as corresponding arteries
 B. Superior vena cava—drains upper part of body
 C. Inferior vena cava—drains lower part of body
 D. Venous sinuses—enlarged venous channels
 E. Hepatic portal system—carries blood from abdominal organs to liver, where it is processed before returning to heart

IV. Physiology of circulation
 A. Capillary exchange
  1. Primary method—diffusion
  2. Medium—tissue fluid
  3. Blood pressure—drives fluid into tissues
  4. Osmotic pressure—pulls fluid into capillary
 B. Regulation of blood flow
  1. Vasodilation—increase in diameter of blood vessel
  2. Vasoconstriction—decrease in diameter of blood vessel
  3. Vasomotor center—in medulla; controls contraction and relaxation of smooth muscle in vessel wall
  4. Precapillary sphincter—regulates blood flow into capillary
  5. Effects
   a. Control of blood distribution
   b. Regulation of blood pressure
 C. Return of blood to heart
  1. Pumping action of heart
  2. Pressure of skeletal muscles on veins

3. Valves in veins
4. Breathing—changes in pressure move blood toward heart
V. Pulse—wave of pressure that travels along arteries as ventricles contract
VI. Blood pressure
 A. Product of cardiac output, vascular resistance
 B. Measured in arm with sphygmomanometer
  1. Systolic pressure—averages 120 mm Hg
  2. Diastolic pressure—averages 80 mm Hg
 C. Abnormal blood pressure
  1. Hypotension—low blood pressure
  2. Hypertension—high blood pressure
   a. Essential hypertension—has no apparent medical cause
   b. May involve renin, enzyme released from kidneys
VII. Disorders involving the blood vessels
 A. Aneurysm—weakness and bulging of a vessel; may burst

 B. Arterial degeneration
  1. Atherosclerosis—fatty deposits in vessel; may result in thrombus
  2. Arteriosclerosis—hardening of arteries with scar tissue or calcium salts; may result in ischemia and cell necrosis
 C. Hemorrhage—massive loss of blood; pressure applied at pressure points will stop blood loss
 D. Shock—inadequate output of blood from heart
 E. Varicose veins—swelling and loss of function in superficial veins, usually in legs and rectum (hemorrhoids)
 F. Phlebitis—inflammation of a vein
  1. Embolus—piece of clot traveling in circulation
  2. Pulmonary embolism—clot lodged in lungs
 G. Other—stasis dermatitis, ulcers, arterial obstruction

QUESTIONS FOR STUDY AND REVIEW

1. Name the three main groups of blood vessels and describe their functions. How has function affected structure?
2. Compare the oxygen content of blood in the pulmonary and systemic vessels.
3. Name the main branches of the aorta.
4. Describe an arterial anastomosis and give several examples.
5. Trace a drop of blood through the shortest possible route from the capillaries of the foot to the capillaries of the head.
6. What are the names and functions of some cranial venous sinuses? Where is the coronary venous sinus and what does it do?
7. What large vessels drain the blood low in oxygen from most of the body into the right atrium? What vessels carry blood high in oxygen into the left atrium?
8. Trace a drop of blood from capillaries in the wall of the small intestine to the right atrium. What is the purpose of going through the liver on this trip?
9. Define *portal system*. Name the largest portal system in the body.
10. What substances diffuse into the tissues from the capillaries? into the capillaries from the tissues?
11. What force pushes fluid out of the capillaries? What force pulls fluid back into the capillaries?
12. What actions help force blood back to the heart?
13. What is meant by *pulse?* Where is the pulse usually determined? If a large part of the body were burned, leaving only the lower extremities accessible for obtaining the pulse, what vessel would you try to use?
14. What are some factors that cause an increase in the pulse rate?
15. List five factors that can change blood pressure.
16. What instrument is used for measuring blood pressure? What are the two values usually obtained called, and what is the significance of each?
17. What are some examples of disorders that cause hypertension of a persistent kind? Of what importance is diastolic blood pressure?

18. What are some symptoms of arteriosclerosis and how are these produced?

19. What is the meaning of *hemorrhage?* What vessels cause the most serious bleeding if they are cut? What are some of the most effective ways of stopping hemorrhage?

20. What is shock, and why is it so dangerous? Name some symptoms of shock and identify four types of shock.

21. In what organs are varicose veins most commonly found? In what situations are they given special names? Give examples.

22. What is the serious complication of phlebitis of lower extremity veins?

23. Explain the difference between the terms in the following pairs:
- **a.** *pulmonary circuit* and *systemic circuit*
- **b.** *arteriole* and *venule*
- **c.** *vasodilation* and *vasoconstriction*
- **d.** *atherosclerosis* and *arteriosclerosis*
- **e.** *hypotension* and *hypertension*

CHAPTER 16

The Lymphatic System and Lymphoid Tissue

Selected Key Terms

The following terms are defined in the Glossary:

chyle

lacteal

lymph

lymphadenitis

lymphadenopathy

lymphangitis

lymphatic duct

lymph node

spleen

thymus

tonsil

Behavioral Objectives

After careful study of this chapter, you should be able to:

- List the three major functions of the lymphatic system
- Explain how lymphatic capillaries differ from blood capillaries
- Name the two main lymphatic ducts and describe the area drained by each
- List the major structures of the lymphatic system and give the locations and functions of each
- Describe the composition and function of the reticulo-endothelial system
- Describe the major disorders of the lymphatic system

Memmler, RL, Cohen, BJ, Wood, DL. *THE HUMAN BODY IN HEALTH AND DISEASE*, 8/e,
© 1996 Lippincott-Raven Publishers

▶ The Lymphatic System

As noted in the preceding chapter, body cells live in tissue fluid, a liquid derived from the bloodstream. Water and dissolved substances, such as oxygen and nutrients, are constantly filtering through capillary walls into the spaces between cells and constantly adding to the volume of tissue fluid. However, under normal conditions, fluid is also constantly removed so that it does not accumulate in the tissues. Part of this fluid simply returns (by diffusion) to the capillary bloodstream, taking with it some of the end products of cellular metabolism, including carbon dioxide and other substances. A second pathway for the drainage of tissue fluid involves the *lymphatic system.* In addition to the blood-carrying capillaries, there are microscopic vessels called *lymphatic capillaries,* which drain away excess tissue fluid that does not return to the blood capillaries. The relation between the circulatory system and the lymphatic system is shown in Figure 16-1. Another important function of the lymphatic capillaries is to absorb protein from the tissue fluid and return it to the bloodstream. As soon as tissue fluid enters the lymphatic capillary, it is called *lymph.* The lymphatic capillaries join to form the larger lymphatic vessels, or ducts, and these vessels (which we shall have a closer look at in a moment) eventually empty into the veins. However, before the lymph reaches the veins, it flows through a series of filters called *lymph nodes,* where bacteria and other foreign particles are trapped and destroyed. Thus, the lymph nodes may be compared in one way with the oil filter in an automobile.

Lymphatic Capillaries

The lymphatic capillaries resemble the blood capillaries in that they are made of one layer of flattened (squamous) epithelial cells, also called *endothelium,* which allows for easy passage of soluble materials and water. Gaps between these endothelial cells allow the entrance of proteins and other relatively large suspended particles. Unlike the capillaries of the bloodstream, the lymphatic capillaries begin blindly; that is, they do not serve to bridge two larger vessels. Instead, one end simply lies within a lake of tissue fluid, and the other communicates with a larger lymphatic vessel (see Fig. 16-1).

In the small intestine are some specialized lymphatic capillaries, called *lacteals* (LAK-te-als), which act as a pathway for the transfer of digested fats into the bloodstream. This process is covered in the chapter on the digestive system (see Chap. 19).

Lymphatic Vessels

The lymphatic vessels are thin walled and delicate and have a beaded appearance because of indentations where valves are located. These valves prevent backflow in the same way as do those found in some veins.

Lymphatic vessels (Fig. 16-2) include *superficial* and *deep* sets. The surface lymphatics are immediately below the skin, often continuing near the superficial veins. The deep vessels are usually larger and accompany the deep veins.

Lymphatic vessels are named according to location. For example, those in the breast are called *mammary* lymphatic vessels, those in the thigh *femoral* lymphatic vessels, and those in the leg *tibial* lymphatic vessels. All the lymphatic vessels form networks, and at certain points they carry lymph into the regional nodes (the nodes that "service" a particular area). For example, nearly all the lymph from the upper extremity and the breast passes through the *axillary lymph nodes,* whereas that from the lower extremity passes through the *inguinal nodes.* Lymphatic vessels carrying lymph away from the regional nodes eventually drain into one of the two terminal vessels, the right lymphatic duct or the thoracic duct, which empty into the bloodstream.

The *right lymphatic duct* is a short vessel about 1.25 cm (1/2 inch) long that receives only the lymph that comes from the upper right quadrant of the body: the right side of the head, neck, and thorax, as well as the right upper extremity. It empties into the right subclavian vein. Its opening into this vein is guarded by two pocket-like semilunar valves to prevent blood from entering the duct. The rest of the body is drained by the thoracic duct.

THE THORACIC DUCT

The *thoracic duct* is much the larger of the two terminal vessels; it is about 40 cm (16 inches) in length. As shown in Figure 16-2, the thoracic duct receives lymph from all parts of the body except those above the diaphragm on the right side. This duct begins in the posterior part of the abdominal cavity, below the attachment of the diaphragm. The first part of this duct is enlarged to form a cistern, or temporary storage pouch, called the *cisterna chyli* (sis-TER-nah KI-li). *Chyle* (kile) is the milky fluid, formed by the combination of fat globules and

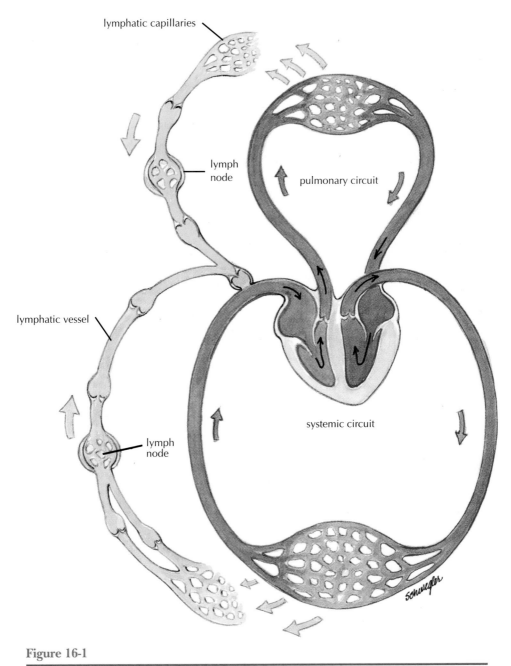

lymphatic capillaries

lymph node

lymphatic vessel

lymph node

pulmonary circuit

systemic circuit

schwegler

Figure 16-1

The lymphatic system in relation to the cardiovascular system.

lymph, that comes from the intestinal lacteals. Chyle passes through the intestinal lymphatic vessels and the lymph nodes of the mesentery, finally entering the cisterna chyli. In addition to chyle, all the lymph from below the diaphragm empties into the cisterna chyli by way of the various clusters of lymph nodes and then is carried by the thoracic duct into the bloodstream.

The thoracic duct extends upward through the diaphragm and along the back wall of the thorax up into the root of the neck on the left side. Here it receives the left jugular lymphatic vessels from the head and neck, the left subclavian vessels from the left upper extremity, and other lymphatic vessels from the thorax and its parts. In addition to the valves along the duct, there are two valves at its open-

How Lymphatic Capillaries Work

 Lymphatic capillaries are more permeable than blood capillaries because the junctions between the cells are wider and also because there is less basement membrane underlying the cells. Proteins and other large molecules can enter the lymphatic capillaries along with fluid.

Why don't substances that get into the lymphatic capillaries flow right out again? The cells that form the capillary wall overlap slightly to function as one-way valves. Pressure outside the vessel forces materials in. Once fluid enters the capillary, pressure against the cells in the capillary wall blocks its return.

ing into the left subclavian vein to prevent the passage of blood into the duct.

Movement of Lymph

The segments of lymphatic vessels located between the valves contract rhythmically, propelling the lymph along. The rate of these contractions is related to the volume of fluid in the vessel—the more fluid, the more rapid the contractions of the vessel. Lymph is also moved by the same mechanisms that promote venous return of blood to the heart. As skeletal muscles contract during movement they compress the lymphatic vessels and drive lymph forward. Changes in pressures within the abdominal and thoracic cavities due to breathing aid the movement of lymph during passage through these body cavities.

▶ Lymphoid Tissue

The above section was just a brief survey of the system of lymph vessels and lymph transport. The lymph nodes were mentioned but they will be described in greater detail now, along with discussion of other organs made of similar tissue.

Lymphoid (LIM-foyd) *tissue* is distributed throughout the body and makes up the specialized organs of the lymphatic system. We will look at some of these organs, but first let us consider some properties of lymphoid tissue to see what characteristics these organs have in common.

Functions

Some of the general functions of lymphoid tissue include the following:

1. Removal of impurities such as carbon particles, cancer cells, pathogenic organisms, and dead blood cells through filtration and phagocytosis
2. Processing of lymphocytes. Some of these lymphocytes produce antibodies, substances in the blood that aid in combating infection; others attack foreign invaders directly. These cells are discussed further in Chapter 17.

Lymph Nodes

The lymph nodes, as we have seen, are designed to filter the lymph once it is drained from the tissues (Fig. 16-3). The lymph nodes are small, rounded masses varying from pinhead size to as long as 2.5 cm (1 inch). Each has a fibrous connective tissue capsule from which partitions (trabeculae) extend into the substance of the node. Inside the node are masses of lymphatic tissue, which include lymphocytes and macrophages, white blood cells active in immunity. At various points in the surface of the node, afferent lymphatic vessels pierce the capsule to carry lymph into the spaces inside the pulp-like nodal tissue. An indented area called the *hilus* (HI-lus) serves as the exit for efferent lymphatic vessels carrying lymph out of the node. At this region other structures, including blood vessels and nerves, connect with the node.

Lymph nodes are seldom isolated. As a rule, they are massed together in groups, the number in each group varying from 2 or 3 to well over 100. Some of these groups are placed deeply, while others are superficial. Those of the most practical importance include the following:

1. *Cervical nodes,* located in the neck, are divided into deep and superficial groups, which drain various parts of the head and neck. They often become enlarged during upper respiratory infections and in certain chronic disorders.
2. *Axillary nodes,* located in the axillae (armpits), may become enlarged following infections of the upper extremities and the breasts. Cancer cells from the breasts often metastasize (spread) to the axillary nodes.
3. *Tracheobronchial* (tra-ke-o-BRONG-ke-al) *nodes* are found near the trachea and around the larger bronchial tubes. In persons living in highly polluted areas, these nodes become so filled with

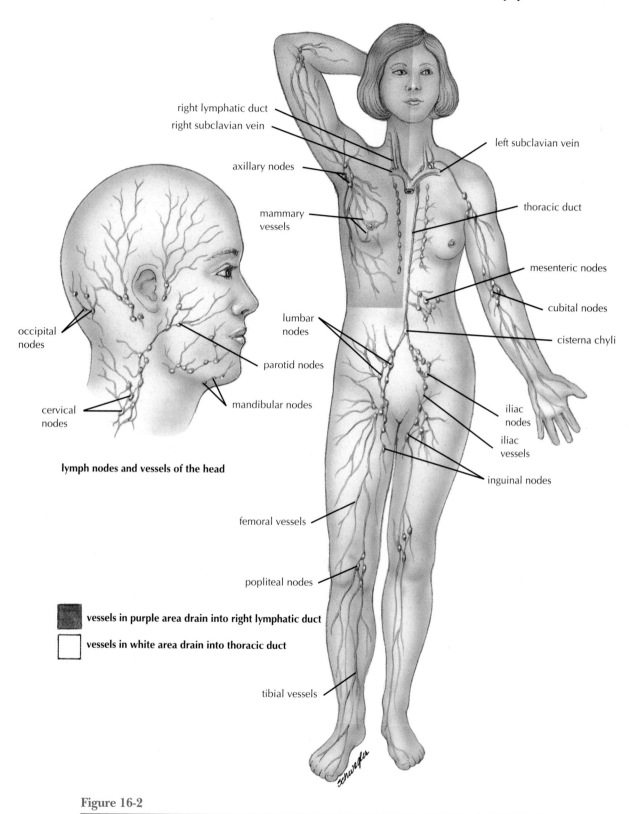

right lymphatic duct

right subclavian vein

axillary nodes

mammary vessels

left subclavian vein

thoracic duct

mesenteric nodes

cubital nodes

lumbar nodes

cisterna chyli

occipital nodes

parotid nodes

iliac nodes

iliac vessels

cervical nodes

mandibular nodes

inguinal nodes

lymph nodes and vessels of the head

femoral vessels

popliteal nodes

■ **vessels in purple area drain into right lymphatic duct**

□ **vessels in white area drain into thoracic duct**

tibial vessels

Figure 16-2

Lymphatic system.

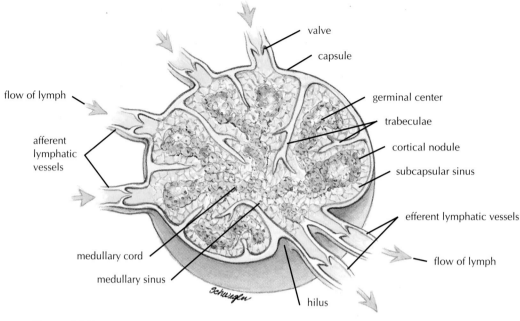

Figure 16-3

Structure of a lymph node.

carbon particles that they are solid black masses resembling pieces of coal.

4. *Mesenteric* (mes-en-TER-ik) *nodes* are found between the two layers of peritoneum that form the mesentery (membrane around the intestines). There are some 100 to 150 of these nodes.

5. *Inguinal nodes,* located in the groin region, receive lymph drainage from the lower extremities and from the external genital organs. When they become enlarged, they are often referred to as *buboes* (BU-bose), from which bubonic plague got its name.

The Tonsils

There are masses of lymphoid tissue that are designed to filter not lymph, but tissue fluid. Found beneath certain areas of moist epithelium that are exposed to the outside, and hence to contamination, these masses include parts of the digestive, urinary, and respiratory tracts. Associated with the latter system are those well-known masses of lymphoid tissue called the *tonsils,* which include:

1. The *palatine* (PAL-ah-tine) *tonsils,* oval bodies located at each side of the soft palate. These are generally meant when one refers to "the tonsils."

2. The *pharyngeal* (fah-RIN-je-al) *tonsil,* commonly referred to as *adenoids* (from a general term that

means "gland-like"). It is located behind the nose on the back wall of the upper pharynx.

3. The *lingual* (LING-gwal) *tonsils,* little mounds of lymphoid tissue at the back of the tongue.

Any or all of these tonsils may become so loaded with bacteria that the pathogens gain the upper hand; removal then is advisable. A slight enlargement of any of them is not an indication for surgery. All lymphoid tissue masses tend to be larger in childhood, so that the physician must consider the patient's age in determining whether these masses are enlarged abnormally. Because the tonsils appear to function in immunity during early childhood, efforts are made *not* to remove them unless absolutely necessary.

The Thymus

Because of its appearance under the microscope, the *thymus* (THI-mus), located in the upper thorax beneath the sternum, has been considered part of the lymphoid system. However, recent studies suggest that this structure has a much more basic function than was originally thought. It now seems apparent that the thymus plays a key role in the development of the immune system before birth and during the first few months of infancy. Certain lymphocytes must mature in the thymus gland before they can perform their functions in the immune system (see Chap. 17). These *T lymphocytes*, or T cells, develop under the effects of the hormone from the thymus gland called *thymosin* (THI-mosin), which also promotes the growth and activity of lymphocytes in lymphoid tissue throughout the body. Removal causes a decrease in the production of T lymphocytes, as well as a decrease in the size of the spleen and of lymph nodes throughout the body. The thymus is most active during early life. After puberty, the tissue undergoes changes; it shrinks in size and is replaced by connective tissue and fat.

The Spleen

The spleen is an organ that contains lymphoid tissue designed to filter blood. It is located in the upper left hypochondriac region of the abdomen and normally is protected by the lower part of the rib cage because it is high up under the dome of the diaphragm. The spleen is a soft, purplish, and somewhat flattened organ about 12.5 to 16 cm (5–6 inches) long and 5 to 7.5 cm (2–3 inches) wide. The capsule of the spleen, as well as its framework, is more elastic than that of the lymph nodes. It contains involuntary muscle, which enables the splenic capsule to contract and to withstand some swelling.

The spleen has an unusually large blood supply, considering its size. The organ is filled with a soft pulp, one of the functions of which is to filter out worn-out red blood cells. The spleen also harbors phagocytes, which engulf bacteria and other foreign particles. Round masses of lymphoid tissue are prominent structures inside the spleen, and it is because of these that the spleen is often classified with the other organs made of lymphoid tissue. Some other category might be better, however, because of the other specialized functions of the spleen. Some of these functions are listed below:

1. Cleansing the blood by filtration and phagocytosis
2. Destroying old, worn-out red blood cells. The iron and other breakdown products of hemoglobin are carried to the liver by the hepatic portal system to be reused or eliminated from the body.
3. Producing red blood cells before birth
4. Serving as a reservoir for blood, which can be returned to the bloodstream in case of hemorrhage or other emergency

Splenectomy (sple-NEK-to-me), or surgical removal of the spleen, is usually tolerated quite well; although the spleen is the largest unit of lymphoid tissue in the body, other lymphoid tissues can take over its functions. The human body has thousands of lymphoid units, and the loss of any one unit or group ordinarily is not a threat to life.

▶ The Reticuloendothelial System

The *reticuloendothelial* (reh-tik-u-lo-en-do-THE-le-al) *system* consists of related cells concerned with the destruction of worn-out blood cells, bacteria, cancer cells, and other foreign substances that are potentially harmful to the body. They include monocytes, which are relatively large white blood cells (see Fig. 13-1) that are formed in the bone marrow and then circulate in the bloodstream to various parts of the body. On entering the tissues, monocytes develop into *macrophages* (MAK-ro-faj-ez), a term that means "big eaters." Some of them are given special names; the *Kupffer's* (KOOP-ferz) *cells,* for example, are located in the lining of the liver sinusoids (blood channels). Other parts of the reticuloendothelial system are found in the spleen, bone marrow, lymph nodes, and brain. Some macrophages are located in the lungs, where they are called *dust cells* because they ingest solid particles that enter the lungs; others are found in soft connective tissues all over the body.

This widely distributed protective system has been called by several other names, including *tissue macrophage system, mononuclear phagocyte system,* and *monocyte–macrophage system.* These names are descriptive of the type of cells found in this system.

▶ Disorders of the Lymphatic System and Lymphoid Tissue

Lymphangitis

In inflammation of the lymphatic vessels, called *lymphangitis* (lim-fan-JI-tis), red streaks can be seen extending along an extremity, usually beginning in the region of an infected and neglected injury. Such inflamed vessels are a danger signal, because the lymph nodes may not be able to stop a serious infection. The next step would be entrance of the pathogens into the bloodstream, causing *septicemia*

Cancer and the Lymphatic System

 Ordinarily, the lymphatic system is one of our primary defenses against disease. In cancer, however, it can be a vehicle for the spread (metastasis) of the disease. Tumor cells that enter the lymphatic vessels can be carried to other parts of the body. In breast cancer, for example, the degree of invasion of nearby lymph nodes is an important factor in diagnosis and treatment. In mastectomies, in addition to removing breast tissue, it is sometimes necessary to remove lymphatic vessels and nodes. This may interfere with normal lymph flow from the arm, resulting in lymphedema and an increased risk of infection.

(sep-tih-SE-me-ah), or blood poisoning. Streptococci often are the invading organisms in such cases.

Elephantiasis

As mentioned in Chapter 5, *elephantiasis* is a great enlargement of the lower extremities resulting from blockage of lymphatic vessels by small worms called *filariae* (fi-LA-re-e). These tiny parasites, carried by insects such as flies and mosquitoes, invade the tissues as embryos or immature forms. They then grow in the lymph channels and thus obstruct the flow of lymph. The swelling of the legs or, as sometimes happens in men, the scrotum, may be so great that the victim becomes incapacitated. This disease is especially common in certain parts of Asia and in some of the Pacific islands. No cure is known.

Lymphadenitis

In *lymphadenitis* (lim-fad-en-I-tis), or inflammation of the lymph nodes, the nodes become enlarged and tender. This condition reflects the body's attempt to combat an infection. Cervical lymphadenitis occurs during measles, scarlet fever, septic sore throat, diphtheria, and, frequently, the common cold. Chronic lymphadenitis may be due to the tubercle bacillus. Infections of the upper extremities cause enlarged axillary nodes, as does cancer of the mammary glands. Infections of the external genitals or the lower extremities may cause enlargement of the inguinal lymph nodes.

Lymphadenopathy

Lymphadenopathy (lim-fad-en-OP-ah-the) is a term meaning "disease of the lymph nodes." Enlarged lymph nodes are a common symptom in a number of infectious and cancerous diseases. An early sign of infection with HIV, the virus that causes AIDS, is a generalized lymphadenopathy. *Infectious mononucleosis* (mon-o-nu-kle-O-sis) is an acute viral infection the hallmark of which is a marked enlargement of the cervical lymph nodes. Mononucleosis is fairly common among college students. Enlarged lymph nodes are commonly referred to as *glands*, as in "swollen glands." However, they do not produce secretions and are not glands.

Splenomegaly

Enlargement of the spleen, known as *splenomegaly* (sple-no-MEG-ah-le), accompanies certain acute infectious diseases, including scarlet fever, typhus fever, typhoid fever, and syphilis. Many tropical parasitic diseases cause splenomegaly. A certain blood fluke (flatworm) that is fairly common among workers in Japan and other parts of Asia causes marked splenic enlargement.

Splenic anemia is characterized by enlargement of the spleen, hemorrhages from the stomach, and accumulation of fluid in the abdomen. In this disease and others of the same nature, splenectomy appears to constitute a cure.

Hodgkin's Disease

Hodgkin's disease is a chronic malignant disorder, most common in young men, that is characterized by enlargement of the lymph nodes. The nodes in the neck particularly, and often those in the armpit, thorax, and groin, enlarge. The spleen may become enlarged as well. Chemotherapy and radiotherapy, either separately or in combination, have been used with good results, affording patients many years of life.

Lymphosarcoma

Lymphosarcoma (lim-fo-sar-KO-mah) is a malignant tumor of lymphoid tissue that is likely to be rapidly fatal. Fortunately, it is not a common disease. Early surgery in combination with appropriate radiotherapy offers the only possible cure at this time. (Note: A *lymphoma* is any tumor, benign or malignant, that occurs in lymphoid tissue.)

SUMMARY

I. Lymphatic system
  A. Functions
  1. Drainage of excess fluid from tissues
  2. Absorption of fats from small intestine
  3. Protection from foreign invaders
  B. Structures
  1. Capillaries—made of endothelium (simple squamous epithelium)
  2. Lymphatic vessels
    a. Superficial
    b. Deep
  3. Right lymphatic duct—drains upper right part of body and empties into right subclavian vein
  4. Thoracic duct—drains remainder of body and empties into left subclavian vein
  C. Movement of lymph

II. Lymphoid tissue—distributed throughout body
  A. General functions
  1. Removal of impurities by filtration and phagocytosis
  2. Processing of lymphocytes of immune system
  B. Specialized functions
  1. Nodes—filtration of lymph
  2. Tonsils—filtration of tissue fluid
  3. Thymus
    a. Processing of T lymphocytes (T cells)
    b. Secretion of thymosin—stimulates T lymphocytes in lymphoid tissue
  4. Spleen
    a. Filtration of blood
    b. Destruction of old red cells
    c. Production of red cells before birth
    d. Storage of blood

III. Reticuloendothelial system—cells throughout body that get rid of impurities

IV. Disorders of the lymphatic system and lymphoid tissue
  A. Lymphangitis—inflammation of lymphatic vessels
  B. Elephantiasis—blockage of lymphatic vessels by parasitic worms
  C. Lymphadenitis—inflammation of lymph nodes that occurs during infection
  D. Lymphadenopathy—disease of lymph nodes
  E. Splenomegaly—enlargement of the spleen
  F. Hodgkin's disease—malignant disease with enlargement of lymph nodes
  G. Lymphosarcoma—malignant tumor of lymphoid tissue

QUESTIONS FOR STUDY AND REVIEW

1. What is lymph? How is it formed?
2. Briefly describe the system of lymph circulation.
3. Describe the lymphatic vessels with respect to design, appearance, and location.
4. Name the two main lymphatic ducts. What part of the body does each drain, and into what blood vessel does each empty?
5. What is the cisterna chyli and what are its purposes?
6. Name two functions of lymphoid tissue.
7. Describe the structure of a typical lymph node.
8. What are the neck nodes called and what are some of the causes of enlargement of these lymph nodes?
9. What parts of the body are drained by vessels entering the axillary nodes and what conditions cause enlargement of these nodes?
10. What parts of the body are drained by lymphatics that pass through the inguinal lymph nodes?
11. What are the different tonsils called and where are they located? What is the purpose of these and related structures?
12. What is the function of the thymus?
13. Give the location of the spleen and name several of its functions.
14. Describe the reticuloendothelial system.
15. What is lymphangitis? lymphadenitis? lymphadenopathy?
16. Describe elephantiasis and its cause.
17. What is splenomegaly and what can cause it?
18. Describe Hodgkin's disease and lymphosarcoma.

CHAPTER 17

Body Defenses, Immunity, and Vaccines

Selected Key Terms

The following terms are defined in the Glossary:

allergy

antibody

antigen

antiserum

attenuated

autoimmunity

B cell

complement

gamma globulin

immunity

immunization

immunodeficiency

inflammation

lymphocyte

macrophage

plasma cell

T cell

toxin.

toxoid

transplant

vaccine

Behavioral Objectives

After careful study of this chapter, you should be able to:

- List the factors that determine the occurrence of infection

- Differentiate between nonspecific and specific body defenses and give examples of each

- Briefly describe the inflammatory reaction

- List several types of inborn immunity

- Define *antigen* and *antibody*

- Compare T cells and B cells with respect to development and type of activity

- Differentiate between natural and artificial acquired immunity

- Differentiate between active and passive immunity

- Define the term *vaccine* and give several examples of vaccines

- Define the term *immune serum* and give several examples of immune sera

- List several disorders of the immune system

Memmler, RL, Cohen, BJ, Wood, DL. *THE HUMAN BODY IN HEALTH AND DISEASE*, 8/e,
© 1996 Lippincott-Raven Publishers

The Body's Defenses Against Disease

C hapter 5 presents a rather frightening list of harmful organisms that surround us in our environment. Fortunately, most of us survive contact with these invaders and even become more resistant to disease in the process. The job of protecting us from these harmful agents belongs in part to certain blood cells and to the lymphatic system, which together make up our *immune system.*

The immune system is part of our general body defenses against disease. Some of these defenses are *nonspecific;* that is, they are effective against any harmful agent that enters the body. Other defenses are referred to as *specific;* that is, they act only against a certain agent and no other.

The Occurrence of Infection

Although the body is constantly exposed to pathogenic invasion, a large number of conditions determine whether an infection will actually occur. Many pathogens have a decided preference for certain body tissues. Some viruses attack only nervous tissue. The poliomyelitis pathogen, for example, may be inhaled or swallowed in large numbers and therefore may be in direct contact with the mucous membrane, yet it causes no apparent disorder of these respiratory or digestive system linings. In contrast, such pathogens as the influenza and cold viruses do attack these mucous membranes.

The respiratory tract is a common *portal of entry* for many pathogens. Other important avenues of entry include the digestive system and the tubes that open into the urinary and reproductive systems. Any break in the skin or in a mucous membrane allows organisms such as staphylococci easy access to the deeper tissues and may lead to infection, whereas unbroken skin or membrane usually is not affected at all. The portal of entry, then, is an important condition influencing the occurrence of infection.

The *virulence* (VIR-u-lens) of an organism, or the power of the organism to overcome the defenses of its host, must also be considered. Virulence has two aspects: one may be thought of as "aggressiveness," or invasive power; the other is the ability of the organism to produce *toxins* (poisons), which damage the body. The virulence of different organisms

varies. The virulence of a specific organism can also change; in other words, an organism can be more dangerous at some times than at others. As a rule, organisms that come from an already infected host are more "vicious" than the same type of organism grown under laboratory conditions.

The *dose* (number) of pathogens that invade the body also has much to do with whether an infection develops. Even if the virulence of a particular organism happens to be low, infection has a better chance of occurring if the number of them that enters the body is large than if the number is small.

Finally, the condition, or *predisposition,* of the individual to infection is also important. Disease organisms are around us all the time. Why does a person occasionally get colds, flu, or other infections? Part of the answer lies in the person's condition, as influenced by general physical and emotional health, nutrition, living habits, and age.

Nonspecific Defenses

The devices that protect the body against disease are usually considered successive "lines of defense," beginning with the relatively simple first, or outer, line and proceeding through progressively more complicated lines until the ultimate defense mechanism—immunity––is reached.

Chemical and Mechanical Barriers

The first line of defense against invaders includes the skin, which serves as a mechanical barrier as long as it remains intact. A serious danger to burn victims is the risk of infection due to destruction of the skin. The mucous membranes that line the passageways leading into the body also act as barriers, trapping foreign material in their sticky secretions. The cilia of the mucous membranes in the upper respiratory tract sweep impurities out of the body.

Body secretions such as tears, perspiration, and saliva wash away microorganisms and may contain acids, enzymes, or other chemicals that destroy invaders. Digestive juices destroy many ingested bacteria and their toxins. Finally, certain reflexes aid in the removal of pathogens. Sneezing and coughing, for instance, tend to remove foreign matter, including microorganisms, from the upper respiratory tract. Vomiting and diarrhea are ways in which toxins and bacteria may be expelled.

Phagocytosis

As previously noted, some white blood cells have the ability to take in and destroy waste and foreign material by the process called *phagocytosis* (see Fig. 13-2). It is mainly neutrophils (granular leukocytes) and macrophages that are active in this process. The macrophages are derived from monocytes (agranular leukocytes). Both they and the neutrophils travel in the blood to sites of infection. Some of the macrophages remain fixed in the tissues, for example in the skin, liver, lungs, lymphoid tissue, and bone marrow. All the macrophages together make up the reticuloendothelial system of protection. Phagocytosis is part of the second line of defense against invaders.

Natural Killer Cells

The *natural killer (NK) cell* is a type of lymphocyte (agranular leukocyte) different from those active in immunity, which are described below. These are cells that can recognize cells with abnormal membranes, such as tumor cells and cells infected with virus, and, as their name indicates, can destroy them on contact. NK cells are found in the lymph nodes, spleen, bone marrow, and blood. They destroy abnormal cells by secreting substances that break down the cell membrane, but the way in which they find their targets is not yet completely understood.

Inflammation

Inflammation is the body's effort to get rid of anything that irritates it (or, if this proves impossible, to limit the harmful effects of the irritant). Inflammation can occur as a result of any irritant, not only bacteria. Friction, fire, chemicals, x-rays, and cuts or blows all can be classed as irritants. If the irritant is due to pathogenic invasion, the resulting inflammation is termed an *infection.* With the entrance of pathogens and their subsequent multiplication, a whole series of defensive processes begins. Called an *inflammatory reaction,* this defense is accompanied by four classic symptoms: heat, redness, swelling, and pain.

When tissues are injured, an inflammatory reaction takes place. *Histamine* (HIS-tah-mene) and other substances are released from the damaged cells and cause the small blood vessels to dilate. More blood then flows into the area, resulting in heat, redness, and swelling.

With the increased blood flow come a vast number of leukocytes. Now a new phenomenon occurs: the walls of the tiny blood vessels become "coarsened" in texture (as does a piece of cloth when it is stretched), the blood flow slows down, and the leukocytes move through these altered walls and into the tissue, where they can get at the irritant directly. Fluid from the blood plasma also leaks out of the vessels into the tissues and begins to clot.

When this response occurs in a local area, it helps prevent the spread of the foreign agent. The mixture of leukocytes and fluid, the *inflammatory exudate,* causes pressure on the nerve endings; this, combined with the increased amount of blood in the vessels, causes the pain of inflammation. As the phagocytes do their work, large numbers of them are destroyed, so that eventually the area becomes filled with dead leukocytes. The mixture of exudate, living and dead white blood cells, pathogens, and destroyed tissue cells is *pus.*

Meanwhile, the lymphatic vessels begin to drain fluid from the inflamed area and carry it toward the lymph nodes for filtration. The regional lymph nodes become enlarged and tender, a sign that they are performing their protective function by working overtime to produce phagocytic cells that "clean" the lymph flowing through them.

Fever

An increase in body temperature above the normal range can be a sign that body defenses are at work. When phagocytes are exposed to infecting organisms, they release substances that act to raise body temperature. Fever boosts the immune system in several ways. It stimulates phagocytes, increases metabolism, and decreases the ability of certain organisms to multiply.

Advertisers would have the public believe that fever is a dangerous symptom that should always be eliminated. Control of fever in itself will do little to alter the course of an illness. The health care worker should, however, always be alert to the development of fever as a possible sign of a serious disorder and should recognize that an increased metabolic rate may have adverse effects on the heart of a weak or elderly ill person.

Interferon

Cells infected with viruses and certain other agents produce a substance that prevents the infection of other cells. This substance was first found in cells infected with influenza virus, and it was called *interferon* because it "interferes" with multiplication of the virus. Interferon, now known to be a group of substances, is also of interest because it nonspecifically stimulates the immune system. It has been used with varying success to boost the immune system in the treatment of malignancies and infections.

The remainder of this chapter is devoted to the immune system, our specific defense against disease.

▶ Immunity

Immunity is the final line of defense against disease. The term can be defined as "the power of an individual to resist or overcome the effects of a particular disease or other harmful agent." Immunity is a selective process; that is, immunity to one disease does not necessarily cause immunity to another. This selective characteristic is called *specificity* (spes-ih-FIS-ih-te).

There are two main categories of immunity: *inborn* (or inherited) *immunity* and *acquired immunity.* Acquired immunity may be obtained by *natural* or *artificial* means; in addition, acquired immunity may be either *active* or *passive.* Figure 17-1 is a summary of the different types of immunity. Refer to this diagram as we investigate each category in turn.

Inborn Immunity

Although certain diseases found in animals may be transmitted to humans, many infections, such as chicken cholera, hog cholera, distemper, and other animal diseases, do not affect human beings. However, the constitutional differences that make human beings immune to these disorders also make them susceptible to others that do not affect the lower animals. Such infections as measles, scarlet fever, diphtheria, and influenza do not seem to affect animals in contact with humans experiencing these illnesses. Thus, both humans and animals have what is called a *species immunity* to many of each other's diseases.

Another form of inborn immunity is *racial immunity.* Some racial groups appear to have a greater inborn immunity to certain diseases than other racial groups. For instance, in the United States, blacks are apparently more immune to poliomye-

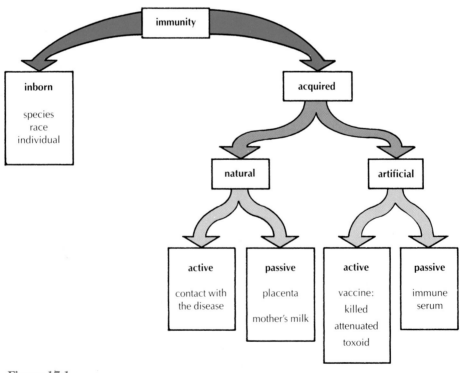

Figure 17-1

Types of immunity.

litis, malaria, and yellow fever than are whites. Of course, it is often difficult to tell how much of this variation in resistance to infection is due to environment and how much is due to inborn traits of the various racial groups.

Some members of a given group have a more highly developed *individual immunity* than other members. Newspapers and magazines sometimes feature the advice of an elderly person who is asked to give her secret for living to a ripe old age. One elderly person may say that he practiced temperance and lived a carefully regulated life with the right amount of rest, exercise, and work, whereas the next may boast of her use of alcoholic beverages, constant smoking, lack of exercise, and other kinds of reputedly unhygienic behavior. However, it is possible that the latter person has lived through the onslaughts of toxins and disease organisms, resisted infection, and maintained health in spite of her habits, rather than because of them, thanks to the resistance factors and immunity to disease she inherited.

Acquired Immunity

Unlike inborn immunity, which is due to inherited factors, acquired immunity develops during an individual's lifetime as that person encounters various specific harmful agents.

ANTIGENS

An *antigen* (AN-te-jen) *(Ag)* is any foreign substance that enters the body and produces an immune response. Most antigens are large protein molecules, but carbohydrates and some lipids may act as antigens. Antigens may be found on the surface of pathogenic organisms, on the surface of red blood cells and tissue cells, on pollens, in toxins, and in foods. The critical feature of any substance described as an antigen is that it stimulates the activity of certain lymphocytes classified as T or B cells.

T CELLS

Both T and B cells come from stem cells in bone marrow, as do all blood cells. They differ, however, in their development and their method of action. Some of the immature stem cells migrate to the thymus gland and become T cells, which constitute about 80% of the lymphocytes in the circulating blood. While in the thymus, these T lymphocytes multiply and become capable of combining with specific foreign antigens, at which time they are described as *sensitized*. These thymus-derived cells pro-

duce an immunity that is said to be *cell-mediated immunity.*

There are several types of T cells, each with different functions. Some of these functions are as follows:

1. To destroy foreign cells directly. These are the *killer (cytotoxic) T cells.*
2. To release substances that stimulate other lymphocytes and macrophages and thereby assist in the destruction of foreign cells. It is these *helper T cells* that are infected and destroyed by the AIDS virus. They have a special surface receptor (CD4) to which HIV, the virus that causes AIDS, attaches.
3. To suppress the immune response in order to regulate it. These *suppressor T cells* may inhibit or destroy active lymphocytes.
4. To remember an antigen and start a rapid response if that antigen is met again. These are *memory T cells.*

The T cell portion of the immune system is generally responsible for defense against cancer cells, certain viruses, and other pathogens that grow within cells (intracellular parasites), as well as for the rejection of tissue transplanted from another person.

Working with the T cells are *macrophages,* cells derived from blood monocytes. For a T cell to react with an antigen, that antigen must be presented to the T cell on the surface of a macrophage in combination with proteins that the T cell can recognize as belonging to the "self." After combining with T cells, the macrophages release substances called *interleukins* (a name that means "between white blood cells"). These stimulate the growth of T cells and have been used medically in efforts to boost the immune system.

B CELLS AND ANTIBODIES

An *antibody (Ab),* also known as an *immunoglobulin (Ig),* is a substance produced in response to an antigen. Antibodies are manufactured by the second type of lymphocyte active in the immune system. The *B cells,* or B lymphocytes, must mature in the fetal liver or in lymphoid tissue before becoming active in the blood. Exposure to an antigen stimulates B cells to multiply rapidly and produce large numbers (clones) of *plasma cells.* These cells produce specific antibodies that circulate in the blood providing the form of immunity described as *humoral immunity* (the term humoral refers to body fluids).

Humoral immunity generally protects against circulating antigens and bacteria that grow outside the cells (extracellular parasites). All antibodies are contained in a portion of the blood plasma called the *gamma globulin* fraction.

Some antibodies remain in the blood to give long-term immunity. In addition, some of the activated B cells do not become plasma cells but, like certain T cells, become memory cells. On repeated contact with an antigen, these cells are ready to produce antibodies immediately. Because of this "immunologic memory," one is usually immune to a childhood disease after having it.

THE ANTIGEN–ANTIBODY REACTION

The antibody that is produced in response to a specific antigen, such as a bacterial cell or a toxin, has a shape that matches some part of that antigen, much in the same way that the shape of a key matches the shape of its lock. For this reason the antibody can bind specifically to the antigen that caused its production and thereby destroy or inactivate it. In the laboratory this can be seen as a settling (precipitation) or a clumping (agglutination) of the antigen–antibody combination, as is seen in the typing of red blood cells.

The destruction of foreign cells sometimes requires the enzymatic activity of a group of nonspecific proteins in the blood together called *complement.*

If descriptions of the immune system seem complex, bear in mind that from infancy on, your immune system is able to protect you from millions of foreign substances, even those that are synthetic and not found in nature. All the while, the system is kept in check so that it does not usually overreact to produce allergies or mistakenly attack and damage your own body tissues.

Naturally Acquired Immunity

Immunity may be acquired naturally through contraction of a specific disease. In this case, antibodies manufactured by the infected person's cells act against the infecting agent or its toxins. Each time a person is invaded by the organisms of a disease, his or her cells may manufacture antibodies that provide immunity against the infection. Such immunity may last for years, and in some cases lasts for life. Because the host is actively involved in the production of antibodies, this type of immunity is called *active immunity.*

Antibodies

Antibodies are substances produced in response to specific antigens. They are all contained in a fraction of the blood plasma known as *gamma globulin.* Because the plasma contains other globulins as well, antibodies have become known as *immunoglobulins (Ig).* Immunologic studies have shown that there are several classes of immunoglobulins that vary in molecular size and in function. Studies of these antibody fractions can be of help in diagnosis. For example, high levels of IgM antibodies, because they are the first to be produced in an immune response, indicate a recent infection. Briefly described, these classes are as follows:

Immunoglobulin	Percentage of Antibodies	Function
IgG	75%	Protect against bacteria and viruses; the only antibodies that can pass through the placenta
IgA	15%	Local protection in mucous membranes
IgM	5%–10%	The first antibodies to be secreted after infection
IgD	Less than 1%	Help to activate B cells
IgE	Less than 0.1%	Active in allergic reactions and parasitic infections

The infection that calls forth the immunity may be a subclinical or inapparent infection that is so mild as to cause no symptoms. Nevertheless, it may stimulate the host's cells to produce an active immunity.

Immunity may be acquired naturally by a fetus through the passage of antibodies from the mother through the placenta. Because these antibodies come from an outside source, this type of immunity is called *passive immunity.* The antibodies obtained in this way do not last as long as actively produced antibodies, but they do help protect the infant for about 6 months, at which time the child's own immune system begins to function. Nursing an infant can lengthen this period of protection owing to the presence of specific antibodies in breast milk and colostrum (first breast secretion). These are the only known examples of naturally acquired passive immunity.

Artificially Acquired Immunity

A person who is not exposed to repeated small doses of a particular organism has no antibodies against that organism and is defenseless against a heavy onslaught of the pathogen. Therefore, artificial measures are usually taken to cause persons' tissues to actively manufacture antibodies. One could inject virulent pathogens into the tissues, but obviously this would be dangerous. Instead, the harmful agent is treated to reduce its virulence and then administered. In this way the tissues are made to produce antibodies without causing a serious illness. This process is known as *vaccination* (vak-sin-A-shun), or *immunization,* and the solution used is called a *vaccine* (vak-SENE). Ordinarily, the administration of a vaccine is a preventive measure designed to provide protection in anticipation of invasion by a certain disease organism.

VACCINES

Vaccines can be made with live organisms or with organisms killed by heat or chemicals. If live organisms are used, they must be nonvirulent for humans, such as the cowpox virus used for smallpox immunization, or they must be treated in the laboratory to weaken them as human pathogens. An organism weakened for use in vaccines is described as *attenuated.* A third type of vaccine is made from a form of the toxin produced by a disease organism. The toxin is altered with heat or chemicals to reduce its harmfulness, but it can still function as an antigen to induce immunity. Such an altered toxin is called a *toxoid.*

EXAMPLES OF VACCINES

Originally, the word *vaccination* meant inoculation against smallpox. (The term even comes from the Latin word for *cow,* referring to cowpox, which is used to vaccinate against smallpox.) According to the World Health Organization, however, smallpox has now been eliminated as a result of widespread immunization programs. Mandatory vaccination has been discontinued because the chance of adverse side effects from the vaccine is thought to be greater than the probability of contraction of the disease.

Because of the seriousness of whooping cough in young infants, early inoculation with whooping cough, or *pertussis* (per-TUS-is), vaccine, made from heat-killed whooping cough bacteria, is recommended. The pertussis vaccine usually is given in conjunction with diphtheria toxoid and tetanus toxoid, all in one mixture. This combination, referred to as *DTP* or *DPT,* may be given as early as the second month of life and should be followed by additional injections at 4, 6, and 15 months and again when the child enters day care, a school, or any other environment in which he or she might be exposed to one of these contagious diseases. Early immunizations now also include inoculation against *Haemophilus influenzae* type B (HIB), which causes life-threatening meningitis, pneumonia, and recurrent ear infections in young children.

Intensive research in viruses has resulted in the development of vaccines for an increasing number of viral diseases. Spectacular results in preventing poliomyelitis have been obtained by the use of a variety of vaccines. The first of these was the Salk vaccine, made with killed poliovirus. Now the more effective and convenient Sabin oral vaccine, made with live attenuated virus, is used. A number of vaccines have been developed against influenza, which is caused by several different strains of virus. MMR, the measles (rubeola), mumps, and rubella (German measles) vaccine, made with live attenuated viruses, is proving to be effective. The use of a rubella vaccine should lower the number of birth defects; rubella itself is very mild but has serious effects on the developing fetus (see Appendix 4, Table 3).

A vaccine against hepatitis B virus has been available since 1982. Initially, limited availability and high cost restricted its use to those at risk for the infection. Advances in the production of the vaccine by recombinant DNA techniques have increased its safety and availability. The U.S. Public Health Service has recommended universal childhood immunization against hepatitis B during the first year of life. A vaccine against hepatitis A virus has been approved and is recommended for travelers and others at high risk of infection.

A new varicella vaccine protects against the childhood disease chicken pox. Although this is usually a mild disease, infection of pregnant women can cause congenital malformation of the fetus. Because varicella is the same virus that causes shingles, vaccination may prevent this late life sequel.

The rabies vaccine is an exception to the rule that a vaccine should be given before the invasion of the disease organism. Rabies is a viral disease transmitted by the bite of wild animals such as raccoons, bats, foxes, and skunks. Mandatory vaccination of domestic animals has practically eliminated this source of rabies in some countries, including the United States, but worldwide a variety of wild and domestic animals are host to the virus. There is no cure for rabies; it is fatal in nearly all cases. However, the disease develops so slowly that affected persons vaccinated after transmission of the organism still have time to develop an active immunity. The vaccine may be given preventively to people who work with animals.

A final word about the long-term efficacy of vaccines: in many cases an active immunity acquired by artificial (or even natural) means does not last a lifetime. Circulating antibodies can decline with time. To help maintain a high titer (level) of antibodies in the blood, repeated inoculations, called *booster shots,* are administered at intervals. The number of booster injections recommended varies with the disease and with the environment or range of exposure of the individual.

The successful completion of childhood immunizations can be a public health concern. Because of recent epidemics in high schools and colleges, several vaccines have been recommended for boosters in this age group.

PASSIVE IMMUNIZATION

It takes several weeks to produce a naturally acquired active immunity and even longer to produce an artificial active immunity through the administration of a vaccine. Therefore, a person who receives a large dose of virulent organisms and has no established immunity to them stands in great danger. To prevent catastrophe, then, the victim must quickly receive a counteracting dose of borrowed antibodies. This is accomplished through the administration of an *immune serum,* or *antiserum.* The immune serum gives short-lived but effective protection against the invaders in the form of an artificially acquired passive immunity.

There are a number of differences between vaccines and antisera. Administration of a vaccine causes antibody formation by the body tissues. In the administration of an antiserum, in contrast, the antibodies are supplied "ready made." However, the passive immunity produced by an antiserum does not last as long as that actively produced by the body tissues.

Sera prepared for immune purposes are often derived from animals, mainly horses. It has been found that the tissues of the horse produce large quantities of antibodies in response to the injection of organisms or their toxins. After repeated injections, the horse is bled according to careful sterile technique; because of the size of the animal it is possible to remove large quantities of blood without causing injury. The blood is allowed to clot, and the serum is removed and packaged in sterile glass vials or other appropriate containers. It is then sent to the hospital, clinic, or doctor's office for use. Immune sera are usually used in emergencies, that is, in situations in which there is no time to wait until an active immunity has developed. Injecting humans with serum derived from animals is not without its problems. The foreign proteins in animal sera may cause an often serious sensitivity reaction, called *serum sickness.* To avoid this problem, human antibody in the form of gamma globulin may be used.

EXAMPLES OF ANTISERA

Some immune sera contain antibodies, known as *antitoxins,* that neutralize toxins but have no effect on the toxic organisms themselves. Certain antibodies act directly on pathogens, engulfing and destroying them or preventing their continued reproduction. Some antisera are obtained from animal sources, others from human sources. Examples of immune sera are listed below:

1. Diphtheria antitoxin, obtained from immunized horses
2. Tetanus immune globulin, effective in preventing lockjaw (tetanus), which is often a complication

of neglected wounds. Because tetanus immune globulin is of human origin, there is less chance of untoward reactions with it than with sera obtained from horses.

3. Immune globulin (human) is given to persons exposed to hepatitis A, measles, polio, or chicken pox. It is also given on a regular basis to persons with congenital (present at birth) immune deficiencies.
4. Hepatitis B immune globulin, used after hepatitis B exposure, is given principally to infants born of mothers who have hepatitis.
5. The immune globulin $Rh_o(D)$ (trade name RhoGAM), a concentrated human antibody given to prevent the formation of Rh antibodies by an Rh-negative mother. It is given during pregnancy with, and following the birth of, an Rh-positive infant (or even in a miscarriage of a presumably Rh-positive fetus) (see Chap. 13). It is also given when Rh transfusion incompatibilities occur.
6. Anti–snakebite sera, or **antivenins** (an-te-VEN-ins), used to combat the effects of bites of certain poisonous snakes
7. Botulism antitoxin, from horses, an antiserum that offers the best hope for botulism victims, although only if given early
8. Rabies antiserum, from humans or horses, used with vaccine to treat victims of bites of rabid animals.

▶ Disorders of the Immune System

Allergy

Allergy involves antigens and antibodies, and its chemical processes are much like those of immunity. Allergy—a broader term for which is *hypersensitivity*—can be defined informally as a tendency to react unfavorably to the presence of certain substances that are normally harmless to most people.

These substances are called **allergens** (AL-er-jens) and, like most antigens, are usually proteins. Examples of typical allergens are pollens, house dust, horse dander (*dander* is the term for the minute scales that are found on hairs and feathers), and certain food proteins. When the tissues of a susceptible person are repeatedly exposed to an allergen—for example, exposure of the nasal mucosa to pollens—these tissues become **sensitized;** that is, antibodies are produced in them. When the next invasion of the allergen occurs, there is an antigen–antibody reaction. Normally this type of reaction takes place in the blood without harm, as in immunity. In allergy, however, the antigen–antibody reaction takes place within the cells of the sensitized tissues, with results that are disagreeable and sometimes dangerous. In the case of the nasal mucosa that has become sensitized to pollen, the allergic manifestation is **hay fever,** with symptoms much like those of the common cold. Many drugs can bring about the allergic state, particularly aspirin, barbiturates, and antibiotics (especially penicillin).

As mentioned, an important allergic manifestation that may occur in response to various sera is serum sickness. Persons who are allergic to the proteins in serum derived from the horse or some other animal show such symptoms as fever, vomiting, joint pain, enlargement of the regional lymph nodes, and hives, or **urticaria** (ur-tih-KA-re-ah). This type of allergic reaction can be severe but is rarely fatal.

The antigen–antibody reaction in sensitive individuals may cause the release of excessive amounts of histamine. The histamine causes dilation and leaking from capillaries as well as contraction of involuntary muscles (*e.g.,* in the bronchi). A group of drugs called *antihistamines* is effective in treating the symptoms of certain allergies. Sometimes it is possible to desensitize an allergic person by means of repeated injections of the offending allergen at short intervals. Unfortunately, this form of protection does not last long.

Some allergic disorders are strongly associated with emotional disturbances. Asthma and **migraine** (MI-grane), or *sick headache,* are two examples of this type. In such disorders the interaction of body and mind is not fully understood.

Autoimmunity

The term **autoimmunity** refers to an abnormal reactivity to one's own tissues. In autoimmunity, the immune system reacts to the body's own cells, described as "self," as if they were foreign substances, or "nonself." Normally the immune system learns before birth to ignore (tolerate) the body's own tissues by eliminating or inactivating those lymphocytes that will attack them. The loss of this immune tolerance may be a cause of rheumatoid arthritis, multiple sclerosis, lupus erythematosus, Graves' disease, glomerulonephritis, and juvenile diabetes, among others.

Immune Deficiency Diseases

An immune deficiency is some failure of the immune system. This failure may involve any part of the system, such as T cells, B cells, or the thymus gland, and it may vary in severity. Such disorders may be congenital (present at birth) or may be acquired as a result of malnutrition, infection, or treatment with x-rays or certain drugs. The disease **AIDS** (*acquired immunodeficiency syndrome*) is a devastating example of an infection that attacks the immune system. Caused by HIV (*human immunodeficiency virus*), which destroys helper T cells, this disease was unknown in the United States before 1980. It first appeared among homosexual men and intravenous drug abusers, but has now spread to heterosexual men and women, teenagers, and young adults, many of whom have ignored warnings against unprotected sexual activity and the use of contaminated needles. The incidence among recipients of blood transfusions has declined thanks to identification of the virus and the testing of donated blood for its presence. So far the disease is uniformly fatal, with patients succumbing to rare diseases such as a parasitic (*Pneumocystis carinii*) pneumonia and an especially malignant skin cancer, **Kaposi's** (KAP-o-seze) **sarcoma**. Some drugs can slow the progress of AIDS, but so far, none can cure it. An obstacle to the development of a vaccine is the tremendous variability of the virus.

Cancer

Cancer cells differ slightly from normal body cells and therefore should be recognized as "nonself" by the immune system. The fact that persons with AIDS and other immune deficiencies develop cancer at a higher than normal rate suggests that this is so. Cancer cells probably form continuously in the body but normally are destroyed by NK cells and the immune system, a process called *immune surveillance* (sur-VAY-lans). As a person ages, cell-mediated immunity declines and cancer is more likely to develop.

Some efforts are being made to treat cancer by stimulating the patient's immune system, a practice called *immunotherapy*. In one approach, T cells have been removed from the patient, activated with interleukin, and then reinjected. This method has given some positive results, especially in treatment of melanoma, a highly malignant form of skin cancer.

AIDS Update

The AIDS virus, HIV, infects a specific subgroup of T helper cells that have surface receptors for the virus. Decrease in the number of cells with this so-called *CD4 receptor* is one of the criteria used to diagnose active cases of AIDS. Apparently the virus can remain in the lymphatic system for varying intervals, sometimes years, before it is triggered to reproduce and cause active AIDS.

Some of the avenues being explored for control of this disease include education in AIDS prevention, treatment with combinations of drugs, and development of a vaccine. There is great interest in people who are long-term survivors of HIV infection. They may hold a clue to how the virus is controlled by the immune system. Researchers are also studying another strain of the virus, HIV-2, which causes a chronic infection. This strain might reveal why HIV-1 is so virulent and also may be used to prepare a vaccine.

▶ Transplants and the Rejection Syndrome

The hope that organs and tissues might be obtained from animals or other human beings to replace injured or incompetent parts of the body has long been under discussion. Much experimental work with *transplantation* (grafting) of organs and tissues in animals has preceded transplant operations in humans. Transplants by grafting of bone marrow, lymphoid tissue, skin, corneas, parathyroid glands, ovaries, kidneys, lungs, heart, liver, and uterus are among those that have been attempted. The natural tendency of every organism to destroy foreign substances, including tissues from another person or any other animal, has been the most formidable obstruction to complete success. This normal antigen–antibody reaction has, in this case, been called the *rejection syndrome*.

In all cases of transplanting or grafting, the tissues of the donor, the person donating the part, should be typed in much the same way that blood is typed when a transfusion is given. Blood types are much fewer in number than tissue substances; thus, the process of obtaining matching blood is much less involved than is the process of obtaining matching tissues. Tissue typing is being done in a number of laboratories, and an effort is being made to obtain donors whose tissues contain relatively few antigens that might cause transplant rejection in a recipient (the person receiving the part).

Because it has been impossible to completely match all the antigens of a donor with those of the recipient, drugs that suppress the immune reaction to the transplanted organ are given to the recipient.

As you must now realize, these drugs also leave the patient unprotected from infection. Because much of the reaction against the foreign material in transplants is caused by the T cells, efforts are being made, using drugs and antibodies, to suppress the action of these lymphocytes without damaging the B cells. B cells produce circulating antibodies and are most important in preventing infections. Success with transplants will increase when methods are found to selectively suppress the immune attack on transplants without destroying the recipient's ability to combat disease.

SUMMARY

I. **Factors that influence occurrence of infection**
  A. Portal of entry of pathogen
  B. Virulence of pathogen
    1. Invasive power
    2. Production of toxins (poisons)
  C. Dose (number) of pathogens
  D. Predisposition of host
II. **Nonspecific defenses against disease**
  A. Chemical and mechanical barriers
    1. Skin
    2. Mucous membranes
    3. Body secretions
    4. Reflexes—coughing, sneezing
  B. Phagocytosis—mainly by neutrophils and macrophages
  C. Natural killer (NK) cells—attack tumor cells and virus-infected cells
  D. Inflammation
  E. Fever
  F. Interferon—substances released from virus-infected cells that prevent infection of other cells and stimulate the immune response
III. **Immunity**—specific defenses against disease
  A. Inborn immunity (inherited)—species, racial, individual
  B. Acquired immunity—obtained during life
    1. Immune response
      a. Antigens—stimulate lymphocyte activity
      b. Antibodies—inactivate pathogens or neutralize toxins
      c. T cells (T lymphocytes)

  (1) Processed in thymus
  (2) Produce cell-mediated immunity
  (3) Macrophages—present antigen to T cells
      d. B cells (B lymphocytes)
  (1) Mature in lymphoid tissue
  (2) Develop into plasma cells that produce circulating antibodies
  (3) Produce humoral immunity
    2. Naturally acquired immunity
      a. Active—acquired through contact with the disease
      b. Passive—acquired from antibodies obtained through placenta and mother's milk
    3. Artificially acquired immunity
      a. Active—immunization with vaccines
      b. Passive—administration of immune serum (antiserum)
IV. **Disorders of the immune system**
  A. Allergy—hypersensitivity to normally harmless substances (allergens)
  B. Autoimmunity—abnormal response to body's own tissues
  C. Immune deficiency disease—failure in the immune system
    1. Hereditary
    2. Acquired (*e.g.,* AIDS)
  D. Cancer—may be partly due to failure of immune system to destroy abnormal cells (surveillance)
V. **Transplantation**
  A. Tissue typing
  B. Suppression of immune system

QUESTIONS FOR STUDY AND REVIEW

1. Name four conditions that determine whether an infection will occur in the body.
2. What are the body's "lines of defense" against disease?
3. Define *phagocytosis*. Name cells active in phagocytosis.
4. What is meant by a *nonspecific defense*? Give several examples.
5. Describe the process of inflammation.
6. Define *immunity* and describe the basic immune process.
7. Give three examples of inborn immunity.
8. What is the basic difference between inborn and acquired immunity?
9. Define *antigen* and *antibody*. Why is the reaction between and antigen and antibody described as specific?
10. How do T and B cells differ? In what ways are they the same?
11. Name four types of T cells.
12. What is the role of macrophages in immunity?
13. Outline the various categories of acquired immunity and give an example of each.
14. What is a vaccine? a toxoid? Give examples of each. What is a booster shot?
15. What is an immune serum? Give examples. Define *antitoxin*.
16. Name two sources of antisera. Why is one sometimes used in preference to another?
17. Define *allergy*. How is the process of allergy like that of immunity, and how do they differ?
18. What is an allergen? Give some examples.
19. Name and describe some typical allergic disorders. Why would the diagnosis of some of these be especially difficult?
20. What does *autoimmunity* mean? Name four disorders caused by autoimmunity.
21. Why are bone marrow transplants sometimes used to treat cases of immune deficiency disease?
22. What does *AIDS* stand for? Name the virus that causes AIDS.
23. What is meant by *rejection syndrome,* and what is being done to offset this syndrome?

Energy: Supply and Use

It is the purpose of the five chapters of this unit to show how oxygen and nutrients are processed, taken up by the body fluids, and used by the cells to yield energy. This unit also describes how the stability of body functions (homeostasis) is maintained and how waste products are excreted.

CHAPTER 18

Respiration

Selected Key Terms

The following items are defined in the Glossary:

alveolus (pl., alveoli)

bronchiole

bronchus (pl., bronchi)

chemoreceptor

diaphragm

emphysema

epiglottis

hemoglobin

hilus (hilum)

hypoxia

larynx

lung

mediastinum

pharynx

pleura

pneumothorax

respiration

surfactant

trachea

ventilation

Behavioral Objectives

After careful study of this chapter, you should be able to:

- Define *respiration* and describe the three phases of respiration
- Name all the structures of the respiratory system
- Explain the mechanism for pulmonary ventilation
- List the ways in which oxygen and carbon dioxide are transported in the blood
- Describe the ways in which respiration is regulated
- List and define five symptoms of abnormal respiration
- Describe four types of respiratory infection
- Name the diseases involved in chronic obstructive pulmonary disease (COPD)

Memmler, RL, Cohen, BJ, Wood, DL. *THE HUMAN BODY IN HEALTH AND DISEASE*, 8/e,
© 1996 Lippincott-Raven Publishers

M̲ost people think of respiration simply as the process by which air moves into and out of the lungs, that is, *breathing*. By scientific definition, respiration is the process by which oxygen is obtained from the environment and delivered to the cells. Carbon dioxide is transported to the outside in a reverse pathway.

Respiration includes three phases:

1. *Pulmonary ventilation,* which is the exchange of air between the atmosphere and the air sacs of the lungs. This is normally accomplished by the inspiration and expiration of breathing.
2. The *diffusion* of gases, which includes the passage of oxygen from the air sacs into the blood and the passage of carbon dioxide out of the blood
3. The *transport* of oxygen to the cells and the transport of carbon dioxide from the cells to the lungs. This is accomplished by the circulating blood.

In the process of *cellular respiration,* oxygen is taken into the cell and used to break down nutrients with the release of energy. Carbon dioxide is the main waste product of cellular respiration.

The respiratory system is an intricate arrangement of spaces and passageways that conduct air into the lungs (Fig. 18-1). These spaces include the *nasal cavities;* the *pharynx* (FAR-inks), which is common to the digestive and respiratory systems; the voice box, or *larynx* (LAR-inks); the windpipe, or *trachea* (TRA-ke-ah); and the *lungs* themselves, with their conducting tubes and air sacs. The entire system might be thought of as a pathway for air between the atmosphere and the blood.

▶ The Respiratory System

The Nasal Cavities

Air makes its initial entrance into the body through the openings in the nose called the *nostrils.* Immediately inside the nostrils, located between the roof of the mouth and the cranium, are the two spaces known as the *nasal cavities.* These two spaces are separated from each other by a partition, the *nasal septum.* The septum and the walls of the nasal cavities are constructed of bone covered with mucous membrane. On the lateral (side) walls of each nasal cavity are three projections called the *conchae*

(KONG-ke). The conchae greatly increase the surface over which air must travel on its way through the nasal cavities.

The lining of the nasal cavities is a mucous membrane, which contains many blood vessels that bring heat and moisture to it. The cells of this membrane secrete a large amount of fluid—up to 1 quart each day. The following changes are produced in the air as it comes in contact with the lining of the nose:

1. Foreign bodies, such as dust particles and pathogens, are filtered out by the hairs of the nostrils or caught in the surface mucus.
2. Air is warmed by the blood in the vascular membrane.
3. Air is moistened by the liquid secretion.

For these protective changes to occur, it is preferable to breathe through the nose than through the mouth.

The *sinuses* are small cavities lined with mucous membrane in the bones of the skull. The sinuses communicate with the nasal cavities, and they are highly susceptible to infection.

The Pharynx

The muscular *pharynx,* or throat, carries air into the respiratory tract and carries foods and liquids into the digestive system. The upper portion, located immediately behind the nasal cavity, is called the *nasopharynx* (na-zo-FAR-inks); the middle section, located behind the mouth, is called the *oropharynx* (o-ro-FAR-inks); and the lowest portion is called the *laryngeal* (lah-RIN-je-al) *pharynx.* This last section opens into the larynx toward the front and into the esophagus toward the back.

The Larynx

The *larynx,* or voice box, (Fig. 18-2) is located between the pharynx and the trachea. It has a framework of cartilage, one of which is the thyroid cartilage that protrudes in the front of the neck. The projection formed by the thyroid cartilage is popularly called the *Adam's apple* because it is considerably larger in the male than in the female. On both sides at the upper end of the larynx are folds of mucous membrane used in producing speech. These are the vocal folds, or *vocal cords* (Fig. 18-3). They are set into vibration by the flow of air from the

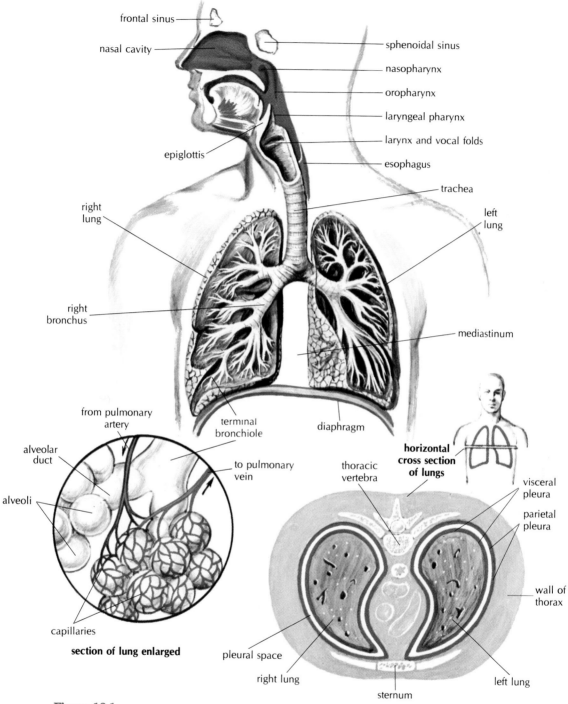

frontal sinus

nasal cavity

sphenoidal sinus

nasopharynx

oropharynx

laryngeal pharynx

larynx and vocal folds

epiglottis

esophagus

trachea

right lung

left lung

right bronchus

mediastinum

from pulmonary artery

terminal bronchiole

to pulmonary vein

alveolar duct

diaphragm

horizontal cross section of lungs

thoracic vertebra

visceral pleura

parietal pleura

alveoli

capillaries

wall of thorax

section of lung enlarged

pleural space

right lung

left lung

sternum

Figure 18-1

Respiratory system.

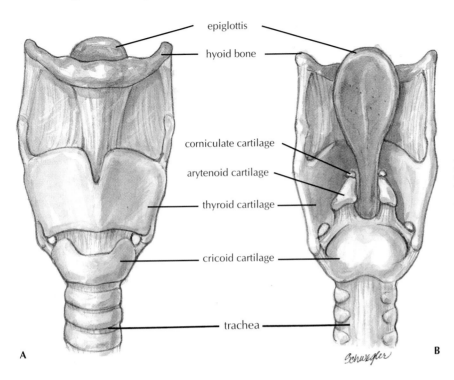

epiglottis

hyoid bone

corniculate cartilage

arytenoid cartilage

thyroid cartilage

cricoid cartilage

trachea

A

B

Figure 18-2

The larynx. **(A)** Anterior view. **(B)** Posterior view.

lungs. A difference in the size of the larynx is what accounts for the difference between male and female voices; because a man's larynx is larger than a woman's, his voice is lower in pitch. The nasal cavities, the sinuses, and the pharynx all serve as resonating chambers for speech, just as the cabinet does for a stereo speaker.

The space between the vocal cords is called the *glottis* (GLOT-is), and the little leaf-shaped cartilage that covers the larynx during swallowing is called the *epiglottis* (ep-ih-GLOT-is). The glottis and epiglottis help keep food out of the remainder of the respiratory tract. As the larynx moves upward and forward during swallowing, the epiglottis moves downward, covering the opening into the larynx. The glottis assists by closing during swallowing. You can feel the larynx move upward toward the epiglottis during this process by placing the flat ends of your fingers on your larynx as you swallow.

The Trachea, or Windpipe

The *trachea* is a tube that extends from the lower edge of the larynx to the upper part of the chest above the heart. It has a framework of cartilages to keep it open. These cartilages, shaped somewhat like a tiny horseshoe or the letter C, are found along the entire length of the trachea. The open sections

in the cartilages are lined up in the back so that the esophagus can bulge into this region during swallowing. The purpose of the trachea is to conduct air between the larynx and the lungs.

The Bronchi

At its inferior end, the trachea divides into two primary, or main stem, *bronchi* (BRONG-ki) which enter the lungs. The right bronchus is considerably larger in diameter than the left and extends downward in a more vertical direction. Therefore, if a foreign body is inhaled, it is likely to enter the right lung. Each bronchus enters the lung at a notch or depression called the *hilus* (HI-lus) or *hilum* (HI-lum). Blood vessels and nerves also connect with the lung in this region.

THE LINING OF THE AIR PASSAGEWAYS

The bronchi and other conducting passageways of the respiratory tract are lined with a special type of epithelium (Fig. 18-4). Basically, it is simple columnar epithelium, but the cells are arranged in such a way that they appear stratified. The tissue is thus described as *pseudostratified* meaning "falsely stratified." These epithelial cells have cilia to filter out impurities and to create movement of fluids within the conducting tubes. The cilia beat to drive

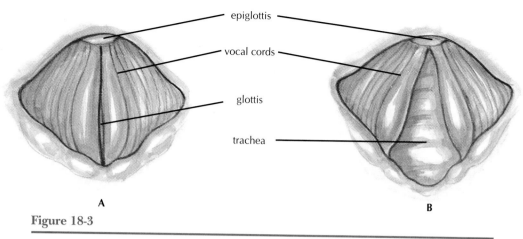

epiglottis

vocal cords

glottis

trachea

A

B

Figure 18-3

The vocal cords viewed from above. **(A)** The glottis in closed position. **(B)** The glottis in open position.

impurities toward the throat where they can be eliminated by coughing, sneezing, or blowing the nose.

The Lungs

The *lungs* are the organs in which the diffusion of gases takes place through the extremely thin and delicate lung tissues. The two lungs, set side by side in the thoracic (chest) cavity, are constructed in the following manner:

Each primary bronchus enters the lung at the hilus and immediately subdivides. The right bronchus divides into three secondary bronchi, each of which enters one of the three lobes of the right lung. The left bronchus gives rise to two secondary bronchi, which enter the two lobes of the left lung. Because the subdivisions of the bronchi resemble the branches of a tree, they have been given the common name *bronchial tree*. The bronchi subdivide

again and again becoming progressively smaller as they branch through lung tissue. The smallest of these conducting tubes are called *bronchioles* (BRONG-ke-oles). The bronchi contain small bits of cartilage, which give firmness to the walls and serve to hold the passageways open so that air can pass in and out easily. However, as the bronchi become smaller, the cartilage decreases in amount. In the bronchioles there is no cartilage at all; what remains is mostly smooth muscle, which is under the control of the autonomic (involuntary) nervous system.

At the end of the *terminal bronchioles,* the smallest subdivisions of the bronchial tree, are the clusters of tiny air sacs in which most gas exchange takes place. These sacs are known as *alveoli* (al-VE-o-li). The wall of each alveolus is made of a single-cell layer of squamous (flat) epithelium. This very thin wall provides easy passage for the gases entering and leaving the blood as it circulates through the mil-

Figure 18-4

Microscopic view of the ciliated epithelium that lines the respiratory passageways (Cormack DH: Essential Histology, plate 4-1D. Philadelphia, JB Lippincott, 1993)

cilia

goblet cell (secretes mucus)

columnar epithelium (pseudo stratified)

lions of tiny capillaries covering the alveoli. Certain cells in the alveolar wall produce *surfactant* (sur-FAK-tant), a substance that prevents the alveoli from collapsing by reducing the surface tension ("pull") of the fluids that line them. There are millions of alveoli in the human lung. The resulting surface area in contact with gases approximates 60 square meters, about three times as much lung tissue as is necessary for life. Because of the many air spaces, the lung is light in weight; normally a piece of lung tissue dropped into a glass of water will float. Figure 18-5 shows a microscopic view of lung tissue.

The pulmonary circuit brings blood to and from the lungs. In the lungs the blood passes through the capillaries around the alveoli, where the gas exchange takes place (see Fig. 18-7).

▶ The Lung Cavities

The lungs occupy a considerable portion of the thoracic cavity, which is separated from the abdominal cavity by the muscular partition known as the *diaphragm.* Each lung is covered by a continuous doubled sac, known as the *pleura.* Each layer of the pleura is named according to its location. The portion of the pleura that is attached to the chest wall is the *parietal pleura,* and the portion that is attached to the surface of the lung is called the *visceral pleura.* Each closed sac completely surrounds the lung, except in the place where the bronchus and blood vessels enter the lung, known as the *root* of the lung. Between the two layers of the pleura is the *pleural space,* in which there is a thin film of fluid that lubricates

Figure 18-5

Lung tissue as seen through a microscope. (Courtesy of Dana Morse Bittus and B. J. Cohen)

the membranes. The effect is the same as two pieces of glass joined by a film of water: that is, they slide easily on each other but strongly resist separation. Thus, the lungs are able to move and enlarge effortlessly in response to changes in the thoracic volume that occur during breathing.

The region between the lungs, the *mediastinum* (me-de-as-TI-num), contains the heart, great blood vessels, esophagus, trachea, and lymph nodes.

▶ Physiology of Respiration

Pulmonary Ventilation

Ventilation is the movement of air into and out of the lungs, as in breathing. There are two phases of ventilation (Fig. 18-6):

1. *Inhalation* is the drawing of air into the lungs.
2. *Exhalation* is the expulsion of air from the lungs.

In *inhalation,* the active phase of breathing, the respiratory muscles contract to enlarge the thoracic cavity. The diaphragm is a strong, dome-shaped muscle attached to the body wall around the base of the rib cage. The contraction and flattening of the diaphragm cause a piston-like downward motion that results in an increase in the vertical dimension of the chest. During quiet breathing, the movement of the diaphragm accounts for most of the increase in thoracic volume. The muscles between the ribs (intercostal muscles) also participate. During exertion, the rib cage is moved further up and out by contraction of muscles in the neck and chest wall.

As the thoracic cavity increases in size, gas pressure within the cavity decreases. When the pressure drops to slightly below atmospheric pressure, air is drawn into the lungs, as by suction.

In *exhalation,* the passive phase of breathing, the muscles of respiration relax, allowing the ribs and diaphragm to return to their original positions. The tissues of the lung are elastic and recoil during exhalation. During forced exhalation, the internal intercostal muscles and the muscles of the abdominal wall contract, pulling the bottom of the rib cage in and down, pushing the abdominal viscera upwards against the relaxed diaphragm.

Air Movement

Air enters the respiratory passages and flows through the ever-dividing tubes of the bronchial tree. As the air traverses this passage, it moves more and more slowly through the great number of bronchial tubes until there is virtually no forward

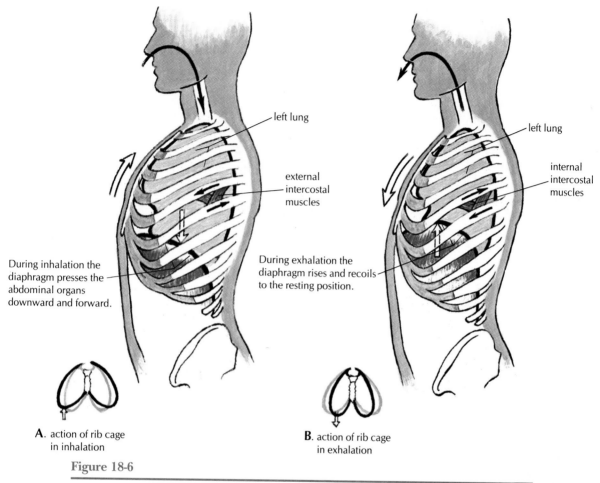

left lung

external
intercostal
muscles

During inhalation the
diaphragm presses the
abdominal organs
downward and forward.

left lung

internal
intercostal
muscles

During exhalation the
diaphragm rises and recoils
to the resting position.

A. action of rib cage
in inhalation

B. action of rib cage
in exhalation

Figure 18-6

(A) Inhalation. **(B)** Exhalation.

flow as it reaches the alveoli. Here the air moves by diffusion, which soon equalizes any differences in the amounts of gases present. Each breath causes relatively little change in the gas composition of the alveoli, but normal continuous breathing ensures the presence of adequate oxygen and the removal of carbon dioxide.

Table 18-1 gives the definitions of and average values for some of the breathing volumes and capacities that are important in any evaluation of respiratory function. A lung *capacity* is a sum of volumes.

Gas Exchanges

The barrier that separates the air in the alveolus from the blood in the capillary is very thin and moist, ideally suited for the exchange of gases by diffusion. Normally, inspired air contains about 21% oxygen

and 0.04% carbon dioxide; expired air has only 16% oxygen and 3.5% carbon dioxide. A two-way diffusion takes place through the walls of the alveoli.

Recall that *diffusion* refers to the movement of molecules from an area in which they are in higher concentration to an area where they are in lower concentration. Blood entering the lung capillaries after returning from the tissues is relatively low in oxygen. Therefore, oxygen diffuses from the alveolus, where its concentration is higher, into the blood. Again, based on relative concentration, carbon dioxide diffuses out of the blood into the air of the alveolus (Fig. 18-7).

Gas Transport

Almost all the oxygen that diffuses into the capillary blood in the lungs is bound to the **hemoglobin** of the red blood cells. A very small percentage is

TABLE 18-1
Lung Volumes and Capacities

Volume	Definition	Average Value
Tidal volume	The amount of air moved into or out of the lungs in quiet, relaxed breathing	500 mL
Residual volume	The volume of air that remains in the lungs after maximum inhalation	1200 mL
Vital capacity	The volume of air that can be expelled from the lungs by maximum exhalation following maximum inhalation	4800 mL
Total lung capacity	The total volume of air that can be contained in the lungs after maximum inhalation	6000 mL
Functional residual capacity	The amount of air remaining in the lungs after normal exhalation	2400 mL

carried in solution in the plasma. The hemoglobin molecule is a large protein with four small iron-containing "heme" regions. The oxygen is bound to these heme portions. Arterial blood (in systemic arteries and pulmonary veins) is 97% saturated with oxygen, whereas venous blood (in systemic veins and pulmonary arteries) is about 70% saturated with oxygen. This 27% difference represents the oxygen that has been taken up by the cells. Note that in clinical practice, gas concentrations are expressed as pressure in millimeters mercury (mmHg), just as blood pressure is reported. Because air is a mixture of gases, each gas exerts a partial pressure that is abbreviated as P_{O_2} and P_{CO_2} for oxygen and carbon dioxide respectively.

To enter the cells, oxygen must separate from hemoglobin. Normally the bond between oxygen and hemoglobin is easily broken, with oxygen being released as blood travels into areas in which the oxygen concentration is relatively low.

The carbon dioxide produced in the tissues is transported to the lungs in three ways:

1. About 10% is dissolved in the plasma and the fluid in red blood cells.
2. About 20% is combined with the protein portion of hemoglobin and plasma proteins.
3. About 70% is transported as an ion, known as a **bicarbonate ion,** which is formed when carbon dioxide dissolves in blood fluids.

The bicarbonate ion is formed slowly in the plasma but much more rapidly inside the red blood cells, where an enzyme called **carbonic anhydrase** in-creases the speed of the reaction. The bicarbonate formed in the red blood cells moves to the plasma and then is carried to the lungs. Here, the process is reversed as bicarbonate reenters the red blood cells and releases carbon dioxide for diffusion into the alveoli and exhalation.

The Air We Breathe

The air we breathe comes under close inspection because of our increasing awareness of air pollution, both indoors and out. The air in many populated areas is polluted by automobile exhaust and industrial smoke. Air in agricultural areas may include smoke from burning fields and drift of aerial pesticide applications far from the intended target. Lumber operations may burn scrap. Wind carries dust and pollen everywhere. Much work is being carried out to decrease these sources of airborne pollution.

Air inside homes or offices often carries a less visible burden of pollution. Many products give off invisible gases that tend to accumulate in modern buildings, which usually are tightly sealed. Unless adequate amounts of clean air are supplied, the occupants often complain of vague illnesses. Attention needs to be given to processing the air not only for temperature, but also for humidity. Most importantly, the air must be filtered so that undesirable pollutants are removed.

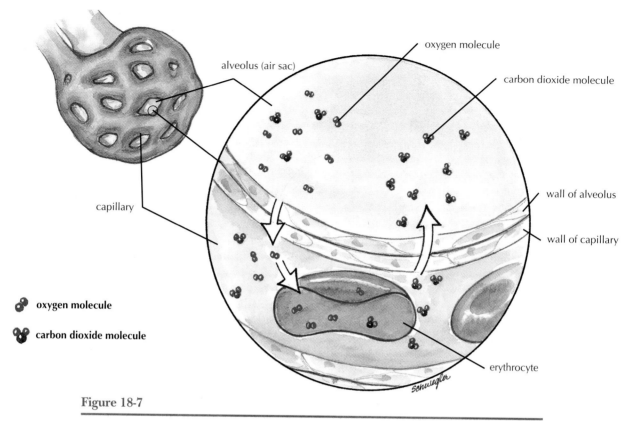

oxygen molecule

carbon dioxide molecule

alveolus (air sac)

oxygen molecule

carbon dioxide molecule

wall of alveolus

wall of capillary

capillary

erythrocyte

Figure 18-7

Diagram showing the diffusion of gas molecules in the lungs.

Carbon dioxide is important in regulating the pH (acid–base balance) of the blood. As a bicarbonate ion is formed from carbon dioxide in the plasma, a hydrogen ion (H+) is also produced. Therefore, the blood becomes more acidic as the amount of carbon dioxide in the blood increases. The exhalation of carbon dioxide shifts the pH of the blood more toward the alkaline (basic) range. The bicarbonate ion is also an important buffer in the blood, acting chemically to help keep the pH of body fluids at a steady pH of 7.4.

Regulation of Respiration

Regulation of respiration is a complex process that must keep pace with moment-to-moment changes in cellular oxygen requirements and carbon dioxide production. Regulation depends primarily on the respiratory control centers located in the medulla and pons of the brain stem. The main control center in the medulla sets the basic pattern of respiration. This can be modified by centers in the pons. Breathing is regulated so that levels of oxygen, carbon dioxide, and acid are kept within normal limits.

From the respiratory center in the medulla, motor nerve fibers extend into the spinal cord. From the cervical (neck) part of the cord, these nerve fibers continue through the ***phrenic*** (FREN-ik) ***nerve*** to the diaphragm. The diaphragm and the other muscles of respiration are voluntary in the sense that they can be regulated by messages from the higher brain centers, notably the cortex. It is possible for a person to deliberately breathe more rapidly or more slowly or to hold his breath and not breathe at all for a time. Usually we breathe without thinking about it, while the respiratory centers in the medulla and pons do the controlling.

Of vital importance in the control of respiration are ***chemoreceptors*** (ke-mo-re-SEP-tors), which, like the receptors for taste and smell, are sensitive to chemicals dissolved in body fluids. The receptors that regulate respiration are found in structures called the *carotid* and *aortic bodies*, as well as outside the medulla of the brain stem. The carotid bodies are located near the bifurcation (forking) of the common carotid arteries in the neck, whereas the aortic bodies are located in the aortic arch. These bodies contain many small blood vessels and sensory neurons, which are sensitive to increases in carbon dioxide and acidity (H+) and to decreases in oxygen supply.

Changes in the composition of the blood generate nerve impulses that are then sent to the brain. Because there is usually an ample reserve of oxygen in the blood, carbon dioxide has the most immediate effect in regulating respiration. When carbon dioxide levels increase, breathing must be increased to blow off the excess. It is only when oxygen levels fall considerably that this gas becomes a controlling factor. The receptor cells outside the medulla respond to the concentration of hydrogen ion in cerebrospinal fluid (CSF) as governed by the amount of carbon dioxide that is dissolved in the blood.

THE EFFECTS OF ABNORMAL VENTILATION

Hyperventilation is deep, rapid respiration that occurs commonly during anxiety attacks. It causes an increase in the oxygen level and a decrease in the carbon dioxide level of the blood. The loss of carbon dioxide increases the pH of the blood (alkalosis), resulting in dizziness and tingling sensations. Breathing may stop because the respiratory control centers are not stimulated. Gradually the carbon dioxide level returns to normal and a regular breathing pattern is resumed. Symptoms of hyperventilation may be relieved by breathing into a paper bag to increase the carbon dioxide level in the blood.

In *hypoventilation,* an insufficient amount of air enters the alveoli. The many possible causes of this condition include respiratory obstruction, lung disease, toxins, damage to the respiratory center, and chest deformity. Hypoventilation results in an increase in the concentration of carbon dioxide in the blood leading to a decrease in blood pH (acidosis). Unless the respiratory center itself is injured or depressed, respiration rate may increase, but breathing is shallow and difficult.

Respiratory Rates

Normal rates of breathing vary from 12 to 20 times per minute for adults. In children, rates may vary from 20 to 40 times per minute, depending on age and size. In infants, the respiratory rate may be more than 40 times per minute. To determine the respiratory rate, the health worker counts the client's breathing for at least 30 seconds, observing in such a way that the person is unaware that a count is being made. Changes in respiratory rates are important in various disorders and should be recorded carefully.

Terms Describing Abnormal Respiration

The following is a list of terms designating various abnormalities of respiration. They are symptoms, not diseases.

1. *Hyperpnea* (hi-perp-NE-ah) refers to an increase in the depth and rate of respiration.
2. *Apnea* (AP-ne-ah) is a temporary cessation of breathing.
3. *Dyspnea* (DISP-ne-ah) is a subjective feeling of difficult or labored breathing.
4. *Cheyne–Stokes* (CHANE-stokes) respiration is a rhythmic variation in the depth of respiratory movements found in certain critically ill persons.
5. *Orthopnea* (or-thop-NE-ah) refers to a difficulty in breathing that is relieved by sitting in an upright position, either against two pillows in bed or in a chair.

Situations that may occur in relation to changes in respiration include the following:

1. *Cyanosis* (si-ah-NO-sis) is a bluish color of the skin and mucous membranes caused by an insufficient amount of oxygen in the blood.
2. *Hypoxia* (hi-POK-se-ah) means a lower than normal level of oxygen in the tissues. The term *anoxia* (ah-NOK-se-ah) is sometimes used instead.
3. *Hypoxemia* (hi-pok-SE-me-ah) refers to a lower than normal concentration of oxygen in arterial blood.
4. *Suffocation* is the cessation of respiration, often the result of a mechanical blockage of the respiratory passages.

▶ Disorders of the Respiratory System

Disorders of the Nasal Cavities and Related Structures
SINUSITIS AND NASAL POLYPS

The sinuses are located close to the nasal cavities and in one case near the ear. Infection may easily travel into these sinuses from the mouth, the nose, and the throat along the mucous membrane lining; the resulting inflammation is called *sinusitis.* Long-standing, or chronic, sinus infection may cause changes in the epithelial cells, resulting in tumor formation. Some of these growths have a grapelike appearance and cause obstruction of the air pathway; these tumors are called *polyps* (POL-ips).

DEVIATED SEPTUM

The partition separating the two nasal cavities is called the *nasal septum.* Because many of us have minor structural defects, it is not surprising that the nasal septum is rarely exactly in the midline. If it is markedly to one side, it is described as a *deviated septum.* In this condition one nasal space may be considerably smaller than the other. If an affected person has an attack of hay fever or develops a cold with accompanying swelling of the mucosa, the smaller nasal cavity may be completely closed. Sometimes the septum is curved in such a way that both nasal cavities are occluded, forcing the individual to breathe through the mouth. Such an occlusion may also prevent proper drainage from the sinuses and aggravate a case of sinusitis.

NOSEBLEED

The most common cause of nosebleed, also called *epistaxis* (ep-e-STAK-sis), is an injury or blow to the nose. Other causes include inflammation and ulceration such as may occur following a persistent discharge from a sinusitis. Growths, including polyps, can also cause epistaxis. Rarely, abnormally high blood pressure causes the vessels in the nasal lining to break, resulting in varying degrees of hemorrhage. To stop a nosebleed, the affected person should remain quiet with the head slightly elevated. Pressure applied to the nostril of the bleeding side and cold compresses over the nose usually are helpful. In some cases, it may be necessary to insert a plug into the bleeding side to encourage adequate clotting. If these methods fail, a physician should be consulted.

Infection
THE SPREAD OF INFECTION

The mucosa of the respiratory tract is one of the most important portals of entry for disease-producing organisms. The transfer of disease organisms from the respiratory system of one human being to that of another occurs most rapidly in crowded places such as schools, theaters, and institutions.

Droplets from one sneeze may be loaded with many billions of disease-producing organisms. To a certain extent the mucous membranes can protect themselves by producing larger quantities of mucus. The runny nose, an unpleasant symptom of the common cold, is an attempt on the part of nature to wash away the pathogens and so protect the deeper tissues from further invasion by the infection. If the resistance of the mucous membrane is reduced, however, the membrane may act as a pathway for the spread of disease. The infection may travel along the membrane into the nasal sinuses, into the middle ear, or into the lung.

Among the infections transmitted through the respiratory passageways are the common cold, diphtheria, chickenpox, measles, influenza, pneumonia, and tuberculosis. Any infection that is confined to the nose and throat is called an *upper respiratory infection* (often called *URI*). Very often, the first symptoms of infectious diseases in children are those of an upper respiratory infection. Such an infection may precede the onset of a serious disease such as rheumatic fever.

The respiratory passageways may become infected one by one as the organisms travel along the lining membrane. The disorder is named according to the part involved, such as pharyngitis (commonly called a *sore throat*), laryngitis, bronchitis, and so forth.

THE COMMON COLD

The common cold is the most widespread of all respiratory diseases—of all communicable diseases, for that matter. More time is lost from school and from work because of the common cold than because of any other disorder. The causative agents include many cold viruses and other infectious organisms. Because there are so many causative organisms, the production of an effective vaccine against colds seems unlikely. Medical science has yet to establish the effectiveness of any method of preventing the common cold.

The symptoms of the common cold are familiar: first the swollen and inflamed mucosa of the nose and the throat, then the copious discharge of watery fluid from the nose, and finally the thick and ropy discharge that occurs when the cold is subsiding. The scientific name for the common cold is *acute coryza* (ko-RI-zah); the word *coryza* can also mean simply "a nasal discharge."

A discharge from the nasal cavities may not only be a symptom of the common cold; it may also stem from sinusitis or be an important forerunner of a more serious disease.

The notion that all nasal mucus is undesirable is wrong. Mucous secretions are required to keep the tissues moist as well as to help protect the cells against invasion by pathogens. The idea that excess secretion is caused by eggs, milk, or any other food in the diet has not been proved, except in the case of an allergy.

INFLUENZA

Influenza, or "flu," is an acute contagious disease characterized by an inflammatory condition of the upper respiratory tract accompanied by generalized aches and pains. It is caused by a virus and may spread to the sinuses and downward to the lungs. Inflammation of the trachea and the bronchi causes the characteristic cough of influenza, and the general infection causes an extremely weakened condition. The great danger of influenza is its tendency to develop into a particularly severe form of pneumonia. At intervals in history, there have been tremendous epidemics of influenza in which millions of people have died. Vaccines have been effective, although the protection is of short duration.

PNEUMONIA

Pneumonia is an inflammation of the lungs in which the air spaces become filled with fluid. A variety of organisms, including staphylococci, pneumococci, streptococci, *Legionella pneumophila* (as in legionnaires' disease), chlamydias, and viruses may be responsible. Many of these pathogens may be carried by a healthy person in the mucosa of the upper respiratory tract. If the person remains in good health, these pathogens may be carried for an indefinite period with no ill effect. However, if the individual's resistance to infection is lowered, the pathogens may invade the tissues and cause disease. Susceptibility to pneumonia is increased in persons with chronic, debilitating illness or chronic respiratory disease, in smokers, and in alcoholics. It is also increased in cases of exposure to toxic gases, immunosuppression, or viral respiratory infections.

There are two main kinds of pneumonia as determined by the extent of lung involvement and other factors. These are:

1. *Lobar pneumonia,* in which an entire lobe of the lung is infected at one time. The organism is usually a pneumococcus, although other pathogens also may cause this disease. The *Legionella* organ-

ism is the causative agent of a severe lobar pneumonia that occurs mostly in localized epidemics.

2. *Bronchopneumonia,* in which the disease process is scattered here and there throughout the lung. The cause may be a staphylococcus, a gram-negative proteus or colon bacillus (not normally pathogenic), or a virus. Bronchopneumonia most often is secondary to an infection or to some agent that has lowered the individual's resistance to disease. This is the most common form of pneumonia.

A characteristic of most types of pneumonia is the formation of a fluid, or *exudate,* in the infected alveoli; this fluid consists chiefly of serum and pus, products of infection. Some red blood cells may be present, as indicated by red streaks in the sputum. Sometimes so many air sacs become filled with fluid that the victim finds it hard to absorb enough oxygen to maintain life.

TUBERCULOSIS

Tuberculosis (TB) is an inflammation caused by the bacillus *Mycobacterium tuberculosis.* Although the tubercle bacillus may invade any tissue in the body, the lung is the usual site. Tuberculosis remains a leading cause of death from communicable disease primarily because of the relatively large numbers of cases among recent immigrants and poor population groups in metropolitan areas. The spread of AIDS has been linked with rising TB, as this viral disease weakens host defenses.

The lungs, with the pleurae, are the organs affected most often by tuberculosis, but other organs may be infected. The lymph nodes in the thorax, especially those surrounding the trachea and the bronchi, frequently are involved. In tuberculous pleurisy (inflammation of the pleura) with fluid formation, fluid may collect in the pleural space; such a collection is called an *effusion* (e-FU-zhun).

Drugs can be used successfully in many cases of tuberculosis, although strains of the TB organism that are resistant to multiple antibiotics have appeared recently. The best results have been obtained by use of a combination of several drugs, with prompt, intensive, and uninterrupted treatment once such a program is begun. Such therapy is usually continued for a minimum of 6 to 18 months; therefore, close supervision by the health care practitioner is important. Adverse drug reactions are rather common, necessitating changes in the drug combinations. Drug treatment of persons with infection that has not progressed to active disease is particularly effective.

Hay Fever and Asthma

Sensitivity to plant pollens, dust, certain foods, and other allergens may lead to *hay fever* or *asthma* or both. Hay fever is characterized by a watery discharge from the eyes and nose. The symptoms of asthma are usually due to a spasm of the involuntary musculature of the bronchial tube walls. This spasm constricts the tubes so that the victim cannot exhale easily. He or she experiences a sense of suffocation and has labored breathing (dyspnea). Much has been written about the role that psychologic factors play in the causation of asthma, but individuals vary considerably, and most cases of asthma involve a multiplicity of problems.

One great difficulty in the treatment of hay fever or asthma is isolation of the particular substance to which the patient is allergic. Usually a number of skin tests are given, but in most cases the results of these are far from conclusive.

Persons with asthma may benefit from treatment with a series of injections to reduce their sensitivity to specific substances.

Emphysema and Bronchitis

Chronic obstructive pulmonary disease (COPD) is the term used to describe several lung disorders, including *chronic bronchitis* and *emphysema*. Most affected persons have symptoms and lung damage characteristic of both diseases. In chronic bronchitis the linings of the airways are chronically inflamed and produce excessive secretions. Emphysema is characterized by dilation and finally destruction of the alveoli.

In COPD, respiratory function is impaired by obstruction to normal air flow, by air trapping and overinflation of parts of the lungs, and by reduced exchange of oxygen and carbon dioxide. In the early stages of these diseases, the small airways are involved, and several years may pass before symptoms become evident. Later the affected person develops dyspnea owing to the difficulty of exhaling air through the obstructed air passages.

The progressive character of these diseases may be reversed when they are detected early in their course and irritants to the respiratory system are eliminated. The major culprit is cigarette smoke, although industrial wastes and air pollution also play a large role. In the popular press the word *emphysema* is used to mean *COPD*.

Atelectasis

Atelectasis (at-e-LEK-tah-sis) is the incomplete expansion of a lung or portion of a lung. This disorder may affect scattered groups of alveoli or an entire lobe, causing its collapse. The affected person will have hypoxia and dyspnea. Obstruction by mucous plugs in COPD, by foreign bodies, or by lung cancer is a cause of this disorder. Another cause may be the insufficient production of surfactant, as in *respiratory distress syndrome* (*RDS*), also called *hyaline membrane disease*, seen in premature newborns. Atelectasis is also seen in cases of external compression of the lung or interference with deep breathing, for example with pain from fractured ribs.

Lung Cancer

Cancer of the lungs is the most common cause of cancer deaths in both men and women. The rate in women is still increasing; the rate in men is stable or slightly decreasing. By far the most important cause of lung cancer is cigarette smoking. Smokers suffer from lung cancer ten times as often as nonsmokers. The risk of getting cancer of the lungs is increased in persons who started smoking early in life, who smoke large numbers of cigarettes daily, and who inhale deeply. Smokers who are exposed to toxic chemicals or particles in the air have an even higher rate of lung cancer. Smoking has also been linked with an increase in COPD and cancers of respiratory passages.

A common form of lung cancer is *bronchogenic* (brong-ko-JEN-ik) *carcinoma,* so called because the malignancy originates in a bronchus. The tumor may grow until the bronchus is blocked, cutting off the supply of air to that lung. The lung then collapses, and the secretions trapped in the lung spaces become infected, with a resulting pneumonia or formation of a lung abscess. Such a lung cancer can spread, causing secondary growths in the lymph nodes of the chest and neck and in the brain and other parts of the body. A treatment that offers a possibility of cure, before secondary growths have had time to form, is complete removal of the lung. This operation is called a *pneumonectomy* (nu-mo-NEK-to-me).

Liquid Ventilation

Researchers have been attempting for years to develop a fluid that could carry high concentrations of oxygen in the body. Such a fluid could substitute for blood in transfusions or be used to carry oxygen into the lungs. Early work on liquid ventilation climaxed dramatically in the mid-1960s when a pioneer in this field submerged a laboratory mouse in a beaker of fluid and the animal survived total immersion for more than 10 minutes. The fluid was a synthetic substance that could hold as much oxygen as does air. A newer version of this fluid, a fluorine-containing chemical known as *PFC,* has been tested to ventilate the collapsed lungs of premature babies. The fluid is less damaging to delicate lung tissue than is air, which has to be pumped in under higher pressure. Other possible uses include treatment of people whose lungs have been damaged by infection or inhalation of toxins.

Malignant tumors of the stomach, breast, and other organs may spread to the lungs as secondary growths (metastases).

Disorders Involving the Pleural Space

Pleurisy (PLUR-ih-se), or inflammation of the pleura, usually accompanies a lung infection—particularly pneumonia or tuberculosis. This condition can be quite painful, because the inflammation produces a sticky exudate that roughens the pleura of both the lung and the chest wall; when the two surfaces rub together during ventilation, the roughness causes acute irritation. The sticking together of two surfaces is called an *adhesion* (ad-HE-zhun). Infection of the pleura also causes an increase in the amount of pleural fluid. This fluid may accumulate in the pleural space in quantities large enough to compress the lung, making the patient unable to obtain enough air. Withdrawal of the fluid by chest tube or syringe may be necessary.

Pneumothorax is an accumulation of air in the pleural space. The lung on the affected side collapses, causing the affected person to have great difficulty breathing. Pneumothorax may be caused by a wound in the chest wall or by rupture of lung air spaces. In a pneumothorax caused by a penetrating wound in the chest wall, an airtight cover over the opening will prevent further air from entering. The remaining lung can then function to provide adequate amounts of oxygen.

Blood in the pleural space, a condition called *hemothorax,* is also caused by penetrating wounds of the chest. In such cases the first priority is to stop the bleeding.

▶ Special Equipment for Respiratory Tract Treatments

The *bronchoscope* (BRONG-ko-skope) is a rigid or flexible fiberoptic tubular instrument for inspection of the primary bronchi and the larger bronchial tubes. The bronchoscope is passed into the respiratory tract by way of the mouth and the pharynx. It may be used to remove foreign bodies or to inspect the bronchi or bronchioles for tumors or other evidence of disease. Children inhale a variety of objects, such as pins, beans, pieces of nuts, and small coins, all of which may be removed with the aid of a bronchoscope. If such things are left in the lung, an abscess or other serious complication may cause death.

Oxygen therapy is used to sustain life when some condition interferes with adequate oxygen supply to the tissues. The oxygen must first have moisture added by being bubbled through water that is at room temperature or heated. Oxygen may be delivered to the patient by mask, catheter, or nasal prongs. Because there is danger of fire when oxygen is being administered, no one in the room should smoke.

A *suction apparatus* is used for removing mucus or other substances from the respiratory tract. Usually, the device takes the form of a drainage bottle, with one tube from it leading to the area to be drained, and another tube from the bottle leading to a suction machine. When the suction is applied, the drainage flows from the patient's respiratory tract into the bottle.

A *tracheostomy* (tra-ke-OS-to-me) *tube* is used if the pharynx or the larynx is obstructed. It is a small metal or plastic tube that is inserted through a cut made in the trachea, and it acts as an artificial airway for ventilation. The procedure for the insertion of such a tube is a *tracheostomy.* The word *tracheotomy* (tra-ke-OT-o-me) refers to the incision in the trachea. Very often, emergency tracheotomies are performed on children who have inhaled something large enough to block the respiratory passages; without this procedure, such children would die in a very short time.

Artificial respiration is resorted to when an individual has temporarily lost the capacity to perform the normal motions of respiration. Such emergencies include cases of smoke asphyxiation, electric shock, and drowning.

Classes are offered by many public agencies in the techniques of mouth-to-mouth respiration and cardiac massage to revive persons experiencing respiratory or cardiac arrest. This technique is known as *cardiopulmonary resuscitation,* or *CPR.*

SUMMARY

I. **Respiration**—supply of oxygen to the cells and removal of carbon dioxide
   A. Pulmonary ventilation—the movement of air into and out of the lungs
   B. Diffusion of gases
   C. Transport of oxygen and carbon dioxide

II. **Respiratory system**
   A. Nasal cavities—filter, warm, and moisten air
   B. Pharynx (throat)—carries air into respiratory tract and food into digestive tract
   C. Larynx (voice box)—contains vocal cords
      1. Epiglottis—covers larynx on swallowing to help prevent food from entering
   D. Trachea (windpipe)
   E. Bronchi—branches of trachea that enter lungs and then subdivide
      1. Bronchioles—smallest subdivisions
   F. Lungs
      1. Pleura—membrane that encloses lung and lines chest wall
      2. Mediastinum—space and organs between lungs
   G. Alveoli—tiny air sacs in lungs

III. **Physiology of respiration**
   A. Pulmonary ventilation
      1. Inhalation—drawing of air into lungs
      2. Exhalation—expulsion of air from lungs
      3. Lung volumes—used to evaluate respiratory function (*e.g.,* vital capacity, total lung capacity, etc.)
   B. Gas exchange
      1. Diffusion of gases from area of higher concentration to area of lower concentration
      2. In lungs—oxygen enters blood and carbon dioxide leaves

      3. In tissues—oxygen leaves blood and carbon dioxide enters
   C. Gas transport
      1. Oxygen—almost totally bound to heme portion of hemoglobin in red blood cells; separates from hemoglobin when oxygen concentration is low (in tissues)
      2. Carbon dioxide—most carried as bicarbonate ion; regulates pH of blood
   D. Regulation of respiration
      1. Normal ventilation
         a. Controlled by centers in medulla and pons
         b. Chemoreceptors detect changes in gas composition of blood
      2. Abnormal ventilation
         a. Hyperventilation—rapid, deep respiration
         b. Hypoventilation—inadequate air in alveoli
   E. Abnormal respiration
      1. Types
         a. Hyperpnea—increase in depth and rate of breathing
         b. Apnea—temporary cessation of breathing
         c. Dyspnea—difficulty in breathing
         d. Cheyne–Stokes—irregularity found in critically ill
         e. Orthopnea—difficulty relieved by upright position
      2. Possible results—cyanosis, hypoxia (anoxia), hypoxemia, suffocation

IV. **Disorders of the respiratory system**
   A. Nasal problems—sinusitis, polyps, deviated septum, nosebleed (epistaxis)

   B. Infection—colds, influenza, pneumonia, tuberculosis
   C. Allergy—hay fever, asthma
   D. Chronic obstructive pulmonary disease (COPD)—involves emphysema and bronchitis
   E. Atelectasis—collapse of lung or portion of lung; seen in respiratory distress syndrome due to lack of surfactant
   F. Lung cancer—smoking a major causative factor
   G. Disorders involving pleural space
      1. Pleurisy—inflammation of pleura
      2. Pneumothorax—air in pleural space
      3. Hemothorax—blood in pleural space
V. **Special equipment for respiratory tract treatment**
   A. Bronchoscope—tube used to examine air passageways and remove foreign bodies
   B. Oxygen therapy
   C. Suction apparatus—removes mucus and other substances
   D. Tracheostomy—artificial airway
      1. Tracheotomy—incision into trachea
   E. Artificial respiration—cardiopulmonary resuscitation (CPR)

QUESTIONS FOR STUDY AND REVIEW

1. What is the definition of *respiration* and what are its three phases?
2. Trace the pathway of air from the outside into the blood.
3. What is the purpose of the cilia on the cells that line the respiratory passageways?
4. What are the advantages of breathing through the nose?
5. Describe the lung cavities and pleura.
6. What muscles are used for inhalation? forceful exhalation?
7. How is oxygen transported in the blood? What causes its release to the tissues?
8. How is carbon dioxide transported in the blood?
9. How does the pH of the blood change as the level of carbon dioxide increases? decreases?
10. Why does hyperventilation cause a cessation of breathing?
11. What are chemoreceptors and how do they function to regulate breathing?
12. Name and describe five types of abnormal respiration.
13. Define five volumes or capacities used to measure breathing.
14. Define four terms related to lack of oxygen.
15. What are *sinusitis, polyps,* and *deviated septum?* What is the effect of each?
16. What are some causes of nosebleed?
17. Describe some possible developments and complications of upper respiratory infections.
18. Describe the common cold and influenza.
19. What changes occur in the lungs in chronic obstructive lung disease?
20. What are the symptoms of atelectasis?
21. What do hay fever and asthma have in common?
22. What are some possible causes of lung cancer?
23. What is pleurisy? What are its complications?
24. Describe pneumothorax. What first aid measure can be used for it?
25. Name and describe three devices used in treatment of the respiratory tract.

CHAPTER 19

Digestion

Selected Key Terms

The following terms are defined in the Glossary:

absorption

bile

chyle

chyme

colon

defecation

deglutition

digestion

duodenum

emulsify

enzyme

esophagus

hydrolysis

lacteal

mastication

peristalsis

peritoneum

saliva

sphincter

ulcer

villi

Behavioral Objectives

After careful study of this chapter, you should be able to:

- Name the two main functions of the digestive system
- Describe the four layers of the digestive tract wall
- Describe the layers of the peritoneum
- Name and describe the organs of the digestive tract
- Name and describe the accessory organs of digestion
- List the functions of each organ involved in digestion
- Explain the role of enzymes in digestion and give examples of enzymes
- Name the digestion products of fats, proteins, and carbohydrates
- Define *absorption*
- Define *villi* and state how villi function in absorption
- Describe how bile functions in digestion
- Name the ducts that carry bile into the digestive tract
- List the main functions of the liver
- Explain the use of feedback in regulating digestion and give several examples
- Describe common disorders of the digestive tract and the accessory organs

Memmler, RL, Cohen, BJ, Wood, DL. *THE HUMAN BODY IN HEALTH AND DISEASE*, 8/e,
© 1996 Lippincott-Raven Publishers

▶ What the Digestive System Does

E very body cell needs a constant supply of nutrients to provide energy and building blocks for the manufacture of body substances. Food as we take it in, however, is too large to enter the cells. It must first be broken down into particles small enough to pass through the cell membrane. This process is known as *digestion.* After digestion, food must be carried to the cells in every part of the body by the circulation. The transfer of food into the circulation is called *absorption.* Digestion and absorption are the two chief functions of the digestive system.

For our purposes the digestive system may be divided into two groups of organs:

1. The *digestive tract,* a continuous passageway beginning at the mouth, where food is taken in, and terminating at the anus, where the solid waste products of digestion are expelled from the body
2. The *accessory organs,* which are necessary for the digestive process but are not a direct part of the digestive tract. They release substances into the digestive tract through ducts.

The Wall of the Digestive Tract

Although modified for specific tasks in different organs, the wall of the digestive tract, from the esophagus to the anus, is similar in structure throughout. Follow the diagram of the small intestine in Figure 19-1 as we describe the layers of this wall from the innermost to the outermost surface. First is the *mucous membrane,* so called because its epithelial layer contains many mucus-secreting cells. The type of epithelium is simple columnar. In Figure 19-2, a microscopic view of the small intestine, the mucus-secreting cells appear as clear areas mixed in with the epithelial cells.

The layer of connective tissue beneath this, the *submucosa,* contains blood vessels and some of the nerves that help regulate digestive activity. Next are two layers of *smooth muscle.* The inner layer has circular fibers, and the outer layer has longitudinal fibers. The alternate contractions of these muscles create the wavelike movement that propels food through the digestive tract and mixes it with digestive juices. This movement is called *peristalsis* (per-ih-STAL-sis).

The outermost layer of the wall consists of fibrous connective tissue. Most of the abdominal organs have an additional layer of *serous membrane* that is part of the peritoneum (per-i-to-NE-um).

The Peritoneum

The abdominal cavity is lined with a thin, shiny serous membrane that also covers most of the abdominal organs (Fig. 19-3). The portion of this membrane that lines the abdomen is called the *parietal peritoneum;* that covering the organs is called the *visceral peritoneum.* In addition to these single-layered portions of the peritoneum there are a number of double-layered structures that carry blood vessels, lymph vessels, and nerves, and sometimes act as ligaments supporting the organs.

The *mesentery* (MES-en-ter-e) is a double-layered portion of the peritoneum shaped somewhat like a fan. The handle portion is attached to the back wall, and the expanded long edge is attached to the small intestine. Between the two layers of membrane that form the mesentery are the blood vessels, lymphatic vessels, and nerves that supply the intestine. The section of the peritoneum that extends from the colon to the back wall is the *mesocolon* (mes-o-KO-lon).

A large double layer of the peritoneum containing much fat hangs like an apron over the front of the intestine. This *greater omentum* (o-MEN-tum) extends from the lower border of the stomach into the pelvic part of the abdomen and then loops back up to the transverse colon. There is also a smaller membrane, called the *lesser omentum,* that extends between the stomach and the liver.

Peritonitis is an inflammation of the peritoneum. It may be a serious complication following infection of one of the organs covered by the peritoneum—often the appendix. The frequency and severity of peritonitis have been greatly reduced by the use of antibiotic drugs. However, the disorder still occurs and can be very dangerous. If the infection is kept in one area, it is said to be *localized peritonitis.* A *generalized peritonitis,* as caused by a ruptured appendix or perforated ulcer, may lead to absorption of so many disease organisms and their toxins as to be fatal. Immediate surgery to repair the rupture as well as medical care are needed.

The Digestive Tract

As we study the organs of the digestive system, locate each in Figure 19-4.

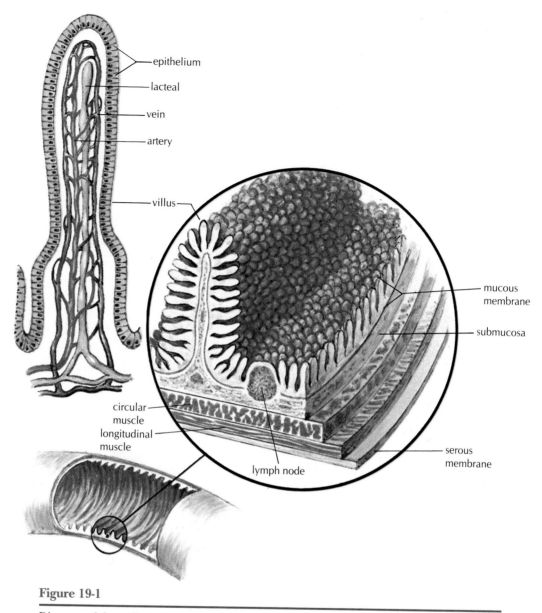

epithelium

lacteal

vein

artery

villus

mucous membrane

submucosa

circular muscle

longitudinal muscle

serous membrane

lymph node

Figure 19-1

Diagram of the wall of the small intestine showing numerous villi. At the left is an enlarged drawing of a single villus.

The digestive tract is a muscular tube extending through the body. It is composed of several parts: the *mouth, pharynx, esophagus, stomach, small intestine,* and *large intestine.* The digestive tract is sometimes called the *alimentary tract,* derived from a Latin word that means "food." It is more commonly referred to as the *gastrointestinal (GI) tract* because of the major importance of the stomach and intestine in the process of digestion.

The Mouth

The mouth, also called the *oral cavity,* is where a substance begins its travels through the digestive tract (Fig. 19-5). The mouth has three digestive functions:

1. To receive food, a process called *ingestion*
2. To prepare food for digestion
3. To begin the digestion of starch.

columnar epithelium

villi

goblet cells
(secrete mucus)

connective
tissue

digestive
glands

smooth
muscle

mucous
glands

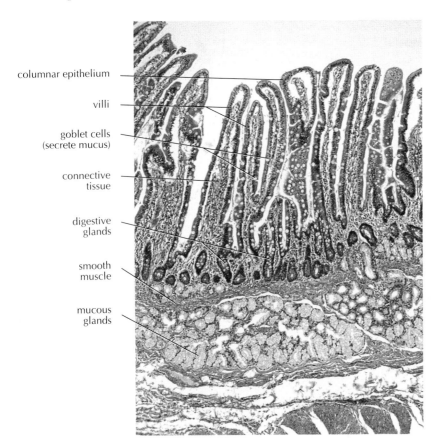

Figure 19-2

Microscopic view of the duodenum (Cormack DH: Essential Histology, plate 13-48. Philadelphia, JB Lippincott, 1993)

Into this space projects a muscular organ, the tongue, which is used for chewing and swallowing, and is one of the principal organs of speech. The tongue has on its surface a number of special organs, called *taste buds*, by means of which taste sensations (bitter, sweet, sour, or salty) can be differentiated.

The oral cavity also contains the teeth. A child between 2 and 6 years of age has 20 teeth; an adult with a complete set of teeth has 32. Among these, the cutting teeth, or **incisors**, occupy the front part of the oral cavity, whereas the larger grinding teeth, the **molars**, are in the back.

DECIDUOUS, OR BABY, TEETH

The first eight **deciduous** (de-SID-u-us) teeth to appear through the gums are the incisors. Later the **canines** (eyeteeth) and molars appear. Usually, the 20 baby teeth have all appeared by the time a child has reached the age of 2 or 2½ years. During the first 2 years the permanent teeth develop within the jawbones from buds that are present at birth. The first permanent tooth to appear is the important 6-year

molar. This permanent tooth comes in before the baby incisors are lost. Because decay and infection of adjacent deciduous molars may spread to and involve new, permanent teeth, deciduous teeth need proper care.

PERMANENT TEETH

As a child grows, the jawbones grow, making space for additional teeth. After the 6-year molars have appeared, the baby incisors loosen and are replaced by permanent incisors. Next, the baby canines (cuspids) are replaced by permanent canines, and finally, the baby molars are replaced by the bicuspids (premolars) of the permanent teeth.

Now the larger jawbones are ready for the appearance of the 12-year, or second, permanent molar teeth. During or after the late teens, the third molars, or so-called *wisdom teeth*, may appear. In some cases the jaw is not large enough for these teeth or there are other abnormalities, so that the third molars may have to be removed. Figure 19-6 shows the parts of a molar.

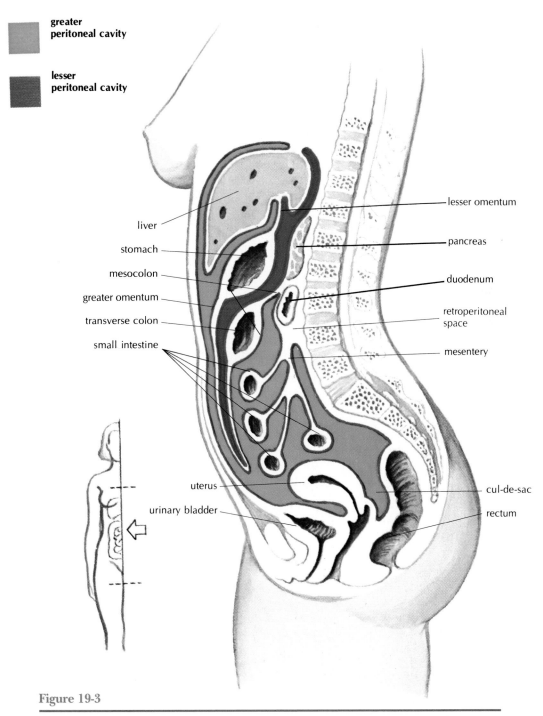

greater
peritoneal cavity

lesser
peritoneal cavity

liver

stomach

mesocolon

greater omentum

transverse colon

small intestine

uterus

urinary bladder

lesser omentum

pancreas

duodenum

retroperitoneal
space

mesentery

cul-de-sac

rectum

Figure 19-3

Diagram of the abdominal cavity showing the peritoneum.

DISEASES OF THE MOUTH AND TEETH

Tooth decay is also termed dental *caries* (KA-reze), which means "rottenness." It has a number of causes, including diet, heredity, mechanical problems, and endocrine disorders. Persons who ingest a lot of sugar are particularly prone to this disease. Because a baby's teeth begin to develop before birth, the diet of the mother during pregnancy is very important in ensuring the formation of healthy teeth in her baby.

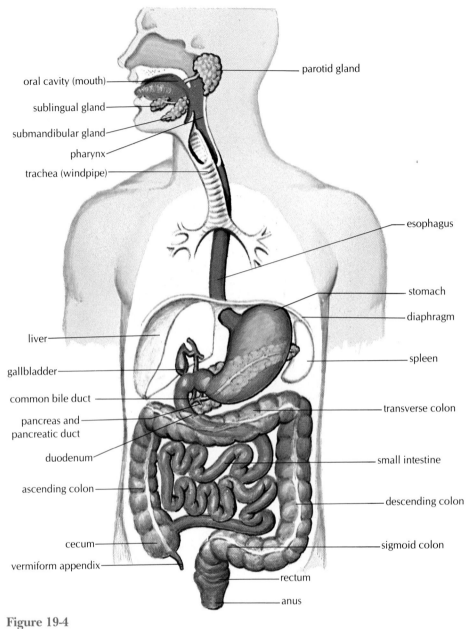

oral cavity (mouth)

sublingual gland

submandibular gland

pharynx

trachea (windpipe)

liver

gallbladder

common bile duct

pancreas and
pancreatic duct

duodenum

ascending colon

cecum

vermiform appendix

parotid gland

esophagus

stomach

diaphragm

spleen

transverse colon

small intestine

descending colon

sigmoid colon

rectum

anus

Figure 19-4

Digestive system.

Any infection of the gum is called *gingivitis* (jin-jih-VI-tis). If such an infection continues and is untreated, it may lead to a more serious condition, *periodontitis* (per-e-o-don-TI-tis), which involves not only the gum tissue but also the supporting bone of the teeth. Loosening of the teeth and destruction of the supporting bone follow unless the process is halted by proper treatment and improved dental hy-

giene. Periodontitis is responsible for nearly 80% of loss of teeth in persons over the age of 35.

Vincent's disease, a kind of gingivitis caused by a spirochete, is most prevalent in teenagers and young adults. Characterized by inflammation, ulceration, and infection of the mucous membranes of the mouth and gums, this disorder is highly contagious, particularly by oral contact.

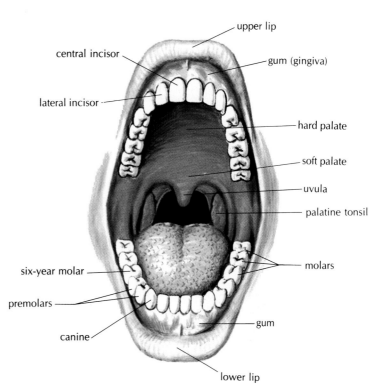

upper lip
central incisor
gum (gingiva)
lateral incisor
hard palate
soft palate
uvula
palatine tonsil
six-year molar
molars
premolars
gum
canine
lower lip

Figure 19-5

The mouth, showing the teeth and tonsils.

Patients on antibiotic therapy are more likely than normal to develop fungal infections of the mouth and tongue because these drugs may destroy the normal bacteria of the mouth and allow other organisms to grow.

Leukoplakia (lu-ko-PLA-ke-ah) is characterized by thickened white patches on the mucous membranes of the mouth. It is common in smokers and may lead to cancer.

THE SALIVARY GLANDS

While food is in the mouth, it is mixed with *saliva,* one purpose of which is to moisten the food and facilitate the processes of chewing, or *mastication* (mas-tih-KA-shun), and swallowing, or *deglutition* (deg-lu-TISH-un). Saliva also helps keep the teeth and mouth clean and reduce bacterial growth.

This watery mixture contains mucus and an enzyme called *salivary amylase* (AM-ih-laze), which begins the digestive process by converting starch to sugar. It is manufactured mainly by three pairs of glands that function as accessory organs:

1. The *parotid* (pah-ROT-id) *glands,* the largest of the group, are located below and in front of the ear.

2. The *submandibular* (sub-man-DIB-u-lar), or *submaxillary* (sub-MAK-sih-ler-e), *glands* are located near the body of the lower jaw.
3. The *sublingual* (sub-LING-gwal) *glands* are under the tongue.

All these glands empty by means of ducts into the oral cavity.

The contagious disease commonly called *mumps* is a viral infection of the parotid salivary glands. This type of *parotitis* (par-o-TI-tis), or inflammation of the parotid glands, may lead to inflammation of the testicles by the same virus. Males affected after puberty are at risk for permanent damage to these sex organs, resulting in sterility. Another complication of mumps that occurs in about 10% of cases is meningitis. Like many contagious diseases, mumps is now preventable by use of a vaccine given to children early in life.

The Pharynx and Esophagus

The *pharynx* (FAR-inks) is commonly referred to as the *throat.* The oral part of the pharynx is visible when you look into an open mouth and depress the tongue. The palatine tonsils may be seen at either side. The pharynx also extends upward to the nasal

Figure 19-6

A molar tooth. (Cohen B: Medical Terminology: An Illustrated Guide, 2nd ed, p. 208. Philadelphia, JB Lippincott, 1994)

cavity and downward to the level of the larynx. The **soft palate** is tissue that forms the back of the roof of the oral cavity. From it hangs a soft, fleshy, V-shaped mass called the **uvula** (U-vu-lah).

In swallowing, a small portion of chewed food mixed with saliva, called a **bolus,** is pushed by the tongue into the pharynx. When the food reaches the pharynx, swallowing occurs rapidly by an involuntary reflex action. At the same time, the soft palate and uvula are raised to prevent food and liquid from entering the nasal cavity, and the tongue is raised to seal the back of the oral cavity. The entrance of the trachea is guarded during swallowing by a leaf-

shaped cartilage, the **epiglottis,** which covers the opening of the larynx. The swallowed food is then moved by peristalsis into the **esophagus** (eh-SOF-ah-gus), a muscular tube about 25 cm (10 inches) long that carries food into the stomach. No additional digestion occurs in the esophagus.

Before joining the stomach, the esophagus must pass through the diaphragm. It passes through a space in the diaphragm called the **esophageal hiatus** (eh-sof-ah-JE-al hi-A-tus). If there is a weakness in the diaphragm at this point, a portion of the stomach or other abdominal organ may protrude through the space, a condition called **hiatal hernia.**

The Stomach

The stomach is an expanded J-shaped organ in the upper left region of the abdominal cavity (Fig. 19-7). In addition to the two muscle layers already described, it has a third, inner oblique (angled) layer that aids in grinding food and mixing it with digestive juices. The left-facing arch of the stomach is the *greater curvature,* whereas the right surface forms the *lesser curvature.* Each end of the stomach is guarded by a muscular ring, or *sphincter* (SFINK-ter), that permits the passage of substances in only one direction.

Between the esophagus and the stomach is the *lower esophageal sphincter (LES).* This valve has also been called the *cardiac sphincter* because it separates the esophagus from the region of the stomach that is close to the heart. We are sometimes aware of the existence of this sphincter; sometimes it does not relax as it should, producing a feeling of being un-able to swallow past that point. Between the distal, or far, end of the stomach and the small intestine is the *pyloric* (pi-LOR-ik) *sphincter.* The region of the stomach leading into this sphincter, the *pylorus* (pi-LOR-us), is important in regulating how rapidly food moves into the small intestine.

The stomach serves as a storage pouch, digestive organ, and churn. When the stomach is empty, the lining forms many folds called *rugae* (RU-je). These folds disappear as the stomach expands. (It may be stretched to hold one half of a gallon of food and liquid.) Special cells in the lining of the stomach secrete substances that mix together to form *gastric juice,* the two main components of which are:

1. Hydrochloric acid (HCl), a strong acid that soft-ens the connective tissue in meat and destroys foreign organisms

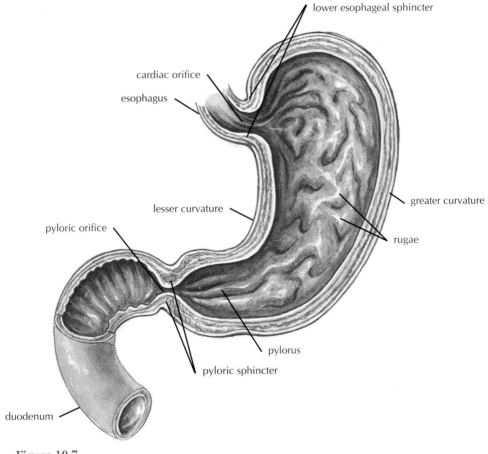

lower esophageal sphincter

cardiac orifice

esophagus

lesser curvature

pyloric orifice

greater curvature

rugae

pylorus

pyloric sphincter

duodenum

Figure 19-7

Longitudinal section of the stomach and a portion of the duodenum showing the interior.

2. Pepsin, a protein-digesting enzyme. This enzyme is produced in an inactive form and is activated only when food enters the stomach and HCl is produced.

The semi-liquid mixture of gastric juice and food that leaves the stomach to enter the small intestine is called *chyme* (kime).

DISORDERS INVOLVING THE STOMACH

A burning sensation in the region of the esophagus and stomach, popularly known as *heartburn,* is often caused by the sudden intake of a large amount of fluid or food resulting in excessive stretching of the lower esophagus. This interferes with the functioning of the lower esophageal sphincter, allowing gastric acid to enter the esophagus. Contrary to popular belief, heartburn is not due to hyperacidity of the stomach contents.

Nausea is an unpleasant sensation that may follow distention or irritation of the lower esophagus or of the stomach as a result of various nervous and mechanical factors. It may be a symptom of interference with the normal forward peristaltic motion of the stomach and intestine and thus may be followed by vomiting. *Vomiting* is the expulsion of gastric (and sometimes intestinal) contents through the mouth by reverse peristalsis. The contraction of the abdominal wall muscles forcibly empties the stomach. Vomiting is frequently caused by overeating or by inflammation of the stomach lining, a condition called *gastritis* (gas-TRI-tis). Gastritis results from irritation of the mucosa by certain drugs, food, or drinks. For example, the long-term use of aspirin, highly spiced foods, or alcohol can lead to gastritis. The nicotine in cigarettes can also cause gastritis.

Flatus (FLA-tus) usually refers to excessive amounts of air (gas) in the stomach or intestine. The resulting condition is referred to as *flatulence* (FLAT-u-lens). In some cases, it may be necessary to insert a tube into the stomach or rectum to aid the patient in expelling flatus.

STOMACH CANCER

Although stomach cancer has become rare in the United States, it is common in many parts of the world, and it is an important disorder because of the high death rate associated with it. Males are more susceptible than females to stomach cancer. The tumor nearly always develops from the epithelial or mucosal lining of the stomach and is often of the type called *adenocarcinoma* (ad-en-o-kar-sih-NO-mah). Sometimes the victim has suffered from long-standing indigestion but has failed to consult a physician until the cancer has spread to other organs, such as the liver, in which there may be several metastases. Persistent indigestion is one of the important warning signs of cancer of the stomach.

PEPTIC ULCER

An ulcer is an area of the skin or mucous membrane in which the tissues are gradually destroyed. An ulcer may be caused by the acid in gastric juice. Peptic ulcers (named for the enzyme pepsin) occur in the mucous membrane of the esophagus, stomach, or duodenum (the first part of the small intestine) and are most common in persons between the ages of 30 and 45. Peptic ulcers in the stomach are *gastric ulcers;* those in the duodenum are *duodenal* (du-o-DE-nal) *ulcers.* Smoking cigarettes and taking aspirin or other anti-inflammatory drugs are major causative factors. Recently it has been found that infection with a bacterium, *Helicobacter pylori,* is a factor in causing peptic ulcers. The organism is associated with inflammation of the stomach and duodenum, and most people with ulcers who show this infection are cured when the organism is eliminated with antibiotics. Drugs that inhibit the secretion of stomach acids are often an effective aid in treatment.

PYLORIC STENOSIS

Normally, the stomach contents are moved through the pyloric sphincter within about 2 to 6 hours after eating. However, some infants, most often males, are born with an obstruction of the pyloric sphincter, a condition called *pyloric stenosis* (steh-NO-sis). Usually, surgery is required in these cases to modify the muscle so food can pass from the stomach into the duodenum.

The Small Intestine

The small intestine is the longest part of the digestive tract. It is known as the small intestine because, although it is longer than the large intestine, it is smaller in diameter, with an average width of about 2.5 cm (1 inch). When relaxed to its full length, the small intestine is about 6 m (20 feet) long. The first 25 cm (10 inches) or so of the small intestine make up the *duodenum* (du-o-DE-num). Beyond the duodenum are two more divisions: the *jejunum* (je-JU-num), which forms the next two fifths of the small intestine, and the *ileum* (IL-e-um), which constitutes the remaining portion.

The wall of the duodenum contains glands that secrete large amounts of mucus to protect the small intestine from the strongly acid chyme entering from the stomach. Cells of the small intestine also secrete enzymes that digest proteins and carbohydrates. In addition, digestive juices from the liver and pancreas enter the small intestine through a small opening in the duodenum. Most of the digestive process takes place in the small intestine under the effects of these juices.

Most absorption of digested food also occurs through the walls of the small intestine. To increase the surface area of the organ for this purpose, the mucosa is formed into millions of tiny, finger-like projections, called *villi* (VIL-li) (see Fig. 19-1), which give the inner surface a velvety appearance. In addition, each epithelial cell has small projecting folds of the cell membrane known as *microvilli*. These create a remarkable increase in the total surface area available in the small intestine for the absorption of nutrients.

The Large Intestine

Any material that cannot be digested as it passes through the digestive tract must be eliminated from the body. In addition, most of the water secreted into the digestive tract for proper digestion must be reabsorbed into the body to prevent dehydration. The storage and elimination of undigested waste and the reabsorption of water are the functions of the large intestine.

The large intestine is about 6.5 cm (2.5 inches) in diameter and about 1.5 m (5 feet) long. The outer longitudinal muscle fibers form three separate bands on the surface. These bands draw up the wall of the organ to give it its distinctive puckered appearance.

The large intestine begins in the lower right region of the abdomen. The first part is a small pouch called the *cecum* (SE-kum). Between the ileum of the small intestine and the cecum is a sphincter, the *ileocecal* (il-e-o-SE-kal) *valve,* that prevents food from traveling backward into the small intestine. Attached to the cecum is a small, blind tube containing lymphoid tissue; it is called the *vermiform* (VER-mih-form) *appendix* (*vermiform* means "wormlike"). Inflammation of this tissue as a result of infection or obstruction is *appendicitis.*

The second portion, the *ascending colon,* extends upward along the right side of the abdomen toward the liver. The large intestine then bends and extends

Surface Area and Transport

Whenever materials pass from one system to another, they must travel through a cellular membrane. A major factor in how much transport can occur per unit of time is the total surface area of the membrane.

The problem of packing a large amount of surface into a small space is solved in the body by folding the membranes. We do the same thing. Imagine trying to store a bed sheet in the closet without folding it!

In the small intestine, where digested food must move into the bloodstream, there is folding of membranes down to the level of the single cells. The organ as a whole is coiled to fit its length into the abdominal cavity. The inner wall of the organ forms circular folds called *plicae circulares.* The projecting villi provide more surface area than a flat membrane would. Even further, the individual cells that line the small intestine have microvilli, tiny fingerlike folds of the cell membrane, that increase surface area tremendously.

Can you name other portions of the digestive tract or other systems that show this folding pattern and explain why it is there?

across the abdomen, forming the *transverse colon.* At this point it bends sharply and extends downward on the left side of the abdomen into the pelvis, forming the *descending colon.* The lower part of the colon bends posteriorly in an S shape and continues downward as the *sigmoid colon.* The sigmoid colon empties into the *rectum,* which serves as a temporary storage area for indigestible or unabsorbable food residue (see Fig. 19-4). Enlargement of the veins in this area constitutes *hemorrhoids.* A narrow portion of the distal large intestine is called the *anal canal.* This leads to the outside of the body through an opening called the *anus* (A-nus).

Large quantities of mucus, but no enzymes, are secreted by the large intestine. At intervals, usually after meals, the involuntary muscles within the walls of the large intestine propel solid waste material, called *feces* or stool, toward the rectum. This material is then eliminated from the body by both voluntary and involuntary muscle actions, a process called *defecation* (def-e-KA-shun).

While the food residue is stored in the large intestine, bacteria that normally live in the colon act on it to produce vitamin K and some of the B-complex vitamins. As mentioned, systemic antibiotic therapy may destroy these bacteria and others living in the large intestine, causing undesirable side effects.

Intestinal Disorders
INFLAMMATION

Difficulties with digestion or absorption may be due to *enteritis* (en-ter-I-tis), an intestinal inflammation. When both the stomach and the small intestine are involved, the illness is called *gastroenteritis* (gas-tro-en-ter-I-tis). The symptoms of gastroenteritis include nausea, vomiting, and diarrhea, as well as acute abdominal pain (colic). The disorder may be caused by a variety of pathogenic organisms, including viruses, bacteria, and protozoa. Chemical irritants such as alcohol, certain drugs (*e.g.,* aspirin), and other toxins have been known to cause this disorder.

DIARRHEA

Diarrhea is a symptom characterized by abnormally frequent watery bowel movements. The danger of diarrhea is dehydration and loss of salts, especially in infants. Diarrhea may result from excess activity of the colon, faulty absorption, or infection. Infections resulting in diarrhea include cholera, dysentery, and food poisoning. Tables 1 and 4 in Appendix 4 list some of the organisms causing these diseases. Such infections often are spread by poor sanitation and by contaminated food, milk, or water. An examination of the stool may be required to establish the cause of diarrhea; such an examination may reveal the presence of pathogenic organisms, worm eggs, or blood.

CONSTIPATION

Millions of dollars are spent each year in an effort to remedy a condition called *constipation.* What is constipation? Many persons erroneously think they are constipated if they go a day or more without having a bowel movement. Actually, what is normal varies greatly; one person may normally have a bowel movement only once every 2 or 3 days, whereas another may normally have more than one movement daily. The term *constipation* is also used to refer to hard stools or difficulty with defecation.

On the basis of its onset, constipation may be classified as acute or chronic. Acute constipation occurs suddenly and may be due to an intestinal obstruction, such as a tumor associated with cancer or an inflammation of the saclike bulges (diverticula) of the intestinal wall as seen in *diverticulitis* (di-ver-tik-u-LI-tis). Laxatives and enemas should be avoided, and a physician should be consulted at once. Chronic constipation, in contrast, has a more gradual onset and may be divided into two groups:

1. *Spastic constipation,* in which the intestinal musculature is overstimulated so that the canal becomes narrowed and the space (lumen) inside the intestine is not large enough to permit the passage of fecal material
2. *Flaccid* (FLAK-sid) *constipation,* which is characterized by a lazy, or *atonic* (ah-TON-ik), intestinal muscle. Elderly persons and persons on bed rest are particularly susceptible to this condition. Often it results from repeated denial of the urge to defecate.

Persons who have sluggish intestinal muscles may be helped by regular bowel habits, moderate exercise, an increase in the ingestion of vegetables and other bulky foods, and an increase in fluid intake.

The chronic use of laxatives and enemas should be avoided. The streams of hot water used in enemas may injure the lining of the intestine by removing the normal protective mucus. In addition, enemas aggravate piles (hemorrhoids). Enemas should be done only on the order of a physician, and sparingly.

CANCER OF THE COLON AND RECTUM

Tumors of the colon and rectum are among the most common types of cancer in the United States. These tumors are usually adenocarcinomas that arise from the mucosal lining. The occurrence of cancer of the colon is evenly divided between the sexes, but malignant tumors of the rectum are more common in men than in women. Tumors may be detected by examination of the rectum and lower colon with an instrument called a *sigmoidoscope* (sig-MOY-do-skope) (named for the sigmoid colon). The presence of blood in the stool may indicate cancer of the bowel or some other gastrointestinal disturbance. A simple chemical test can detect extremely small quantities of blood in the stool, referred to as *occult* ("hidden") *blood.* Early detection and treatment are the key to increasing survival rates.

An instrument used for examining the interior of a tube or hollow organ is called an ***endoscope*** (EN-do-skope), and the examination is known as ***endoscopy*** (en-DOS-ko-pe). With a ***fiberoptic*** (fi-ber-OP-tik) ***endoscope,*** which contains flexible bundles of glass or plastic that propagate light, transmitted images show the lining of internal organs. Such endoscopic examinations are used to detect structural abnormalities, bleeding, ulcers, inflammation, and tumors in such organs as the esophagus, stomach, and small and large intestine.

▶ The Accessory Structures

The Liver

The liver, often referred to by the word root *hepat,* is the largest glandular organ of the body (Fig. 19-8). It is located in the upper right portion of the abdominal cavity under the dome of the diaphragm. The lower edge of a normal-sized liver is level with the lower margin of the ribs. The human liver is the same reddish brown color as the animal liver seen in the supermarket. It has a large right lobe and a smaller left lobe; the right lobe includes two inferior smaller lobes. The liver is supplied with blood through two vessels: the portal vein and the hepatic artery. These vessels deliver about 1½ quarts of blood to the liver every minute. The hepatic artery carries oxygenated blood, whereas the portal system of veins carries blood that is rich in the end products of digestion. This most remarkable organ has so many functions that only some of its major activities can be listed here:

1. The storage of glucose (simple sugar) in the form of ***glycogen,*** an animal starch. When the blood sugar level falls below normal, liver cells convert glycogen to glucose and release it into the bloodstream; this serves to restore the normal concentration of blood sugar.
2. The formation of blood plasma proteins, such as albumin, globulins, and clotting factors
3. The synthesis of ***urea*** (u-RE-ah), a waste product of protein metabolism. Urea is released into the blood and transported to the kidneys for elimination.
4. The modification of fats so they can be used more efficiently by cells all over the body

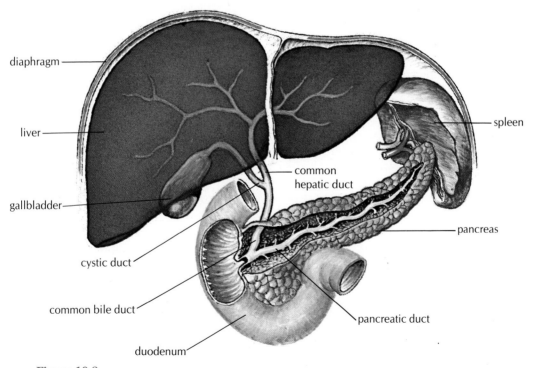

Figure 19-8

Accessory organs of digestion.

5. The manufacture of bile
6. The destruction of old red blood cells. The pigment released from these cells in both the liver and the spleen is eliminated in the bile. This pigment (bilirubin) gives the stool its characteristic dark color.
7. The **detoxification** (de-tok-sih-fih-KA-shun) (removal of the poisonous properties) of harmful substances such as alcohol and certain drugs
8. The storage of some vitamins and iron.

The main digestive function of the liver is the production of *bile.* The salts contained in bile act like a detergent to *emulsify* fat, that is, to break up fat into small droplets that can be acted on more effectively by digestive enzymes. Bile also aids in the absorption of fat from the small intestine. Bile leaves the lobes of the liver by two ducts that merge to form the *common hepatic duct.* After collecting bile from the gallbladder, this duct, now called the *common bile duct,* delivers bile into the duodenum. These and the other accessory ducts are shown on Figure 19-8.

Diseases of the Liver
HEPATITIS

Inflammation of the liver, called *hepatitis* (hep-ah-TI-tis), may be caused by drugs, alcohol, or infection (see Table 3 of Appendix 4). The known viruses that cause hepatitis are named A through E. These vary in route (pathway) of infection, severity, and complications. All types of hepatitis are marked by hepatic cell destruction and such symptoms as loss of appetite, jaundice, and liver enlargement. In most patients, the liver cells regenerate with little residual damage. The types of hepatitis and their primary routes of transmission are:

Hepatitis A: commonly transmitted in fecal matter and contaminated food and water
Hepatitis B: transmitted by direct exchange of blood or body fluids
Hepatitis C: transmitted primarily by direct exchange of blood with evidence of limited sexual exchange
Hepatitis D: transmitted by direct exchange of blood. It occurs only in those with hepatitis B infection.
Hepatitis E: transmitted by fecal contamination of water. Most cases have been linked to epidemics in the Middle East and Asia.

There is no specific treatment for hepatitis.

Hepatitis B

Although there are five known types of hepatitis virus, type B is the most prevalent. Formerly called *serum hepatitis*, it is usually transmitted by use of improperly sterilized needles and syringes. More than 50% of intravenous drug users show signs of hepatitis B infection, and health care workers are at risk because of possible exposure to infected blood. Though not as commonly spread by these means, the virus has also been found in feces and body fluids. All donated blood is now tested for hepatitis B virus, so that most cases of infection by transfusion of blood or blood products today are caused by other types of the virus. Type B hepatitis is difficult to treat. Some victims remain chronically infected for life and can continue to spread the disease. In some carriers, chronic infection can lead to cirrhosis and liver cancer. There is a vaccine available for hepatitis B, and it is now recommended for childhood immunization.

CIRRHOSIS

Cirrhosis (sih-RO-sis) of the liver is a chronic disease in which active liver cells are replaced by inactive connective (scar) tissue. The most common type of cirrhosis is alcoholic (portal) cirrhosis. Alcohol has a direct damaging effect on liver cells that is compounded by malnutrition. Destruction of the liver cells hampers the portal circulation, causing blood to accumulate in the spleen and gastrointestinal tract and causing fluid (ascites) to accumulate in the peritoneal cavity.

JAUNDICE

Damage to the liver or blockage in any of the bile ducts may cause bile pigment to accumulate in the blood. As a result, the stool may become pale in color and the skin and sclera of the eyes may become yellowish; this symptom is called *jaundice* (JAWN-dis). Jaundice may also be caused by excess destruction of red blood cells. In addition, it is often seen in newborns, in whom the liver is immature and not yet functioning efficiently.

CANCER

Hepatic metastases are common in cases that begin as cancer in one of the organs of the abdominal cavity; the tumor cells are carried from the intestine or another organ through the veins that go to the liver.

The Gallbladder

The gallbladder is a muscular sac on the inferior surface of the liver that serves as a storage pouch for bile. Although the liver may manufacture bile continuously, the body is likely to need it only a few times a day. Consequently, bile from the liver flows into the hepatic ducts and then up through the *cystic* (SIS-tik) *duct* connected with the gallbladder. When chyme enters the duodenum, the gallbladder contracts, squeezing bile through the cystic duct and into the common bile duct leading to the duodenum.

The most common disease of the gallbladder is the formation of stones, or *cholelithiasis* (ko-le-lih-THI-ah-sis). Stones are formed from the substances contained in bile, mainly cholesterol. They may remain in the gallbladder or may lodge in the bile ducts, causing extreme pain. Cholelithiasis is usually associated with inflammation of the gallbladder, or *cholecystitis* (ko-le-sis-TI-tis).

The Pancreas

The pancreas is a long gland that extends from the duodenum to the spleen. The pancreas produces enzymes that digest fats, proteins, carbohydrates, and nucleic acids. The protein-digesting enzymes are produced in inactive forms, which must be converted to active forms in the small intestine by other enzymes. The pancreas also produces large amounts of alkaline fluid, which neutralizes the chyme in the small intestine, thus protecting the lining of the digestive tract. These juices collect in a main duct that joins the common bile duct or empties into the duodenum near the common bile duct. Most persons also have an additional smaller duct that opens into the duodenum.

Because they are usually confined to proper channels, pancreatic enzymes do not damage body tissues. However, if the bile ducts become blocked, pancreatic enzymes back up into the pancreas. Also, in some cases of gallbladder disease, infection may extend to the pancreas and cause abnormal activation of the pancreatic enzymes. In either circumstance, the pancreas suffers destruction by its own juice, and the outcome can be fatal; this condition is known as *acute pancreatitis.*

The pancreas also functions as an endocrine gland, producing the hormones insulin and glucagon that regulate sugar metabolism. These secretions of the islet cells are released directly into the blood.

▶ The Process of Digestion

Although the different organs of the digestive tract are specialized for digesting different types of food, the basic chemical process of digestion is the same for fats, proteins, and carbohydrates. In every case this process requires enzymes. Enzymes are proteins that speed the rate of chemical reactions but are not themselves changed or used up in these reactions. An enzyme is highly specific; that is, it acts only in a certain type of reaction involving a certain type of food molecule. For example, the carbohydrate-digesting enzyme amylase only splits starch into the disaccharide (double sugar) maltose. Another enzyme is required to split maltose into two molecules of the monosaccharide (simple sugar) glucose. Other enzymes split fats into their building blocks, glycerol and fatty acids, and still others split proteins into their building blocks, amino acids.

Because water is added to these molecules as they are split by enzymes, the process is called *hydrolysis* (hi-DROL-ih-sis), which means "splitting by means of water." You can now understand why a large amount of water is needed in digestion. Water is needed not only to produce digestive juices and to dilute food so that it can move more easily through the digestive tract, but also is used in the chemical process of digestion itself.

Let us see what happens to a mass of food from the time it is taken into the mouth to the moment that it is ready to be absorbed.

In the mouth the food is chewed and mixed with saliva, softening it so that it can be swallowed easily. Salivary amylase initiates the process of digestion by changing some of the starches into sugars.

When the food reaches the stomach, it is acted upon by gastric juice, which contains hydrochloric acid and enzymes. The hydrochloric acid has the important function of liquefying the food. In addition, it activates the enzyme pepsin, which is secreted by the cells of the gastric lining in an inactive form. Once activated by hydrochloric acid, pepsin works to digest protein; nearly every type of protein in the diet is first digested by this enzyme. Two other enzymes are also secreted by the stomach, but they are of no importance in adults. The food, gastric juice, and mucus, the latter of which is also secreted by cells of the gastric lining, are mixed to form the semi-liquid substance chyme. Chyme is moved from the stomach to the small intestine for further digestion.

In the duodenum the chyme is mixed with the greenish yellow bile delivered from the liver and the gallbladder through the common bile duct. Rather than containing enzymes, the bile contains salts that split (emulsify) fats into smaller particles to allow the powerful secretion from the pancreas to act on them most efficiently. The pancreatic juice contains a number of enzymes, including the following:

1. *Lipase.* Following the physical division of fats into tiny particles by the action of bile, the powerful pancreatic lipase does almost all the digesting of fats. In this process fats are usually broken down into two simpler compounds, glycerol (glycerine) and fatty acids, which are more readily absorbable. If pancreatic lipase is absent, fats are expelled with the feces in undigested form.
2. *Amylase.* This enzyme changes starch to sugar.
3. *Trypsin* (TRIP-sin). This enzyme splits proteins into amino acids, which are small enough to enter the bloodstream.

The intestinal juice contains a number of enzymes, including three that act on complex sugars to transform them into the simpler form in which they are absorbed. These are *maltase, sucrase,* and *lactase.* It must be emphasized that most of the chemical changes in foods occur in the intestinal tract because of the pancreatic juice, which could probably adequately digest all foods even if no other digestive juice were produced. When pancreatic juice is absent, serious digestive disturbances always occur.

Table 19-1 summarizes the main substances used in digestion. Note that, except for HCl, sodium bicarbonate, and bile salts, all the substances listed are enzymes.

Absorption

The means by which the digested food reaches the blood is known as *absorption.* Most absorption takes place through the mucosa of the small intestine by means of the villi (see Fig. 19-1). Within each villus are a small artery and a small vein bridged with capillaries. Simple sugars, amino acids, some simple fatty acids, and water are absorbed into the blood through the capillary walls in the villi. From here they pass by way of the portal system to the liver, to be stored or released and used as needed.

Most fats have an alternative method of reaching the blood. Instead of entering the blood capillaries, they are absorbed by lymphatic capillaries in the villi that are called **lacteals**. The fat droplets give the lymph a milky appearance (the word *lacteal* means "like milk"). The mixture of lymph and fat globules that is drained from the small intestine after a quantity of fat has been digested is called *chyle* (kile). It circulates in the lymph and eventually enters the bloodstream near the heart.

Minerals and vitamins taken in with food are also absorbed from the small intestine. The minerals and some of the vitamins dissolve in water and are absorbed directly into the blood. Other vitamins are incorporated in fats and are absorbed along with the fats. Some B vitamins and vitamin K are produced by the action of bacteria in the colon and are absorbed from the large intestine.

Control of Digestion

As food moves through the digestive tract, its rate of movement and the activity of each organ it passes through must be carefully regulated. If food moves too slowly or digestive secretions are inadequate, the body will not get enough nourishment. If food moves too rapidly or excess secretions are produced, digestion may be incomplete or the lining of the digestive tract may be damaged.

There are two types of control over digestion: nervous and hormonal. Both illustrate the principles of feedback control. The nerves that control digestive activity are located in the submucosa and between the muscle layers of the organ walls. Instructions for action come from the autonomic (visceral) nervous system. In general, parasympathetic stimulation increases activity and sympathetic stimulation decreases activity. Excess sympathetic stimulation, as in stress, can block the movement of food through the digestive tract and inhibit secretion of the mucus that is so important in protecting the lining of the digestive tract.

The hormones involved in regulation of digestion are produced by the digestive organs themselves. Let us look at some examples of these controls.

The sight, smell, thought, taste, or feel of food in the mouth stimulates, through the nervous system, the secretion of saliva and the release of gastric juice. Once in the stomach, food stimulates the release into the blood of the hormone *gastrin,* which

TABLE 19-1
Digestive Juices Produced by the Organs of the
Digestive Tract and the Accessory Organs

Organ	Main Digestive Juices Secreted	Action
Salivary glands	Salivary amylase*	Begins starch digestion
Stomach	Hydrochloric acid (HCl)	Breaks down proteins
	Pepsin*	Begins protein digestion
Small intestine	Peptidases*	Digest proteins to amino acids
	Lactase, maltase, sucrase*	Digest disaccharides to monosaccharides
Pancreas	Sodium bicarbonate	Neutralizes HCl
	Amylase*	Digests starch
	Trypsin*	Digests protein to amino acids
	Lipases*	Digest fats to fatty acids and glycerol
	Nucleases*	Digest nucleic acids
Liver	Bile salts	Emulsify fats

*Enzymes

promotes stomach secretions and movement. When chyme enters the duodenum, nerve impulses inhibit movement of the stomach so that food will not move too rapidly into the small intestine. This action is a good example of negative feedback. At the same time, hormones released from the duodenum feed back to the stomach to reduce its activity. Two of these hormones, *secretin* (se-KRE-tin) and *cholecystokinin* (ko-le-sis-to-KI-nin), have the additional function of stimulating the liver, pancreas, or gallbladder to release their juices. Table 19-2 lists some of the hormones that help regulate the digestive process.

Hunger and Appetite

Hunger is the desire for food and can be satisfied by the ingestion of a filling meal. Hunger is regulated by centers in the hypothalamus that can be modified by input from higher brain centers. Therefore, cultural factors and memories of past food intake can influence hunger. Strong, mildly painful contractions of the empty stomach may stimulate a feeling of hunger. Messages received by the hypothalamus reduce hunger as food is chewed and swallowed and begins to fill the stomach. The short-term regulation of food intake works to keep the amount

TABLE 19-2
Hormones Active in Digestion

Hormone	Source	Action
Gastrin	Stomach	Stimulates release of gastric juice
Gastric-inhibitory peptide (GIP)	Small intestine	Inhibits release of gastric juice
Secretin	Duodenum	Stimulates release of water and bicarbonate from pancreas; stimulates release of bile from liver; inhibits the stomach
Cholecystokinin (CCK)	Duodenum	Stimulates release of digestive enzymes from pancreas; stimulates release of bile from gallbladder; inhibits the stomach

of food taken in within the limits of what can be processed by the intestine. The long-term regulation of food intake maintains appropriate blood levels of certain nutrients.

Appetite differs from hunger in that, although it is basically a desire for food, it often has no relationship to the need for food. Even after an adequate meal that has relieved hunger, a person may still have an appetite for additional food. A chronic loss of appetite, called *anorexia* (an-o-REK-se-ah), may be due to a great variety of physical and mental disorders. Because the hypothalamus and the higher brain centers are involved in the regulation of hunger, it is possible that emotional and social factors contribute to the development of anorexia.

Anorexia nervosa is a psychologic disorder that mainly afflicts young women. In a desire to be excessively thin, affected persons literally starve themselves, sometimes to the point of death. A related disorder, *bulimia* (bu-LIM-e-ah), is also called the *binge–purge syndrome*. Affected individuals take in huge quantities of food at one time and then induce vomiting or take large doses of laxatives to prevent absorption of the food.

SUMMARY

I. Digestive system
- **A.** Functions—digestion and absorption
- **B.** Structure of wall—mucosa, submucosa, smooth muscle, serosa
- **C.** Peritoneum—serous membrane that lines abdominal cavity and folds over organs
  - **1.** Peritonitis—peritoneal inflammation

II. Digestive tract
- **A.** Mouth
  - **1.** Functions
    - **a.** Ingestion of food
    - **b.** Mastication of food
    - **c.** Lubrication of food
    - **d.** Digestion of starch with salivary amylase
  - **2.** Structures
    - **a.** Tongue
    - **b.** Deciduous (baby) teeth—20 (incisors, canines, molars)
    - **c.** Permanent teeth—32 (incisors, canines, premolars, molars)
  - **3.** Diseases of mouth and teeth—caries, gingivitis, periodontitis, Vincent's disease, leukoplakia, mumps
- **B.** Pharynx (throat)—moves portion of food (bolus) into esophagus by reflex swallowing (deglutition)
- **C.** Esophagus—long muscular tube that carries food to stomach by peristalsis
- **D.** Stomach
  - **1.** Functions
    - **a.** Storage of food
    - **b.** Breakdown of food by churning
    - **c.** Liquefaction of food with hydrochloric acid (HCl) to form chyme
    - **d.** Digestion of protein with enzyme pepsin
  - **2.** Diseases—gastritis, ulcer, cancer
- **E.** Small intestine
  - **1.** Functions
    - **a.** Digestion of food
    - **b.** Absorption of food through villi (small projections of intestinal lining)
  - **2.** Divisions—duodenum, jejunum, ileum
- **F.** Large intestine
  - **1.** Functions
    - **a.** Storage and elimination of waste (defecation)
    - **b.** Reabsorption of water
  - **2.** Divisions—cecum; ascending, transverse, descending, and sigmoid colons; rectum; anus
  - **3.** Disorders—enteritis, diarrhea, constipation (spastic and flaccid), diverticulitis, appendicitis, cancer

III. Accessory organs
- **A.** Liver
  - **1.** Functions
    - **a.** Storage of glucose
    - **b.** Formation of blood plasma proteins
    - **c.** Synthesis of urea
    - **d.** Modification of fats
    - **e.** Manufacture of bile
    - **f.** Destruction of old red blood cells
    - **g.** Detoxification of harmful substances
    - **h.** Storage of vitamins and iron

      2. Diseases—hepatitis, cirrhosis, jaundice, cancer

  B. Gallbladder

      1. Function—storage of bile until needed for digestion

      2. Diseases—gallstones (cholelithiasis), inflammation (cholecystitis)

  C. Pancreas

      1. secretes powerful digestive juice

      2. secretes neutralizing fluid

IV. **Process of digestion**

  A. Products

      1. Simple sugars (monosaccharides) from carbohydrates

      2. Amino acids from proteins

      3. Glycerol and fatty acids from fats

  B. Absorption—movement of digested food into the circulation

  C. Control

      1. Nervous control

         a. Parasympathetic system—generally increases activity

         b. Sympathetic system—generally decreases activity

      2. Hormonal control

         a. Stimulation of digestive activity

         b. Feedback to inhibit responses

QUESTIONS FOR STUDY AND REVIEW

1. Trace the path of a mouthful of food through the digestive tract.
2. What is the peritoneum? Name the two layers and describe their locations. Name four double-layered peritoneal structures.
3. Differentiate between deciduous and permanent teeth with respect to kinds and numbers.
4. What is the pharynx? What is above and below it?
5. What is peristalsis? Name some structures in which it occurs.
6. Name two purposes of the acid in gastric juice.
7. Name the principal digestive enzymes. Where is each formed? What does each do?
8. Where does absorption occur, and what structures are needed for absorption?
9. What types of digested materials are absorbed into the blood?
10. What types of digested materials are absorbed into the lymph?
11. Name the accessory organs of digestion and the functions of each.
12. Name five nondigestive functions of the liver.
13. Give examples of negative feedback in the control of digestion.
14. Name several hormones that regulate digestion.
15. Name several diseases that involve the organs of digestion.
16. Describe what happens in the formation of a peptic ulcer.
17. What is an important symptom of stomach cancer?
18. How is hunger regulated?

CHAPTER 20

Metabolism, Nutrition, and Body Temperature

Selected Key Terms

The following terms are defined in the Glossary:

anabolism

catabolism

fever

glucose

glycogen

hypothermia

malnutrition

metabolic rate

mineral

oxidation

pyrogen

vitamin

Behavioral Objectives

After careful study of this chapter, you should be able to:

- Differentiate between catabolism and anabolism
- Differentiate between the anaerobic and aerobic phases of catabolism and give the end products and the relative amount of energy released from each
- Explain the role of glucose in metabolism
- Compare the energy contents of fats, proteins, and carbohydrates
- Define *essential amino acid*
- Explain the roles of minerals and vitamins in nutrition and give examples of each
- Define *metabolic rate*
- Name several factors that affect the metabolic rate
- Explain how heat is produced in the body
- List the ways in which heat is lost from the body
- Describe the role of the hypothalamus in regulating body temperature
- Define *fever*
- Describe some adverse effects of excessive heat and cold

Memmler, RL, Cohen, BJ, Wood, DL. *THE HUMAN BODY IN HEALTH AND DISEASE*, 8/e, © 1996 Lippincott-Raven Publishers

▶ Metabolism

The end products of digestion are used for all the cellular activities of the body which, all together, make up *metabolism.* These activities fall into two categories:

1. *Catabolism,* which is the breakdown of complex compounds into simpler compounds. It includes the digestion of food into usable small molecules and the release of energy from these molecules within the cell.
2. *Anabolism,* which is the building of simple compounds into substances needed for cellular activities and the growth and repair of tissues.

Through the steps of catabolism and anabolism there is a constant turnover of body materials as energy is consumed, cells function and grow, and waste products are generated.

How Cells Obtain Energy From Food

Energy is released from nutrients in a series of reactions called *cellular respiration,* each step of which requires a specific enzyme as a catalyst. Early studies on cellular respiration were done with *glucose* as the starting compound. Glucose is a simple sugar that is the main energy source for the body. The first steps in the breakdown of glucose do not require oxygen, that is, they are *anaerobic.* This phase of catabolism occurs in the cytoplasm of the cell. It yields a very small amount of energy which is used to make ATP (adenosine triphosphate), the energy compound of the cells. Because the breakdown is incomplete, some organic end product results. In muscle cells operating briefly under anaerobic conditions, lactic acid accumulates while the cells build up an oxygen debt, as described in Chapter 8. However, our cells must use nutrients more completely to generate enough energy for survival. Further oxygen-requiring, or *aerobic,* reactions occur within the mitochondria of the cell and transfer most of the energy remaining in the nutrients to ATP.

In the course of these aerobic reactions, carbon dioxide is released and oxygen is combined with hydrogen from the food molecules to form water. The carbon dioxide is eliminated by the lungs in breathing. Because of the type of chemical reactions involved, and because oxygen is used in the final steps, this process of cellular respiration is called *oxidation.* Although the oxidation of food is often compared to the burning of fuel, this comparison is inaccurate. Burning results in a sudden and often wasteful release of energy in the form of heat and light. In contrast, metabolic oxidation occurs in steps, and much of the energy released is stored as ATP for later use by the cells; some of the energy is released as heat, which is used to maintain body temperature, as discussed later in this chapter.

METABOLIC RATE

Metabolic rate refers to the rate at which energy is released from food in the cells. It is affected by a person's size, body fat, sex, age, activity, and hormones, especially thyroid hormone (thyroxine). *Basal metabolism* is the amount of energy needed to maintain life functions while the body is at rest. Metabolic rate is high in children and adolescents and decreases with age.

The Use of Nutrients

As noted, the main source of energy in the body is glucose. Other carbohydrates in the diet are converted to glucose for metabolism. Glucose circulates in the blood to provide energy for all the cells. Reserves of glucose are stored in liver and muscle cells as *glycogen* (GLI-ko-jen), a compound built from glucose molecules. When it is needed for energy, glycogen is broken down to glucose. Glycerol and fatty acids (from fat digestion) and amino acids (from protein digestion) can also be used for energy, but they enter the breakdown process at different points. Fat in the diet yields more than twice as much energy as protein or carbohydrate (it is more "fattening"); fat yields about 9 kilocalories (kcal) of energy per gram, whereas protein and carbohydrate each yield about 4 kcal per gram. Food that is ingested in excess of need is converted to fat and stored in adipose tissue.

Before oxidation, the amino acids must have their nitrogen (amine) groups removed. This occurs in the liver, where the nitrogen groups are then formed into urea by combination with carbon dioxide. The bloodstream transports the urea to the kidneys to be eliminated.

Anabolism

Food molecules are built into body materials by other metabolic steps, all of which are catalyzed by enzymes. Ten of the 20 amino acids needed to build proteins can be manufactured internally by metabolic reactions. The remaining 10 amino acids cannot be made by the body and therefore must be taken in with the diet; these are the ***essential amino acids.*** Most animal proteins supply all of the essential amino acids and are described as ***complete proteins.*** Vegetables may be lacking in one or more of the essential amino acids. Persons on strict vegetarian diets must learn to combine foods, such as beans with rice or wheat, to obtain all the essential amino acids. The few essential fatty acids that are needed are easily obtained through a healthful, balanced diet.

Minerals and Vitamins

In addition to needing fats, proteins, and carbohydrates, the human body requires minerals and vitamins. ***Minerals*** are elements needed for body structure, fluid balance, and such activities as muscle contraction, nerve impulse conduction, and blood clotting. A list of the main minerals needed in a proper diet is given in Table 20-1. Some additional minerals not listed are also required for good health. Those needed in extremely small amounts are referred to as ***trace elements.***

Vitamins are organic substances needed in very small quantities. Vitamins are parts of enzymes or other essential substances used in cell metabolism, and vitamin deficiencies lead to a variety of nutritional diseases. A list of vitamins is given in Table 20-2.

◗ Some Practical Aspects of Nutrition

Good nutrition is absolutely essential for the maintenance of health. If any vital food material is miss-

TABLE 20-1
Minerals

Mineral	Functions	Sources	Deficiencies
Potassium (K)	Nerve and muscle activity	Fruits, meats, seafood, milk	Muscular and neurologic disorders
Sodium (Na)	Body fluid balance; nerve impulse conduction	Most foods and table salt	Weakness, cramps, diarrhea, dehydration
Calcium (Ca)	Formation of bones and teeth, blood clotting, nerve conduction, muscle contraction	Dairy products, eggs, green vegetables, legumes (peas and beans)	Rickets, tetany, bone demineralization
Phosphorus (P)	Formation of bones and teeth; found in ATP, nucleic acids, cell membrane	Beef, egg yolk, dairy products	Bone demineralization, abnormal metabolism
Iron (Fe)	Oxygen carrier (hemoglobin)	Meat, eggs, spinach, prunes, cereals, legumes	Anemia, dry skin, indigestion
Iodine (I)	Thyroid hormones	Seafood, iodized salt	Hypothyroidism, goiter
Magnesium (Mg)	Catalyst for enzyme reactions, carbohydrate metabolism	Green vegetables, grains	Spasticity, arrhythmia, vasodilation
Manganese (Mn)	Catalyst in actions of calcium and phosphorus	Legumes, nuts, cereals, green leafy vegetables	Possible reproductive disorders
Copper (Cu)	Necessary for absorption and oxidation of vitamin C and iron and in formation of hemoglobin	Liver, fish, oysters, legumes, nuts	Anemia
Cobalt (Co)	Part of vitamin B_{12} involved in blood cell production and in synthesis of insulin	Animal products	Pernicious anemia
Zinc (Zn)	Promotes carbon dioxide transport and energy metabolism; found in enzymes	Many foods	Alopecia (baldness) possibly related to diabetes
Fluorine (F)	Prevents tooth decay	Water and many foods	Dental caries

TABLE 20-2
Vitamins

Vitamins	Functions	Sources	Deficiencies
Retinol (A)	Required for healthy epithelial tissues and for eye pigments	Yellow vegetables, fish, liver oils, dairy products	Night blindness; dry, scaly skin
Thiamin (B_1)	Required for enzymes involved in oxidation of food	Pork, cereal, grains, meats, legumes	Beriberi, a disease of nerves (neuritis)
Riboflavin (B_2)	Needed for enzymes required for oxidation of food	Milk, eggs, kidney, liver, green leafy vegetables	Skin and tongue disorders
Niacin (nicotinic acid)	Involved in oxidation of food	Yeast, lean meat, liver, grains, legumes,	Pellagra with dermatitis, diarrhea, mental disorders
Pyridoxine (B_6)	Involved in metabolism of various food substances and transport of amino acids	Liver, milk, cereals, eggs, yeast, vegetables	Irritability, convulsions, muscle twitching, skin disorders
Pantothenic acid	Essential for normal growth	Yeast, liver, eggs	Sleep disturbances, lack of coordination, skin lesions
Cyanocobalamin (B_{12})	Production of blood cells (hemopoiesis)	Meat, liver, milk, eggs	Pernicious anemia
Biotin	Involved in fat and glycogen formation, amino acid metabolism	Peanuts, liver, tomatoes, eggs	Lack of coordination, dermatitis, fatigue
Folic acid and folates	Required for amino acid metabolism, maturation of red blood cells	Vegetables, liver	Anemia, digestive disorders, neural tube defects in the embryo
Ascorbic acid (C)	Maintains healthy skin and mucous membranes; involved in synthesis of collagen	Citrus fruits, green vegetables	Scurvy, poor bone and wound healing
Calciferol (D)	Aids in absorption of calcium from intestinal tract	Fish liver oils, dairy products	Rickets, bone deformities
Tocopherol (E)	Protects cell membranes	Seeds, green vegetables	In question

ing from the diet, the body will suffer from ***malnutrition.*** One commonly thinks of a malnourished person as one who does not have enough to eat, but malnutrition can also occur from eating too much of the wrong foods. The simplest and perhaps best advice for maintaining a healthful diet is to eat a wide variety of fresh, wholesome foods daily. The U.S. Department of Agriculture (USDA) presently recommends a distribution of food in the diet as shown in Figure 20-1. The "Food Guide Pyramid" emphasizes grains, fruits, and vegetables with limited intake of fats, oils, and sugars.

Because proteins, unlike carbohydrates and fats, are not stored, protein foods should be taken in on a regular basis. The average U.S. diet, however, contains more than enough protein and too much fat, especially animal fat. It is currently recommended that calories from the three types of food be distributed as follows:

Fat	30%
Carbohydrate	58%
Protein	12%

The carbohydrates should be mainly complex, naturally occurring carbohydrates with a minimum of refined sugars. Less than half of the fats in the diet should be ***saturated fats.*** These are fats, mostly from animal sources, that are solid at room temperature, such as butter and lard. Also included in this group are fats that are artificially hardened (hydrogenated), such as vegetable shortening and margarine, which it is best to avoid. Evidence shows that these manufactured fat products may be just as harmful, if not more so, than natural saturated fats.

Food as Good Medicine

Does an apple a day keep the doctor away? This adage reflects what people have long known—that the foods we eat have a profound effect on health. That apple has fiber, which is known to reduce the risk of colon cancer by diluting carcinogens in the bowel and reducing their contact with the bowel wall. It is also lacking in fats, which should be controlled in the diet, especially saturated fats. People with diets rich in fruits and vegetables have lower incidences of diabetes, cardiovascular disease, and several types of cancer. Vegetables in the mustard family, such as broccoli, cabbage, and kale, appear to contain compounds that inhibit the growth of cancer cells. Studies are now underway to identify which components of the various vegetables may provide this protection.

Unsaturated fats are mainly from plants. They are liquid at room temperature and are generally referred to as *oils*. Diets high in saturated fats are associated with a higher than normal incidence of cancer, heart disease, and cardiovascular problems, although the relation between these factors is not clear.

A weight loss diet should follow the same guidelines as given above with a reduction in portion sizes. Exercise is also recommended.

The need for mineral and vitamin supplements to the diet is a subject of controversy. Some researchers maintain that adequate amounts of these substances can be obtained from a varied, healthful diet; others hold that pollution, depletion of the soils, and the storage, refining, and processing of foods make supplementation beneficial. Most agree, however, that children, elderly persons, pregnant and lactating women, and teenagers, who often do not get enough of the proper foods, would profit from additional minerals and vitamins.

When required, supplements should be selected by a physician or nutritionist to fit the particular needs of the individual. Megavitamin dosages may cause unpleasant reactions and in some cases are hazardous. Vitamins A and D have both been found to cause serious toxic effects; a relatively small excess of vitamin D may result in the appearance of dangerous symptoms.

The subject of allergies has received much attention recently. Some persons develop clear allergic (hypersensitive) symptoms if they eat certain foods. The most common food allergens are wheat, yeast, milk, and eggs, but almost any food might cause an allergic reaction in a given individual. Although food allergies may be responsible for symptoms with no other known cause, most persons can probably eat all types of foods.

Body Temperature

Heat is an important byproduct of the many chemical activities constantly going on in tissues all over the body. Heat is always being lost through a variety of outlets. However, because of a number of regulatory devices, under normal conditions body temperature remains constant within quite narrow limits. Maintenance of a constant temperature despite both internal and external influences is one phase of homeostasis, the tendency of all body processes to maintain a normal state despite forces that tend to alter them.

Heat Production

Heat is produced when oxygen combines with nutrients in the cells. Thus, heat is a byproduct of the cellular reactions that generate energy. The amount of heat produced by a given organ varies with the kind of tissue and its activity. While at rest, muscles may produce as little as 25% of total body heat, but when a number of muscles contract, heat production may be multiplied hundreds of times owing to the increase in metabolic rate. Under basal conditions (at rest), the abdominal organs, particularly the liver, produce about 50% of total body heat. While the body is at rest, the brain produces 15% of body heat, and an increase in nervous tissue activity produces very little increase in heat production. The largest amount of heat, therefore, is produced in the muscles and the glands. Although it would seem from this description that some parts of the body would tend to become much warmer than others, the circulating blood distributes the heat fairly evenly.

The rate at which heat is produced is affected by a number of factors, which may include activity level, hormone production, food intake, and age. Hormones, such as thyroxine from the thyroid gland and epinephrine (adrenaline) from the medulla of

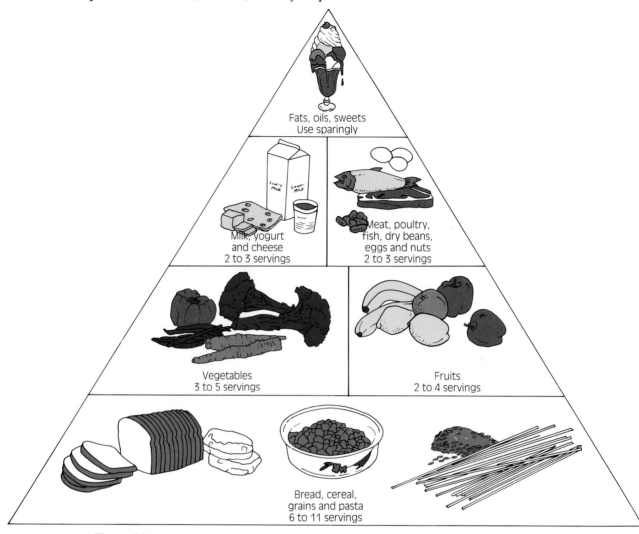

Figure 20-1

Guide to daily food choices. (Taylor C, Lillis C, LeMone P: Fundamentals of Nursing: The Art and Science of Nursing Care, 2nd ed, p. 805. Philadelphia, JB Lippincott, 1993)

the adrenal gland, may increase the rate of heat production. The intake of food is also accompanied by increased heat production. The reasons for this are not entirely clear, but a few causes are known. More fuel is poured into the blood with food intake and is therefore more readily available for cellular metabolism. The glandular structures and the muscles of the digestive system generate additional heat as they set to work. This does not account for all the increase, however, nor does it account for the much greater increase in metabolism following a meal containing a large amount of protein. Whatever the reasons, the intake of food definitely increases the chemical activities that go on in the body and thus adds to heat production.

Heat Loss

More than 80% of heat loss occurs through the skin. The remaining 15% to 20% is dissipated by the respiratory system and with the urine and feces. Networks of blood vessels in the dermis (deeper part) of the skin can bring considerable quantities of blood near the surface so that heat can be dissipated to the outside. This can occur in several ways. Heat can be transferred to the surrounding air in the process of *conduction.* Heat also travels from its source in the form of heat waves or rays, a process termed *radiation.* If the air is moving so that the layer of heated air next to the body is constantly being carried away and replaced with cooler air (as by an electric fan), the process is known as *convection.* Finally,

Obesity

Obesity is usually defined as being 20% or more above normal weight. Its causes are complex and include social, psychologic, and hereditary components. Obesity has been linked with high blood pressure, diabetes, cardiovascular problems, and breast cancer.

People in the United States spend millions each year in the effort to lose weight. Unfortunately, for most people, reducing diets do not work, especially if they are viewed as temporary measures. Usually the lost weight is regained—and more. When faced with very low calorie diets, the body's metabolic rate actually declines in an effort to save energy, making weight loss even more difficult. For most people, a varied diet eaten in moderation and regular exercise are the surest ways to avoid obesity.

heat loss may be produced by **evaporation,** the process by which liquid changes to the vapor state. Rub some alcohol on your arm; it evaporates rapidly, and in so doing uses so much heat from the skin that your arm feels cold. Perspiration does the same thing, although not as quickly. The rate of heat loss through evaporation depends on the humidity of the surrounding air. When this exceeds 60% or so, perspiration does not evaporate so readily, making one feel generally miserable unless one can resort to some other means of heat loss such as convection caused by a fan.

If the temperature of the surrounding air is lower than that of the body, excessive heat loss is prevented by both natural and artificial means. Clothing checks heat loss by trapping "dead air" in both its material and its layers. This noncirculating air is a good insulator. An effective natural insulation against cold is the layer of fat under the skin. Even when skin temperature is low, this fatty tissue prevents the deeper tissues from losing much heat. On the average, this layer is slightly thicker in females than in males. Naturally there are individual variations, but as a rule the degree of insulation depends on the thickness of this layer of subcutaneous fat.

Other factors that play a part in heat loss include the volume of tissue compared with the amount of skin surface. A child loses heat more rapidly than an

adult. Such parts as fingers and toes are affected most by exposure to cold because they have a great amount of skin compared with total tissue volume.

Temperature Regulation

Given that body temperature remains almost constant in spite of wide variations in the rate of heat production or loss, there must be mechanisms for regulating temperature. Many areas of the body take part in this process, but the most important heat-regulating center is the area inside the brain, located just above the pituitary gland, the **hypothalamus.** Some of the cells in the hypothalamus control the production of heat in the body tissues, whereas another group of cells controls heat loss. This control comes about in response to the heat brought to the brain by the blood as well as to nerve impulses from temperature receptors in the skin. If these two factors indicate that too much heat is being lost, impulses are sent quickly from the hypothalamus to the autonomic (involuntary) nervous system, which in turn causes constriction of the skin blood vessels to reduce heat loss. Other impulses are sent to the muscles to cause shivering, a rhythmic contraction of many body muscles, which results in increased heat production. Furthermore, the output of epinephrine may be increased if conditions call for it. Epinephrine increases cell metabolism for a short period, and this in turn increases heat production. Also, the smooth muscle around the hair roots contracts, forming "goose-flesh," a reaction that conserves heat in many animals but only little heat in humans.

If there is danger of overheating, the hypothalamus will transmit impulses that (1) stimulate the sweat glands to increase their activity and (2) dilate the blood vessels in the skin so that there is increased blood flow with a correspondingly greater loss of heat. The hypothalamus may also encourage the relaxation of muscles and thus minimize the production of heat in these organs.

Muscles are especially important in temperature regulation because variations in the amount of activity in these large masses of tissue can readily increase or decrease the total amount of heat produced. Because muscles form roughly one third of the bulk of the body, either an involuntary or a purposeful increase in the activity of this big group of organs can form enough heat to offset a considerable decrease in the temperature of the environment.

Very young and very old persons are limited in their ability to regulate body temperature when exposed to extremes in environment. The body temperature of a newborn infant will decrease if the infant is exposed to a cool environment for a long period. Elderly persons also are not able to produce enough heat to maintain body temperature in a cool environment. Heat loss mechanisms are not fully developed in the newborn; the elderly do not lose as much heat from their skin. Both groups should be protected from extreme temperatures.

Normal Body Temperature

The normal temperature range obtained by either a mercury or an electronic thermometer may extend from 36.2°C to 37.6°C (97°F to 100°F). Body temperature varies with the time of day. Usually it is lowest in the early morning, because the muscles have been relaxed and no food has been taken in for several hours. Temperature tends to be higher in the late afternoon and evening because of physical activity and consumption of food.

Normal temperature also varies in different parts of the body. Skin temperature obtained in the axilla (armpit) is lower than mouth temperature, and mouth temperature is a degree or so lower than rectal temperature. It is believed that, if it were possible to place a thermometer inside the liver, it would register a degree or more higher than rectal temperature. The temperature within a muscle might be even higher during its activity.

Although the Fahrenheit scale is used in the United States, in most parts of the world temperature is measured with the *Celsius* (SEL-se-us) thermometer. On this scale the ice point is at 0° and the normal boiling point of water is at 100°, the interval between these two points being divided into 100 equal units. The Celsius scale is also called the *centigrade scale* (think of 100 cents). See Appendix 2 for a comparison of the Celsius and Fahrenheit scales and formulas for converting from one to the other.

Abnormal Body Temperature
FEVER

Fever is a condition in which the body temperature is higher than normal. An individual with a fever is described as *febrile* (FEB-ril). Usually the presence of fever is due to an infection, but there can be many other causes, such as malignancies, brain injuries, toxic reactions, reactions to vaccines, and diseases involving the CNS. Sometimes emotional upsets can bring on a fever. Whatever the cause, the effect is to reset the body's thermostat in the hypothalamus.

Curiously enough, fever usually is preceded by a chill—that is, a violent attack of shivering and a sensation of cold that such measures as blankets and heating pads seem unable to relieve. Owing to these reactions heat is being generated and stored in the body, and when the chill subsides, the body temperature is elevated.

The old adage that a fever should be starved is completely wrong. During a fever there is an increase in metabolism that is usually proportional to the amount of fever. In addition to the use of available sugar and fat, there is an increase in the use of protein, and during the first week or so of a fever there is definite evidence of destruction of body protein. A high-calorie diet with plenty of protein is therefore desirable.

When a fever ends, sometimes the drop in temperature to normal occurs very rapidly. This sudden fall in temperature is called the *crisis,* and it is usually accompanied by symptoms indicating rapid heat loss: profuse perspiration, muscular relaxation, and dilation of blood vessels in the skin. A gradual drop in temperature, in contrast, is known as *lysis.* A drug that reduces fever is described as *antipyretic* (an-ti-pi-RET-ik).

The mechanism of fever production is not completely understood, but we might think of the hypothalamus as a thermostat that is set higher during fever than normally. This change in the heat-regulating mechanism often follows the injection of a foreign protein or the entrance into the bloodstream of bacteria or their toxins. Substances that produce fever are called *pyrogens* (PI-ro-jens). Up to a point, fever may be beneficial, because it steps up phagocytosis (the process by which white blood cells surround, engulf, and digest bacteria and other foreign bodies), inhibits the growth of certain organisms, and increases cellular metabolism, which may help recovery from disease.

EXTREME OUTSIDE TEMPERATURES

The body's heat-regulating devices are efficient, but there is a limit to what they can accomplish. High outside temperature may overcome the body's heat loss mechanisms, in which case body temperature will rise and cellular metabolism and accompanying heat production will increase. When body temperature goes up, the affected individual is apt to suffer from a series of disorders: heat cramps are followed by heat exhaustion, and then, if that is untreated, heat stroke.

In *heat cramps,* there is localized muscle cramping of the extremities and occasionally of the abdomen. The condition will abate with rest in a cool environment and adequate fluids.

With further heat retention and more fluid loss, *heat exhaustion* occurs. Symptoms of this disorder include headache, tiredness, vomiting, and a rapid pulse. There may be a decrease in circulating blood volume and lowered blood pressure. Heat exhaustion may also be treated by rest and fluid replacement.

Heat stroke (also called *sunstroke*) is a medical emergency. Heat stroke can be recognized by a body temperature of up to 41°C (105°F); hot, dry skin; and CNS symptoms, including confusion, dizziness, and loss of consciousness. The body has responded to the loss of fluid from the circulation by reducing blood flow to the skin and sweat glands. It is important to lower the sunstroke victim's body temperature immediately by removing the individual's clothing, placing him or her in a cool environment, and cooling him or her with cold water or ice. The patient should be treated with appropriate fluids containing the vital electrolytes, including sodium, potassium, calcium, and chloride. Supportive medical care is also necessary, because heat stroke can cause fatal complications.

The body is no more capable of coping with prolonged exposure to cold than it is to prolonged exposure to heat. If, for example, the body is immersed in cold water for a time, the water (a better heat conductor than air) removes more heat from the body than can be replaced, and body temperature falls. This can happen too, of course, in cold air, particularly when clothing is inadequate. The main effects of an excessively low body temperature, termed *hypothermia* (hi-po-THER-me-ah), are uncontrolled

shivering, lack of coordination, and decreased heart and respiratory rates. Speech becomes slurred, and there is overpowering sleepiness, which may lead to coma and death. Outdoor activities in cool, not necessarily cold, weather cause many unrecognized cases of hypothermia. Wind, fatigue, and depletion of water and energy stores all play a part. Once the body is cooled below a certain point, cellular metabolism is slowed and heat production is inadequate for maintaining a normal body temperature. The person must then be warmed by heat from an outside source. The best first aid measure is to remove the person's clothing and put him or her in a warmed sleeping bag with an unclothed companion until shivering stops. Administration of hot, sweetened fluids also helps.

Exposure to cold, particularly to moist cold, may result in *frostbite,* which can cause permanent local tissue damage. The areas most likely to be affected by frostbite are the face, ears, and extremities. The causes of damage include the formation of ice crystals and the reduction of blood supply to the area. Necrosis (death) of the tissues with gangrene can result. The very young, the very old, and those who suffer from disease of the circulatory system are particularly susceptible to cold injuries.

A frostbitten area should *never* be rubbed; rather, it should be rapidly thawed by immersion in warm water or by contact with warm bare skin. The affected area should be treated very gently; a person with frostbitten feet should not be permitted to walk. Persons with cold-damaged extremities frequently have some lowering of body temperature. Warming of the whole person should not be neglected during warming of the affected part.

Hypothermia is employed in certain types of surgery. In such cases the hypothalamus is depressed by drugs and the body temperature reduced to as low as 25°C (77°F) before the operation is begun. In the case of heart surgery, further cooling to 20°C (68°F) is accomplished as the blood goes through the heart–lung machine. This method has been successful even with tiny infants suffering from congenital heart abnormalities.

SUMMARY

I. **Metabolism**—life-sustaining reactions that occur in the living cell
  A. Catabolism—breakdown of complex compounds into simpler compounds
    1. Cellular respiration—a series of reactions in which food is oxidized for energy
      a. Anaerobic phase—does not require oxygen
        (1) Location—cytoplasm
        (2) Yield—small amount of energy
        (3) End product—organic (*i.e.*, lactic acid)
      b. Aerobic phase—requires oxygen
        (1) Location—mitochondria
        (2) Yield—almost all remaining energy in food
        (3) End products—carbon dioxide and water
    2. Metabolic rate—rate at which energy is released from food in the cells
      a. Basal metabolism—amount of energy needed to maintain life functions while at rest
    3. Use of nutrients
      a. Glucose—main energy source
      b. Fats—highest energy yield
      c. Proteins—used for energy after removal of nitrogen
  B. Anabolism—building of simple compounds into materials needed by the body
  C. Minerals—elements needed for body structure and cell activities
    1. Trace elements—elements needed in extremely small amounts
  D. Vitamins—organic substances needed in small amounts

II. **Practical aspects of nutrition**
  A. Malnutrition—too little food or inadequate amounts of specific foods
  B. Components of healthy diet
    1. USDA Food Guide Pyramid
    2. Less than 15% of total daily calories in form of saturated fats

III. **Body temperature**
  A. Heat production—most heat produced in muscles and glands; distributed by the circulation
  B. Heat loss
    1. Skin
    2. Urine
    3. Feces
    4. Breathing
  C. Temperature regulation
    1. Hypothalamus—main temperature regulating center; responds to temperature of blood and temperature receptors in skin
    2. Results of decrease in body temperature
      a. Constriction of blood vessels in skin
      b. Shivering
      c. Increased release of epinephrine
    3. Results of increase in body temperature
      a. Dilation of skin vessels
      b. Sweating
      c. Relaxation of muscles
  D. Normal body temperature—ranges from 36.2°C to 37.6°C; varies with time of day and location measured
  E. Abnormal body temperature
    1. Fever—higher than normal body temperature resulting from infection, injury, toxin, damage to CNS, etc.
      a. Pyrogen—substance that produces fever
      b. Antipyretic—drug that reduces fever
    2. Extreme outside temperatures
      a. Excessive heat—heat cramps, heat exhaustion, heat stroke
      b. Excessive cold
        (1) Hypothermia—low body temperature
          (a) Results—coma and death
          (b) Uses—surgery
        (2) Frostbite—reduction of blood supply to areas such as face, ears, toes, fingers
          (a) Results—necrosis and gangrene

QUESTIONS FOR STUDY AND REVIEW

1. Define *cellular respiration.*
2. In what part of the cell does anaerobic respiration occur? What are its end products?
3. In what part of the cell does aerobic respiration occur? What are its end products?
4. Define *metabolic rate.* What factors affect the metabolic rate?
5. Approximately how many kilocalories are released from a gram of butter as compared with a gram of egg white? with a gram of sugar?
6. Gelatin is not a complete protein. What are the dangers of eating flavored gelatin as a sole source of protein?
7. Name the foods that are emphasized in the USDA Food Guide Pyramid; which are to be eaten sparingly?
8. How is heat produced in the body? What structures produce the most heat during increased activity?
9. Name four factors that affect heat production.
10. How is heat lost from the body?
11. Name four ways in which heat escapes to the environment.
12. In what ways is heat kept in the body?
13. Name the main temperature regulator and describe what it does when the body is too hot and when it is too cold. What parts do muscles play?
14. What is the normal body temperature range? How does it vary with respect to the time of day and the part of the body?
15. Define *fever,* name some aspects of fever's course, and list some of fever's beneficial and detrimental effects.
16. Name and describe two consequences of excessive outside heat. Why do these conditions occur? What is the prime emergency measure for sunstroke?
17. What is hypothermia? Under what circumstances does it usually occur? List some of its effects. In what types of surgery is hypothermia induced?
18. Name and describe two common injuries resulting from cold. What happens in the body to bring these conditions about?
19. Differentiate between the terms in each of the following pairs:
    a. *catabolism* and *anabolism*
    b. *aerobic* and *anaerobic*
    c. *mineral* and *vitamin*
    d. *saturated fats* and *unsaturated fats*
    e. *crisis* and *lysis*

CHAPTER 21

Body Fluids

Selected Key Terms

The following terms are defined in the Glossary:

acidosis

alkalosis

ascites

buffer

dehydration

edema

effusion

electrolyte

extracellular

interstitial

intracellular

pH

Behavioral Objectives

After careful study of this chapter, you should be able to:

- Compare intracellular and extracellular fluids
- List four types of extracellular fluids
- Name the systems that are involved in water balance
- Define *electrolytes* and describe their importance
- Describe three methods for regulating the pH of body fluids
- Describe five disorders involving body fluids
- Describe fluids used in therapy

Memmler, RL, Cohen, BJ, Wood, DL. *THE HUMAN BODY IN HEALTH AND DISEASE*, 8/e,
© 1996 Lippincott-Raven Publishers

The normal proportion of body water varies from 50% to 70% of a person's weight. It is highest in the young and in thin, muscular individuals. As the amount of fat increases, and as a person ages, the percentage of water in the body decreases. Water is important to living cells as a solvent, as a transport medium, and as a participant in metabolic reactions.

Various electrolytes (salts), nutrients, gases, waste, and special substances such as enzymes and hormones are dissolved or suspended in body water. The composition of body fluids is an important factor in homeostasis. Whenever the volume or chemical makeup of these fluids deviates even slightly from normal, disease results. (See Appendix 3 for normal values). The constancy of body fluids is maintained in ways that include the following:

1. The thirst mechanism, which maintains the volume of water at a constant level
2. Kidney activity, which regulates the volume and composition of body fluids (see Chap. 22)
3. Hormones, which serve to regulate fluid volume and electrolytes
4. Regulators of pH (acidity), including buffers, respiration, and kidney function.

Fluid Compartments

Although body fluids have much in common no matter where they are located, there are some important differences between fluid inside and fluid outside cells. Accordingly, fluids are grouped into two main compartments:

1. *Intracellular fluid* is contained within the cells. About two thirds to three fourths of all body fluids are in this category.
2. *Extracellular fluid* includes all body fluids outside of cells. In this group are included the following:
   a. *Blood plasma,* which constitutes about 4% of an individual's body weight
   b. *Interstitial* (in-ter-STISH-al) *fluid,* or more simply, tissue fluid. This fluid is located in the spaces between the cells of tissues all over the body. It is estimated that tissue fluid constitutes about 15% of body weight.
   c. *Lymph*
   d. *Fluid in special compartments,* such as cerebrospinal fluid, the aqueous and vitreous humors of the eye, serous fluid, and synovial fluid. Together these make up about 1% to 3% of total body fluids. Fluids are not locked into one compartment. There is a constant interchange between compartments as fluids are transferred across semipermeable cell membranes by diffusion and osmosis (Fig. 21-1).

Intake and Output of Water

In a person whose health is normal, the quantity of water taken in (intake) is approximately equal to the quantity lost (output). The quantity of water consumed in a day varies considerably. The average adult in a comfortable environment takes in about 2500 mL of water daily. About half of this quantity comes from drinking water and other beverages, and about half comes from foods—fruits, vegetables, and soups.

Water is constantly being lost from the body by the following routes:

1. The *skin.* Although sebum and keratin help prevent dehydration, water is constantly evaporating from the surface of the skin. Larger amounts of water are lost from the skin in the form of sweat when it is necessary to cool the body.
2. The *lungs* expel water along with carbon dioxide.
3. The *intestinal tract* eliminates water along with the feces.
4. The *kidneys* excrete the largest quantity of water lost each day. About 1 to 1.5 liters of water are eliminated daily in the urine.

In many disorders it is important for the health care team to know whether a patient's intake and output are approximately equal; in such a case a 24-hour intake–output record is kept. The intake record includes *all* the liquid the patient has consumed in the form of water, beverages, and food, as well as the quantity and kinds of foods that have been eaten in each 24-hour period. The output record includes the quantity of urine excreted in the same 24-hour period, as well as an estimation of fluid losses due to fever, vomiting, diarrhea, bleeding, wound discharge, or other causes.

Sense of Thirst

The control center for the sense of thirst is located in the hypothalamus. This center plays a major role in the regulation of total fluid volume. A decrease in fluid volume or an increase in the concentration of body fluids stimulates the thirst center

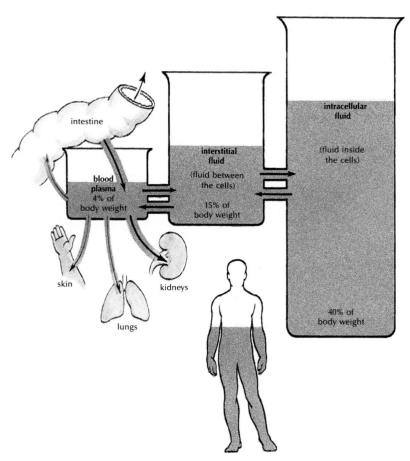

Figure 21-1

Main fluid compartments showing the relative percentage of body fluid in each.

and thus causes an individual to drink water or other fluids containing large amounts of water. Dryness of the mouth also causes a sensation of thirst. Excessive thirst, such as that caused by excessive urine loss in cases of diabetes, is called *polydipsia* (pol-e-DIP-se-ah).

Electrolytes and Their Functions

Electrolytes are important constituents of body fluids. These are compounds that separate into positively and negatively charged ions and carry an electric current in solution. A few of the most important ions are reviewed below. The first three are positive ions (cations), and the last two are negative ions (anions).

1. *Sodium* is chiefly responsible for maintaining osmotic balance and body fluid volume. It is the main positive ion in extracellular fluids. Sodium is required for nerve impulse conduction and is important in maintaining acid–base balance.

How Much to Drink

Most people engaging in routine activities drink 6 to 8 cups of fluid daily to meet their needs. This may be in the form of water, juices, or milk. Beverages containing alcohol or caffeine should not be included in this total because they act as diuretics and increase water loss. The thirst center of the brain should stimulate enough drinking to balance fluids, but this is not always the case. When exercising vigorously, especially in hot weather, the body can dehydrate rapidly. People may not drink enough to replace needed fluids. In addition, if plain water is consumed, the dilution of body fluids may depress the thirst center. Athletes exercising very strenuously may need to drink beverages with some carbohydrates for energy and also some electrolytes to keep fluids in balance.

2. *Potassium* is also important in the transmission of nerve impulses and is a major positive ion in intracellular fluids. Potassium is involved in cellular enzyme activities, and it helps regulate the chemical reactions by which carbohydrate is converted to energy and amino acids are converted to protein.

3. *Calcium* is required for bone formation, muscle contraction, nerve impulse transmission, and blood clotting.

4. *Phosphate* is essential in the metabolism of carbohydrates, bone formation, and acid–base balance. Phosphates are found in the cell membrane and in the nucleic acids (DNA and RNA).

5. *Chloride* is essential for formation of the hydrochloric acid of the gastric juice.

Electrolytes must be kept in the proper concentration in both intracellular and extracellular fluids. The maintenance of water and electrolyte balance is one of the most difficult problems for health workers in caring for patients. Although some electrolytes are lost in the feces and through the skin as sweat, the job of balancing electrolytes is left mainly to the kidneys, as described in Chapter 22.

Several hormones are involved in this process. Aldosterone produced by the adrenal cortex promotes the reabsorption of sodium (and water) and the elimination of potassium. In *Addison's disease,* in which the adrenal cortex does not produce enough aldosterone, there is a loss of sodium and water and an excess of potassium. Calcium and phosphate levels are regulated by hormones from the parathyroid and thyroid glands. Parathyroid hormone increases blood calcium levels by causing the bones to release calcium and by causing the kidneys to reabsorb calcium. The thyroid hormone calcitonin lowers blood calcium by causing calcium to be deposited in the bones.

The Acid–Base Balance

The pH scale is a measure of how acidic or basic (alkaline) a solution is. Body fluids are slightly alkaline at approximately pH 7.4. They must be kept within a narrow range of pH, or damage, even death, will result. Several systems act together to maintain acid–base balance. These are:

1. *Buffer systems.* Buffers are substances that prevent sharp changes in hydrogen ion (H+) concentration and thus maintain a relatively constant pH. They do so by accepting or releasing these ions as needed. The main buffer systems in the body are bicarbonate buffers, phosphate buffers, and proteins, such as hemoglobin and plasma proteins.

2. *Respiration.* The role of respiration in controlling pH was described in Chapter 18. Recall that the release of carbon dioxide from the lungs acts to make the blood more alkaline (increase the pH) by reducing the amount of carbonic acid formed. In contrast, a reduction in the release of carbon dioxide will act to make the blood more acidic (decrease the pH).

3. *Kidney function.* The kidneys serve to regulate pH by reabsorbing or eliminating hydrogen ions as needed. The activity of the kidneys is described in Chapter 22.

If shifts in pH cannot be controlled, the following conditions result:

A drop in the pH of body fluids produces a condition called *acidosis* (as-ih-DO-sis). Acidosis may result, for example, from respiratory obstruction, lung disease, kidney failure, or prolonged diarrhea, the latter of which drains the alkaline contents of the intestine. It may also result from inadequate carbohydrate metabolism, as occurs in diabetes mellitus, ingestion of a low-carbohydrate diet, or starvation. In these cases the body metabolizes too much fat and protein from food or body materials, leading to the production of excess acid. When acidosis results from the accumulation of ketone bodies, as in diabetes, the condition is more accurately described as *ketoacidosis.*

Alkalosis (al-kah-LO-sis) results from an increase in pH. Its possible causes include hyperventilation (the release of too much carbon dioxide), ingestion of antacids, and prolonged vomiting with loss of stomach acids.

Disorders Involving Body Fluids

Edema is the accumulation of excessive fluid in the intercellular spaces. Some causes of edema are as follows:

1. Interference with normal fluid return to the heart, as caused by congestive heart failure or blockage in the venous or lymphatic systems. A backup of fluid in the lungs, *pulmonary edema,* is a serious potential consequence of congestive heart failure.

2. Protein loss or ingestion of too little dietary protein over an extended period. The decrease in

Sodium and Potassium Imbalances

The concentrations of sodium and potassium in body fluids are important measures of water and electrolyte balance. An excess of sodium in body fluids is termed *hypernatremia* (hi-per-nah-TRE-me-ah), taken from the Latin name for sodium, *natrium.* This condition accompanies dehydration. *Hyponatremia*, a deficiency of sodium in body fluids, can come from water intoxication, heart failure, kidney failure, cirrhosis of the liver, pH imbalance, or endocrine disorders.

The term *hyperkalemia* (hi-per-kah-LE-me-ah) is taken from the Latin name for potassium, *kalium.* It refers to excess potassium in body fluids, which may result from kidney failure and other causes. *Hypokalemia*, or low potassium in body fluids, may result from taking diuretics that cause potassium to be lost along with water. It may also result from pH imbalance or secretion of too much aldosterone from the adrenal cortex.

---

protein lowers the osmotic pressure of the blood and reduces the return of fluid to the circulation. This results in accumulation of fluid in the tissues.

3. Kidney failure, a common clinical cause of edema, resulting from the inability of the kidneys to eliminate adequate amounts of urine.
4. Increased loss of fluid through the capillaries, as caused by injury, allergic reaction, or certain infections.

Water intoxication involves dilution of body fluids in both the intracellular and extracellular compartments. Transport of water into the cells results in swelling. In the brain, the swelling of cells may lead to convulsions, coma, and finally death. Causes of water intoxication include an excess of antidiuretic hormone (ADH) and intake of excess fluids by mouth or by intravenous injection.

Dehydration (de-hi-DRA-shun), a severe deficit of body fluids, will result in death if it is prolonged. The causes include vomiting, diarrhea, drainage from burns or wounds, excessive perspiration, and too little fluid intake, as in cases of damage to the thirst mechanism. In such cases it may be necessary to administer intravenous fluids to correct fluid and electrolyte imbalances.

Effusion (e-FU-zhun) is the escape of fluid into a cavity or a space. An example is pleural effusion, fluid within the pleural space; in this condition fluid compresses the lung so that normal breathing is not possible. Tuberculosis, cancer, and some infections may give rise to effusion.

Effusion into the pericardial sac may occur in autoimmune disorders such as rheumatoid arthritis and lupus erythematosus, because of infections, or for other unknown causes. This may interfere with normal heart contractions and can cause death.

Ascites (ah-SI-teze) is effusion with accumulation of fluid within the abdominal cavity. It may occur in disorders of the liver, kidneys, and heart, as well as in cancers and infection.

▶ Fluid Therapy

Fluids are administered into a vein under a wide variety of conditions to help maintain normal body functions when natural intake is not possible. Fluids are also administered to correct specific fluid and electrolyte imbalances in cases of losses due to disease or injury.

The first fluid started in emergencies is normal saline, which contains 0.9% sodium chloride, a concentration equal to that of plasma. Being isotonic, this type of solution will not change the distribution in the body fluid compartments.

Another common fluid is dextrose (glucose) 5% in water. This is slightly hypotonic when it is infused. The amount of sugar contained in a liter of such fluid is equal to 170 calories. The sugar is soon used up, resulting in a fluid that is effectively pure water. Frequently the client will receive 5% dextrose in ½ normal saline. This solution is hypertonic when infused, but becomes hypotonic once the sugar is used. Both of these solutions increase the plasma fluid volume. Small amounts of potassium chloride are often added to the above fluids to replace this electrolyte lost in the course of vomiting or diarrhea.

Ringer's lactate solution contains sodium, potassium, calcium, chloride, and lactate. In this formulation, the electrolyte concentrations are equal to normal plasma values. The lactate is metabolized to bicarbonate, which acts as a buffer. This fluid is given when the need is for additional plasma volume with the electrolyte concentration equal to that of the blood.

Serum albumin 25% contains the plasma protein albumin in a concentration five times normal. This hypertonic solution will draw fluid from the interstitial space into the circulation.

Fluids containing varied concentrations of dextrose, sodium chloride, potassium, and other elec-

trolytes and substances are manufactured. These fluids are used to correct specific imbalances. Nutritional solutions containing concentrated sugar, protein, and fat are available for administration in cases when oral intake will not be possible for an extended period.

SUMMARY

I. Fluid compartments
  A. Intracellular fluid—contained within the cells
  B. Extracellular fluid—outside the cells
    1. Blood plasma
    2. Interstitial (tissue) fluid
    3. Lymph
    4. Fluid in special compartments
II. Water balance
  A. Output—through skin, lungs, intestinal tract, kidneys
  B. Intake—through food and beverages
    1. Thirst center—in hypothalamus
III. Electrolytes—release ions in solution
  A. Regulation
    1. Kidneys—main regulators
    2. Hormones
      a. Aldosterone—reabsorption of sodium, excretion of potassium
      b. Parathyroid hormone—increase in blood calcium level
      c. Calcitonin—decrease in blood calcium level
IV. Acid–base balance—normal pH is 7.4
  A. Mechanisms to regulate pH
    1. Buffers—maintain constant pH
    2. Kidney—regulates amount of hydrogen ion excreted

    3. Respiration—release of carbon dioxide increases pH (less acid); retention of carbon dioxide decreases pH (more acid)
  B. Abnormal pH
    1. Acidosis—decrease in pH due to respiratory disorder, kidney failure, diarrhea, or diabetes mellitus
    2. Alkalosis—increase in pH due to hyperventilation, prolonged vomiting, or ingestion of antacids
V. Disorders of body fluids
  A. Edema—accumulation of fluid in tissues
  B. Water intoxication—dilution of body fluids
  C. Dehydration—deficiency of fluid
  D. Effusion—escape of fluid into a cavity or space
  E. Ascites—accumulation of fluid in abdominal cavity
VI. Fluid therapy
  A. Purpose
    1. Correct fluid balance
    2. Correct electrolyte balance
    3. Provide nourishment
  B. Commonly used types
    1. Normal saline
    2. 5% dextrose
    3. Ringer's lactate
    4. 25% serum albumin

QUESTIONS FOR STUDY AND REVIEW

1. List four ways in which body fluids are regulated.
2. In a healthy person, what is the ratio of fluid intake to output?
3. Name five common ions in the body. How is each used?
4. Name three hormones involved in electrolyte balance and explain what each does.
5. Name the three main buffer systems in the body.
6. How does the respiratory system help regulate pH?
7. Describe five disorders involving body fluids.
8. List several reasons for administering intravenous fluids.
9. When is normal saline used for treatment?
10. List the contents of some other fluids commonly used for treatment.
11. Compare the terms in each of the following pairs.
    a. *intracellular fluid* and *extracellular fluid*
    b. *acidosis* and *alkalosis*

CHAPTER 22

The Urinary System

Selected Key Terms

The following terms are defined in the Glossary:

angiotensin

antidiuretic hormone (ADH)

cystitis

dialysis

erythropoietin

excretion

glomerular filtrate

glomerulonephritis

glomerulus

hemodialysis

kidney

micturition

nephron

renin

urea

ureter

urethra

urinalysis

urinary bladder

urine

Behavioral Objectives

After careful study of this chapter, you should be able to:

- List the systems that eliminate waste and name the substances eliminated by each
- Describe the parts of the urinary system and give the functions of each
- Trace the path of a drop of blood as it flows through the kidney
- Describe a nephron
- Describe the functions of renin and angiotensin
- Name the four processes involved in urine formation and describe the functions of each
- Identify the role of ADH in urine formation
- List the common disorders of the urinary system
- List five signs of chronic renal failure
- Explain the principle and the purpose of kidney dialysis
- Name three normal and six abnormal constituents of urine

Memmler, RL, Cohen, BJ, Wood, DL. *THE HUMAN BODY IN HEALTH AND DISEASE*, 8/e,
© 1996 Lippincott-Raven Publishers

The urinary system is also called the *excretory system* because one of its main functions is to remove waste products from the blood and eliminate them from the body. Other functions of the urinary system include regulation of the volume of body fluids and balance of the pH and electrolyte composition of these fluids.

Although the focus of this chapter is the urinary system, certain aspects of other systems will also be discussed because body systems work interdependently to maintain a balance, or homeostasis. The chief excretory mechanisms of the body, and some of the substances that they eliminate, are listed below:

1. The *urinary system* excretes water, waste products containing nitrogen, and salts. These are all constituents of the urine.
2. The *digestive system* eliminates water, some salts, bile, and the residue of digestion, all of which are contained in the feces. The liver is important in eliminating the products of red blood cell destruction and in breaking down certain drugs and toxins.
3. The *respiratory system* eliminates carbon dioxide and water. The latter appears as vapor, as can be demonstrated by breathing on a windowpane.
4. The skin, or *integumentary system,* excretes water, salts, and very small quantities of nitrogenous wastes. These all appear in perspiration, though evaporation of water from the skin may go on most of the time without our being conscious of it.

Organs of the Urinary System

The main parts of the urinary system, shown in Figure 22-1, are as follows:

1. Two *kidneys.* These organs extract wastes from the blood, balance body fluids, and form urine.
2. Two *ureters* (U-re-ters). These tubes conduct urine from the kidneys to the urinary bladder.
3. A single *urinary bladder.* This reservoir receives and stores the urine brought to it by the two ureters.
4. A single *urethra* (u-RE-thrah). This tube conducts urine from the bladder to the outside of the body for elimination.

The Kidneys

Location of the Kidneys

The two kidneys lie against the muscles of the back in the upper abdomen. They are up under the dome of the diaphragm and are protected by the lower ribs and the rib (costal) cartilages. Each kidney is enclosed in a membranous capsule that is made of fibrous connective tissue; this is loosely adherent to the kidney itself. In addition, there is a crescent of fat called the *adipose capsule* around the perimeter of the organ. This capsule is one of the chief supporting structures of the kidney. The kidneys, as well as the ureters, lie behind the peritoneum. Thus, they are not in the peritoneal cavity but rather in an area known as the *retroperitoneal* (ret-ro-per-ih-to-NE-al) *space.*

The blood supply to the kidney is illustrated in Figure 22-2. Blood is brought to the kidney by a short branch of the abdominal aorta called the *renal artery.* After entering the kidney, the renal artery subdivides into smaller and smaller branches, which eventually make contact with the functional units of the kidney, the *nephrons* (NEF-ronz) (Figs. 22-3 and 22-4). Blood leaves the kidney by vessels that finally merge to form the *renal vein*, which carries blood into the inferior vena cava for return to the heart.

Structure of the Kidneys

The kidney is a somewhat flattened organ about 10 cm (4 inches) long, 5 cm (2 inches) wide, and 2.5 cm (1 inch) thick. On the inner, or medial, border there is a notch called the *hilus,* at which region the renal artery, the renal vein, and the ureter connect with the kidney. The outer, or lateral, border is convex (curved outward), giving the entire organ a bean-shaped appearance.

The kidney is divided into three regions: the renal cortex, the renal medulla, and the renal pelvis (see Fig. 22-3). The renal *cortex* is the outer portion of the kidney. The renal *medulla* contains the tubes that collect urine. These tubes form a number of cone-shaped structures called *pyramids.* The tips of the pyramids point toward the *renal pelvis*, a funnel-shaped basin that forms the upper end of the ureter. Cuplike extensions of the renal pelvis surround the tips of the pyramids and collect urine; these extensions are called *calyces* (KA-lih-seze) (sing., *calyx* [KA-liks]). The urine that collects in the pelvis then passes down the ureters to the bladder.

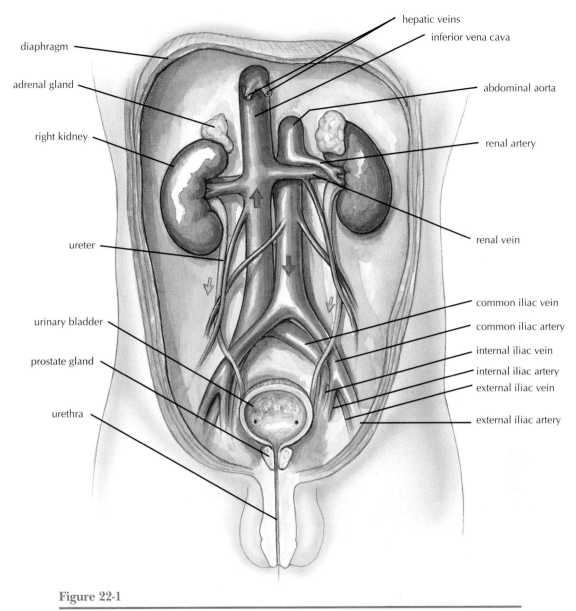

diaphragm

adrenal gland

right kidney

ureter

urinary bladder

prostate gland

urethra

hepatic veins

inferior vena cava

abdominal aorta

renal artery

renal vein

common iliac vein

common iliac artery

internal iliac vein

internal iliac artery

external iliac vein

external iliac artery

Figure 22-1

Urinary system, with blood vessels.

The kidney is a glandular organ; that is, most of the tissue is epithelium with just enough connective tissue to serve as a framework. As is the case with most organs, the most fascinating aspect of the kidney is too small to be seen with the naked eye. This basic unit of the kidney, where the kidney's work is actually done, is the nephron (see Fig. 22-4). The nephron is essentially a tiny coiled tube with a bulb at one end. This bulb, called ***Bowman's capsule,*** surrounds a cluster of capillaries called the ***glomerulus*** (glo-MER-u-lus) (pl., *glomeruli* [glo-MER-u-li]). Each kidney contains about 1 million nephrons; if all these coiled tubes were separated, straightened out, and laid end to end, they would span some 120 kilometers (75 miles)! Figure 22-5 is a microscopic view of kidney tissue showing several glomeruli, each surrounded by a Bowman's capsule. It also shows sections through the tubular portions of the nephrons.

A small blood vessel, called the ***afferent arteriole,*** supplies the glomerulus with blood; another small

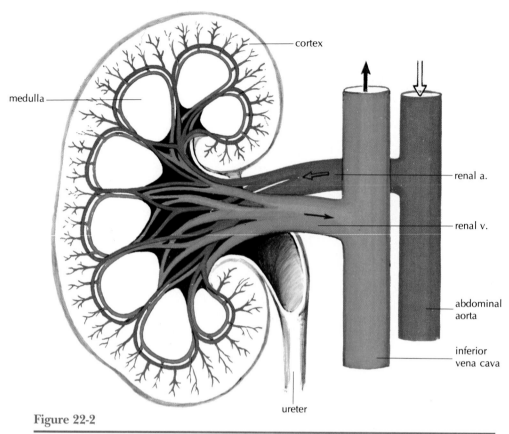

Figure 22-2

Blood supply and circulation of the kidney.

vessel, called the *efferent arteriole,* carries blood from the glomerulus to capillaries that surround the tubular portion of the nephron. These small vessels, *peritubular capillaries,* are named for their location around the nephron.

The tubular part of the nephron consists of several portions. The coiled portion leading from Bowman's capsule is called the *proximal convoluted* (KON-vo-lu-ted) *tubule (PCT).* The tubule then uncoils to form a hairpin-shaped segment called the *loop of Henle.* Continuing from the loop, the tubule coils once again into the *distal convoluted tubule (DCT),* so called because it is farther along the tubule from Bowman's capsule than is the PCT. The distal end of each tubule empties into a collecting tubule, which then continues through the medulla toward the renal pelvis. The glomerulus, Bowman's capsule, and the proximal and distal convoluted tubules of most nephrons are within the renal cortex. The loops of Henle extend varying distances into the medulla.

The distal convoluted tubule curls back toward the glomerulus between the afferent and efferent arterioles. At the point at which the distal convoluted tubule contacts the afferent arteriole, there are specialized glandular cells that form the *juxtaglomerular* (juks-tah-glo-MER-u-lar) *apparatus* (Fig. 22-6), abbreviated as *JG apparatus.* The name means "near the glomerulus," which describes the location of the cells in this structure. The function of the JG apparatus is described below.

Functions of the Kidneys

The kidneys are involved in the following processes:

1. Excretion of unwanted substances such as waste products from cell metabolism, excess salts, and toxins. One product of amino acid metabolism is nitrogen-containing waste material, a chief form of which is *urea* (u-RE-ah). The kidneys provide a specialized mechanism for the elimination of these nitrogenous (ni-TROJ-en-us) wastes.

pyramids of medulla

cortex

nephron

calyx

renal pelvis

capsule

ureter

Figure 22-3

Longitudinal section through the kidney showing its internal structure and a much enlarged diagram of a nephron. There are more than 1 million nephrons in each kidney.

2. Maintenance of water balance. Although the amount of water consumed in a day can vary tremendously, the kidneys can adapt to these variations so that the volume of body water remains remarkably stable from day to day. Water is constantly lost in many ways: from the skin, from the respiratory system during exhalation, and from the intestinal tract. Normally, the amount of water taken in or produced (intake) is approximately equal to the amount lost (output).

3. Regulation of the acid–base balance of body fluids. Acids are constantly being produced by cell metabolism, and certain foods can cause acids or bases to be formed in the body. Bases in the form of antacids, such as baking soda, may also be ingested. However, if the body is to function normally, a certain critical proportion of acids and bases must be maintained at all times.

4. Production of *renin* (RE-nin), an enzyme which is important in the regulation of blood pressure. If blood pressure falls too low for effective filtration through the glomerulus, the cells of the juxtaglomerular (JG) apparatus release renin into the blood. Here, the enzyme activates a protein called *angiotensin* (an-je-o-TEN-sin) that causes blood vessels to constrict, thus raising blood pressure.

5. Production of the hormone *erythropoietin* (eh-rith-ro-POY-eh-tin) *(EPO),* which stimulates the production of red blood cells in the red bone marrow. The hormone is produced when the kidneys do not get enough oxygen. EPO made by genetic engineering is available to reduce the severe anemia that accompanies the end stage of kidney failure.

Renal Physiology
GLOMERULAR FILTRATION

The process of urine formation begins in the glomerulus and Bowman's capsule. The membranes that form the walls of the glomerular capillaries are

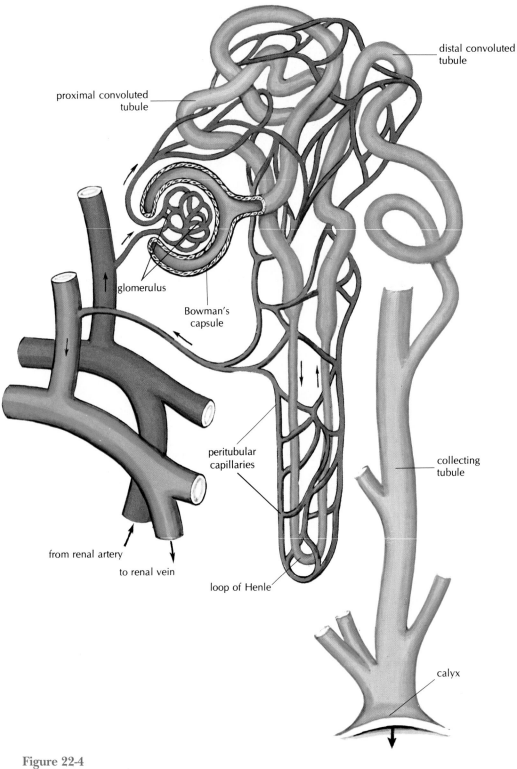

distal convoluted
tubule

proximal convoluted
tubule

glomerulus

Bowman's
capsule

peritubular
capillaries

from renal artery

to renal vein

loop of Henle

collecting
tubule

calyx

Figure 22-4

Simplified diagram of a nephron.

Figure 22-5

Microscopic view of the kidney. (Courtesy of Dana Morse Bittus and B. J. Cohen)

glomerulus

renal tubules

Bowman's capsule

sieve-like and permit the free flow of water and soluble materials through them. Like other capillary walls, these are impermeable (im-PER-me-abl) to blood cells and large protein molecules, so these substances remain in the blood (Fig. 22-7).

Because the diameter of the afferent arteriole is slightly larger than the diameter of the efferent arteriole, blood can enter the glomerulus more easily than it can leave. Thus, the pressure of the blood in the glomerulus is about three to four times as high

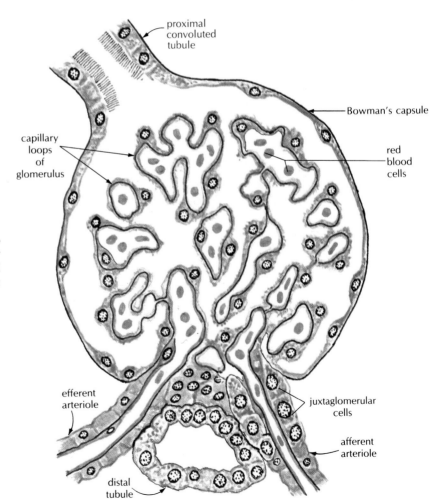

proximal convoluted tubule

Bowman's capsule

capillary loops of glomerulus

red blood cells

efferent arteriole

juxtaglomerular cells

afferent arteriole

distal tubule

Figure 22-6

Structure of the juxtaglomerular apparatus. Note how the distal convoluted tubule contacts the arterioles.

blood

blood

afferent
arteriole

efferent
arteriole

glomerulus
(tuft of capillaries)

endothelium
(capillary wall)

simple
squamous
epithelium

filtrate

water molecules

soluble molecules

insoluble molecules

secreting
epithelium

convoluted
tubule

Figure 22-7

Diagram showing the process of filtration in the formation of urine. The high pressure
inside the capillaries of the glomerulus forces dissolved substances (but not plasma proteins) and much water into the space inside Bowman's capsule. The smaller caliber of
the efferent vessel as compared with that of the larger afferent vessel causes this pressure.

as it is in other body capillaries. To understand this effect, think of placing your thumb over the end of a garden hose as water comes through. Because the diameter of the opening is made smaller, water is forced out under higher pressure. As a result of this increased pressure in the glomerulus, materials are constantly being "squeezed" out of the blood into Bowman's capsule. This process is known as *glomerular filtration.* The fluid that enters Bowman's capsule, called the *glomerular filtrate,* begins its journey along the tubular system of the nephron. In addition to water and the normal soluble substances in the blood, other substances, such as drugs, may also be filtered and become part of the glomerular filtrate.

TUBULAR REABSORPTION

About 160 to 180 liters of filtrate are formed each day in the kidneys. However, only 1 to 1.5 liters of urine are eliminated daily. Clearly, most of the water that enters the nephron is not excreted with the urine but rather is returned to the circulation. In addition to water, many substances that are needed by the body, such as nutrients and ions, also pass into the nephron as part of the filtrate. These must be returned to the body as well. So the process of filtration that occurs in Bowman's capsule is followed by a process of *reabsorption.* As the filtrate travels through the tubular system of the nephron, water and other needed substances leave the tubule by diffusion and active transport and enter the surrounding tissue fluids. They then enter the blood in the peritubular capillaries and return to the circulation. Most of the urea and other nitrogenous waste material is kept within the tubule to be eliminated with the urine.

TUBULAR SECRETION

Before the filtrate leaves the body as urine, a final adjustment in composition is made by the process of *tubular secretion.* In this process, some substances are actively moved from the blood into the nephron. Of particular importance is the secretion of potassium ions and hydrogen ions. The long term regulation of acid-base balance is carried out by the kidney through tubular secretion of hydrogen ions.

CONCENTRATION OF THE URINE

The amount of water that is eliminated with the urine is regulated by a complex mechanism within the nephron that is influenced by *antidiuretic hor-*

The Transport Maximum

The kidney works very efficiently to return valuable substances to the blood following glomerular filtration. However, the carriers that are needed for active transport of these substances can become overloaded. There is a limit to the amount of each substance that can be reabsorbed in a given time period. The limit of this rate of reabsorption is called the *transport maximum (Tm)* for that substance, and it is measured in milligrams (mg) per minute.

If a substance is present in excess in the blood, it may exceed its transport maximum and then, because it cannot be totally reabsorbed, some will be excreted in the urine. For example, if the concentration of glucose in the blood exceeds 180 mg/mL, glucose will begin to appear in the urine, a condition called *glucosuria* (glu-ko-SU-re-ah). The most common cause of glucosuria is uncontrolled diabetes mellitus.

mone (ADH) from the posterior pituitary gland. The process is called the *counter-current mechanism* because it involves fluid traveling in opposite directions within the loop of Henle. The counter-current mechanism is illustrated in Figure 22-8. Its essentials are described below.

As the filtrate passes through the loop of Henle, salts, especially sodium, are actively pumped out by the cells of the nephron, with the result that the interstitial fluid of the medulla becomes increasingly concentrated. Because the nephron is not very permeable to water at this point, the fluid within the nephron becomes increasingly dilute. As the fluid passes through the more permeable DCT and through the collecting tubule, water is drawn out by the concentrated fluids around the nephron and returned to the blood. (Remember, according to the laws of diffusion, water follows salt.) In this manner the urine becomes more concentrated and its volume is reduced.

The role of ADH is to make the walls of the DCT and collecting tubule more permeable to water so that more water will be reabsorbed and less will be excreted with the urine. The release of ADH is regulated by a feedback system. As the blood becomes more concentrated, the hypothalamus causes more

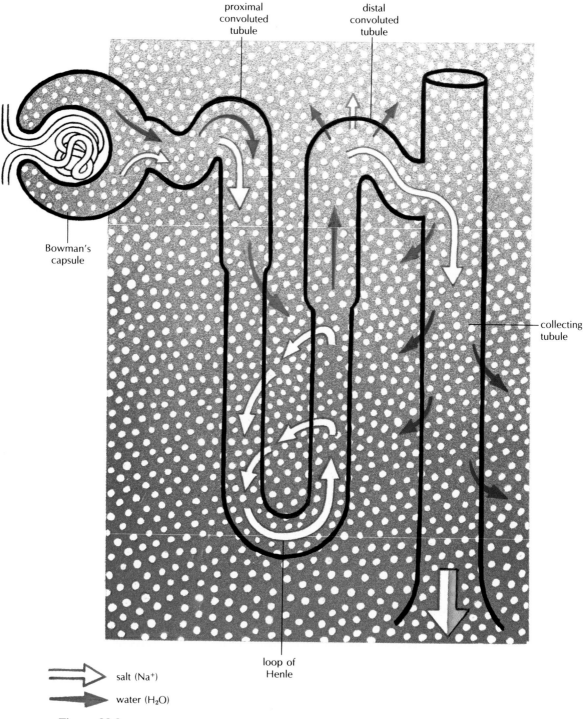

proximal
convoluted
tubule

distal
convoluted
tubule

Bowman's
capsule

collecting
tubule

loop of
Henle

⟹ salt (Na⁺)

⟶ water (H₂O)

Figure 22-8

Loop of Henle, where the proportions of waste and water in urine are regulated according to
the body's constantly changing needs. The concentration of urine is determined by means of
intricate exchanges of water and salt, constituting the counter-current mechanism.

ADH to be released from the posterior pituitary; as the blood becomes more dilute, less ADH is released. In the disease diabetes insipidus there is inadequate secretion of ADH from the hypothalamus. This results in the elimination of large amounts of very dilute urine accompanied by excessive thirst.

To summarize the above processes involved in urine formation:

1. Glomerular filtration allows all diffusible materials to pass from the blood into the nephron.
2. Tubular reabsorption moves useful substances back into the blood while keeping waste products in the nephron to be eliminated in the urine.
3. Tubular secretion moves additional substances from the blood into the nephron for elimination. Movement of hydrogen ions is one means by which the pH of body fluids is balanced.
4. The counter-current mechanism concentrates the urine and reduces the volume excreted. The pituitary hormone ADH allows more water to be reabsorbed from the nephron.

Although this story seems quite complex, and there appears to be a great deal of back-and-forth exchanges, these processes together allow the kidney to "fine tune" body fluids. As the filtrate makes its slow journey through the twists and turns of the nephron, there is ample time for exchanges to take place between the kidney tubules and the circulating blood.

▶ The Ureters

The two ureters are long, slender, muscular tubes that extend from the kidney basin down to and through the lower part of the urinary bladder. Their length naturally varies with the size of the individual and so may be anywhere from 25 cm to 32 cm (10–13 inches) long. Nearly 2.5 cm (1 inch) of its lower part enters the bladder by passing obliquely (at an angle) through the bladder wall. The ureters are entirely extraperitoneal, being located behind and, at the lower part, below the peritoneum.

The wall of the ureter includes a lining of epithelial cells, a relatively thick layer of involuntary muscle, and finally an outer coat of fibrous connective tissue. The lining is continuous with that of the renal pelvis and the bladder. The muscles of the ureters are capable of the same rhythmic contraction (peristalsis) found in the digestive system.

Urine is moved along the ureter from the kidneys to the bladder by gravity and by peristalsis at frequent intervals. Because of the oblique direction of the last part of each ureter through the lower bladder wall, compression of the ureters by the full bladder prevents backflow of urine.

▶ The Urinary Bladder

Characteristics of the Bladder

When it is empty, the urinary bladder (Fig. 22-9) is located below the parietal peritoneum and behind the pubic joint. When it is filled, it pushes the peritoneum upward and may extend well into the abdominal cavity proper. The urinary bladder is a temporary reservoir for urine, just as the gallbladder is a storage sac for bile.

The bladder wall has many layers. It is lined with mucous membrane; the lining of the bladder, like that of the stomach, is thrown into folds called *rugae* when the organ is empty. Beneath the mucosa is a layer of connective tissue. Then follows a three-layered coat of involuntary muscle tissue that is capable of great stretching. Finally, there is an incomplete coat of peritoneum that covers only the upper portion of the bladder. When the bladder is empty, the muscular wall becomes thick and the entire organ feels firm. As the bladder fills, the muscular wall becomes thinner, and the organ may increase from a length of 5 cm (2 inches) up to as much as 12.5 cm (5 inches) or even more. A moderately full bladder holds about 470 mL (1 pint) of urine.

In the floor of the bladder is the *trigone* (TRI-gone), a triangle formed by the openings of the two ureters and the urethra (see Fig. 22-9). As the bladder fills with urine, it expands upward leaving the trigone at the base stationary. This prevents the stretching of the ureteral openings and the possible back flow of urine into the ureters.

Urination

The process of expelling (voiding) urine from the bladder is called *urination* or *micturition* (mik-tu-RISH-un). Near the outlet of the bladder, a circle of smooth muscle forms the *internal sphincter*, which contracts involuntarily to prevent emptying of the bladder. As the bladder fills, stretch receptors send impulses to a center in the lower part of the spinal cord. From this center, motor impulses are sent out

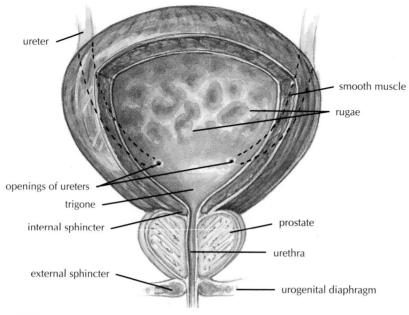

ureter

smooth muscle

rugae

openings of ureters

trigone

internal sphincter

prostate

urethra

external sphincter

urogenital diaphragm

Figure 22-9

Interior of the urinary bladder, shown in the male. The trigone is a triangle in the floor of the bladder marked by the openings of the ureters and the urethra.

to the bladder musculature and the organ is emptied. In the infant this emptying is automatic (a reflex action). As a child matures, higher brain centers gain control over the reflex action and over a voluntary external sphincter that is located below the internal sphincter. The time of urination can then be voluntarily controlled unless the bladder becomes too full.

The Urethra

The *urethra,* the tube that extends from the bladder to the outside, is the means by which the bladder is emptied. The urethra differs in men and women; in the male it is part of the reproductive system and the urinary system, and it is much longer than is the female urethra.

The urethra in the female is a thin-walled tube about 4 cm (1.5 inches) long. It is located behind the pubic joint and is embedded in the muscle of the front wall of the vagina. The external opening, called the *urethral meatus,* is located just in front of the vaginal opening between the labia minora.

The male urethra is about 20 cm (8 inches) long. Early in its course, it passes through the prostate

gland, where it is joined by the two ducts carrying the male sex cells. From here it leads through the *penis* (PE-nis), the male organ of copulation, to the outside. The male urethra, then, serves the dual purpose of conveying the sex cells and draining the bladder, whereas the female urethra performs only the latter function.

Disorders of the Urinary System

Kidney Disorders

Kidney disorders may be acute or chronic. Acute conditions usually arise suddenly, most frequently as the result of infection with inflammation of the nephrons. These diseases commonly run a course of a few weeks and are followed by complete recovery. Chronic conditions arise slowly and often are progressive with gradual loss of kidney function.

Acute glomerulonephritis (glo-mer-u-lo-nef-RI-tis) is the most common disease of the kidneys. This condition usually occurs in children about 1 to 4 weeks after a streptococcal infection of the throat. Antibodies formed in response to the streptococci attach to the glomerular membrane and injure it. These damaged glomeruli allow protein, especially albumin, to filter into Bowman's capsule and ultimately to appear in the urine (albuminuria). They also allow red blood cells to filter into the urine (hematuria). Usually, the patient recovers without permanent kidney damage. Sometimes, however, particularly in adult patients, the disease becomes chronic, with a gradual decrease in the number of functioning nephrons leading to chronic renal failure.

Pyelonephritis (pi-el-o-nef-RI-tis), an inflammation of the kidney pelvis and the tissue of the kidney itself, may be either acute or chronic. In acute pyelonephritis, the inflammation results from a bacterial infection. Bacteria most commonly reach the kidney by ascending along the lining membrane from an infection in the lower part of the urinary tract (see Fig. 23-12). More rarely, bacteria are carried to the kidney by the blood. Acute pyelonephritis is often seen in persons with partial obstruction of urine flow with stagnation (urinary stasis). It is most likely to occur in pregnant women and in men with enlarged prostates, because the prostate surrounds the first portion of the urethra in males. Usually, the disease responds to the administration of antibiotics, fluid replacement, rest, and fever control. Chronic pyelonephritis, a more serious disease, is frequently seen in patients with urinary tract blockage. It may be caused by persistent or repeated bacterial infections. Progressive damage of kidney tissue in this condition eventually leads to chronic renal failure.

Hydronephrosis (hi-dro-nef-RO-sis) is the distention of the renal pelvis and calyces with accumulated fluid caused by obstruction of urine flow. The obstruction may occur at any level in the urinary tract. The most common causes of obstruction, in addition to pregnancy and an enlarged prostate, are a kidney stone that has formed in the pelvis and dropped into the ureter, a tumor that presses on a ureter, and scars due to inflammation. Removal of the obstruction within a few weeks, before the kidney is damaged, may result in complete recovery. If the obstruction is not removed, the kidney will be permanently damaged.

Acute renal failure may result from a medical or surgical emergency or from toxins that damage the tubules. This condition is characterized by a sudden, serious decrease in kidney function and may be fatal without immediate medical treatment.

Chronic renal failure results from a gradual loss of nephrons. As more and more nephrons are destroyed, the kidneys gradually lose the ability to per-form their normal functions. As the disease progresses, nitrogenous waste products accumulate to high levels in the blood, causing a condition known as *uremia* (u-RE-me-ah).

A few of the characteristic signs and symptoms of chronic renal failure are the following:

1. *Dehydration* (de-hi-DRA-shun). Excessive loss of body fluid may occur early in renal failure when the kidneys cannot concentrate the urine and large amounts of water are eliminated.
2. *Edema* (eh-DE-mah). Accumulation of fluid in the tissue spaces may occur late in chronic renal disease when the kidneys cannot eliminate water in adequate amounts.
3. *Hypertension* may occur as the result of fluid overload and the increased production of renin.
4. *Anemia* occurs when the kidneys cannot produce the hormone erythropoietin to activate red blood cell production in bone marrow.
5. *Uremia.* If levels of nitrogenous waste products in the blood are very high, urea can be changed into ammonia in the stomach and intestine and cause ulcerations and bleeding.

Tumors of the kidneys usually grow rather slowly, but occasionally rapidly invading types are found. Blood in the urine and dull pain in the kidney region are warnings that should be heeded at once.

Angiotensin

Angiotensin is a hormone that has a powerful effect on blood pressure. It is generated through the action of renin, an enzyme from the kidney that normally acts to keep blood pressure high enough for effective filtration. In the blood, renin promotes formation of a substance called *angiotensin I*. Under the effects of another enzyme, angiotensin I is converted to *angiotensin II*. This compound raises blood pressure by causing constriction of blood vessels. It also stimulates the release of the hormone aldosterone and antidiuretic hormone (ADH), both of which act to increase blood pressure. A class of drugs used in the treatment of hypertension are the *ACE inhibitors* (angiotensin-converting enzyme inhibitors). They control blood pressure by blocking the production of angiotensin II.

Immediate surgery may be lifesaving. A *polycystic* (pol-e-SIS-tik) *kidney* is one in which many fluid-containing sacs develop in the active tissue and gradually, by pressure, destroy the functioning parts. This disorder runs in families, and until now treatment has not proved very satisfactory, except for the use of dialysis machines or kidney transplants.

Kidney stones, or *calculi* (KAL-ku-li), are made of substances, such as uric acid and calcium salts, that precipitate out of the urine instead of remaining in solution. They usually form in the renal pelvis, although the bladder can be another site of formation. The causes of this precipitation include infection of the urinary tract and stagnation of the urine. These stones may vary in size from tiny grains resembling bits of gravel up to large masses that fill the kidney pelvis and extend into the calyces. The latter are described as *staghorn calculi*. There is no way of dissolving these stones, since substances that would be able to do so would also destroy the kidney tissue. Sometimes instruments can be used to crush small stones and thus allow them to be expelled with the urine, but more often surgical removal is required. Also in use is a device called a *lithotriptor* (LITH-o-trip-tor), literally a "stone-cracker," which employs external shock waves to shatter kidney stones. The procedure is called *lithotripsy* (LITH-o-trip-se).

DIALYSIS MACHINES
AND KIDNEY TRANSPLANTS

Dialysis (di-AL-ih-sis) means "the separation of dissolved molecules based on their ability to pass through a semipermeable membrane." Molecules that can pass through the membrane will move from an area of greater concentration to one of lesser concentration. In patients who have defective kidney function, the accumulation of urea and other nitrogenous waste products can be reduced by passage of the patient's blood through a dialysis machine. The principle of "molecules leaving the area of greater concentration" thus operates to remove the excess products from the blood. The fluid in the dialysis machine, the dialysate, can be adjusted to regulate the flow of substances out of the blood.

There are two methods of dialysis in use: hemodialysis (blood dialysis) and peritoneal dialysis (dialysis in the abdominal cavity). In hemodialysis the dialysis membrane is made of cellophane or other synthetic material. In peritoneal dialysis the surface area of the peritoneum acts as the membrane. Dialysis fluid is introduced into the peritoneal cavity and then periodically removed along with waste products.

A 1973 amendment to the Social Security Act provides federal financial assistance for persons who have chronic renal disease and require dialysis. Most hemodialysis is performed in freestanding clinics. Treatment time has been reduced; a typical schedule involves 2 to 3 hours, 3 times a week. Access to the bloodstream has been made safer and easier through surgical establishment of a permanent exchange site (shunt). Peritoneal dialysis has been improved and simplified so that patients are able to manage treatment at home.

Many hundreds of kidney transplants have been performed successfully in recent years. Kidneys have so much extra functioning tissue that the loss of one kidney normally poses no problem. Records show that the likelihood that a transplant will be successful is greatest when a living donor who is closely related to the patient is used. However, organs from deceased donors have also proved satisfactory in many cases. The problem of tissue rejection (the rejection syndrome) is discussed in Chapter 17.

Disorders of the Ureters

Abnormalities in structure of the ureter include double portions at the kidney pelves and constricted or abnormally narrow parts, called *strictures* (STRICK-tures). Narrowing of the ureter may be caused by abnormal pressure from tumors or other masses outside the tube. Obstruction of the ureters may be due to stones from the kidneys, or to kinking of the tube due to a dropping of the kidney, a condition known as *renal ptosis* (TO-sis).

The passage of a small stone along the ureter causes excruciating pain. Relief of this pain, called *renal colic*, usually requires morphine or an equally powerful drug. The first "barber surgeons," operating without benefit of anesthesia, were permitted by their patients to cut through the skin and the muscles of the back to remove stones from the ureters. "Cutting for stone" in this way was relatively successful, despite the lack of sterile technique, because the approach through the back avoided the peritoneal cavity, and the serious risk of peritonitis. Modern surgery employs a variety of instruments for removal of stones from the ureter, including endoscopes similar to those described in Chapter 19. The transurethral route through the urethra and urinary bladder and then into the ureters, as well as entrance through the skin and muscles of the back, may be used to remove calculi from the kidney pelvis or from the ureters.

Disorders Involving the Bladder

A full (distended) bladder lies in an unprotected position in the lower abdomen, and a blow may rupture it, necessitating immediate surgical repair. Blood in the urine is a rather common symptom of infection or tumors, which may involve the bladder. Inflammation of the bladder, called *cystitis* (sis-TI-tis), is 10 times as frequent in women as in men. This may be due at least in part to the very short urethra in the female compared with that of the male. Usually, bacteria (*e.g.,* colon bacilli) ascend from the outside through the urethra into the bladder (see Fig. 23-12). Pain, urgency, and frequency of urination are common symptoms.

Another type of cystitis, called *interstitial cystitis,* may cause pelvic pain with discomfort before and after urination. The tissues below the mucosa are involved. The disease can be diagnosed only with the use of a cystoscope (a kind of endoscope). Because no bacteria are involved, antibiotics are not effective treatment and may even be harmful.

Obstruction by an enlarged prostate gland or from a pregnancy may lead to stagnation and cystitis. Reduction of a person's general resistance to infection, as in diabetes, may also lead to cystitis. The danger of cystitis is that the infection may ascend to other parts of the urinary tract.

Tumors of the bladder, which are most prevalent in men older than 50, include benign papillomas and various kinds of cancer. About 90% of bladder tumors arise from the epithelial lining. Possible causes include toxins (particularly certain aniline dyes), chronic infestations (schistosomiasis), heavy cigarette smoking, and the presence of urinary stones, which may develop and increase in size within the bladder. Cystoscopic examinations and biopsies should be done as soon as blood in the urine (hematuria) is detected. Removal before the tumor invades the muscle wall gives the best prognosis.

Disorders of the Urethra

Congenital anomalies (defects present at birth) involve the urethra and other parts of the urinary tract. The opening of the urethra to the outside may be too small, or the urethra itself may be narrowed. Occasionally it happens that an abnormal valve-like structure is located at the point at which the urethra enters the bladder. If it is not removed surgically,

such a valve-like fold of tissue can cause a back pressure of the urine with serious consequences. There is also a condition in the male in which the urethra opens on the undersurface of the penis instead of at the end. This is called *hypospadias* (hi-po-SPA-de-as).

Urethritis, which is characterized by inflammation of the mucous membrane and the glands of the urethra, is much more common in the male than in the female. It is often due to gonococci or chlamydia, although many other bacteria may be responsible for the infection.

"Straddle" injuries to the urethra are common in men. This type of injury occurs when, for example, a man walking along a raised beam slips and lands with the beam between his legs. Such an accident may catch the urethra between the hard surfaces of the beam and the pubic arch and rupture the urethra. In accidents in which the bones of the pelvis are fractured, rupture of the urethra is fairly common.

The Effects of Aging

Even without kidney disease, aging causes the kidneys to lose some of their ability to concentrate urine. With aging, progressively more water is needed to excrete the same amount of waste. Older persons find it necessary to drink more water than young persons, and they eliminate larger amounts of urine (polyuria) even at night (nocturia). Beginning at about age 40 there is a decrease in the number and size of the nephrons. Often more than 50% of them are lost before age 80. There may be an increase in blood urea nitrogen (BUN) without serious symptoms. Elderly persons are more susceptible than young persons to infections of the urinary system. Childbearing may cause damage to the musculature of the pelvic floor, resulting in urinary tract problems in later years. Enlargement of the prostate, common in older men, may cause obstruction and a back pressure in the ureters and kidneys. If this condition is untreated, it will cause permanent damage to the kidneys. Age changes may predispose to but do not cause incontinence. Most elderly persons (60% in nursing homes and up to 90% living independently) have no incontinence.

▶ The Urine

Normal Constituents

Urine is a yellowish liquid that is about 95% water and 5% dissolved solids and gases. The amount of these dissolved substances is indicated by its *specific gravity.* The specific gravity of pure water, used as a standard, is 1.000. Because of the dissolved materials it contains, urine has a specific gravity that normally varies from 1.002 (very dilute urine) to 1.040 (very concentrated urine). When the kidneys are diseased, they lose the ability to concentrate urine, and the specific gravity no longer varies as it does when the kidneys function normally.

Some of the dissolved substances normally found in the urine are the following:

1. *Nitrogenous waste products,* including urea, uric acid, and *creatinine* (kre-AT-ih-nin)
2. *Electrolytes,* including sodium chloride (as in common table salt) and different kinds of sulfates and phosphates. Electrolytes are excreted in appropriate amounts to keep their blood concentration constant.
3. *Yellow pigment,* which is derived from certain bile compounds. Pigments from foods and drugs also may appear in the urine.

Abnormal Constituents

Examination of urine, called a *urinalysis* (u-rin-AL-ih-sis) *(UA),* is one of the most important parts of an evaluation of a person's physical state. A routine urinalysis includes observation of color and turbidity (cloudiness) as well as measurement of pH and specific gravity. A variety of abnormal components are also tested for. Among the most significant abnormal substances found in the urine are the following:

1. *Glucose* is usually an important indicator of a disease known as *diabetes mellitus,* in which blood sugar is not adequately oxidized (metabolized) in the body cells. The excess glucose, which cannot be reabsorbed, is excreted in the urine. The presence of glucose in the urine is known as *glycosuria* (gli-ko-SU-re-ah).
2. *Albumin.* The presence of this protein, which is normally retained in the blood, may indicate a kidney disorder such as glomerulonephritis. Albumin in the urine is known as *albuminuria* (al-bu-mih-NU-re-ah).
3. *Blood* in the urine is usually an important indicator of urinary system disease, including nephritis. Blood in the urine is known as *hematuria* (hem-ah-TU-re-ah).
4. *Ketones* (KE-tones) are produced when fats are incompletely oxidized; ketones in the urine are seen in diabetes mellitus and starvation.
5. *White blood cells* (pus) are evidence of infection; they can be seen by microscopic examination of a centrifuged specimen. Pus in the urine is known as *pyuria* (pi-U-re-ah).
6. *Casts* are molds formed in the microscopic kidney tubules; they usually indicate disease of the nephrons.

More extensive tests on urine may include analysis for drugs, enzymes, hormones, and other metabolites as well as cultures for microorganisms. Normal values for common urine tests are given in Appendix 3, Table 1.

SUMMARY

I. Urinary system
  A. Main function is excretion—removal and elimination of waste materials from blood
   1. Other systems that eliminate waste
     a. Digestive system—eliminates undigested food, water, salts, bile
     b. Respiratory system—eliminates carbon dioxide, water
     c. Skin—eliminates water, salts, nitrogen waste
II. Kidneys—located in upper abdomen, against the back
  A. Structure
   1. Cortex—outer portion

  2. Medulla—inner portion
  3. Pelvis
    a. Upper end of ureter
    b. Calyces—cuplike extensions that receive urine
  4. Nephron
    a. Functional unit of kidney
    b. Parts—Bowman's capsule, proximal convoluted tubule, Henle's loop, distal convoluted tubule
  5. Blood supply
    a. Renal artery
     (1) Afferent arteriole—enters Bowman's capsule

(2) Efferent arteriole—leaves Bowman's capsule

(3) Glomerulus—coil of capillaries in Bowman's capsule

(4) Peritubular capillaries—surround nephron

b. Renal vein

B. Functions

1. Excretion of waste, excess salts, toxins
2. Water balance
3. Regulation of pH
4. Production of renin—enzyme that regulates blood pressure via angiotensin
5. Production of erythropoietin—hormone that stimulates red blood cell production

C. Physiology

1. Filtration—pressure in glomerulus forces water and soluble substances out of blood and into Bowman's capsule

   a. Glomerular filtrate—material that leaves blood and enters the nephron

2. Reabsorption—most of filtrate leaves nephron by diffusion and active transport and returns to blood through peritubular capillaries

3. Tubular secretion—materials moved from blood into nephron for excretion

4. Counter-current mechanism—method for concentrating urine based on movement of ions out of nephron

   a. ADH

   (1) Hormone from posterior pituitary

   (2) Promotes reabsorption of water

III. Ureters—carry urine from the kidneys to the bladder

IV. Urinary bladder—stores urine until it is eliminated

A. Micturition—urination

1. Reflex control—impulses from central nervous system in response to stretching of bladder wall

2. Voluntary control—external sphincter around urethra

V. Urethra—carries urine out of body

A. Female urethra—4.0 cm long; opens in front of vagina

B. Male urethra—20 cm long; carries both urine and semen

VI. Disorders of urinary system

A. Kidney disorders

1. Acute glomerulonephritis—damages glomeruli
2. Pyelonephritis—inflammation of kidney and renal pelvis
3. Hydronephrosis—distention with obstructed fluids
4. Renal failure

   a. Acute or chronic

   b. Signs of chronic renal failure—dehydration, edema, hypertension, anemia, uremia

   c. Dialysis—removes unwanted substances from blood when kidneys fail

5. Calculi—kidney stones

B. Cystitis—inflammation of bladder; most common in females

C. Urethritis—inflammation of urethra

VII. Urine

A. Normal constituents—water, nitrogenous waste, electrolytes, pigments

B. Abnormal constituents—glucose, albumin, blood, ketones, white blood cells, casts

QUESTIONS FOR STUDY AND REVIEW

1. Name the body systems that have excretory functions.
2. Name the organs of the urinary system.
3. Where are the kidneys located?
4. Describe the external appearance of the kidneys and tell what tissues form most of the kidney structure.
5. Describe the structure of a nephron.
6. Name the blood vessels associated with the nephron.
7. Compare the afferent arteriole and the efferent arteriole.
8. What happens to the glomerular filtrate as it passes through the nephron?

9. What results from inadequate secretion of ADH? What disease is associated with this condition?
10. What is the function of the juxtaglomerular apparatus?
11. What structures empty into the kidney pelvis and what drains the pelvis?
12. What is micturition? How is it controlled?
13. Describe the female urethra and tell how it differs from the male urethra in structure and function.
14. What are some of the infections that involve the kidney and what parts are most often affected?
15. What are calculi and what are some of their causes?
16. What is meant by the word *dialysis* and how is this principle used for persons with kidney failure? What kinds of membranes are used for peritoneal dialysis? for hemodialysis?
17. What types of donors are best for kidney transplants?
18. What is inflammation of the bladder called? Why is it more common in women than in men?
19. What are some of the effects of aging on the urinary system?
20. What is urea? Where and how is it formed?
21. What tests are done in a routine urinalysis?
22. Tell the possible significance of six abnormal constituents of the urine.

UNIT **VII**

Perpetuating Life

The last unit includes three chapters on the structures and functions related to reproduction and inheritance. The reproductive system is not necessary for the continuation of the life of the individual, but rather is needed for the continuation of the human species. The germ cells and their genes have been studied intensively during recent years and are part of the rapidly developing science of heredity.

CHAPTER 23

The Male and Female Reproductive Systems

Behavioral Objectives

After careful study of this chapter, you should be able to:

- Name the male and female gonads and describe the function of each
- State the purpose of meiosis
- List the accessory organs of the male and female reproductive tracts and cite the function of each
- Describe the composition and function of semen
- Draw and label a spermatozoon
- List in the correct order the hormones produced during the menstrual cycle and cite the source of each
- Describe the functions of the main male and female sex hormones
- Explain how negative feedback regulates reproductive function in both males and females
- Briefly describe the major disorders of the male and female reproductive tracts
- Describe the changes that occur during and after menopause
- Define *contraception* and cite the main methods of contraception in use

Memmler, RL, Cohen, BJ, Wood, DL. *THE HUMAN BODY IN HEALTH AND DISEASE,* 8/e,
© 1996 Lippincott-Raven Publishers

The chapters in this unit deal with what is certainly one of the most interesting and mysterious attributes of life: the ability to reproduce. The simplest forms of life, one-celled organisms, usually need no partner to reproduce; they simply divide by themselves. This form of reproduction is known as *asexual* (nonsexual) reproduction.

In most animals, however, reproduction is *sexual,* meaning that there are two kinds of individuals, males and females, each of which has specialized cells designed specifically for the perpetuation of the species. These specialized sex cells are known as *germ cells,* or *gametes* (GAM-etes). In the male they are called *spermatozoa* (sper-mah-to-ZO-ah), and in the female they are called *ova* (O-vah). Germ cells are characterized by having half as many chromosomes as are found in any other cell in the body. During their formation they go through a special process of cell division, called *meiosis* (mi-O-sis), that halves the number of chromosomes. In humans, meiosis reduces the chromosome number from 46 to 23.

Although the reproductive apparatuses of males and females are different, the organs of both sexes may be divided into two groups: primary and accessory.

1. The primary organs are the *gonads* (GO-nads), or sex glands; they produce the germ cells and manufacture hormones. The male gonads are the *testes* (TES-teze), and the female gonads are the *ovaries* (O-vah-reze).
2. The *accessory organs* include a series of ducts that provide for the transport of germ cells, as well as various exocrine glands.

▶ The Male Reproductive System

The Testes

The male gonads (testes) are normally located outside the body proper, suspended between the thighs in a sac called the *scrotum* (SKRO-tum) (Fig. 23-1). The testes are egg-shaped organs measuring about 3.7 to 5 cm (1.5–2 inches) in length and approximately 2.5 cm (1 inch) in each of the other two dimensions. During embryonic life the testes develop from tissue near the kidney. A month or two before birth, the testes normally descend (move downward) through the *inguinal* (ING-gwih-nal)

canal in the abdominal wall into the scrotum. Each testis then remains suspended by a *spermatic cord* (Fig. 23-2) that extends through the inguinal canal. This cord contains blood vessels, lymphatic vessels, nerves, and the tube (ductus deferens) that transports spermatozoa away from the testis. Each gland must descend completely if it is to function normally; to produce spermatozoa, the testes must be kept at the temperature of the scrotum, which is lower than that of the abdominal cavity.

The bulk of the specialized tissue of the testes consists of tiny coiled *seminiferous* (seh-mih-NIF-er-us) *tubules.* Cells in the walls of these tubules produce spermatozoa. Between the tubules are the specialized *interstitial* (in-ter-STISH-al) *cells* that secrete the male sex hormone *testosterone* (tes-TOS-teh-rone). Figure 23-3 is a microscopic view of the testis in cross section showing the seminiferous tubules, interstitial cells, and developing spermatozoa.

After being secreted by the testes, testosterone is absorbed directly into the bloodstream. This hormone has two functions. The first is maintenance of the reproductive structures, including development of the spermatozoa. A second involves the development of *secondary sex characteristics,* traits that characterize males and females but are not directly concerned with reproduction. In males they include a deeper voice, broader shoulders, narrower hips, a greater percentage of muscle tissue, and more body hair than are found in females.

The Duct System

The tubes that carry the spermatozoa (also referred to simply as *sperm cells* or *sperm*) begin with the tubules inside the testis itself. From these tubes the cells are collected by a greatly coiled tube 6 meters (20 feet) long, called the *epididymis* (ep-ih-DID-ih-mis), which is located inside the scrotal sac. While they are temporarily stored in the epididymis, the sperm cells mature and become motile, that is, able to move, or "swim," by themselves. The epididymis finally extends upward as the *ductus deferens* (DEF-er-enz), also called the *vas deferens.* The ductus deferens, contained in the spermatic cord, continues through the inguinal canal into the abdominal cavity. Here, it separates from the remainder of the spermatic cord and curves behind the urinary bladder. The ductus deferens then joins with the duct of the *seminal vesicle* (VES-ih-kl) on the same side to form the *ejaculatory* (e-JAK-u-lah-to-re) *duct.* The two

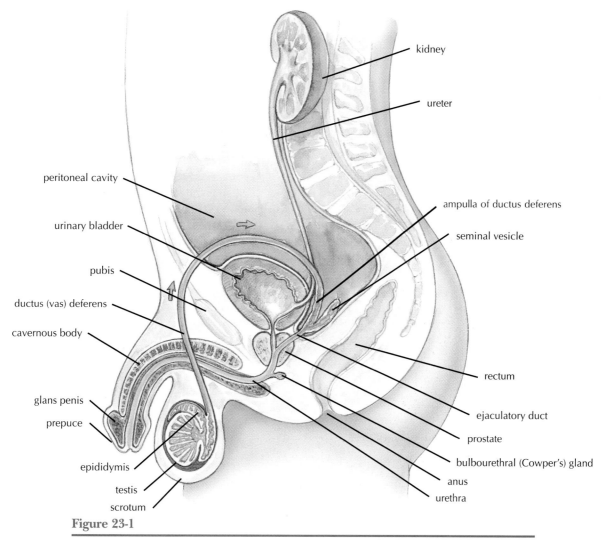

kidney

ureter

peritoneal cavity

ampulla of ductus deferens

urinary bladder

seminal vesicle

pubis

ductus (vas) deferens

cavernous body

rectum

glans penis

ejaculatory duct

prepuce

prostate

bulbourethral (Cowper's) gland

epididymis

anus

testis

urethra

scrotum

Figure 23-1

Male genitourinary system. The arrows indicate the course of sperm cells through the duct system.

ejaculatory ducts enter the body of the prostate gland where they empty into the urethra.

Formation of Semen

Semen (SE-men) is the mixture of spermatozoa and various secretions that is expelled from the body. The secretions serve to nourish and transport the spermatozoa, neutralize the acidity of the vaginal tract, and lubricate the reproductive tract during sexual intercourse. The glands discussed below contribute secretions to the semen.

THE SEMINAL VESICLES

The seminal vesicles are tortuous muscular tubes with small outpouchings. They are about 7.5 cm (3 inches) long and are attached to the connective tissue at the back of the urinary bladder. The glandular lining produces a thick, yellow, alkaline secretion containing large quantities of simple sugar and other substances that provide nourishment for the sperm. The seminal fluid forms a large part of the volume of the semen.

THE PROSTATE GLAND

The prostate gland lies immediately below the urinary bladder, where it surrounds the first part of the urethra. Ducts from the prostate carry its secretions into the urethra. The thin, alkaline prostatic secretion helps neutralize the acidity of the vaginal tract and enhance the motility of the spermatozoa.

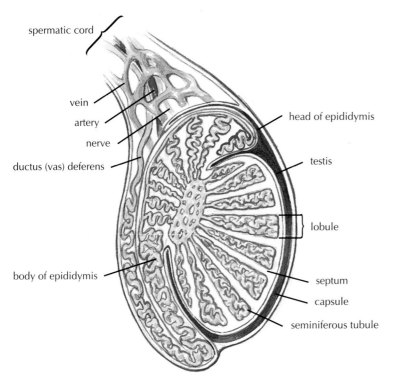

spermatic cord

vein

artery

nerve

ductus (vas) deferens

body of epididymis

head of epididymis

testis

lobule

septum

capsule

seminiferous tubule

Figure 23-2

Structure of the testis, also showing the epididymis and spermatic cord.

The prostate gland is also supplied with muscular tissue, which upon signal from the nervous system contracts to aid in the expulsion of the semen from the body.

MUCUS-PRODUCING GLANDS

The largest of the mucus-producing glands in the male reproductive system are *Cowper's,* or *bulbourethral* (bul-bo-u-RE-thral), *glands,* a pair of pea-sized organs located in the pelvic floor just below the prostate gland. The ducts of these glands extend about 2.5 cm (1 inch) from each side and empty into the urethra before it extends within the penis. Other very small glands secrete mucus into the urethra as it passes through the penis. The mucus from all of these glands serves mainly as a lubricant.

The Urethra and the Penis

The male urethra, as we saw earlier, serves the dual purpose of conveying urine from the bladder and carrying the reproductive cells and their accompanying secretions to the outside. The ejection of semen into the receiving canal (vagina) of the female is made possible by the *erection,* or stiffening and enlargement, of the penis, through which the longest part of the urethra extends. The penis is

primitive spermatozoa

mature spermatozoa

seminiferous tubule

interstitial cells

Figure 23-3

Microscopic view of the testis. (Courtesy of Dana Morse Bittus and B. J. Cohen) made of a sponge-like tissue containing many blood spaces that are relatively empty when the organ is flaccid but fill with blood and distend when the penis is erect. The spongy tissue is subdivided into three segments, each called a *corpus* (meaning "body") (Fig. 23-4). A single, ventrally located *corpus spongiosum* contains the urethra. On either side is a larger *corpus cavernosum* (pl., *corpora cavernosa*). The penis and scrotum are referred to as the *external genitalia* of the male.

Reflex centers in the spinal cord order the contraction of the smooth muscle tissue in the prostate gland, followed by the contraction of skeletal muscle in the pelvic floor. This provides the force needed for *ejaculation* (e-jak-u-LA-shun), the expulsion of semen through the urethra to the outside.

The Spermatozoa

Spermatozoa are tiny individual cells (Fig. 23-5). They are so small that at least 200 million are contained in the average ejaculation. Spermatozoa are continuously manufactured in the testes. They develop within the seminiferous tubules with the aid of special cells called *Sertoli,* or "nurse," *cells.* These cells help nourish and protect the developing sperm cells.

The individual sperm cell has an oval head that is largely a nucleus containing chromosomes. Covering the head like a cap is the *acrosome* (AK-ro-some), which contains enzymes that help the spermatozoon penetrate the ovum. The whiplike tail (flagellum) enables the sperm cell to make its way through the various passages until it reaches the ovum of the female. A middle region contains many mitochondria to provide energy for movement. Out of the millions of spermatozoa in an ejaculation, only one, if any, fertilizes the ovum. The remainder of the cells live only a few hours, up to a maximum of 3 days.

▶ Hormonal Control of Male Reproduction

The activities of the testes are under the control of two hormones produced by the anterior pituitary gland (hypophysis). One of these, *follicle stimulating hormone (FSH)*, stimulates the Sertoli cells and promotes the formation of spermatozoa. Testosterone is also needed in this process. The other hormone is *luteinizing hormone (LH)*, called *interstitial cell–stimulating hormone (ICSH)* in males. It stimulates the interstitial cells to produce testosterone. These pituitary hormones are named for their activity in female reproduction, although they are chemically the same in both males and females.

The pituitary gland is regulated by a region of the brain just above it called the *hypothalamus.* Starting at puberty, the hypothalamus begins to secrete hormones that trigger the release of FSH and LH. These hormones are secreted continuously in the male. The activity of the hypothalamus is in turn regulated by a negative feedback mechanism involving testosterone. As the level of testosterone in the blood increases, the hypothalamus secretes less releasing hormone; as the level of testosterone decreases, the hypothalamus secretes more releasing hormone (see Fig. 12-2).

The Effects of Aging on Male Reproduction

A gradual decrease in the production of testosterone and spermatozoa begins as early as age 20

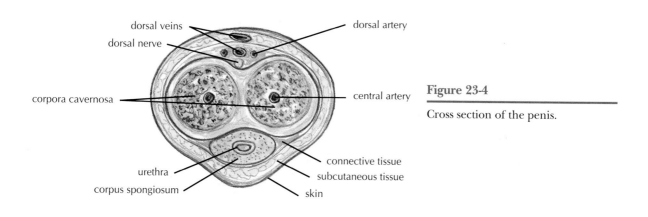

Figure 23-4

Cross section of the penis.

dorsal veins

dorsal nerve

dorsal artery

corpora cavernosa

central artery

urethra

corpus spongiosum

connective tissue

subcutaneous tissue

skin

acrosome

head

nucleus

neck

mitochondria

middle piece

tail
(flagellum)

Figure 23-5

Diagram of a human spermatozoon showing major structural features. (Chaffee EE, Lytle IM: Basic Physiology and Anatomy, 4th ed, p. 549. Philadelphia, JB Lippincott, 1980)

and continues throughout life. Secretions from the prostate and seminal vesicles decrease in amount and become less viscous. In a few men (less than 10%) sperm cells remain late in life, even to age 80.

▶ Disorders of the Male Reproductive System

Infertility

Infertility means significantly lower than normal ability to reproduce. If the inability is complete, the condition is termed *sterility*. The proportion of infertile marriages due to defects involving the male has been estimated variously from 40% to 50%. The

tubules of the testes are very sensitive to x-rays, infections, toxins, and malnutrition, all of which bring about degenerative changes. Such damage may cause a decrease in the numbers of spermatozoa produced, leading to a condition called *oligospermia* (ol-ih-go-SPER-me-ah). Adequate numbers of sperm are required to disperse the coating around the ovum so that one sperm can fertilize it. Absence of or an inadequate number of male sex cells is a significant cause of infertility.

Intentional sterilization of the male may be accomplished by an operation called a *vasectomy* (vah-SEK-to-me). In this procedure a portion of the ductus deferens on each side is removed and the cut end closed to keep the spermatozoa from reaching the urethra. These tiny cells are absorbed without injury. The vasectomized male retains the ability to produce hormones and all other seminal secretions, as well as the ability to perform the sex act, but no fertilization can occur.

Cryptorchidism

Cryptorchidism (kript-OR-kid-izm), which means "hidden testes," is a disorder characterized by failure of the testes to descend into the scrotum. Unless corrected in childhood, this condition results in sterility. Undescended testes also are particularly subject to tumor formation. Most testes that are undescended at birth descend spontaneously by age 1.

In Vitro Fertilization

The term *in vitro* literally means "in glass." It is used to describe tests or procedures done in the laboratory. In vitro fertilization refers to fertilization of an egg cell outside the mother's body in a laboratory dish. The first such procedure was done successfully in 1978.

The mother must be given hormones to cause ovulation of several eggs. These are then withdrawn with a needle and fertilized with the father's sperm. After a few divisions the fertilized egg is placed in the uterus for development. Additional fertilized eggs can be frozen to repeat the procedure in case of failure or for later pregnancies. Other possibilities for in vitro fertilization include using donor eggs, donor sperm cells, or both.

Surgical correction is the usual remedy in the remaining cases.

Inguinal Hernia

Hernia (HER-ne-ah), or *rupture,* refers to the abnormal protrusion of an organ, or part of an organ, through the wall of the cavity in which it is normally contained. Hernias most often occur where there is a weak area in the abdominal wall, for example, at the inguinal canal. In this region of the lower abdomen, the testis pushes its way through the muscles and connective tissues of the abdominal wall, carrying with it the blood vessels and other structures that will form the spermatic cord. In the normal adult this area is fairly well reinforced with connective tissue, and there is no direct connection between the abdominal cavity and the scrotal sac. However, like some other regions in which openings permit the passage of various structures to and from the abdominal cavity, this area constitutes a weak place at which a hernia may occur.

Infections

Infections of various kinds may involve the male reproductive organs, but by far the most common are chlamydial and gonococcal infections. These sexually transmitted (venereal) diseases are manifested by a discharge from the urethra, which may be accompanied by burning and pain, especially during urination. The infection may travel along the mucous membrane into the prostate gland and into the epididymis; if both sides are affected and enough scar tissue is formed to destroy the tubules, sterility may result.

Another common sexually transmitted disease is an unpleasant, persistent infection called **genital herpes.** Caused by a virus, this disorder is characterized by fluid-filled vesicles (blisters) on and around the genital organs.

The sexually transmitted disease *syphilis* is caused by a spirochete (*Treponema pallidum).* Because syphilis spreads quickly in the bloodstream, it is regarded as a systemic disorder (see Appendix 4, Table 1). The incidence of syphilis has been rising in recent years to reach its highest level since the development of antibiotics, especially among poor minority populations. This increase is alarming because the genital ulcers caused by syphilis increase the chances of infection with the AIDS virus.

Other infectious agents that sometimes invade the reproductive organs include the tubercle bacillus and various staphylococci. The testes may be involved in mumps, with a resulting *orchitis* (or-KI-is), or inflammation of the testes.

Tumors

Tumors may develop in the male reproductive organs, most commonly the prostate, and are quite common in elderly men. Such growths may be benign or malignant. Both cause such pressure on the urethra that urination becomes difficult. Back pressure often destroys kidney tissue and permits stagnation of urine in the bladder with a resulting tendency to infection. The man with benign prostate enlargement may respond to medication to shrink the prostate. However, if kidney function is threatened, surgery is performed to reduce the obstruction.

Cancer of the prostate is a common disorder in men over 50. It is frequently detected as a nodule during rectal examination. Early detection is now possible with an annual blood test for prostate-specific antigen (PSA). This protein is increased in prostate cancer, although it may increase in other prostate disorders as well. Surgical treatment very early in the disease may be associated with the best prognosis. Unfortunately, this cancer frequently spreads to the bones with a fatal outcome.

Testicular cancer affects young to middle-aged adults, frequently causing death due to widespread metastasis by way of the lymphatic system. Early detection with regular self-examination allows for effective treatment and possible preservation of fertility.

Phimosis

Phimosis (fi-MO-sis) is a tightness of the foreskin (prepuce) so that it cannot be drawn back. This may be remedied through circumcision, by which part or all of the foreskin is removed. This operation is often done on very young male infants as a routine measure, either for hygienic reasons or because of religious principles.

▶ The Female Reproductive System

The Ovaries

In the female the counterparts of the testes are the two *ovaries,* where the female sex cells, or *ova,* are formed (Fig. 23-6). The ovaries are small, somewhat flattened oval bodies measuring about 4 cm (1.6 inches) in length, 2 cm (0.8 inch) in width, and 1 cm (0.4 inch) in depth. Like the testes, the ovaries descend, but only as far as the pelvic portion of the abdomen. Here, they are held in place by ligaments, including the broad ligament, the ovarian ligament, and others, that attach them to the uterus and the body wall.

The outer layer of each ovary is made of a single layer of epithelium. Beneath this layer the ova are produced. The ova begin a complicated process of maturation, or "ripening," which takes place in small fluid-filled clusters of cells called *ovarian follicles* (o-VA-re-an FOL-ih-kls) or *Graafian* (GRAF-e-an) *follicles* (Fig. 23-7). The cells of the ovarian follicle walls secrete estrogen. When an ovum has ripened, the ovarian follicle ruptures, and the ovum is discharged from the surface of the ovary and makes its way to the nearest *oviduct* (O-vih-dukt). The oviducts are two tubes, one of which is in the vicinity of each ovary. The rupture of a follicle allowing the escape of the egg cell is called *ovulation* (ov-u-LA-shun).

The ovaries of a newborn female contain a large number of potential ova. Each month during the reproductive years, several ripen, but usually only one is released. Once this ovum has been expelled, the follicle is transformed into a solid glandular mass called the *corpus luteum* (LU-te-um), which means "yellow body." This structure secretes both estrogen and progesterone. Commonly, the corpus luteum shrinks and is replaced by scar tissue. When a pregnancy occurs, however, this structure remains active. Often as a result of normal ovulation, the corpus luteum will persist and form a small ovarian cyst (fluid-filled sac). This condition usually resolves without treatment.

The Oviducts

The tubes that transport the ova in the female reproductive system, the oviducts, are also known as *uterine* (U-ter-in) *tubes,* or *fallopian* (fah-LO-pe-an) *tubes.* They are small, muscular structures, nearly 12.5 cm (5 inches) long, extending from a point near the ovaries to the uterus (womb). There is no direct connection between the ovaries and these tubes. The ova are swept into the oviducts by a current in the peritoneal fluid produced by the small, fringe-like extensions called *fimbriae* (FIM-bre-e) that are located at the edges of the abdominal openings of the tubes.

Unlike the spermatozoon, the ovum cannot move by itself. Its progress through the oviduct toward the uterus depends on the sweeping action of cilia in the lining of the tubes and on peristalsis of the tubes. It takes about 5 days for an ovum to reach the uterus from the ovary.

The Uterus

The organ to which the oviducts lead is the *uterus* (U-ter-us), and it is within this structure that the fetus grows to maturity.

The uterus is a pear-shaped, muscular organ about 7.5 cm (3 inches) long, 5 cm (2 inches) wide, and 2.5 cm (1 inch) deep. The upper portion rests on the upper surface of the urinary bladder; the lower portion is embedded in the pelvic floor between the bladder and the rectum. The wider upper portion is called the body, or *corpus;* the lower, narrower part is the *cervix* (SER-viks), or neck. The small, rounded part above the level of the tubal entrances is known as the *fundus* (FUN-dus) (see Fig. 23-6). The cavity inside the uterus is shaped somewhat like a capital T, but it is capable of changing shape and dilating as a fetus develops. The cervix leads to the *vagina* (vah-JI-nah), the lower part of the birth canal, which opens to the outside of the body.

The lining of the uterus is a specialized epithelium known as *endometrium* (en-do-ME-tre-um), and it is this layer that is involved in menstruation.

The *broad ligaments* support the uterus, extending from each side of the organ to the lateral body wall. Along with the uterus, these two portions of peritoneum form a partition dividing the female pelvis into anterior and posterior areas. The ovaries are suspended from the broad ligaments, and the oviducts lie within the upper borders. Blood vessels that supply these organs are found between the layers of the broad ligament.

The Vagina

The vagina is a muscular tube about 7.5 cm (3 inches) long connecting the uterine cavity with the outside. It receives the cervix, which dips into the upper vagina in such a way that a circular recess is formed, giving rise to areas known as *fornices* (FOR-nih-seze). The deepest of these spaces, behind the cervix, is the *posterior fornix* (FOR-niks) (Fig. 23-8). This recess in the posterior vagina lies adjacent to the lowest part of the peritoneal cavity, a narrow passage between the uterus and the rectum named the

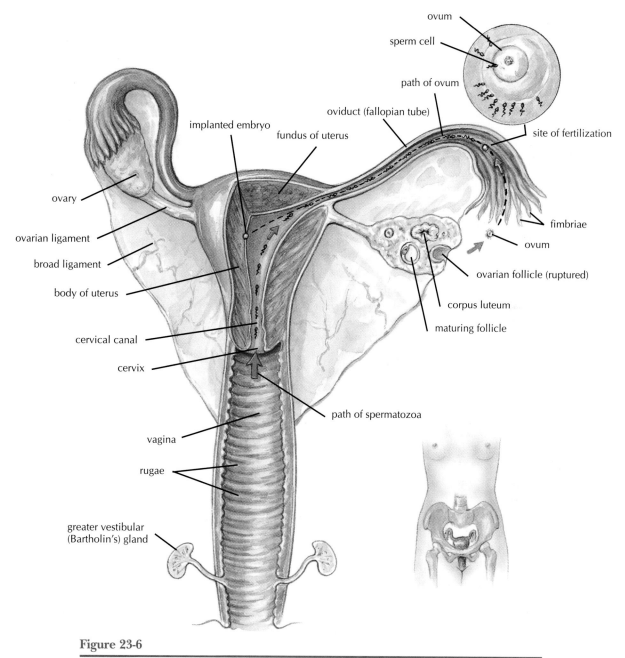

Figure 23-6

Female reproductive system.

cul-de-sac (from a French term meaning "bottom of the sack"). A rather thin layer of tissue separates the posterior fornix from this region, so that abscesses or tumor cells in the peritoneal cavity can sometimes be detected by vaginal examination.

The lining of the vagina is a wrinkled mucous membrane something like that found in the stom-ach. The folds (rugae) permit enlargement so that childbirth usually does not tear the lining. In addi-tion to being a part of the birth canal, the vagina is the organ that receives the penis during sexual in-tercourse. A fold of membrane called the *hymen* may sometimes be found at or near the vaginal (VAJ-ih-nal) canal opening.

ovum

ovarian
(Graafian)
follicle

Figure 23-7

Microscopic view of the ovary showing egg cells developing within ovarian follicles. (Courtesy of Dana Morse Bittus and B. J. Cohen)

THE GREATER VESTIBULAR GLANDS

Just above and to each side of the vaginal opening are the mucus-producing *greater vestibular* (ves-TIB-u-lar), or *Bartholin's, glands.* These glands open into an area near the vaginal opening known as the *vestibule.* They may become infected, then painfully swollen, and finally, abscessed. A surgical incision to promote drainage may be required.

The Vulva and the Perineum

The external parts of the female reproductive system form the *vulva* (VUL-vah), which includes two pairs of lips, or *labia* (LA-be-ah); the *clitoris* (KLIT-o-ris), which is a small organ of great sensitivity; and related structures (Fig. 23-9). Although the entire pelvic floor in both the male and female is properly called the *perineum* (per-ih-NE-um) (see Fig. 8-10), those who care for pregnant women usually refer to the limited area between the vaginal opening and the anus as the perineum.

▶ The Menstrual Cycle

In the female, as in the male, reproductive function is controlled by hormones from the pituitary gland as regulated by the hypothalamus. Female activity differs, however, in that it is cyclic, that is, shows regular patterns of increases and decreases in hormone levels. These changes are regulated by hormonal feedback.

The length of the menstrual cycle varies between 22 and 45 days in normal women, but 28 days is taken as an average, with the first day of menstrual flow being considered the first day of the cycle (Fig. 23-10).

At the start of each cycle, under the influence of FSH produced by the pituitary, a follicle begins to develop in the ovary. This follicle produces increasing amounts of *estrogen* as the ovum matures. (*Estrogen* is the term used for a group of related hormones, the most active of which is estradiol.) The estrogen is carried in the bloodstream to the uterus, where it starts preparing the endometrium for a possible pregnancy. This preparation includes thickening of the endometrium and elongation of the glands that produce the uterine secretion. Estrogen in the blood also acts as a feedback messenger to inhibit the release of FSH and stimulate the release of LH from the pituitary (see Fig. 12-2).

About 1 day before ovulation, there is an *LH surge,* a sharp rise in LH. This hormone causes ovulation and transforms the ruptured follicle into the corpus luteum. The corpus luteum produces some estrogen and large amounts of *progesterone.* Under the influence of these hormones, the endometrium continues to thicken, the glands and blood vessels increasing in size. The rising levels of estrogen and progesterone feed back to inhibit the release of FSH and LH from the pituitary (Fig. 23-11). During this time the ovum makes its journey to the uterus by way of the oviduct. If the ovum is not fertilized while passing through the uterine tube, it dies within 2 to 3 days and then disintegrates.

If fertilization does not occur, the corpus luteum degenerates and the levels of estrogen and progesterone decrease. Without the hormones to support growth, the endometrium degenerates. Small hemorrhages appear in this lining, producing the bleeding known as *menstrual flow.* Bits of endometrium break away and accompany the flow of blood. The average duration of this discharge is 2 to 6 days.

oviduct
(fallopian tube)

ovary

peritoneal
cavity

uterus

round
ligament

pubis

clitoris

labium minus

labium majus

sacrum

uterosacral
ligament

cul-de-sac

posterior
fornix

cervix

rectum

urinary
bladder

urethra

vagina

anus

Figure 23-8

Female reproductive system, as seen in sagittal section.

Before the flow ceases, the endometrium begins to repair itself through the growth of new cells. The low levels of estrogen and progesterone allow the release of FSH from the anterior pituitary. This causes a new ovum to begin to ripen within the ovaries, and the cycle begins anew.

The activity of ovarian hormones as negative feedback messengers is the basis of hormonal methods of contraception (birth control). The hormones act to inhibit the release of FSH and LH from the pituitary, resulting in a menstrual period but no ovulation (Table 23-1 on page 373).

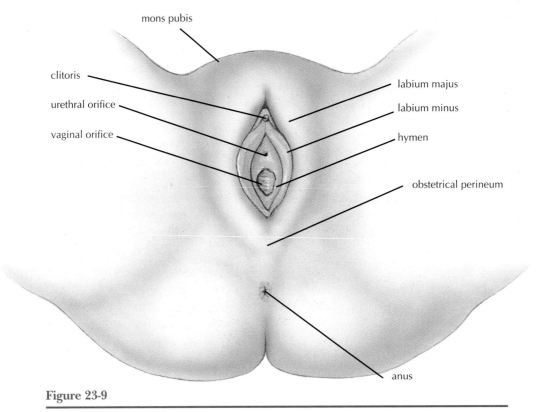

mons pubis

clitoris

urethral orifice

vaginal orifice

labium majus

labium minus

hymen

obstetrical perineum

anus

Figure 23-9

The external female genitalia.

▶ Disorders of the Female Reproductive System

Leukorrhea

A nonbloody vaginal discharge, usually whitish in color, is called *leukorrhea* (lu-ko-RE-ah). In many cases it is merely the colorless mucus produced by the cervical glands. Leukorrhea is not a disease, but it can be a sign of irritation or infection involving the cervix or vagina.

Menstrual Disorders

Absence of menstrual flow is known as *amenorrhea* (ah-men-o-RE-ah). This condition can be symptomatic of such a disorder as insufficient hormone secretion or congenital abnormality of the reproductive organs. Psychologic factors often play a part in cessation of the menstrual flow. For example, any significant change in a woman's general state of health, or a change in her pattern of living, such as a shift in working hours, can cause her to miss a period. The most common cause of amenorrhea (except, of course, for menopause) is pregnancy.

Dysmenorrhea (dis-men-o-RE-ah) means painful or difficult menstruation, and in young women it may be due to immaturity of the uterus. Dysmenorrhea is frequently associated with cycles in which ovulation has occurred. Often the pain can be relieved by drugs that block prostaglandins, since some prostaglandins are known to cause painful uterine contractions. In many cases women have been completely relieved of menstrual cramps by their first pregnancies. Apparently, enlargement of the cervical opening remedies the condition. Artificial dilation of the cervical opening may alleviate dysmenorrhea for several months. Often such health measures as sufficient rest, a well-balanced diet, and appropriate exercises remedy the disorder. During the attack, the application of heat over the abdomen usually relieves the pain, just as it may ease other types of muscular cramps.

Premenstrual syndrome (PMS), also called *premenstrual tension,* is a condition in which nervousness, irritability, and depression precede the menstrual period. It is thought to be due to fluid

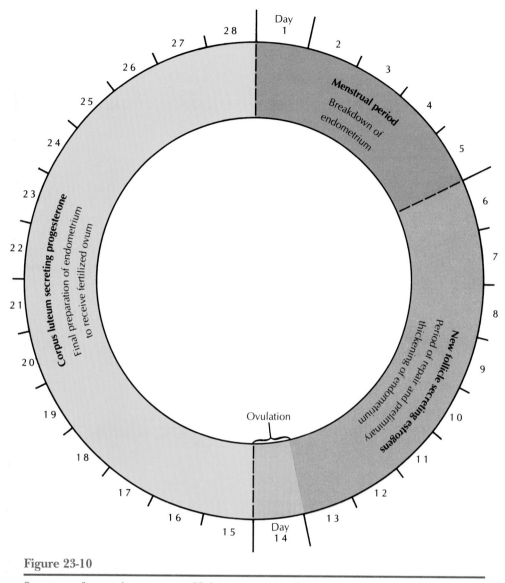

Day 1

28

27

26

25

24

23

22

21

20

19

18

17

16

Day 15

2

3

4

5

6

7

8

9

10

11

12

13

Day 14

Menstrual period
Breakdown of
endometrium

New follicle secreting estrogens
Period of repair and preliminary
thickening of endometrium

Ovulation

Corpus luteum secreting progesterone
Final preparation of endometrium
to receive fertilized ovum

Figure 23-10

Summary of events in an average 28-day menstrual cycle.

retention in various tissues, including the brain. Sometimes a low-salt diet and appropriate medication for 2 weeks before the menses prevent this disorder. This treatment may also avert dysmenorrhea. PMS treatment centers in some areas of the United States are proving helpful for many women.

Abnormal uterine bleeding includes excessive menstrual flow, too-frequent menstruation, and nonmenstrual bleeding. Any of these may cause serious anemias and deserve careful medical attention. Nonmenstrual bleeding may be an indication of a tumor, possibly cancer.

Benign Tumors

Fibroids, which are more correctly called *myomas,* are common tumors of the uterus. Studies indicate that about 50% of women who reach the age of 50 have one or more of these growths in the walls of the uterus. Very often these tumors are small, and usually they remain benign and produce no symptoms. They develop between puberty and menopause and ordinarily stop growing after a woman has reached the age of 50. In some cases these growths interfere with pregnancy, and in a patient under 40, the surgeon may simply remove the tumor and leave

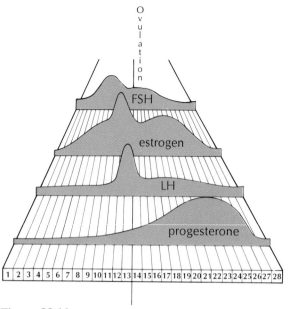

Figure 23-11

Hormones in the menstrual cycle. (Redrawn from Djerassi C: Fertility awareness: Jet-age rhythm method? Science 1990;248:1061)

the uterus fairly intact. Normal pregnancies have occurred after such surgery.

Fibroids may become so large that pressure on adjacent structures causes grave disorders. In some cases invasion of blood vessels near the uterine cavity causes serious hemorrhages. For these and other reasons, it may be necessary to remove the entire uterus or a large part of it. Surgical removal of the uterus is called a *hysterectomy* (his-ter-EK-to-me).

Malignant Tumors

Cancer of the breast is the most commonly occurring malignant disease in women. The tumor is usually a painless, movable mass that is often noticed by the woman and all too frequently ignored. However, in recent years there has been increasing emphasis on the importance of regular self-examination of the breasts. Most breast lumps are discovered by women themselves. Any lump, no matter how small, should be reported to a physician immediately. The *mammogram,* x-ray study of the breast, has improved the detection of early breast cancer. Mammograms are generally recommended yearly after the age of 45 unless there is a history of breast cancer in the family, in which case earlier studies are recommended.

The treatment of breast cancer consists of surgery with follow-up therapy of radiation, chemotherapy, or both. Surgical treatment by removal of the lump or a segment of the breast is most common. Removal of the entire breast and dissection of the lymph nodes in the axilla (armpit) is called *modified radical mastectomy* (mas-TEK-to-me). Studies are underway to determine which combinations of therapy are most effective in each stage.

The most common cancer of the female reproductive tract is cancer of the endometrium (the lining of the uterus). This type of cancer usually affects women during or after menopause. It is seen most frequently in women who have had few pregnancies, abnormal bleeding, or cycles in which ovulation did not occur. Symptoms include an abnormal discharge or irregular bleeding. The usual methods of treatment include surgery and irradiation. Early, aggressive treatment of this, as of all cancers, can save the patient's life.

Ovarian cancer is a leading cause of cancer deaths in women. Although most ovarian cysts are not malignant, they should always be investigated for possible malignant change. Ovarian malignancies tend to progress rapidly, so that careful staging (see Chap. 4) is important if a patient's life is to be saved. Ovarian cancer is the second most common reproductive tract cancer in the female, usually occurring in women between the ages of 40 and 65. Early surgery, irradiation, and especially chemotherapy have proved to be effective in many cases.

Cancer of the cervix, the third most common cancer of the female reproductive tract, is most frequent in women from 30 to 50 years of age. Appropriate screening allows the discovery and treatment of many early cases. Although no specific cause has been identified, statistics indicate that the risk of development of cervical cancer is strongly increased by factors such as first sexual intercourse at an early age, many sexual partners, and infection, as with genital herpes or a virus that causes genital warts. The decline in the death rate from this type of cancer is directly related to use of the *Papanicolaou* (pap-ah-nik-o-LAH-o) *test,* also known as the *Pap test* or *Pap smear.* The Pap smear is a microscopic examination of cells obtained from scrapings of the cervix and swabs of the cervical canal. All women should be encouraged to have this test every year. Even girls under 18 should be tested if they are sexually active.

The incidences of the various types of cancer should not be confused with the death rates for each type. Owing to education of the public and increas-

TABLE 23-1
Main Methods of Contraception Currently in Use

Method	Description	Advantages	Disadvantages
Surgical			
Vasectomy/tubal ligation	Cutting and tying of tubes carrying gametes	Nearly 100% effective; involves no chemical or mechanical devices	Not usually reversible; rare surgical complications
Hormonal			
Birth control pill	Estrogen and progesterone or progesterone alone taken orally to prevent ovulation	Highly effective; requires no last-minute preparation	Alters physiology; possible serious side effects
Birth control shot	Injection of synthetic progesterone every 3 months to prevent ovulation	Highly effective; lasts for 3 to 4 months	Alters physiology; possible side effects include menstrual irregularity, amenorrhea
Birth control implants	Devices containing synthetic progesterone implanted under skin to prevent ovulation	Highly effective; lasts for 5 years	Alters physiology; possible side effects include menstrual irregularity, amenorrhea; expensive
Barrier			
Male condom	Sheath that fits over erect penis and prevents release of semen	Easily available; does not affect physiology; protects against sexually transmitted disease (STD)	Must be applied just before intercourse; may slip or tear
Female condom	Sheath that fits in vagina, held in place with rings	Easily available; protects against STD	More expensive than male condom; must be applied before intercourse
Diaphragm (with spermicide)	Rubber cap that fits over cervix and prevents entrance of sperm	Does not affect physiology; some protection against STD	Must be inserted before intercourse; requires fitting by physician
Other			
Spermicide	Chemicals used to kill sperm; best when used in combination with a barrier method	Easily available, does not affect physiology; some protection against STD	Local irritation; must be used just before intercourse
Fertility awareness	Abstinence during fertile part of cycle as determined by menstrual history, basal body temperature, or quality of cervical mucus	Does not affect physiology; accepted by certain religions	High failure rate; requires careful record keeping

ingly better methods of diagnosis and treatment, some forms of cancer have a higher cure rate than others. For example, cancer of the breast appears much more often in women than does cancer of the lung, but more women now die each year from lung cancer than from breast cancer.

Infections

The common sexually transmitted diseases that involve the male reproductive system also attack the female genital organs (Fig. 23-12); the most common are chlamydial infections, gonorrhea, and genital herpes. Syphilis also occurs in women, and the

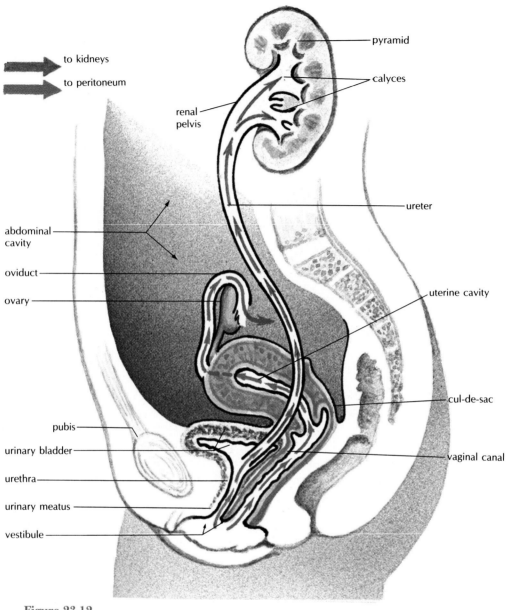

to kidneys

to peritoneum

pyramid

calyces

renal pelvis

ureter

abdominal cavity

oviduct

ovary

uterine cavity

cul-de-sac

pubis

urinary bladder

vaginal canal

urethra

urinary meatus

vestibule

Figure 23-12

Pathway of infection from outside to the peritoneum and into the urinary system.

fetus of an untreated syphilitic mother may be still-born; an infant born alive may have highly infectious lesions of the palms or the soles as well as other manifestations of the disease.

The incidence of **genital warts,** caused by papillomavirus, has increased in recent years. These infections have been linked to cancer of the reproductive tract, especially cancer of the uterine cervix.

Salpingitis (sal-pin-JI-tis) means "inflammation of a tube." However, the term is used most often to refer to disease of the uterine tubes. Most cases of infection of the uterine tubes are caused by gonococci or by the bacterium *Chlamydia trachomatis,* but other bacteria can bring it about. Salpingitis may cause sterility by obstructing the tubes, thus preventing the passage of ova.

Pelvic inflammatory disease (**PID**) is due to extension of infections from the reproductive organs into the pelvic cavity, and it often involves the peritoneum. (See the red arrow pathway in Fig. 23-12.)

Gonococci or chlamydia are usually the initial cause of infection, but most cases of PID involve multiple organisms.

Toxic shock syndrome (TSS), is a serious bacterial infection usually due to certain staphylococci. Symptoms include a sudden fever, rash, diarrhea, vomiting, and hypotension, which may lead to shock. Most cases in the 1980s were associated with the use of superabsorbent tampons. Changes in the composition of these tampons have resulted in a reduction of cases due to this cause. Toxic shock syndrome also occurs in males and in nonmenstruating females, in which cases the cause is unknown.

Infertility

Infertility is much more difficult to diagnose and evaluate in the female than in the male. Whereas a microscopic examination of properly collected semen may be all that is required to determine the presence of abnormal or too-few sperm cells in the male, no such simple study can be made in the female. Infertility in women, as in men, may be relative or absolute. Causes of female infertility include infections, endocrine disorders, psychogenic factors, and abnormalities in the structure and function of the reproductive organs themselves. In all cases of apparent infertility, the male partner should be investigated first, because the procedures for determining lack of fertility in the male are much simpler and less costly than those in the female, as well as being essential for the evaluation.

❱ Menopause

Menopause (MEN-o-pawz), often called *change of life,* is that period at which menstruation ceases altogether. It ordinarily occurs between the ages of 45 and 55 and is caused by a normal decline in ovarian function. The ovary becomes chiefly scar tissue and no longer produces ova or appreciable amounts of estrogen. Eventually the uterus, oviducts, vagina, and vulva all become somewhat atrophied.

Although menopause is an entirely normal condition, its onset sometimes brings about effects that are temporarily disturbing. The decrease in estrogen levels can cause such nervous symptoms as irritability, "hot flashes," and dizzy spells.

Although still a subject of controversy, the use of estrogen replacement therapy is considered to be of overall benefit for menopausal women. Estrogen therapy has proved effective in alleviating hot flashes and the sensitivity of a thinning vaginal mucosa. Some women may require hormone therapy to prevent or halt osteoporosis (bone weakening). It has also been found to reduce the risk of heart attacks, which generally increase among women after menopause. The longer the duration of hormone treatment, the greater is the risk of cancer of the uterus (endometrial cancer). Reducing the dosages and the duration of the therapy lessens this risk, as does giving the estrogen in combination with progesterone (progestin) to prevent overgrowth of the endometrium. Other factors that must be considered involve the risk of thrombosis and embolism, which is highest among women who smoke. Some studies have also shown an increased risk of breast cancer in the wake of long-term estrogen therapy. A woman who has a family history of cancer of the breast or of the uterus should avoid such treatment.

STDs

Sexually transmitted disease (STD), formerly known as venereal disease (VD), is infection spread through intimate sexual contact. Examples are gonorrhea, syphilis, human papilloma virus (HPV), genital herpes, hepatitis B, and chlamydia. These diseases can lead to sterility, cancer of the cervix, and increased susceptibility to AIDS. A trend toward participation in sex at earlier ages exposes people to more sexual contacts and an increased risk of infection.

Women are more likely to contract STDs than are males. The same mechanisms that transport sperm cells through the female reproductive tract also move infectious organisms. Also, symptoms in women may be less apparent than in males. There needs to be more awareness of STDs and more fast, cheap, and accurate tests for diagnosis.

The main preventive measure at present is the use of a condom. Efforts to produce safe and effective drugs to kill STD-causing germs are in progress. A vaccine is available for hepatitis B, and immunizations for other STDs are under study.

▶ Contraception

Contraception is defined as the use of artificial methods to prevent fertilization of the ovum or implantation of the fertilized ovum. Table 23-1 presents a brief description of the main contraceptive methods currently in use, along with some advantages and disadvantages of each. The list is given in rough order of decreasing effectiveness.

The spread of AIDS and other sexually transmitted diseases has stimulated interest in the development of better condoms, including a female condom, as these not only prevent conception but also reduce the chances of infection.

Most other advances involve new methods for administering birth control hormones. Capsules of synthetic progesterone implanted under the skin are effective for 5 years and avoid the side effects of estrogen. This same drug may be given by injection at 3-month intervals.

Although not strictly a contraceptive, the drug *RU 486* is taken after conception to terminate an early pregnancy. It blocks the action of progesterone, causing the uterus to shed its lining and release the fertilized egg. It must be combined with administration of prostaglandins to expel the uterine tissue.

SUMMARY

I. **Structure of reproductive tract**
   A. Primary organs—gonads
      1. Testes in male; ovaries in female
      2. Germ cells (sperm and eggs)—produced in gonads by meiosis, which reduces chromosome number from 46 to 23
   B. Accessory organs—ducts and exocrine glands

II. **Male reproductive system**
   A. Structure
      1. Testes
         a. Products
            (1) Spermatozoa—produced in seminiferous tubules
               (a) Head—mainly nucleus
               (b) Middle region—has mitochondria
               (c) Tail—for locomotion
               (d) Acrosome—covers head (aids in penetration of ovum)
            (2) Testosterone—produced in interstitial cells (between tubules)
               (a) Development of spermatozoa
               (b) Production of secondary sex characteristics
      2. Ducts
         a. Epididymis—stores spermatozoa until ejaculation
         b. Ductus deferens (vas deferens)
            (1) Transports spermatozoa

            (2) Part of spermatic cord that suspends testis
         c. Ejaculatory duct—receives secretions from seminal vesicle and ductus deferens
         d. Urethra—carries semen through penis
      3. Penis—copulatory organ
      4. Scrotum—sac that contains testes
   B. Semen
      1. Functions
         a. Nourishment and transportation of spermatozoa
         b. Lubrication of reproductive tract
         c. Neutralization of female reproductive tract
      2. Formation
         a. Seminal vesicles
         b. Prostate
         c. Bulbourethral glands
   C. Hormonal control of reproduction
      1. Hypothalamus—produces releasing hormone, causing anterior pituitary to release FSH and LH
      2. FSH—stimulates formation of spermatozoa
      3. LH (ICSH)—stimulates production of testosterone
      4. Testosterone—regulates continuous hormone production by acting as negative feedback messenger to hypothalamus

D. Disorders
  1. Infertility—decreased ability to reproduce
     a. Sterility—complete inability to reproduce
  2. Cryptorchidism—failure of testes to descend into scrotum
  3. Inguinal hernia—caused by weakness in abdominal wall at inguinal canal
  4. Infections
     a. Gonorrhea and chlamydial infections—main sexually transmitted diseases
     b. Others—genital herpes, syphilis
  5. Tumors

III. Female reproductive tract
  A. Structure
    1. Ovaries—produce ova and hormones
    2. Oviducts—carry ovum to uterus by means of cilia in lining and by peristalsis
    3. Uterus—site for development of fertilized egg
       a. Endometrium—lining of uterus
       b. Cervix—narrow region at bottom
       c. Broad ligament—supports uterus, ovaries, oviducts
    4. Vagina—canal for copulation, birth, menstrual flow
       a. Hymen—membrane that covers opening of vagina
       b. Greater vestibular glands—secrete mucus
    5. Vulva—external female genitalia
       a. Labia—two liplike pairs of tissue
       b. Clitoris—sensitive tissue
       c. Perineum—pelvic floor
  B. Menstrual cycle
    1. First hormone is FSH from anterior pituitary, which starts ripening of ovum within follicle
    2. Production of estrogen by follicle
       a. Development of endometrium for possible pregnancy
       b. Feedback to hypothalamus to inhibit FSH and start production of LH
    3. LH—surge occurs 24 hours before ovulation
       a. Ovulation—caused by LH on day 14 of 28-day cycle
       b. Conversion of follicle to corpus luteum by LH
    4. Production of progesterone (and some estrogen) by corpus luteum
       a. Continued development of endometrium
       b. Feedback to inhibit release of LH
    5. No fertilization
       a. Degeneration of corpus luteum
       b. Drop in hormone levels
       c. Menstruation
  C. Disorders
    1. Leukorrhea—vaginal discharge; sign of irritation or infection
    2. Menstrual disorders
       a. Amenorrhea—absence of menstrual flow
       b. Dysmenorrhea—painful or difficult menstruation
       c. Premenstrual syndrome
    3. Malignant tumors—cancer of breast, endometrium, ovary, cervix most common
    4. Infections
       a. Salpingitis—inflammation of uterine tubes
       b. Pelvic inflammatory disease
       c. Toxic shock syndrome
    5. Infertility

IV. Menopause—cessation of menstruation
  A. Characteristics
    1. Degeneration of reproductive organs
    2. Possibility of disturbing side effects
    3. Increased risk of osteoporosis and heart disease
  B. Therapy—estrogen replacement

V. Contraception—use of methods to prevent fertilization or implantation of ovum

QUESTIONS FOR STUDY AND REVIEW

1. In what fundamental respect does reproduction in some single-celled animals such as the ameba differ from that in most animals?
2. Name the sex cells of both the male and the female.
3. Name all the parts of the male reproductive system and describe the function of each.
4. What is cryptorchidism? Why does this condition cause sterility?
5. Name and describe a disorder of the male reproductive system that is common in elderly men. What are some other effects of aging in the male?
6. Name the principal parts of the female reproductive system and describe the function of each.
7. Describe the process of ovulation.
8. Beginning with the first day of the menstrual flow, describe the events of one complete cycle, including the role of the various hormones.
9. Distinguish between dysmenorrhea and amenorrhea.
10. What is the most common malignant disease in women? For early detection, what should a woman do at regular intervals?
11. What are the two most common cancers of the uterus?
12. What is a Pap test? Why is it important?
13. What are some causes of infertility in both males and females?
14. What is menopause? What causes it? What are some of the changes that take place in the body as a result?
15. What are the values of estrogen replacement therapy after menopause? What are some of the dangers of such therapy?
16. Define *contraception*. Describe methods of contraception that involve (1) barriers; (2) chemicals; (3) hormones; (4) prevention of implantation.

CHAPTER 24

Development
and Birth

Selected Key Terms

The following terms are
defined in the Glossary:

abortion

amniotic sac

embryo

fertilization

fetus

gestation

implantation

lactation

parturition

placenta

umbilical cord

zygote

Behavioral Objectives

After careful study of this chapter, you should be able to:

- Describe fertilization and the early development of the fertilized egg
- Describe the structure and function of the placenta
- Briefly describe the four stages of labor
- Cite the advantages of breastfeeding
- Describe several disorders associated with pregnancy

Memmler, RL, Cohen, BJ, Wood, DL. *THE HUMAN BODY IN HEALTH AND DISEASE,* 8/e,
© 1996 Lippincott-Raven Publishers

▶ Pregnancy

Pregnancy begins with fertilization of an ovum and ends with delivery of the fetus and afterbirth. During this 38-week period of development, known as *gestation* (jes-TA-shun), all body tissues differentiate from a single fertilized egg. Along the way, many changes occur in both the mother and the growing offspring.

Fertilization and the Start of Pregnancy

When semen is deposited in the vagina, the many spermatozoa immediately wriggle about in all directions, some traveling into the uterus and oviducts (see Fig. 23-6). If an egg cell is present in the oviduct, many spermatozoa cluster around it. Using enzymes they dissolve the coating around the ovum so that eventually one sperm cell can penetrate the cell membrane. The nuclei of the sperm and egg then combine.

The result of this union is a single cell with the full human chromosome number of 46. This new cell, called a *zygote* (ZI-gote), can divide and grow into a new individual. The zygote divides rapidly into two cells and then four cells, and soon a ball of cells is formed. During this time, the ball of cells is traveling toward the uterine cavity, pushed along by the cilia and by the peristalsis of the tube. After reaching the uterus, the little ball of cells burrows into the greatly thickened uterine lining, where it is soon completely covered and implanted.

Development and Functions of the Placenta

Following *implantation,* a group of cells within the ball becomes an *embryo* (EM-bre-o). The outer cells form projections, called *villi,* that invade the uterine wall and maternal blood channels (venous sinuses). This eventually leads to the formation of the *placenta* (plah-SEN-tah), a flat, circular organ that consists of a spongy network of blood-filled lakes and capillary-containing villi.

The embryo, later called the *fetus* (FE-tus), is connected to the developing placenta by a stalk of tissue that eventually becomes the *umbilical* (um-BIL-ih-kal) *cord.* The cord contains two arteries that carry blood from the fetus to the placenta and one vein that carries blood from the placenta to the fetus

(Fig. 24-1). The placenta serves as the organ of nutrition, respiration, and excretion for the embryo through exchanges that occur between the blood of the embryo and the blood of the mother across the capillaries of the villi.

Another function of the placenta is endocrine in nature. Beginning soon after implantation, some of the embryonic cells produce a hormone called *human chorionic gonadotropin* (ko-re-ON-ik gon-ah-do-TRO-pin) *(HCG).* This hormone stimulates the corpus luteum of the ovary, prolonging its life span (to 11 or 12 weeks) and causing it to secrete increasing amounts of progesterone and estrogen. Progesterone is essential for the maintenance of pregnancy. It promotes endometrial secretion to nourish the embryo, and it decreases the ability of the uterine muscle to contract, thus preventing the embryo from being expelled from the body. During pregnancy progesterone also helps prepare the breasts for the secretion of milk. Estrogen promotes enlargement of the uterus and breasts. By the 11th or 12th week of pregnancy, the corpus luteum is no longer needed; by this time, the placenta itself can secrete adequate amounts of progesterone and estrogen.

Development of the Embryo

For the first eight weeks of life the developing offspring is referred to as an *embryo* (Fig. 24-2). The beginnings of all body systems are established during this period. The heart and the brain are among the first organs to develop. A primitive nervous system begins to form in the third week. The heart and blood vessels originate during the second week, and the first heartbeat appears during week 4, at the same time that other muscles begin to develop. By the end of the first month, the embryo is about 0.62 cm (0.25 inches) long, with four small swellings at the sides called *limb buds,* which will develop into the four extremities. At this time the heart produces a prominent bulge at the front of the embryo. By the end of the second month, the embryo takes on an appearance that is recognizably human. The study of embryonic development is *embryology* (em-bre-OL-o-je).

The Fetus

The term *fetus* is used for the developing individual from the beginning of the third month until birth. During this period the organ systems continue to grow and mature. For study, the period of devel-

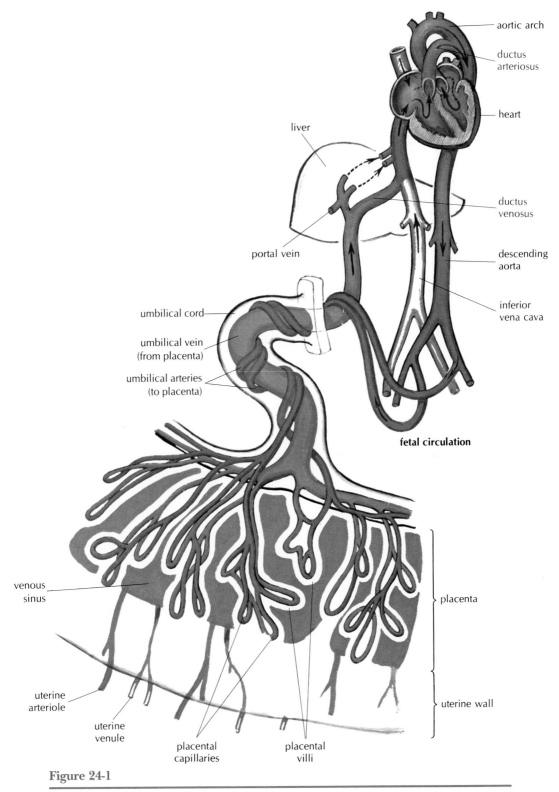

aortic arch

ductus arteriosus

heart

liver

ductus venosus

portal vein

descending aorta

inferior vena cava

umbilical cord

umbilical vein (from placenta)

umbilical arteries (to placenta)

fetal circulation

venous sinus

placenta

uterine arteriole

uterine venule

placental capillaries

placental villi

uterine wall

Figure 24-1

Fetal circulation and placenta.

umbilical
stalk

28 days

6 weeks

8 weeks

umbilical cord

3 months

5 months

Figure 24-2

Development of an embryo into a fetus.

opment (gestation), may be divided into three equal segments or *trimesters.* The most rapid growth occurs during the second trimester (months 4–6). By the end of the fourth month, the fetus is almost 15 cm (6 inches) long, and its external genitalia are sufficiently developed to reveal its sex. By the seventh month, the fetus is usually about 35 cm (14 inches) long and weighs about 1.1 kg (2.4 lbs). At the end of pregnancy, the normal length of the fetus is 45 to 56 cm (18–22.5 inches), and the weight varies from 2.7 to 4.5 kg (6–10 lb).

The *amniotic* (am-ne-OT-ik) *sac,* which is filled with a clear liquid known as *amniotic fluid,* surrounds the fetus and serves as a protective cushion for it (Fig. 24-3). Popularly called the *bag of waters,* the amniotic sac ruptures at birth. During development the skin of the fetus is protected by a layer of cheese-like material called *vernix caseosa* (VER-niks KA-se-o-sa).

The Origin of Tissues

 Within the first weeks of development, the cells of the fertilized egg differentiate into three layers that will give rise to all the tissues of the body. These three *germ layers* are the *ectoderm* (outermost), *endoderm* (innermost), and *mesoderm* (middle).

The cells of the ectoderm eventually form all nervous tissue, the epidermis of the skin, hair, nails, and skin glands. The endoderm forms the lining of the respiratory and digestive tracts and some glands. The mesoderm gives rise to all muscle and connective tissue, including bone, cartilage, and blood, and to serous membranes.

Most organs eventually contain cells that originate from all three of these layers. For example, the stomach has nerves (ectoderm), muscle and connective tissue (mesoderm), and an epithelial lining (endoderm). No matter where they appear, all body cells trace their origins to one of these three primitive germinal layers.

The Mother

The total period of pregnancy, from fertilization of the ovum to birth, is about 266 days, also given as 280 days or 40 weeks from the last menstrual period (LMP). During this time, the mother must supply all the food and oxygen for the fetus and eliminate its waste materials. To support the additional demands of the growing fetus, the mother's metabolism changes markedly, with increased demands being made on several organ systems:

1. The heart pumps more blood to supply the needs of the uterus and its contents.
2. The lungs provide more oxygen to supply the fetus by increasing the rate and depth of respiration.
3. The kidneys excrete nitrogenous wastes from the fetus and from the mother's body.
4. Nutritional needs are increased to provide for the growth of maternal organs (uterus and breasts) and growth of the fetus, as well as for preparation for labor and the secretion of milk.

Nausea and vomiting are common discomforts in early pregnancy. The specific cause is not known, but these symptoms may be due to the great changes in hormone levels that occur at this point. The nausea and vomiting usually last for only a few weeks. Frequency of urination and constipation are often present during the early stages of pregnancy and then usually disappear. They may reappear late in pregnancy as the head of the fetus drops from the abdominal region down into the pelvis, pressing on the rectum and the urinary bladder.

▶ Childbirth

The mechanisms that trigger the beginning of uterine contractions for childbirth are still unknown. However, it is recognized that the uterine muscle becomes increasingly sensitive to oxytocin (from the pituitary gland) late in pregnancy. Once labor is started, stimuli from the cervix and vagina produce reflex secretion of this hormone, which in turn increases the uterine contractions.

The process by which the fetus is expelled from the uterus is known as *labor* and *delivery;* it also may be called *parturition* (par-tu-RISH-un). It is divided into four stages:

1. The *first stage* begins with the onset of regular contractions of the uterus. With each contraction, the cervix becomes thinner and the opening larger. Rupture of the amniotic sac may occur at any time, with a gush of fluid from the vagina.
2. The *second stage* begins when the cervix is completely dilated and ends with the delivery of the baby. This stage involves the passage of the fetus, usually head first, through the cervical canal and the vagina to the outside.
3. The *third stage* begins after the child is born and ends with the expulsion of the *afterbirth.* The afterbirth includes the placenta, the membranes of the amniotic sac, and the umbilical cord, except for a small portion remaining attached to the baby's *umbilicus* (um-BIL-ih-kus), or navel.
4. The *fourth stage* begins with the expulsion of the afterbirth and constitutes a period in which bleeding is controlled. Contraction of the uterine muscle acts to close off the blood vessels leading to the placental site.

To prevent the pelvic floor tissues from being torn during childbirth, as often happens, the physician may cut the mother's perineum just before her infant is born and then repair this clean cut immewall of uterus

placenta

umbilical cord

amniotic sac

amniotic fluid

fetus

urinary bladder

pubis

rectum

anus

urethra vagina perineum

cervix

Figure 24-3

Midsagittal section of a pregnant uterus with intact fetus.

diately after childbirth; such an operation is called an *episiotomy* (eh-piz-e-OT-o-me). The area between the vagina and the anus that is cut in an episiotomy is referred to as the *surgical* or *obstetrical perineum* (see Fig. 23-9).

Multiple Births

Until recently, statistics indicated that twins occurred in about 1 of every 80 to 90 births, varying somewhat in different countries. Triplets occurred much less frequently, usually once in several thousand births, while quadruplets occurred very rarely indeed. The birth of quintuplets represented a historic event unless the mother had taken fertility drugs. Now these fertility drugs, usually gonadotropins, are given quite frequently, and the number of multiple births has increased significantly. Multiple fetuses tend to be born prematurely and therefore have a high death rate. However, better care of infants and newer treatments have resulted in more living multiple births than ever.

Twins originate in two different ways and on this basis are divided into two groups:

1. *Fraternal twins* are formed as a result of the fertilization of two different ova by two spermatozoa. Two completely different individuals, as distinct from each other as brothers and sisters of different ages, are produced. Each fetus has its own placenta and surrounding sac.
2. *Identical twins* develop from a single zygote formed from a single ovum fertilized by a single spermatozoon. Obviously, they are always the same sex and carry the same inherited traits. Sometime during the early stages of development the embryonic cells separate into two units. Usually there is a single placenta, although there must be a separate umbilical cord for each individual.

Other multiple births may be fraternal, identical, or combinations of these. The tendency to multiple births seems to be hereditary.

The Mammary Glands and Lactation

The *mammary glands,* or the breasts of the female, are accessories of the reproductive system. They are designed to provide nourishment for the baby after its birth, and the secretion of milk at this time is known as *lactation* (lak-TA-shun).

The mammary glands are constructed in much the same manner as the sweat glands. Each of these glands is divided into a number of lobes composed of glandular tissue and fat, and each lobe is further subdivided. The secretions from the lobes are conveyed through *lactiferous* (lak-TIF-er-us) *ducts,* all of which converge at the nipple (Fig. 24-4).

The mammary glands begin developing during puberty, but they do not become functional until the end of a pregnancy. The hormone *prolactin (PRL),*

produced by the anterior lobe of the pituitary, stimulates the secretory cells of the mammary glands. The first of the mammary gland secretions is a thin liquid called *colostrum* (ko-LOS-trum). It is nutritious but has a somewhat different composition from milk. Secretion of milk begins within a few days and can continue for several months as long as milk is frequently removed by the suckling baby or by pumping. Stimulation of the breast by the suckling infant causes the release of *oxytocin* from the posterior pituitary gland. This hormone promotes the ejection or *letdown* of milk.

The digestive tract of the newborn baby is not ready for the usual mixed diet of an adult. Mother's milk is more desirable for the young infant than milk from other animals for several reasons, some of which are listed below:

1. Infections that may be transmitted by foods exposed to the outside air are avoided by nursing.
2. Both breast milk and colostrum contain maternal antibodies that help protect the baby against pathogens.
3. The proportions of various nutrients and other substances in human milk are perfectly suited to the human infant. Substitutes are not exact imitations of human milk. Nutrients are present in more desirable amounts if the mother's diet is well balanced.
4. The psychologic and emotional benefits of nursing are of infinite value to both the mother and the infant.

Disorders of Pregnancy

Ectopic Pregnancy
A pregnancy that develops in a location outside the uterine cavity is said to be an *ectopic* (ek-TOP-ik) *pregnancy, ectopic* meaning "out of normal place." The most common type is the *tubal ectopic pregnancy,* in which the embryo grows in the oviduct. This structure, which is not designed by nature for pregnancy, is not able to expand to contain the growing embryo and may rupture. Ectopic pregnancy may threaten the mother's life if it does not receive prompt surgical treatment.

Placenta Previa
The placenta is usually attached to the upper part of the uterus. In *placenta previa* (PRE-ve-ah) the pla-

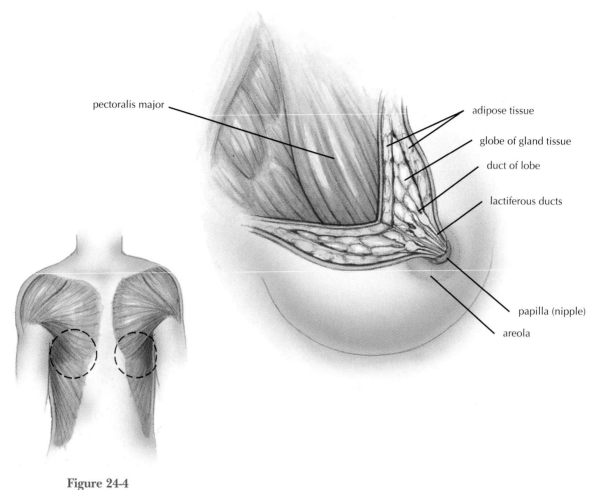

Figure 24-4

Section of the breast.

centa becomes attached at or near the internal open-ing of the cervix. The normal softening and dilation of the cervix that occur in later pregnancy will sepa-rate part of the placenta from its attachment. The result will be painless bleeding and interference with the fetal oxygen supply. Sometimes placenta previa necessitates *cesarean* (se-ZAR-re-an) *section,* which is delivery of the fetus by way of an incision made in the abdominal wall and the wall of the uterus. A cesarean section (C-section) may be required for a variety of other conditions as well.

Placental Abruption

Sometimes the placenta separates from the wall of the uterus prematurely, often after the 20th week of pregnancy, causing hemorrhage. This disorder,

known as ***abruptio placentae*** (ab-RUP-she-o plah-SEN-te), occurs most often in women over 35 who have had several pregnancies (multigravidas). This is a common cause of bleeding during the second half of pregnancy and may require the termination of pregnancy to save the mother's life.

THE USE OF ULTRASOUND

A valuable diagnostic tool that avoids the use of undesirable x-rays is a form of vibrational energy called *ultrasound.* By this method soft tissues can be visualized and such abnormalities as ectopic preg-nancies and placenta previa can be accurately de-lineated. The relatively inexpensive ultrasound equipment has been found useful in detecting ab-normalities of the fetus, tumors, and other disorders of the reproductive system.

The Use of Ultrasound in Obstetrics

Ultrasonography (ul-trah-son-OG-rah-fe) is a safe, painless, and noninvasive method for studying soft tissue. It has proved extremely valuable for monitoring pregnancies and deliveries. For many parents, the first picture in their baby's album is a prenatal ultrasound scan.

An ultrasound image, called a *sonogram,* is made by sending high-frequency sound waves into the body. Each time a wave meets an interface between two tissues of different densities, an echo is produced. An instrument called a *transducer* converts the reflected sound waves into electrical energy, and a computer is used to generate an image on a viewing screen.

Ultrasound scans can be used in obstetrics to diagnose pregnancy, judge the age of the fetus, and determine the location of the placenta. The technique can also show the presence of excess amniotic fluid and fetal abnormalities.

Pregnancy-Induced Hypertension

A serious disorder that can develop in the latter part of pregnancy is *preeclampsia* (pre-eh-KLAMP-se-ah) also called *toxemia of pregnancy.* Symptoms include hypertension, protein in the urine (proteinuria), general edema, and sudden weight gain. The causes of this disorder are unknown. However, it is most common in women whose nutritional state is poor and who have received little or no health care during pregnancy. If the condition remains untreated, it may lead to *eclampsia* (eh-KLAMP-se-ah) with the onset of kidney failure, convulsions, and coma during pregnancy or after delivery. The result may be the death of both the mother and the infant.

Lactation Disturbances

Disturbances in lactation may be due to a variety of reasons, including the following:

1. Malnutrition or anemia, which may prevent lactation entirely
2. Emotional disturbances, which may affect lactation (as they may other glandular activities)
3. Abnormalities of parts of the mammary glands or injuries to these organs, which may cause interference with their functioning

4. *Mastitis* (mas-TI-tis), which means "inflammation of the breast," is due to infection. Antibiotic treatment usually allows for the continuation of nursing.

Live Births and Fetal Deaths

A pregnancy may end before its full term has been completed. The term *live birth* is used if the baby breathes or shows any evidence of life such as heartbeat, pulsation of the umbilical cord, or movement of voluntary muscles. An *immature* or *premature* infant is one born before the organ systems are mature. Infants born before the 37th week of gestation or weighing less than 2500 grams (5.5 lb) are considered *preterm.*

Loss of the fetus is classified according to the duration of the pregnancy:

1. The term *abortion* refers to loss of the embryo or fetus before the 20th week or weight of about 500 grams (1.1 lb). This loss can be either spontaneous or induced.
   a. *Spontaneous abortions* occur naturally with no interference. The most common causes are related to an abnormality of the embryo or fetus. Other causes include abnormality of the mother's reproductive organs, infections, or chronic disorders such as kidney disease or hypertension. *Miscarriage* is the lay term for spontaneous abortions.
   b. *Induced abortions* occur as a result of artificial or mechanical interruption of pregnancy. *Therapeutic abortions* are abortions performed by physicians as a form of treatment for a variety of reasons. With more liberal use of this type of abortion, there has been a dramatic decline in the incidence of death related to illegal abortion.
2. The term *fetal death* refers to loss of the fetus after the eighth week of pregnancy. *Stillbirth* refers to the delivery of an infant that is lifeless.

Immaturity is a leading cause of death in the newborn. After the 20th week of pregnancy, the fetus is considered *viable,* that is, able to live outside the uterus. A fetus expelled before the 28th week or a weight of 1000 grams (2.2 lb) has a less than 50% chance of survival. One born at a point closer to the full 40 weeks stands a much better chance of survival. Increasing numbers of immature infants are being saved because of advances in neonatal intensive care.

Postpartum Disorders
PUERPERAL INFECTION

Childbirth-related deaths are often due to infections. *Puerperal* (pu-ER-per-al) *infections* (those related to childbirth) were once the cause of death in as many as 10% to 12% of women going through labor. Cleanliness and sterile techniques have improved the chances of avoiding such outcomes of pregnancies. Nevertheless, in the United States, puerperal infection still develops in about 6% of ma-

ternity patients. Antibiotics have dramatically improved the chances of recovery of both the mother and the child.

CHORIOCARCINOMA

A very malignant tumor that is made of placental tissue is the *choriocarcinoma* (ko-re-o-kar-sih-NO-mah). It spreads rapidly, and if the mother isn't treated, it may be fatal within 3 to 12 months. With the use of modern chemotherapy, the outlook for cure is very good. If metastases have developed, irradiation and other forms of treatment may be necessary.

SUMMARY

I. **Pregnancy**
  A. Fertilization
    1. Occurs in oviduct
    2. Zygote (fertilized egg)—formed by fusion of egg and sperm nuclei
  B. Early pregnancy
    1. Embryo (developing egg)—travels into uterus and implants
    2. Placenta
      a. Formation—formed by cells around embryo and lining of uterus; connected to embryo by umbilical cord
      b. Functions
        (1) Provision of nourishment to embryo
        (2) Gas exchange with embryo
        (3) Removal of waste from embryo
        (4) Production of chorionic gonadotropin—maintains corpus luteum until 11–12 weeks of pregnancy
    3. Amniotic sac—fluid-filled sac around embryo that serves as protective cushion
    4. Fetus—embryo becomes fetus in third month
II. **Childbirth**
  A. Four stages
    1. Contractions
    2. Delivery of baby
    3. Expulsion of afterbirth

    4. Contraction of uterus
  B. Multiple births—incidence increased because of fertility drugs
III. **Lactation**
  A. Hormones
    1. Prolactin
    2. Oxytocin
  B. Advantages
    1. Reduces infections
    2. Provides best form of nutrition
    3. Transfers antibodies
    4. Emotional satisfaction
IV. **Disorders of pregnancy and delivery**
  A. Ectopic pregnancy—pregnancy outside of uterus
  B. Placenta previa—improper attachment of placenta to uterus
  C. Placental abruption—separation of placenta from uterus
  D. Hypertension—preeclampsia, eclampsia
  E. Abortion—loss of embryo or fetus before week 20 of pregnancy; may be spontaneous or induced
    1. Fetal death—loss of fetus after 8 weeks of pregnancy
  F. Postpartum disorders
    1. Puerperal infection
    2. Choriocarcinoma—malignant tumor of placental tissue

QUESTIONS FOR STUDY AND REVIEW

1. Distinguish among the following: zygote, embryo, fetus.
2. Name two auxiliary structures that are designed to serve the fetus. What is their origin and what are their functions?
3. Is blood in the umbilical arteries relatively low or high in oxygen? in the umbilical vein?
4. Describe some of the changes that take place in the mother's body during pregnancy.
5. What is the major event of each of the four stages of labor and delivery?
6. What is an episiotomy? Why is it done?
7. Distinguish between fraternal and identical twins.
8. Name two hormones involved in lactation. Where do they originate and what do they do?
9. List some of the advantages associated with breastfeeding a baby.
10. Define *ectopic pregnancy; placenta previa; preeclampsia.*
11. Define *abortion; premature infant; preterm infant.*

CHAPTER 25

Heredity and Hereditary Diseases

Selected Key Terms

The following terms are defined in the Glossary:

amniocentesis

carrier

chromosome

congenital

dominant

gene

genetic

heredity

karyotype

meiosis

mutagenic

mutation

pedigree

progeny

recessive

sex-linked

trait

Behavioral Objectives

After careful study of this chapter, you should be able to:

- Briefly describe the mechanism of gene function
- Explain the difference between dominant and recessive genes
- Describe what is meant by a *carrier* of a genetic trait
- Define *meiosis* and explain the function of meiosis in reproduction
- Explain how sex is determined in humans
- Describe what is meant by the term *sex linked* and list several sex-linked traits
- List several factors that may influence the expression of a gene
- Define *mutation*
- Differentiate among congenital, genetic, and hereditary disorders and give several examples of each
- List several factors that may produce genetic disorders
- Define *karyotype* and explain how karyotypes are used in genetic counseling
- Briefly describe several methods used to treat genetic disorders

Memmler, RL, Cohen, BJ, Wood, DL. *THE HUMAN BODY IN HEALTH AND DISEASE,* 8/e,
© 1996 Lippincott-Raven Publishers

Often we are struck by the resemblance of a baby to one or both of its parents, yet rarely do we stop to consider *how* various traits are transmitted from parents to offspring. This subject—heredity—has fascinated humans for thousands of years; the Old Testament contains numerous references to heredity (although, of course, the word was unknown in biblical times). However, it was not until the 19th century that methodic investigation into heredity was begun. At that time an Austrian monk, Gregor Mendel, discovered through his experiments with garden peas that there was a precise pattern in the appearance of differences among parents and their progeny. Mendel's most important contribution to the understanding of heredity was the demonstration that there are independent units of hereditary influence. Later, these independent units were given the name *genes.*

Genes and Chromosomes

Genes are actually segments of DNA (deoxyribonucleic acid) contained in the threadlike chromosomes within the nucleus of each cell. Genes act by controlling the manufacture of enzymes, which are necessary for all the chemical reactions that occur within the cell. When body cells divide by the process of mitosis, the DNA that makes up the chromosomes is duplicated and distributed to the daughter cells so that each daughter cell gets exactly the same kind and number of chromosomes as were in the original cell. Each chromosome (aside from the Y chromosome, which determines sex) may carry thousands of genes, and each gene carries the code for a specific trait (characteristic). These traits constitute the physical, biochemical, and physiologic makeup of every cell in the body.

In humans, every cell except the sex cells contains 46 chromosomes. These chromosomes exist in pairs. One member of each pair was received at the time of fertilization from the father of the offspring, and one was received from the mother. These paired chromosomes, except for the pair that determines sex, are alike in size and appearance and carry genes for the same traits. Thus, the genes for each trait exist in pairs.

Dominant and Recessive Genes
Another of Mendel's discoveries was that genes can be either dominant or recessive. A ***dominant*** gene is one that expresses its effect in the cell re-

The Human Genome Project

As techniques for studying the exact structure of DNA advanced, scientists began to consider the possibility of identifying every single gene in the human cell. The Human Genome Project was under way by the late 1980s. Scientists from around the world are currently working together to produce a complete map of the human chromosomes. By the end of the project, the exact location and the chemical code of all 100,000 human genes will be known.

So far, about 4000 genes for specific functions have been found. Some of these are genes for hereditary disorders, including cystic fibrosis, Huntington's disease, a form of muscular dystrophy, and a hereditary kidney disease. Areas of specific chromosomes that are involved in some cancers of the breast, ovary, and colon and in some forms of leukemia have been pinpointed. However, it is important to note that the development of cancer is very complex, involving not only specific genes but also gene interactions and environmental factors.

gardless of whether the gene at the same site on the matching chromosome is the same as or different from the dominant gene. The gene need be received from only one parent to be expressed in the offspring. The effect of a ***recessive*** gene will not be evident unless the gene at that site in the matching chromosome of the pair is also recessive. Thus, a recessive trait will appear only if the recessive genes for that trait are received from both parents. For example, genes for dark eyes are dominant, whereas genes for light eyes are recessive. Light eyes will appear in the offspring only if genes for light eyes are received from both parents.

Clearly, a recessive gene will not be expressed if it is present in the cell together with a dominant gene. However, it can be passed on to offspring and may thus appear in future generations. An individual who shows no evidence of a trait but has a recessive gene for that trait is described as a ***carrier*** of the gene.

Distribution of Chromosomes to Offspring
The reproductive cells (ova and spermatozoa) are produced by a special process of cell division called ***meiosis.*** This process divides the chromosome

number in half so that each reproductive cell has 23 chromosomes. Moreover, the division occurs in such a way that each cell receives one member of each chromosome pair that was present in the original cell. The separation occurs at random, meaning that either member of the original pair may be included in a given germ cell. Thus, the maternal and paternal sets of chromosomes get mixed up and redistributed at this time, leading to increased variety within the population.

Sex Determination

The two chromosomes that determine the sex of the offspring, unlike the other 22 pairs of chromosomes, are not matched in size and appearance. The female X chromosome is larger than most other chromosomes and carries genes for other characteristics in addition to that for sex. The male Y chromosome is smaller than other chromosomes and mainly determines sex. A female has two X chromosomes in each body cell; a male has one X and one Y. By the process of meiosis, each male sperm cell receives either an X or a Y chromosome, whereas every egg cell can have only an X chromosome. If a sperm with an X chromosome fertilizes an ovum, the resulting infant will be female; if a sperm with a Y chromosome fertilizes an ovum, the resulting infant will be male (Fig. 25-1).

SEX-LINKED TRAITS

Any trait that is carried on a sex chromosome is said to be *sex-linked*. Because the Y chromosome carries few traits aside from sex determination, most sex-linked traits are carried on the X chromosome and are described as *X-linked*. Examples are hemophilia, certain forms of baldness, and red–green color blindness. Sex-linked traits appear almost exclusively in males. The reason for this is that most of these traits are recessive, and if a recessive gene is located on the X chromosome in a male it cannot be masked by a matching dominant gene. Thus, a male who has only one recessive gene for the trait will exhibit the characteristic, whereas a female must have two recessive genes to show the trait.

▶ Hereditary Traits

Some observable hereditary traits are skin, eye, and hair color and facial features. Also influenced by genetics are less clearly defined traits such as weight, body build, life span, and susceptibility to disease.

Some human traits, including the traits involved in many genetic diseases, are determined by a single pair of genes; most, however, are the result of two or more gene pairs acting together in what is termed *multifactorial inheritance.* This accounts for the wide range of variations within populations in such characteristics as coloration, height, and weight, all of which are determined by more than one pair of genes.

Gene Expression

The effect or expression of a gene may be influenced by a variety of factors, including the sex of the individual and the presence of other genes. For example, the genes for certain types of baldness and certain types of color blindness may be inherited by either males or females, but the traits appear mostly in males under the effects of male sex hormone.

Environment also plays a part in the expression of genes. One inherits a potential for a given size, for example, but one's actual size is additionally influenced by such factors as nutrition, development, and general state of health. The same is true of life span and susceptibility to diseases.

Gene Mutation

As a rule, chromosomes replicate exactly during cell division. Occasionally, however, for reasons not yet clearly understood, the genes or chromosomes change. This change may be in a single gene or in the number of chromosomes. Alternatively, it may consist of chromosomal breakage, in which there is loss or rearrangement of gene fragments. Such changes are termed genetic *mutations.* Mutations may occur spontaneously or may be induced by some agent, such as ionizing radiation or chemicals, described as a *mutagenic agent.*

If the mutation occurs in an ovum or a sperm cell involved in reproduction, the altered trait will be inherited by the offspring. The vast majority of harmful mutations never are expressed because the affected fetus dies and is spontaneously aborted. Most remaining mutations are so inconsequential that they have no visible effect. Beneficial mutations, on the other hand, tend to survive and increase as a population evolves.

▶ Genetic Diseases

Any disorders that involve the genes may be said to be genetic, but they are not always hereditary, that

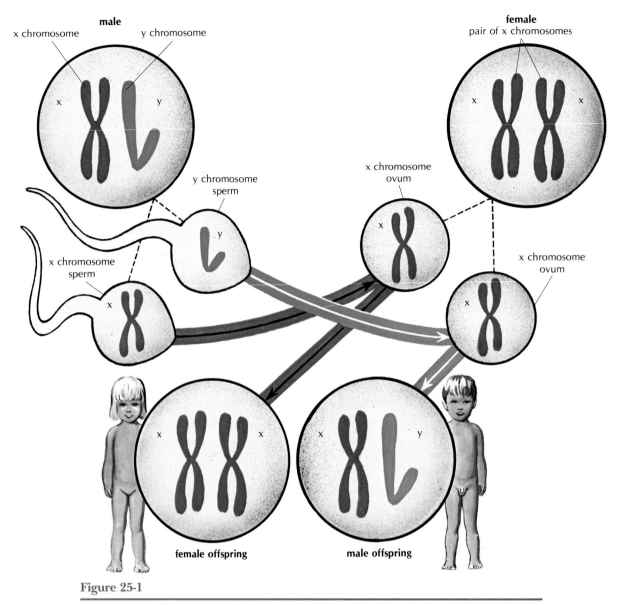

Figure 25-1

If an X chromosome from a male unites with an X chromosome from a female, the child is female; if a Y chromosome from a male unites with an X chromosome from a female, the child is male.

is, passed from parent to offspring in the reproductive cells. Noninherited genetic disorders may begin during maturation of the sex cells or even during development of the embryo.

Advances in genetic research have made it possible to identify the causes of many hereditary disorders and to develop methods of genetic screening. Persons who are "at risk" for having a child with a genetic disorder, as well as fetuses and newborns in

whom the presence of an abnormality may be suspected, can have chromosome studies done to identify genetic abnormalities.

Congenital Versus Hereditary Diseases

Before we discuss hereditary diseases, we need to distinguish them from other congenital diseases. To illustrate, let us assume that two infants are born within seconds in adjoining delivery rooms of the same hospital. It is noted that one infant has a club-foot, a condition called *talipes* (TAL-ih-pes); the second infant has a rudimentary extra finger attached to the fifth finger of each hand, a condition called *polydactyly* (pol-e-DAK-til-e). Are both conditions hereditary? both congenital? Is either hereditary? We can answer these questions by defining the key terms, *congenital* and *hereditary*. *Congenital* means present at the time of birth; *hereditary* means genetically transmitted or transmissible. Thus, one condition may be both congenital and hereditary; another, congenital yet not hereditary.

Hereditary conditions are usually evident at birth or soon thereafter. However, certain inherited disorders, such as adult polycystic kidney and Huntington's disease, a nervous disorder, do not manifest themselves until about midlife (40 to 50 years of age). In the case of our earlier examples, the club-foot is congenital, but not hereditary, having resulted from severe distortion of the developing extremities during intrauterine growth; the extra fingers are hereditary, a familial trait that appears in another relative, a grandparent perhaps, or a parent, and that is evident at the time of birth.

Although causes of congenital deformities and birth defects often are not known, in some cases they are known and can be avoided. For example, certain infections and toxins may be transmitted from the mother's blood by way of the placenta to the circulation of the fetus. Some of these cause serious developmental disorders in affected babies. German measles (rubella) is a contagious viral infection that is ordinarily a mild disease, but if maternal infection occurs during the first 3 or 4 months of pregnancy, the fetus has a 40% chance of developing defects of the eye (cataracts), of the ear (deafness), and of the brain and heart. Infection can be prevented by appropriate immunizations.

Ionizing radiation and various toxins may damage the genes, and sometimes the disorders they produce are transmissible. Environmental agents, such as mercury and some chemicals used in industry (for example, certain phenols and PCB), as well as some drugs, notably LSD, are known to disrupt genetic organization.

Intake of alcohol and cigarette smoking by a pregnant woman often cause growth retardation and low birth weight in her infant. Smaller than normal infants do not do as well as babies of average weight. Some congenital heart defects have been associated with a condition called *fetal alcohol syndrome.* Total abstinence from alcohol and cigarettes is strongly recommended during pregnancy.

Examples of Genetic Diseases

The best known example of a genetic disorder that is not hereditary is the most common form of **Down syndrome,** which results from the presence of an extra chromosome per cell. This abnormality arises during formation of a sex cell. The disorder is usually recognizable at birth by the afflicted child's distinctive facial features. The face is round, with close-set eyes that slant upward at the outside. The head is small and grows at an unusually slow rate. The nose is flat, the tongue is large and protruding, and the muscles and joints are lax. Intellectual function is impaired in Down children; however, the amount of skill they can gain depends on the severity of the disease and their family and school environments. Down syndrome is usually not inherited, although there is a hereditary form of the disorder. In most cases, both parents are normal, as are the child's siblings. The likelihood of having a baby with Down syndrome increases dramatically in women past the age of 35, and may be the result of defects in the germ cells, either male or female, due to age.

Most genetic diseases are *familial* or *hereditary;* that is, they are passed on from parent to child by way of the egg and sperm. In the case of a dominant trait, one parent usually carries the abnormal gene that gives rise to the disorder, as is the case in Huntington's disease. The disturbance appears in the parent and any of the offspring who receive the defective gene. If the trait is a recessive one, as is the case in the majority of inheritable disorders, a defective gene must come from each parent. Important inheritable diseases that are carried as recessive traits include diabetes mellitus, cystic fibrosis, and sickle cell anemia.

As stated earlier, genes control the production of specific enzymes. In the case of **PKU,** or *phenylketonuria* (fen-il-ke-to-NU-re-ah), the lack of a certain enzyme prevents the proper metabolism of *phenylalanine* (fen-il-AL-ah-nin), one of the common amino acids. The phenylalanine accumulates in the infant's blood. If the condition remains untreated, it leads to mental retardation before the age of 2 years. PKU screening is done routinely on newborn infants.

Sickle cell disease is described in Chapter 13 where it is stated that the disease is found almost exclusively in members of the black race. By contrast, *cystic fibrosis* is most common in whites; in fact, it is the most frequently inherited disease in white populations. It is characterized by excessively thickened secretions of the linings of the bronchi, the intestine, and the ducts of the pancreas. Such secretions in the ducts of the pancreas prevent the flow of pancreatic juice to the small intestine. Obstruction and blockage of these vital organs follows, in association with frequent respiratory infections, with uncontrollable intestinal losses, particularly of fats (and the vitamins these fats carry), as well as massive salt loss. Treatment includes oral administration of pancreatic enzymes and special pulmonary exercises. Cystic fibrosis was once fatal by the time of adolescence, but now, with appropriate care, life expectancies are extending into the third decade. Recently, the gene responsible for the disease was identified, raising hope of better diagnosis, treatment, and perhaps even correction of the defective gene that causes the disorder.

Another group of heritable muscle disorders is known collectively as the *progressive muscular atrophies.* *Atrophy* (AT-ro-fe) means wasting due to decrease in the size of a normally developed part. The absence of normal muscle movement in the infant proceeds within a few months to extreme weakness of the respiratory muscles, until ultimately the infant is unable to breathe adequately. Most afflicted babies die within several months. The name *floppy baby syndrome,* as the disease is commonly called, provides a vivid description of its effects.

Albinism is an inherited disorder that is carried by recessive genes. It is of particular interest because of the easily recognizable appearance it lends to the affected person. The skin and hair color are strikingly white and do not darken with age. The skin is abnormally sensitive to sunlight and may appear wrinkled. People with albinism are especially susceptible to skin cancer and to some severe visual disturbances, such as myopia (nearsightedness) and abnormal sensitivity to light (photophobia).

Other inherited disorders include *osteogenesis* (os-te-o-JEN-eh-sis) *imperfecta,* or *brittle bones,* in which multiple fractures may occur during and shortly after fetal life, and a disorder of skin, muscles, and bones called *neurofibromatosis* (nu-ro-fi-bro-mah-TO-sis). In the latter condition multiple masses, often on stalks (pedunculated), grow along nerves all over the body.

▶ Treatment and Prevention of Genetic Diseases

The list of genetic diseases is so lengthy (over 3000) that many pages of this book would be needed simply to enumerate them. Moreover, the list continues to grow as sophisticated research techniques and advances in biology make it clear that various diseases of previously unknown origin are genetic—some hereditary, others not. Can we identify which are inherited genetic disorders and which are due to environmental factors? Can we prevent the occurrence of any of them?

Genetic Counseling

It is possible to prevent genetic disorders in many cases and even to treat some of them. The most effective method of preventing genetic disease is through genetic counseling, a specialized field of health care. Genetic counseling centers use a team approach of medical, nursing, laboratory, and social service professionals to advise and care for the clients.

THE FAMILY HISTORY

An accurate and complete family history of both prospective parents is necessary. It includes information about relatives with respect to age, onset of

Correction of Genetic Disorders

Knowledge of disease genes and how they work leads to the possibility of testing people for genetic defects and perhaps treating these disorders. Some attempts have already been made to supplement failed genes with healthy ones. Two children with hereditary failure of the immune system have received genes for the enzyme they were missing, and these genes were found to be working. One woman with congenital high cholesterol had healthy genes inserted into her liver and has returned to normal. Healthy genes have been inserted into lung cells of a patient with cystic fibrosis to produce a protein that is defective in these cases. Such "gene therapy" is experimental and extremely costly, but marks a first step in the correction of genetic disorders.

a specific disease, health status, and cause of death. The families' ethnic origins may be relevant, since some genetic diseases predominate in certain ethnic groups. Hospital and physician records are studied, as are photographs of family members. The ages of the prospective parents are factors, as is parental or ancestral relationship (for example, marriage between first cousins). The complete, detailed family history, or tree, is called a *pedigree.* Pedigrees are used to determine the pattern of inheritance of a genetic disease within a family. They may also indicate whether a given member of the family is a carrier of the disease.

KARYOTYPE

Abnormalities in the number of chromosomes and some abnormalities within the chromosomes can be detected by analysis of the **karyotype** (KAR-e-o-tipe), a word derived from *karyon,* which means "nucleus." A karyotype is produced by growing cells in a special medium and arresting cell division at the stage called *metaphase.* The chromosomes, visible under the microscope, are then photographed, and the photographs are cut out and arranged in groups according to their size and form (Fig. 25-2). Abnormalities in the number and structure of chromosomes—called *chromosomal errors*—can thereby be detected. Special stains are used to reveal certain changes in fine structure within the chromosomes.

A technique that enables the geneticist to prepare a karyotype of an unborn fetus is **amniocentesis** (am-ne-o-sen-TE-sis). During this procedure, a small amount of the amniotic fluid that surrounds the fetus is withdrawn. Fetal skin cells in the amniotic fluid are removed, grown (cultured), and separated for study. A karyotype is prepared, and the chromosomes are examined. The amniotic fluid is also analyzed for biochemical abnormalities. With these methods almost 200 genetic diseases can be detected before birth.

A newer method for obtaining fetal cells for study involves sampling of the chorionic villi through the cervix. The chorionic villi are hairlike projections of the membrane that surrounds the embryo in early pregnancy. The method is called **chorionic villus sampling.** Samples may be taken between 8 and 10 weeks of pregnancy, and the cells obtained may be analyzed immediately. In contrast, amniocentesis cannot be done before the 14th to 16th week of pregnancy, and test results are not available for about 2 weeks.

COUNSELING THE PROSPECTIVE PARENTS

Armed with all the available pertinent facts, as well as with knowledge of the risk of recurrence, the counselor is equipped to inform the prospective parents of the possibility of their having genetically abnormal offspring. The couple may then elect to have no children, to have an adoptive family, to consider artificial insemination, to terminate the pregnancy, or to accept the risk.

Progress in Medical Treatment

The mental and physical ravages of many genetic diseases are largely preventable, provided the diseases are diagnosed and treated early in the individual's life. Some of these diseases respond well to dietary control. One such disease, called *maple syrup urine disease,* responds to very large doses of thiamine along with control of the intake of certain amino acids. The disastrous effects of Wilson's disease, in which abnormal accumulations of copper in the tissue cause tremor, rigidity, uncontrollable stagger, and finally extensive liver damage, can be prevented by a combination of dietary and drug therapy.

Phenylketonuria is perhaps the best-known example of dietary management of inherited disease. If the disorder is undiagnosed and untreated, 98% of affected patients will be severely mentally retarded by 10 years of age; in contrast, if the condition is diagnosed and treated before the baby reaches the age of 6 months and treatment is maintained until age 10, mental deficiency will be prevented or at least minimized. A simple blood test for PKU is now done routinely in hospitals throughout the United States. The test is done immediately after birth and should be repeated 24 to 48 hours after the infant has received protein.

Klinefelter's (KLINE-fel-terz) *syndrome,* which occurs in about 1 in 600 males, is a common cause of underdevelopment of the gonads with resulting infertility. Victims of this disorder have abnormal sex chromosome patterns, usually an extra X chromosome. Instead of the typical male XY pattern, the cells contain an XXY combination owing to failure of the sex chromosomes to separate during cell division. Treatment of this disorder includes the use of hormones and psychotherapy.

In the future we can anticipate greatly improved methods of screening, diagnosis, and treatment of genetic diseases. There have been reports of fetuses being treated with vitamins or hormones following prenatal diagnosis of a genetic disorder. Ahead lies

Figure 25-2

(*Top*) Metaphase spread of normal male chromosomes. The chromosomes have doubled and are ready to divide. The bands are produced by staining. (*Bottom*) Karyotype. The chromosomes are arranged in matching pairs according to size and other characteristics. (Courtesy of Wenda S. Long, Thomas Jefferson University, Philadelphia) the possibility of treating or correcting genetic disorders through genetic engineering—introducing genetically altered cells to produce missing factors such as enzymes or hormones, or even correcting faulty genes in the victim's cells. For the present, it is important to educate the public about the availability of screening methods for both parents and offspring. People should also be made aware of the damaging genetic effects of radiation, drugs, and other toxic substances.

SUMMARY

I. Genes
   A. Hereditary units
   B. Segments of DNA in chromosomes within nucleus
   C. Control manufacture of enzymes
   D. Types
      1. Dominant gene—always expressed
      2. Recessive gene—expressed only if gene received from both parents
         a. Carrier—person with recessive gene that is not apparent but can be passed to offspring

II. Chromosomes
   A. 46 in each human body cell (except sex cells)
   B. Pairs—matched in size and appearance; one member of each pair from each parent
   C. X and Y chromosomes—determine sex
   D. Meiosis—form of cell division
      1. Production of sex cells
      2. Halving of chromosome number from 46 to 23
      3. Random separation of maternal and paternal chromosomes into different daughter cells
   E. Sex chromosomes—females XX, males XY
      1. X chromosome—larger than most others; carries traits in addition to sex determination
      2. Sex-linked traits—appear mostly in males (e.g. hemophilia, baldness, red–green color blindness)

III. Hereditary traits—genes determine physical, biochemical, and physiologic characteristics of every cell
   A. Some traits determined by single gene pairs
   B. Most traits determined by multiple gene pairs (multifactorial inheritance)
      1. Produces a range of variations in a population
      2. Examples: height, weight, coloration, susceptibility to disease
   C. Factors affecting gene expression
      1. Sex
      2. Presence of other genes

      3. Environment
   D. Mutation
      1. Change in genes or chromosomes
      2. May be passed to offspring if occurs in germ cells
      3. Mutagenic agents
         a. Ionizing radiation
         b. Drugs
         c. Toxins
         d. Chemicals

IV. Genetic diseases—disorders involving genes
   A. Hereditary disorders—passed from parent to offspring in sex cells
   B. Congenital disorders—present at birth; may or may not be hereditary
   C. Examples of genetic diseases
      1. Down syndrome—results from extra chromosome
      2. PKU (phenylketonuria)—inability to metabolize phenylalanine
      3. Cystic fibrosis—common in white populations
      4. Progressive muscular atrophies
      5. Albinism—lack of pigment
      6. Others: Huntington's disease, sickle cell anemia, osteogenesis imperfecta, neurofibromatosis

V. Treatment and prevention
   A. Genetic counseling
      1. Family history to establish pedigree
      2. Karyotype—analysis of chromosomes
         a. Amniocentesis—withdrawal of amniotic fluid at 14–16 weeks of pregnancy
         b. Chorionic villus sampling—done at 8–10 weeks of pregnancy
      3. Counseling of prospective parents about risks
   B. Progress in medical treatment
      1. Dietary control—maple syrup urine disease, Wilson's disease, PKU
      2. Hormone therapy—Klinefelter's disease
      3. Fetal therapy
      4. Correction of faulty genes—experimental

QUESTIONS FOR STUDY AND REVIEW

1. How many chromosomes are there in a human body cell? in a human sex cell?
2. What process results in the distribution of chromosomes to the offspring?
3. What sex chromosomes are present in a male? in a female?
4. From which parent does a male child receive an X-linked gene?
5. Explain how two normal parents can give birth to a child with a recessive hereditary disease.
6. Explain the great variation in the color of skin, hair, and eyes in humans.
7. Explain how a hereditary disease can suddenly appear in a family with no history of the disease.
8. A baby is born with syphilis. Is this congenital or hereditary? Explain.
9. List several mutagenic agents.
10. Could a karyotype be used to diagnose Down syndrome? Klinefelter's syndrome?
11. How is an amniocentesis performed and what can be learned from it?
12. How is a pedigree used in genetic counseling?
13. What is PKU and how should this disease be treated?
14. What are some characteristics of albinism and what are the risks associated with this disorder?
15. What heritable disorder is most common among black persons and how does it manifest itself?
16. What is the most common heritable disease among white persons and what are some of its symptoms?

Suggestions for Further Study

Amenta PS: Histology and Human Anatomy, 6th ed. New York, Piccin, 1991

Beck WS: Hematology, 5th ed. Cambridge, MA, MIT Press, 1994

Bishop ML, Duben-Engelkirk JL, Fody EP: Clinical Chemistry, 2nd ed. Philadelphia, JB Lippincott, 1992

Burton G: Microbiology for the Health Sciences, 4th ed. Philadelphia, JB Lippincott, 1992

Cormack DH: Esssential Histology. Philadelphia, JB Lippincott, 1992

Carroll HJ, Oh MS: Water, Electrolyte and Acid-Base Metabolism, 2nd ed. Philadelphia, JB Lippincott, 1989

Diseases. Springhouse, PA, Springhouse Corporation, 1993

Dorland's Illustrated Medical Dictionary, 28th ed. Philadelphia, WB Saunders, 1994

Fischbach FT: A Manual of Laboratory and Diagnostic Tests, 4th ed. Philadelphia, JB Lippincott, 1992

Foley JF et al: Current Therapy in Cancer. Philadelphia, WB Saunders, 1994

Ganong WF: Review of Medical Physiology, 16th ed. San Mateo CA, Appleton and Lange, 1993

Gosling JA, Harris PF, Whitmore IV, Humpherson JR, Willan PLT: Atlas of Human Anatomy. Philadelphia, JB Lippincott, 1987

Grant's Atlas of Anatomy, 9th ed. Baltimore, Williams and Wilkins, 1991

Guyton AC: Basic Neuroscience: Anatomy and Physiology, 2nd ed. Philadelphia, WB Saunders, 1991

Guyton AC: Medical Physiology, 8th ed. Philadelphia, WB Saunders, 1990

Hamilton H, Rose MB (eds): Diagnostics, 2nd ed. Springhouse, PA, Springhouse Corporation, 1992

Haskell C: Cancer Treatment, 3rd ed. Philadelphia, WB Saunders, 1990

Hole JW: Essentials of Human Anatomy and Physiology, 5th ed. Boston, WC Brown, 1994

Hollinshead WH, Rosse C: Textbook of Anatomy, 4th ed. Philadelphia, JB Lippincott, 1985

Holman SR: Essentials of Nutrition for the Health Professions. Philadelphia, JB Lippincott, 1987

Johnson JY, Smith-Temple AJ: Nurse's Guide to Clinical Procedures. Philadelphia, JB Lippincott, 1994

Kozier B: Fundamentals of Nursing, 4th ed. Reading, MA, Addison-Wesley, 1991

Lotspeich-Steininger CA, Steine-Martin EA, Koepke JA: Clinical Hematology. Philadelphia, JB Lippincott, 1992

Melloni J et al: Melloni's Illustrated Review of Human Anatomy. Philadelphia, JB Lippincott, 1988

Moore K: Clinically Oriented Anatomy, 3rd ed. Baltimore, Williams and Wilkins, 1992

Morton PJ et al: Nurses Clinical Guide to Health Assessment. Springhouse, PA, Springhouse Corporation, 1990

Pennington J: Bowes and Church's Food Values of Portions Commonly Used, 16th ed. Philadelphia, JB Lippincott, 1993

Procedures. Springhouse, PA, Springhouse Corporation, 1992

Rakel RE: Conn's Current Therapy, 1994. Philadelphia, WB Saunders (published yearly)

Rakel RE: Textbook of Family Practice, 4th ed. Philadelphia, WB Saunders, 1990

Rodman MJ: Clinical Pharmacology in Nursing, 2nd ed. Philadelphia, JB Lippincott, 1984

Ross MH et al: Histology: A Text and Atlas, 3rd ed. Baltimore, Williams and Wilkins, 1994

Rothwell NV: Understanding Genetics, A Molecular Approach. New York, John Wiley and Sons, 1993

Rubin E, Farber J: Essential Pathology, 2nd ed. Philadelphia, JB Lippincott, 1995

Scherer JC: Introductory Clinical Pharmacology, 4th ed. Philadelphia, JB Lippincott, 1992

Taylor C, Lillis CA, LeMone P: Fundamentals of Nursing, 2nd ed. Philadelphia, JB Lippincott, 1993

Tortora GJ, Grabowski SR: Principles of Anatomy and Physiology, 7th ed. New York, Harper Collins, 1993

Volk WA, Benjamin DC, Kadner R, Parsons JT: Essentials of Medical Microbiology, 5th ed. Philadelphia, JB Lippincott, 1995

Williams PL et al (eds): Gray's Anatomy, 37th ed. New York, Churchill Livingstone, 1989

Glossary

Abduction (ab-DUK-shun) Movement away from the midline; turning outward

Abortion (ah-BOR-shun) Loss of an embryo or fetus before the 20th week of pregnancy

Abscess (AB-ses) Area of tissue breakdown; a localized space in the body containing pus and liquefied tissue

Absorption (ab-SORP-shun) Transfer of digested food molecules from the digestive tract into the circulation

Accommodation (ah-kom-o-DA-shun) Coordinated changes in the eye that enable one to focus on near and far objects

Acetylcholine (as-e-til-KO-lene) **(Ach)** Neurotransmitter; released at synapses within the nervous system and at the neuromuscular junction

Acid (AH-sid) Substance that can donate a hydrogen ion to another substance

Acidosis (as-ih-DO-sis) Condition that results from a decrease in the pH of body fluids

Acquired immunodeficiency syndrome (AIDS) Viral disease that attacks the immune system, specifically the T helper lymphocytes with CD4 receptors

Acrosome (AK-ro-some) Caplike structure over the head of the sperm cell that helps the sperm to penetrate the ovum

ACTH See Adrenocorticotropic hormone

Actin (AK-tin) One of the two contractile proteins in muscle cells, the other being myosin

Action potential Sudden change in the electric charge on a cell membrane; nerve impulse

Active transport Movement of a substance into or out of a cell in a direction opposite that in which it would normally flow by diffusion

Acute (ah-KUTE) Referring to a severe but short-lived disease or condition

Adduction (ad-DUK-shun) Movement toward the midline; turning outward

Adenosine triphosphate (ah-DEN-o-sene tri-FOS-fate) **(ATP)** Energy-storing compound found in all cells

ADH See Antidiuretic hormone

Adhesion (ad-HE-zhun) Holding together of two surfaces or parts; band of connective tissue between parts that are normally separate; molecular attraction between contacting bodies

Adipose (AD-ih-pose) Referring to a type of connective tissue that stores fat

Adrenal (ah-DRE-nal) Endocrine gland located above the kidney; suprarenal gland

Adrenaline (ah-DREN-ah-lin) See Epinephrine

Adrenocorticotropic (ah-dre-no-kor-tih-ko-TRO-pik) **hormone (ACTH)** Hormone produced by the pituitary that stimulates the adrenal cortex

Aerobic (air-O-bik) Requiring oxygen

Afferent (AF-fer-ent) Carrying toward a given point, such as a sensory neuron that carries nerve impulses toward the central nervous system

Agglutination (ah-glu-tih-NA-shun) Clumping of cells due to an antigen–antibody reaction

AIDS See Acquired immunodeficiency syndrome

Albumin (al-BU-min) Protein in blood plasma and other body fluids; helps maintain the osmotic pressure of the blood

Albuminuria (al-bu-mih-NU-re-ah) Presence of albumin in the urine, usually as a result of a kidney disorder

Aldosterone (al-DOS-ter-one) Hormone released by the adrenal cortex that promotes the reabsorption of sodium and water in the kidneys

Alkalosis (al-kah-LO-sis) Condition that results from an increase in the pH of body fluids

Allergen (AL-er-jen) Substance that causes hypersensitivity; substance that induces allergy

Allergy (AL-er-je) Tendency to react unfavorably to a certain substance that is normally harmless to most people; hypersensitivity

Alveolus (al-VE-o-lus) Small sac or pouch; usually a tiny air sac in the lungs through which gases are exchanged between the outside air and the blood; tooth socket; pl., alveoli

Amino (ah-ME-no) **acid** Building block of protein

Amniocentesis (am-ne-o-sen-TE-sis) Removal of fluid and cells from the amniotic sac for prenatal diagnostic tests

Amniotic (am-ne-OT-ik) **sac** Fluid-filled sac that surrounds and cushions the developing fetus

Amphiarthrosis (am-fe-ar-THRO-sis) Slightly movable joint

Anabolism (ah-NAB-o-lizm) Metabolic building of simple compounds into more complex substances needed by the body

Anaerobic (an-air-O-bik) Not requiring oxygen

Analgesic (an-al-JE-zik) Relieving pain; a pain-relieving agent that does not cause loss of consciousness

Anastomosis (ah-nas-to-MO-sis) Communication between two structures such as blood vessels

Anatomy (ah-NAT-o-me) Study of body structure

Anemia (ah-NE-me-ah) Abnormally low level of hemoglobin or red cells in the blood, resulting in inadequate delivery of oxygen to the tissues

Anesthesia (an-es-THE-ze-ah) Loss of sensation, particularly of pain

Aneurysm (AN-u-rizm) Bulging sac in the wall of a vessel

Angiotensin (an-je-o-TEN-sin) Substance formed in the blood by the action of the enzyme renin from the kidneys. It increases blood pressure by causing constriction of the blood vessels and stimulating the release of aldosterone from the adrenal cortex.

Angina (an-JI-nah) Severe choking pain; disease or condition producing such pain. Angina pectoris is suffocating pain in the chest usually caused by lack of oxygen supply to the heart muscle.

Anion (AN-i-on) Negatively charged particle (ion)

Anorexia (an-o-REK-se-ah) Loss of appetite. Anorexia nervosa is a psychologic condition in which a person may become seriously, even fatally, weakened from lack of food.

Anoxia (ah-NOK-se-ah) See Hypoxia

ANS See Autonomic nervous system

Antagonist (an-TAG-o-nist) Muscle that has an action opposite that of a given movement; substance that opposes the action of another substance

Anterior (an-TE-re-or) Toward the front or belly surface; ventral

Antibody (AN-te-bod-e) Substance produced in response to a specific antigen

Antidiuretic (an-ti-di-u-RET-ik) **hormone (ADH)** Hormone released from the posterior pituitary gland that increases the reabsorption of water in the kidneys, thus decreasing the volume of urine excreted

Antigen (AN-te-jen) **(Ag)** Foreign substance that produces an immune response

Antiserum (an-te-SE-rum) Serum containing antibodies; may be given for the purpose of providing passive immunity

Aorta (a-OR-tah) Large artery that carries blood out of the left ventricle of the heart

Aphasia (ah-FA-ze-ah) Loss or defect in language communication. Loss of the ability to speak or write is *expressive aphasia*; loss of understanding of written or spoken language is *receptive aphasia*.

Aponeurosis (ap-o-nu-RO-sis) Broad sheet of fibrous connective tissue that attaches muscle to bone or to other muscle

Appendicular (ap-en-DIK-u-lar) **skeleton** Part of the skeleton that includes the bones of the upper extremities, lower extremities, shoulder girdle, and hips

Arachnoid (ah-RAK-noyd) Middle layer of the meninges

Areolar (ah-RE-o-lar) Referring to loose connective tissue

Arrhythmia (ah-RITH-me-ah) Abnormal rhythm of the heartbeat

Arteriole (ar-TE-re-ole) Vessel between a small artery and a capillary

Arteriosclerosis (ar-te-re-o-skle-RO-sis) Hardening of the arteries

Artery (AR-ter-e) Vessel that carries blood away from the heart

Arthritis (arth-RI-tis) Inflammation of the joints

Ascites (ah-SI-teze) Abnormal collection of fluid in the abdominal cavity

Asepsis (a-SEP-sis) Condition in which no pathogens are present; adj., aseptic

Astigmatism (ah-STIG-mah-tizm) Visual defect due to an irregularity in the curvature of the cornea or the lens

Ataxia (ah-TAK-se-ah) Lack of muscular coordination; irregular muscular action

Atherosclerosis (ath-er-o-skle-RO-sis) Hardening of the arteries due to the deposit of yellowish, fat-like material in the lining of these vessels

Atom (AT-om) Fundamental unit of a chemical element

Atopic dermatitis (ah-TOP-ik der-mah-TI-tis) Skin condition that may involve redness, blisters, pimples, scaling, and crusting; eczema

ATP See Adenosine triphosphate

Atrioventricular (a-tre-o-ven-TRIK-u-lar) **(AV) node** Part of the conduction system of the heart

Atrium (A-tre-um) One of the two upper chambers of the heart; adj., atrial

Atrophy (AT-ro-fe) Wasting or decrease in size of a part

Attenuated (ah-TEN-u-a-ted) Weakened

Autoimmunity (aw-to-ih-MU-nih-te) Abnormal reactivity to one's own tissues

Autonomic (aw-to-NOM-ik) **nervous system (ANS)** The part of the nervous system that controls smooth muscle, cardiac muscle, and glands; motor portion of the visceral or involuntary nervous system

AV node See Atrioventricular node

Axial (AK-se-al) **skeleton** The part of the skeleton that includes the skull, spinal column, ribs, and sternum

Axilla (ak-SIL-ah) Hollow beneath the arm where it joins the body; armpit

Axon (AK-son) Fiber of a neuron that conducts impulses away from the cell body

Bacillus (bah-SIL-us) Rod-shaped bacterium; pl., bacilli (bah-SIL-i)

Bacterium (bak-TE-re-um) Type of microorganism; pl., bacteria (bak-TE-re-ah)

Bacteriostasis (bak-te-re-o-STA-sis) Condition in which bacterial growth is inhibited but the organisms are not killed

Basal ganglia (BA-sal GANG-le-ah) Gray masses in the lower part of the forebrain that aid in muscle coordination

Base Substance that can accept a hydrogen ion (H+); substance that donates a hydroxide ion (OH−)

Basophil (BA-so-fil) Granular white blood cell that shows large, dark blue cytoplasmic granules when stained with basic stain

B cell Agranular white blood cell that produces antibodies in response to an antigen; B lymphocyte

Benign (be-NINE-) Describing a tumor that does not spread; not recurrent nor becoming worse

Bile Substance produced in the liver that emulsifies fats

Biopsy (BI-op-se) Removal of tissue or other material from the living body for examination, usually under the microscope

Blood urea nitrogen (BUN) Amount of nitrogen from urea in the blood; test to evaluate kidney function

Bradycardia (brad-e-KAR-de-ah) Heart rate of less than 60 beats per minute

Brain stem Portion of the brain that connects the cerebrum with the spinal cord; contains the midbrain, pons, and medulla oblongata

Bronchiole (BRONG-ke-ole) Microscopic terminal branch of a bronchus

Bronchus (BRONG-kus) Large air passageway in the lung; pl., bronchi (BRONG-ki)

Buffer (BUF-er) Substance that prevents sharp changes in the pH of a solution

BUN See Blood urea nitrogen

Bursa (BER-sah) Small, fluid-filled sac found in an area subject to stress around bones and joints; pl., bursae (BER-se)

Bursitis (ber-SI-tis) Inflammation of a bursa

Cancer (KAN-ser) Tumor that spreads to other tissues; a malignant neoplasm

Capillary (CAP-ih-lar-e) Microscopic vessel through which exchanges take place between the blood and the tissues

Carbohydrate (kar-bo-HI-drate) Simple sugar or compound made from simple sugars linked together, such as starch or glycogen

Carbon dioxide (di-OX-ide) **(CO₂)** The gaseous waste product of cellular metabolism

Carcinogen (kar-SIN-o-jen) Cancer-causing substance

Carcinoma (kar-sih-NO-mah) Malignant growth of epithelial cells; a form of cancer

Cardiopulmonary resuscitation (CPR) Method to restore heartbeat and breathing by mouth-to-mouth resuscitation and closed chest cardiac massage

Caries (KA-reze) Tooth decay

Carrier Individual who has a gene that is not expressed but that can be passed to offspring

Cartilage (KAR-tih-lij) Type of hard connective tissue

CAT See Computed tomography

Catabolism (kah-TAB-o-lizm) Metabolic breakdown of substances into simpler substances; includes the digestion of food and the oxidation of nutrient molecules for energy

Cataract (KAT-ah-rakt) Opacity of the eye lens or lens capsule

Catheter (KATH-eh-ter) Tube that can be inserted into a vessel or cavity; may be used to remove fluid, such as urine or blood; v., catheterize

Cation (KAT-i-on) Positively charged particle (ion)

Cecum (SE-kum) Small pouch at the beginning of the large intestine

Cell Basic unit of life

Cell membrane Outer covering of a cell; regulates what enters and leaves cell; plasma membrane

Cellular respiration Series of reactions by which nutrients are oxidized for energy within the cell

Central nervous system (CNS) Part of the nervous system that includes the brain and spinal cord

Centrifuge (SEN-trih-fuje) Instrument for separating materials in a mixture based on density

Centriole (SEN-tre-ole) Rod-shaped body near the nucleus of a cell; functions in cell division

Cerebellum (ser-eh-BEL-um) Small section of the brain located under the cerebral hemispheres; functions in coordination, balance, and muscle tone

Cerebral (SER-e-bral) **cortex** The very thin outer layer of gray matter on the surface of the cerebral hemispheres

Cerebrospinal (ser-e-bro-SPI-nal) **fluid (CSF)** Fluid that circulates in and around the brain and spinal cord

Cerebrovascular (ser-e-bro-VAS-ku-lar) **accident (CVA)** Condition involving bleeding from the brain or obstruction of blood flow to brain tissue, usually as a result of hypertension or atherosclerosis; stroke

Cerebrum (SER-e-brum) Largest part of the brain; composed of the cerebral hemispheres

Cerumen (seh-RU-men) Earwax; adj., Ceruminous (seh-RU-min-us)

Cervix (SER-vix) Constricted portion of an organ or part, such as the lower portion of the uterus; neck.; adj., cervical

Chemoreceptor (ke-mo-re-SEP-tor) Receptor that detects chemical changes

Chemotherapy (ke-mo-THER-ah-pe) Treatment of a disease by administration of a chemical agent

Chlamydia (klah-MID-e-ah) A type of very small bacterium that can exist only within a living cell; members of this group cause inclusion conjunctivitis, trachoma, sexually transmitted diseases, and respiratory diseases

Cholesterol (ko-LES-ter-ol) An organic fat-like compound found in animal fat, bile, blood, myelin, liver, and other parts of the body

Choroid (KO-royd) Pigmented middle layer of the eye

Chromosome (KRO-mo-some) Dark-staining, threadlike body in the nucleus of a cell; contains genes that determine hereditary traits

Chronic (KRON-ik) Referring to a disease that is not severe but is continuous or recurring

Chyle (kile) Milky-appearing fluid absorbed into the lymphatic system from the small intestine. It consists of lymph and droplets of digested fat.

Chyme (kime) Mixture of partially digested food, water, and digestive juices that forms in the stomach

Cilia (SIL-e-ah) Hairs or hairlike processes such as eyelashes or microscopic extensions from the surface of a cell; sing., cilium

Circumduction (ser-kum-DUK-shun) Circular movement at a joint

Cirrhosis (sih-RO-sis) Chronic disease, usually of the liver, in which active cells are replaced by inactive scar tissue

CNS See Central nervous system

Coagulation (ko-ag-u-LA-shun) Clotting, as of blood

Coccus (KOK-us) A round bacterium; pl., cocci (KOK-si)

Cochlea (KOK-le-ah) Coiled portion of the inner ear that contains the organ of hearing

Collagen (KOL-ah-jen) Flexible white protein that gives strength and resilience to connective tissue such as bone and cartilage

Colloidal (kol-OYD-al) **suspension** Mixture in which suspended particles do not dissolve but remain distributed in the solvent because of their small size (*e.g.,* cytoplasm); colloid

Colon (KO-lon) Main portion of the large intestine

Complement (KOM-ple-ment) Group of blood proteins that help antibodies to destroy foreign cells

Compound Substance composed of two or more chemical elements

Computed tomography (to-MOG-rah-fe) **(CT)** Imaging method in which multiple x-ray views taken from different angles are analyzed by computer to show a cross section of an area; used to detect tumors and other abnormalities; also called *computed axial tomography* (*CAT*)

Congenital (con-JEN-ih-tal) Present at birth

Conjunctiva (kon-junk-TI-vah) Membrane that lines the eyelid and covers the anterior part of the sclera (white of the eye)

Contraception (con-trah-SEP-shun) Prevention of fertilization of an ovum or implantation of a fertilized ovum; birth control

Cornea (KOR-ne-ah) Clear portion of the sclera that covers the front of the eye

Coronary (KOR-on-ar-e) Referring to the heart or to the arteries supplying blood to the heart

Corpus callosum (kal-O-sum) Thick bundle of myelinated nerve cell fibers, deep within the brain, that carries nerve impulses from one cerebral hemisphere to the other

Corpus luteum (LU-te-um) Yellow body formed from ovarian follicle after ovulation; produces progesterone

Cortex (KOR-tex) Outer layer of an organ such as the brain, kidney, or adrenal gland

Covalent (KO-va-lent) **bond** Chemical bond formed by the sharing of electrons between atoms

CPR See Cardiopulmonary resuscitation

CSF See Cerebrospinal fluid

CT See Computed tomography

Cutaneous (ku-TA-ne-us) Referring to the skin

Cyanosis (si-ah-NO-sis) Bluish color of the skin and mucous membranes resulting from insufficient oxygen in the blood

Cystitis (sis-TI-tis) Inflammation of the urinary bladder

Cytology (si-TOL-o-je) Study of cells

Cytoplasm (SI-to-plazm) Substance that fills the cell and holds the organelles

Defecation (def-e-KA-shun) Act of eliminating undigested waste from the digestive tract

Degeneration (de-jen-er-A-shun) Breakdown, as from age, injury, or disease

Deglutition (deg-lu-TISH-un) Act of swallowing

Dehydration (de-hi-DRA-shun) Excessive loss of body fluid

Dendrite (DEN-drite) Fiber of a neuron that conducts impulses toward the cell body

Deoxyribonucleic (de-OK-se-ri-bo-nu-kle-ik) **acid (DNA)** Genetic material of the cell; makes up the chromosomes in the nucleus of the cell

Dermatitis (der-mah-TI-tis) Inflammation of the skin

Dermis (DER-mis) True skin; deeper part of the skin

Dextrose (DEK-strose) Glucose; simple sugar

Diabetes mellitus (di-ah-BE-teze mel-LI-tus) Disease in which glucose is not oxidized adequately in the tissues for energy because of insufficient insulin

Diagnosis (di-ag-NO-sis) Identification of an illness

Dialysis (di-AL-ih-sis) Method for separating molecules in solution based on differences in their ability to pass through a semipermeable membrane; method for removing nitrogen waste products from the body, as by hemodialysis or peritoneal dialysis

Diaphragm (DI-ah-fram) Dome-shaped muscle under the lungs that flattens during inhalation; separating membrane or structure

Diaphysis (di-AF-ih-sis) Shaft of a long bone

Diarthrosis (di-ar-THRO-sis) Freely movable joint; synovial joint

Diastole (di-AS-to-le) Relaxation phase of the cardiac cycle; adj., diastolic (di-as-TOL-ik)

Diencephalon (di-en-SEF-ah-lon) Region of the brain between the cerebral hemispheres and the midbrain; contains the thalamus, hypothalamus, and pituitary gland

Diffusion (dih-FU-zhun) Movement of molecules from a region where they are in higher concentration to a region where they are in lower concentration

Digestion (di-JEST-yun) Process of breaking down food into absorbable particles

Dilation (di-LA-shun) Widening of a part, such as the pupil of the eye, a blood vessel, or the uterine cervix; dilatation

Disease Illness; abnormal state in which part or all of the body does not function properly

Distal (DIS-tal) Farther from the origin of a structure or from a given reference point

DNA See Deoxyribonucleic acid

Dominant (DOM-ih-nant) Referring to a gene that is always expressed if present

Dorsal (DOR-sal) Toward the back; posterior

Duct Tube or vessel

Ductus deferens (DEF-er-enz) Tube that carries sperm cells from the testis to the urethra; vas deferens

Duodenum (du-o-DE-num) First portion of the small intestine

Dura mater (DU-rah MA-ter) Outermost layer of the meninges

Dyspnea (DISP-ne-ah) Difficult or labored breathing

ECG See Electrocardiograph

Echocardiograph (ek-o-KAR-de-o-graf) Instrument to study the heart by means of ultrasound; the record produced is an echocardiogram

Eclampsia (eh-KLAMP-se-ah) Serious and sometimes fatal condition involving kidney failure and convulsions that can develop from preeclampsia (toxemia of pregnancy)

Eczema (EK-ze-mah) See atopic dermatitis

Edema (eh-DE-mah) Accumulation of fluid in the tissue spaces

EEG See Electroencephalograph

Effector (ef-FEK-tor) Muscle or gland that responds to a stimulus; effector organ

Efferent (EF-fer-ent) Carrying away from a given point, such as a motor neuron that carries nerve impulses away from the central nervous system

Effusion (eh-FU-zhun) Escape of fluid into a cavity or space; the fluid itself

Ejaculation (e-jak-u-LA-shun) Expulsion of semen through the urethra

EKG See Electrocardiograph

Electrocardiograph (e-lek-tro-KAR-de-o-graf) **(EKG, ECG)** Instrument to study the electric activity of the heart; record made is an electrocardiogram.

Electroencephalograph (e-lek-tro-en-SEF-ah-lo-graf) **(EEG)** Instrument used to study electric activity of the brain; record made is an electroencephalogram

Electrolyte (e-LEK-tro-lite) Compound that forms ions in solution; substance that conducts an electric current in solution

Electron (e-LEK-tron) Negatively charged particle located in an orbital outside the nucleus of an atom

Element (EL-eh-ment) One of the substances from which all matter is made; substance that cannot be decomposed into a simpler substance

Embolism (EM-bo-lizm) The condition of having an embolus

Embolus (EM-bo-lus) Blood clot or other obstruction in the circulation

Embryo (EM-bre-o) Developing offspring during the first 2 months of pregnancy

Emesis (EM-eh-sis) Vomiting

Emphysema (em-fih-SE-mah) Pulmonary disease characterized by dilation and destruction of the alveoli

Emulsify (e-MUL-sih-fi) To break up fats into small particles; n., emulsification

Endocardium (en-do-KAR-de-um) Membrane that lines the heart chambers and covers the valves

Endocrine (EN-do-krin) Referring to a gland that secretes directly into the bloodstream

Endometrium (en-do-ME-tre-um) Lining of the uterus

Endoplasmic reticulum (en-do-PLAS-mik re-TIK-u-lum) **(ER)** Network of membranes in the cytoplasm of a cell

Endosteum (en-DOS-te-um) Thin membrane that lines the marrow cavity of a bone

Endothelium (en-do-THE-le-um) Epithelium that lines the heart, blood vessels, and lymphatic vessels

Enzyme (EN-zime) Organic catalyst; speeds the rate of a reaction but is not changed in the reaction

Eosinophil (e-o-SIN-o-fil) Granular white blood cell that shows beadlike, bright pink cytoplasmic granules when stained with acid stain; acidophil

Epicardium (ep-ih-KAR-de-um) Membrane that forms the outermost layer of the heart wall and is continuous with the lining of the pericardium; visceral pericardium

Epidemic (ep-ih-DEM-ik) Occurrence of a disease among many people in a given region at the same time

Epidermis (ep-ih-DER-mis) Outermost layer of the skin

Epiglottis (ep-e-GLOT-is) Leaf-shaped cartilage that covers the larynx during swallowing

Epilepsy (EP-ih-lep-se) Chronic disorder of the nervous system involving abnormal electrical activity of the brain; characterized by seizures of varying severity

Epimysium (ep-ih-MIS-e-um) Sheath of fibrous connective tissue that encloses a muscle

Epinephrine (ep-ih-NEF-rin) Neurotransmitter and hormone; released from neurons of the sympathetic nervous system and from the adrenal medulla; adrenaline

Epiphysis (eh-PIF-ih-sis) End of a long bone

Episiotomy (eh-piz-e-OT-o-me) Cutting of the perineum between the vaginal opening and the anus to reduce the tearing of tissue in childbirth

Epithelium (ep-ih-THE-le-um) One of the four main types of tissue; forms glands, covers surfaces, and lines cavities; adj., epithelial

ER See Endoplasmic reticulum

Eruption (e-RUP-shun) Raised skin lesion; rash

Erythema (er-eh-THE-mah) Redness of the skin

Erythrocyte (eh-RITH-ro-site) Red blood cell

Erythropoietin (eh-rith-ro-POY-eh-tin) Hormone released from the kidney that stimulates the production of red blood cells in the bone marrow

Esophagus (eh-SOF-ah-gus) Tube that carries food from the throat to the stomach

Estrogen (ES-tro-jen) Group of female sex hormones that promotes development of the uterine lining and maintains secondary sex characteristics

Etiology (e-te-OL-o-je) Study of the cause of a disease or the theory of its origin

Eustachian (u-STA-shun) **tube** Tube that connects the middle ear cavity to the throat; auditory tube

Excretion (eks-KRE-shun) Removal and elimination of metabolic waste products from the blood

Exocrine (EK-so-krin) Referring to a gland that secretes through a duct

Extracellular (EK-strah-sel-u-lar) Outside the cell

Fascia (FASH-e-ah) Band or sheet of fibrous connective tissue

Fascicle (FAS-ih-kl) Small bundle, as of muscle cells or nerve cell fibers

Feces (FE-seze) Waste material discharged from the large intestine; excrement; stool

Feedback Return of information into a system so that it can be used to regulate that system

Fertilization (fer-til-ih-ZA-shun) Union of an ovum and a spermatozoon

Fetus (FE-tus) Developing offspring from the third month of pregnancy until birth

Fever (FE-ver) Abnormally high body temperature

Fibrin (FI-brin) Blood protein that forms a blood clot

Filtration (fil-TRA-shun) Movement of material through a semipermeable membrane under mechanical force

Fissure (FISH-ure) Deep groove

Flaccid (FLAK-sid) Flabby, limp, soft

Flagellum (flah-JEL-lum) Long whiplike extension from a cell used for locomotion; pl., flagella

Flatus (FLA-tus) Gas in the digestive tract

Flexion (FLEK-shun) Bending motion that decreases the angle between bones at a joint

Follicle (FOL-lih-kl) Sac or cavity, such as the ovarian follicle or hair follicle

Follicle-stimulating hormone (FSH) Hormone produced by the anterior pituitary that stimulates development of ova in the ovary and spermatozoa in the testes

Fontanelle (fon-tah-NEL) Area in the infant skull where bone formation has not yet occurred; "soft spot"

Foramen (fo-RA-men) Opening or passageway, as into or through a bone; pl., foramina (fo-RAM-in-ah)

Fossa (FOS-sah) Hollow or depression, as in a bone; pl., fossae (FOS-se)

Fovea (FO-ve-ah) Small pit or cup-shaped depression in a surface; the fovea centralis near the center of the retina is the point of sharpest vision

FSH See Follicle-stimulating hormone

Fungus (FUN-gus) Type of plantlike microorganism; yeast or mold; pl., fungi (FUN-ji)

Gamete (GAM-ete) Reproductive cell; ovum or spermatozoon

Gamma globulin (GLOB-u-lin) Protein fraction in the blood plasma that contains antibodies

Ganglion (GANG-le-on) Collection of nerve cell bodies located outside the central nervous system.

Gangrene (GANG-grene) Death of tissue accompanied by bacterial invasion and putrefaction

Gastrointestinal (gas-tro-in-TES-tih-nal) **(GI)** Pertaining to the stomach and intestine or the digestive tract as a whole

Gene Hereditary unit; portion of the DNA on a chromosome

Genetic (jeh-NET-ik) Pertaining to the genes or heredity

Gestation (jes-TA-shun) Period of development from conception to birth

GH See Growth hormone

GI See Gastrointestinal

Gingiva (JIN-jih-vah) Tissue around the teeth; gum

Glaucoma (glaw-KO-mah) Disorder involving increased fluid pressure within the eye

Glomerular (glo-MER-u-lar) **filtrate** Fluid and dissolved materials that leave the blood and enter the kidney nephron through Bowman's capsule

Glomerulonephritis (glo-mer-u-lo-nef-RI-tis) Kidney disease often resulting from antibodies to a streptococcal infection

Glomerulus (glo-MER-u-lus) Cluster of capillaries in Bowman's capsule of the nephron

Glucose (GLU-kose) Simple sugar; main energy source for the cells; dextrose

Glycogen (GLI-ko-jen) Compound built from glucose molecules that is stored for energy in liver and muscles

Golgi (GOL-je) **apparatus** System of membranes in the cell that formulates special substances

Gonad (GO-nad) Sex gland; ovary or testis

Gonadotropin (gon-ah-do-TRO-pin) Hormone that acts on a reproductive gland (ovary or testis; *e.g.,* FSH, LH)

Gram (g) Basic unit of weight in the metric system

Gray matter Nervous tissue composed of unmyelinated fibers and cell bodies

Growth hormone (GH) Hormone produced by anterior pituitary that promotes growth of tissues; somatotropin

Gustatory (GUS-tah-to-re) Pertaining to the sense of taste (gustation)

Gyrus (JI-rus) Raised area of the cerebral cortex; pl., gyri (JI-ri)

Helminth (HEL-minth) Worm

Hematocrit (he-MAT-o-krit) **(Hct)** Volume percentage of red blood cells in whole blood; packed cell volume

Hematoma (he-mah-TO-mah) Tumor or swelling filled with blood

Hematuria (hem-ah-TU-re-ah) Blood in the urine

Hemodialysis (he-mo-di-AL-ih-sis) Removal of impurities from the blood by their passage through a semipermeable membrane

Hemoglobin (he-mo-GLO-bin) **(Hb)** Iron-containing protein in red blood cells that transports oxygen

Hemolysis (he-MOL-ih-sis) Rupture of red blood cells; v., hemolyze (HE-mo-lize)

Hemopoiesis (he-mo-poy-E-sis) Production of blood cells; hematopoiesis

Hemorrhage (HEM-eh-rij) Loss of blood

Hemostasis (he-mo-STA-sis) Stoppage of bleeding

Heparin (HEP-ah-rin) Substance that prevents blood clotting; anticoagulant

Hepatitis (hep-ah-TI-tis) Inflammation of the liver

Heredity (he-RED-ih-te) Transmission of characteristics from parent to offspring via the genes; the genetic make-up of the individual

Hernia (HER-ne-ah) Protrusion of an organ or tissue through the wall of the cavity in which it is normally enclosed

Hilum (HI-lum) See hilus

Hilus (HI-lus) Area where vessels and nerves enter or leave an organ

Histamine (HIS-tah-mene) Substance released from tissues during an antigen–antibody reaction

Histology (his-TOL-o-je) Study of tissues

HIV See Human immunodeficiency virus

Homeostasis (ho-me-o-STA-sis) State of balance within the body; maintenance of body conditions within set limits

Hormone Secretion of an endocrine gland; chemical messenger that has specific regulatory effects on certain other cells

Human immunodeficiency virus (HIV) The virus that causes AIDS

Hydrolysis (hi-DROL-ih-sis) Splitting of large molecules by the addition of water, as in digestion

Hyperglycemia (hi-per-gli-SE-me-ah) Abnormal increase in the amount of glucose in the blood

Hypersensitivity (hi-per-SEN-sih-tiv-ih-te) Exaggerated reaction of the immune system to a substance that is normally harmless to most people; allergy

Hypertension (hi-per-TEN-shun) High blood pressure

Hypertonic (hi-per-TON-ik) Describing a solution that is more concentrated than the fluids within a cell

Hypertrophy (hy-PER-tro-fe) Enlargement or overgrowth of an organ or part

Hypoglycemia (hi-po-gli-SE-me-ah) Abnormal decrease in the amount of glucose in the blood

Hypophysis (hi-POF-ih-sis) Pituitary gland

Hypotension (hi-po-TEN-shun) Low blood pressure

Hypothalamus (hi-po-THAL-ah-mus) Region of the brain that controls the pituitary and maintains homeostasis

Hypothermia (hi-po-THER-me-ah) Abnormally low body temperature

Hypotonic (hi-po-TON-ik) Describing a solution that is less concentrated than the fluids within a cell

Hypoxemia (hi-pok-SE-me-ah) Lower than normal concentration of oxygen in arterial blood

Hypoxia (hi-POK-se-ah) Lower than normal level of oxygen in the tissues

ICSH Interstitial cell-stimulating hormone; see Luteinizing hormone

Ileum (IL-e-um) The last portion of the small intestine

Immunity (ih-MU-nih-te) Power of an individual to resist or overcome the effects of a particular disease or other harmful agent

Immunization (ih-mu-nih-ZA-shun) Use of a vaccine to produce immunity

Immunodeficiency (im-u-no-de-FISH-en-se) Any failure of the immune system

Implantation (im-plan-TA-shun) The embedding of the fertilized egg into the lining of the uterus

Infarct (IN-farkt) Area of tissue damaged from lack of blood supply caused by blockage of a vessel

Inferior (in-FE-re-or) Below or lower

Inferior vena cava (VE-nah KA-vah) Large vein that drains the lower part of the body and empties into the right atrium of the heart

Infertility (in-fer-TIL-ih-te) Decreased ability to reproduce

Inflammation (in-flah-MA-shun) Response of tissues to injury; characterized by heat, redness, swelling, and pain

Insertion (in-SER-shun) End of a muscle attached to a movable part

Integument (in-TEG-u-ment) Skin; adj., integumentary

Intercellular (in-ter-SEL-u-lar) Between cells

Interstitial (in-ter-STISH-al) Between; pertaining to spaces or structures in an organ between active tissues

Intracellular (in-trah-SEL-u-lar) Within a cell

Ion (I-on) Charged particle formed when an electrolyte goes into solution

Iris (I-ris) Circular colored region of the eye around the pupil

Ischemia (is-KE-me-ah) Lack of blood supply to an area

Islets (I-lets) Groups of cells in the pancreas that produce hormones; islets of Langerhans (LAHNG-er-hanz)

Isometric (i-so-MET-rik) **contraction** Muscle contraction in which there is no change in muscle length but an increase in muscle tension, as in pushing against an immovable force

Isotonic (i-so-TON-ik) Describing a solution that has the same concentration as the fluid within a cell

Isotonic contraction Muscle contraction in which the tone within the muscle remains the same but the muscle shortens to produce movement

Isotope (I-so-tope) Form of an element that has the same atomic number as another but a different atomic weight

Jaundice (JAWN-dis) Yellowish discoloration of the skin that is usually due to the presence of bile in the blood

Jejunum (je-JU-num) Second portion of the small intestine

Joint Area of junction between two or more bones; articulation

Juxtaglomerular (juks-tah-glo-MER-u-lar) **(JG) apparatus** Structure in the kidney composed of cells of the afferent arteriole and distal convoluted tubule that secretes the enzyme renin when blood pressure decreases below a certain level

Karyotype (KAR-e-o-type) Picture of the chromosomes arranged according to size and form.

Keratin (KER-ah-tin) Protein that thickens and protects the skin; makes up hair and nails

Kidney (KID-ne) An organ of excretion

Lacrimal (LAK-rih-mal) Referring to tears or the tear glands

Lactation (lak-TA-shun) Secretion of milk

Lacteal (LAK-te-al) Capillary of the lymphatic system; drains digested fats from the villi of the small intestine

Lactic (LAK-tik) **acid** Organic acid that accumulates in muscle cells functioning without oxygen

Larynx (LAR-inks) Structure between the pharynx and trachea that contains the vocal cords; voice box

Laser (LA-zer) Device that produces a very intense beam of light

Lateral (LAT-er-al) Farther from the midline; toward the side

Lens Biconvex structure of the eye that changes in thickness to accommodate for near and far vision; crystalline lens

Lesion (LE-zhun) Wound or local injury

Leukemia (lu-KE-me-ah) Malignant blood disease characterized by abnormal development of white blood cells

Leukocyte (LU-ko-site) White blood cell

Ligament (LIG-ah-ment) Band of connective tissue that connects a bone to another bone; thickened portion or fold of the peritoneum that supports an organ or attaches it to another organ

Lipid (LIP-id) Type of organic compound, one example of which is a fat

Liter (LE-ter) **(L)** Basic unit of volume in the metric system.

Lumen (LU-men) The central opening of an organ or vessel

Lung An organ of respiration

Luteinizing (LU-te-in-i-zing) **hormone** Hormone produced by the anterior pituitary that induces ovulation and formation of the corpus luteum in females; in males it stimulates cells in the testes to produce testosterone and is called *interstitial cell-stimulating hormone* (*ICSH*)

Lymph (limf) Fluid in the lymphatic system

Lymphadenitis (lim-fad-en-I-tis) Inflammation of the lymph nodes

Lymphadenopathy (lim-fad-en-OP-ah-the) Any disorder of the lymph nodes

Lymphangitis (lim-fan-JI-tis) Inflammation of the lymphatic vessels

Lymphatic duct Vessel of the lymphatic system

Lymph node Mass of lymphoid tissue along the path of a lymphatic vessel that filters lymph and harbors white blood cells active in immunity

Lymphocyte (LIM-fo-site) Agranular white blood cell that functions in immunity

Lysosome (LI-so-some) Cell organelle that contains digestive enzymes

Macrophage (MAK-ro-faj) Large phagocytic cell that develops from a monocyte

Macula (MAK-u-lah) Spot; flat, discolored spot on the skin such as a freckle or measles lesion; small yellow spot in the retina of the eye that contains the fovea, the point of sharpest vision. Also called *macule*

Magnetic resonance imaging (MRI) A method for studying tissue based on nuclear movement following exposure to radio waves in a powerful magnetic field

Malignant (mah-LIG-nant) Describing a tumor that spreads; describing a disorder that tends to become worse and cause death

Malnutrition (mal-nu-TRISH-un) State resulting from lack of food, lack of an essential component of the diet, or faulty use of food in the diet

Mastectomy (mas-TEK-to-me) Removal of the breast; mammectomy

Mastication (mas-tih-KA-shun) Act of chewing

Medial (ME-de-al) Nearer the midline of the body

Mediastinum (me-de-as-TI-num) Region between the lungs and the organs and vessels it contains

Medulla (meh-DUL-lah) Inner region of an organ; marrow

Medulla oblongata (ob-long-GAH-tah) Part of the brain stem that connects the brain with the spinal cord

Megakaryocyte (meg-ah-KAR-e-o-site) Very large cell that gives rise to blood platelets

Meiosis (mi-O-sis) Process of cell division that halves the chromosome number in the formation of the reproductive cells

Melanin (MEL-ah-nin) Dark pigment found in skin, hair, parts of the eye, and certain parts of the brain

Membrane Thin sheet of tissue

Mendelian (men-DE-le-en) **laws** Principles of heredity discovered by an Austrian monk named Gregor Mendel

Meninges (men-IN-jeze) Three layers of fibrous membranes that cover the brain and spinal cord

Menopause (MEN-o-pawz) Time at which menstruation ceases

Menses (MEN-seze) The monthly flow of blood from the female reproductive tract

Mesentery (MES-en-ter-e) The membranous peritoneal ligament that attaches the small intestine to the dorsal abdominal wall

Mesocolon (mes-o-KO-lon) Peritoneal ligament that attaches the colon to the dorsal abdominal wall

Metabolic rate Rate at which energy is released from nutrients in the cells

Metabolism (meh-TAB-o-lizm) The physical and chemical processes by which an organism is maintained

Metastasis (meh-TAS-tah-sis) Spread of tumor cells; pl., metastases (meh-TAS-tah-seze)

Meter (ME-ter) **(m)** Basic unit of length in the metric system

Microbiology (mi-kro-bi-OL-o-je) Study of microscopic organisms

Micrometer (MI-kro-me-ter) **(μm)** 1/1000th of a millimeter; micron; also an instrument for measuring through a microscope (pronounced mi-KROM-eh-ter)

Microorganism (mi-kro-OR-gan-izm) Microscopic organism

Micturition (mik-tu-RISH-un) Act of urination; voiding of the urinary bladder

Midbrain Upper portion of the brainstem

Mineral (MIN-er-al) Inorganic substance; in the diet, an element needed in small amounts for health

Mitochondria (mi-to-KON-dre-ah) Cell organelles that manufacture ATP with the energy released from the oxidation of nutrients

Mitosis (mi-TO-sis) Type of cell division that produces two daughter cells exactly like the parent cell

Mitral (MI-tral) **valve** The valve between the left atrium and left ventricle of the heart; bicuspid valve

Mixture Blend of two or more substances

Molecule (MOL-eh-kule) Particle formed by chemical bonding of two or more atoms; smallest subunit of a compound

Monocyte (MON-o-site) Agranular white blood cell active in phagocytosis

MRI See Magnetic resonance imaging

Mucosa (mu-KO-sah) Lining membrane that produces mucus; mucous membrane

Mucus (MU-kus) Thick protective fluid secreted by mucous membranes and glands; adj., mucous

Murmur Abnormal heart sound

Mutagenic (mu-tah-JEN-ik) Causing mutation

Mutation (mu-TA-shun) Change in a gene or a chromosome

Myalgia (mi-AL-je-ah) Muscular pain

Myelin (MI-el-in) Fatty material that covers and insulates the axons of some neurons

Myocardium (mi-o-KAR-de-um) Middle layer of the heart wall; heart muscle

Myoglobin (MI-o-glo-bin) Compound that stores oxygen in muscle cells

Myometrium (mi-o-ME-tre-um) The muscular layer of the uterus

Myopia (mi-O-pe-ah) Nearsightedness

Myosin (MI-o-sin) One of the two contractile proteins in muscle cells, the other being actin

Necrosis (neh-KRO-sis) Tissue death

Negative feedback A self-regulating system in which the result of an action is the control over that action; a method for keeping body conditions within a normal range and maintaining homeostasis

Neoplasm (NE-o-plazm) Abnormal growth of cells; tumor; adj., neoplastic

Nephron (NEF-ron) Microscopic functional unit of the kidney

Nerve Bundle of neuron fibers outside the central nervous system

Nerve impulse Electric charge that spreads along the membrane of a neuron; action potential

Neuralgia (nu-RAL-je-ah) Pain in a nerve

Neurilemma (nu-rih-LEM-mah) Thin sheath that covers certain peripheral axons; aids in regeneration of the axon

Neuroglia (nu-ROG-le-ah) Supporting and protective cells of the central nervous system; glial cells

Neuromuscular junction Point at which a nerve fiber contacts a muscle cell

Neuron (NU-ron) Conducting cell of the nervous system

Neurotransmitter (nu-ro-TRANS-mit-er) Chemical released from the ending of an axon that enables a nerve impulse to cross a synaptic junction

Neutrophil (NU-tro-fil) Phagocytic granular white blood cell; polymorph; poly; PMN; seg

Node Small mass of tissue such as a lymph node; space between cells in the myelin sheath

Norepinephrine (nor-epi-ih-NEF-rin) Neurotransmitter similar to epinephrine; noradrenaline

Nucleotide (NU-kle-o-tide) Building block of DNA and RNA

Nucleus (NU-kle-us) Largest organelle in the cell, containing the DNA, which directs all cell activities; group of neurons in the central nervous system; in chemistry, the central part of an atom

Olfactory (ol-FAK-to-re) Pertaining to the sense of smell (olfaction)

Oncology (on-KOL-o-je) Study of tumors

Ophthalmic (of-THAL-mik) Pertaining to the eye

Organ (OR-gan) Body part containing two or more tissues functioning together for specific purposes

Organelle (or-gan-EL) Specialized subdivision within a cell

Organic (or-GAN-ik) Referring to compounds found in living things and containing carbon, hydrogen, and oxygen

Organism (OR-gan-izm) Individual plant or animal; any organized living thing

Organ of Corti (KOR-te) Receptor for hearing located in the cochlea of the internal ear

Origin (OR-ih-jin) Source; beginning; end of a muscle attached to a nonmoving part

Osmosis (os-MO-sis) Movement of water through a semipermeable membrane

Osmotic (os-MOT-ik) **pressure** Tendency of a solution to draw water into it; is directly related to the concentration of the solution

Ossicle (OS-ih-kl) One of three small bones of the middle ear: malleus, incus, or stapes

Ossification (os-ih-fih-KA-shun) Process of bone formation

Osteoblast (OS-te-o-blast) Bone-forming cell

Osteoclast (OS-te-o-clast) Cell that breaks down bone

Osteocyte (OS-te-o-site) Mature bone cell; maintains bone but does not divide

Osteoporosis (os-te-o-po-RO-sis) Abnormal loss of bone tissue with tendency to fracture

Ovary (O-vah-re) Female reproductive gland

Ovulation (ov-u-LA-shun) Release of a mature ovum from a follicle in the ovary

Ovum (O-vum) Female reproductive cell or gamete; pl., ova

Oxidation (ok-sih-DA-shun) Chemical breakdown of nutrients for energy

Oxygen (OK-sih-jen) **(O_2)** The gas needed to completely break down nutrients for energy within the cell

Oxygen debt Amount of oxygen needed to reverse the effects produced in muscles functioning without oxygen

Pacemaker Sinoatrial (SA) node of the heart; group of cells or artificial device that sets the rate of heart contractions

Pancreas (PAN-kre-as) Large, elongated gland behind the stomach; produces digestive enzymes and hormones (*e.g.,* insulin)

Papilla (pah-PIL-ah) Small nipple-like projection or elevation

Paracentesis (par-eh-sen-TE-sis) Puncture through the wall of a cavity, usually to remove fluid or promote drainage

Parasite (PAR-ah-site) Organism that lives on or within another (host) at the other's expense

Parasympathetic nervous system Craniosacral division of the autonomic nervous system

Parathyroid (par-ah-THI-royd) Any of four small glands embedded in the capsule enclosing the thyroid gland; produces hormone that regulates calcium in the blood

Parietal (pah-RI-eh-tal) Pertaining to the wall of a space or cavity

Parturition (par-tu-RISH-un) Childbirth; labor

Pathogen (PATH-o-jen) Disease-causing organism; adj., pathogenic (path-o-JEN-ik)

Pathology (pah-THOL-o-je) Study of disease

Pathophysiology (path-o-fiz-e-OL-o-je) Study of the physiologic basis of disease

Pedigree (PED-ih-gre) Family history; used in the study of heredity

Pelvic inflammatory disease (PID) Ascending infection that involves the pelvic organs; common causes are gonorrhea and chlamydia

Pelvis (PEL-vis) Basin-like structure, such as the lower portion of the abdomen or the upper flared portion of the ureter (renal pelvis)

Penis (PE-nis) Male organ of urination and sexual intercourse

Pericardium (per-ih-KAR-de-um) Fibrous sac lined with serous membrane that encloses the heart

Perichondrium (per-ih-KON-dre-um) Layer of connective tissue that covers cartilage

Perineum (per-ih-NE-um) Pelvic floor; external region between the anus and genital organs

Periosteum (per-e-OS-te-um) Connective tissue membrane covering a bone

Peripheral (peh-RIF-er-al) Located away from a center or central structure

Peripheral nervous system (PNS) All the nerves and nervous tissue outside the central nervous system

Peristalsis (per-ih-STAL-sis) Wavelike movements in the wall of an organ or duct that propel its contents forward

Peritoneum (per-ih-to-NE-um) Serous membrane that lines the abdominal cavity and forms outer layer of abdominal organs; forms supporting ligaments for some organs

pH Symbol indicating hydrogen ion (H+) concentration; scale that measures the relative acidity and alkalinity (basicity) of a solution

Phagocytosis (fag-o-si-TO-sis) Engulfing of large particles through the cell membrane

Pharynx (FAR-inks) Throat; passageway between the mouth and esophagus

Phlebitis (fleh-BI-tis) Inflammation of a vein

Physiology (fiz-e-OL-o-je) Study of the function of living organisms

Pia mater (PI-ah MA-ter) Innermost layer of the meninges

PID See Pelvic inflammatory disease

Pineal (PIN-e-al) **body** Gland-like organ in the brain that is regulated by light; involved in sleep–wake cycles

Pinocytosis (pi-no-si-TO-sis) Intake of small particles and droplets by the cell membrane

Pituitary (pih-TU-ih-tar-e) **gland** Endocrine gland located under and controlled by the hypothalamus; releases hormones that control other glands; hypophysis

Placenta (plah-SEN-tah) Structure that nourishes and maintains the developing individual during pregnancy

Plasma (PLAZ-mah) Liquid portion of the blood

Plasma cell Cell that produces antibodies; derived from a B cell

Platelet (PLATE-let) Cell fragment that forms a plug to stop bleeding and acts in blood clotting; thrombocyte

Pleura (PLU-rah) Serous membrane that lines the chest cavity and covers the lungs

Plexus (PLEK-sus) Network of vessels or nerves

Pneumothorax (nu-mo-THO-raks) Accumulation of air in the pleural space

PNS See Peripheral nervous system

Polyp (POL-ip) Protruding growth, often grapelike, from a mucous membrane

Pons (ponz) Area of the brain between the midbrain and medulla; connects the cerebellum with the rest of the central nervous system

Portal system Venous system that carries blood to a second capillary bed before it returns to the heart

Posterior (pos-TE-re-or) Toward the back; dorsal

Preeclampsia (pre-eh-KLAMP-se-ah) Disorder of unknown cause that develops late in pregnancy, with hypertension, proteinuria, general edema, and sudden weight gain; if untreated it may lead to eclampsia; toxemia of pregnancy

Prime mover Muscle that performs a given movement; agonist

Progeny (PROJ-eh-ne) Offspring

Progesterone (pro-JES-ter-one) Hormone produced by the corpus luteum and placenta; maintains the lining of the uterus for pregnancy

Prognosis (prog-NO-sis) Prediction of the probable outcome of a disease based on the condition of the patient and knowledge about the disease

Prophylaxis (pro-fih-LAK-sis) Prevention of disease

Proprioceptor (pro-pre-o-SEP-tor) Sensory receptor that aids in judging body position and changes in position; located in muscles, tendons, and joints

Prostaglandins (pros-tah-GLAN-dinz) Group of hormones produced by many cells; these hormones have a variety of effects

Protein (PRO-tene) Organic compound made of amino acids; contains nitrogen in addition to carbon, hydrogen, and oxygen (some contain sulfur or phosphorus)

Prothrombin (pro-THROM-bin) Clotting factor; converted to thrombin during blood clotting

Proton (PRO-ton) Positively charged particle in the nucleus of an atom

Protozoon (pro-to-ZO-on) Animal-like microorganism; pl., protozoa

Proximal (PROK-sih-mal) Nearer to point of origin or to a reference point

Pulse Wave of increased pressure in the vessels produced by contraction of the heart

Pupil (PU-pil) Opening in the center of the eye through which light enters

Pyrogen (PI-ro-jen) Substance that produces fever

Receptor (re-SEP-tor) Specialized cell or ending of a sensory neuron that can be excited by a stimulus; also a site in the cell membrane to which a special substance (*e.g.*, hormone, antibody) may attach

Recessive (re-SES-iv) Referring to a gene that is not expressed if a dominant gene for the same trait is present

Reflex (RE-flex) A simple, rapid, automatic response involving few neurons

Reflex arc (ark) A pathway through the nervous system from stimulus to response; commonly involves a receptor, sensory neuron, central neuron(s), motor neuron, and effector

Refraction (re-FRAK-shun) Bending of light rays as they pass from one medium to another of a different density

Renin (RE-nin) Enzyme released from the juxtaglomerular apparatus of the kidneys that indirectly increases blood pressure by activating angiotensin

Resorption (re-SORP-shun) Loss of substance, such as that of bone or a tooth

Respiration (res-pih-RA-shun) Exchange of oxygen and carbon dioxide between the outside air and body cells

Reticuloendothelial (reh-tik-u-lo-en-do-THE-le-al) **system** Protective system consisting of highly phagocytic cells in body fluids and tissues, such as the spleen, lymph nodes, bone marrow, and liver

Retina (RET-ih-nah) Innermost layer of the eye; contains light-sensitive cells (rods and cones)

Retroperitoneal (ret-ro-per-ih-to-NE-al) Behind the peritoneum, as are the kidneys, pancreas, and abdominal aorta

Ribonucleic (RI-bo-nu-kle-ik) **acid (RNA)** Substance needed for protein manufacture in the cell

Ribosome (RI-bo-some) Small body in the cytoplasm of a cell that is a site of protein manufacture

Rickettsia (rih-KET-se-ah) Extremely small oval to rod-shaped bacterium that can grow only within a living cell

RNA See Ribonucleic acid

Roentgenogram (rent-GEN-o-gram) Film produced by means of x-rays

Rugae (RU-je) Folds in the lining of an organ such as the stomach or urinary bladder; sing., ruga (RU-gah)

Saliva (sah-LI-vah) Secretion of the salivary glands; moistens food and contains an enzyme that digests starch

SA node See Sinoatrial node

Sarcoma (sar-KO-mah) Malignant tumor of connective tissue; a form of cancer

Sclera (SKLE-rah) Outermost layer of the eye; made of tough connective tissue; "white" of the eye

Scrotum (SKRO-tum) Sac in which testes are suspended

Sebum (SE-bum) Oily secretion that lubricates the skin; adj., **sebaceous** (se-BA-shus)

Semen (SE-men) Mixture of sperm cells and secretions from several glands of the male reproductive tract

Semicircular canal Bony canal in the internal ear that contains receptors for the sense of dynamic equilibrium; there are three semicircular canals in each ear

Sensory adaptation The gradual loss of sensation when sensory receptors are exposed to continuous stimulation

Sepsis (SEP-sis) Presence of pathogenic microorganisms or their toxins in the bloodstream or other tissues; adj., septic

Septicemia (sep-tih-SE-me-ah) Presence of pathogenic organisms or their toxins in the bloodstream; blood poisoning

Septum (SEP-tum) Dividing wall, as between the chambers of the heart or the nasal cavities

Serosa (se-RO-sah) Serous membrane; epithelial membrane that secretes a thin, watery fluid

Serum (SE-rum) Liquid portion of blood without clotting factors; thin, watery fluid; adj., serous (SE-rus)

Sex-linked Referring to a gene carried on a sex chromosome, usually the X chromosome

Sexually transmitted disease (STD) Disease acquired through sexual relations; venereal disease (VD)

Shock Pertaining to the circulation: inadequate output of blood by the heart

Sign Manifestation of a disease as noted by an observer

Sinoatrial (si-no-A-tre-al) **(SA) node** Tissue in the upper wall of the right atrium that sets the rate of heart contractions; pacemaker of the heart

Sinusoid (SI-nus-oyd) Enlarged capillary that serves as a blood channel

Solute (SOL-ute) Substance that is dissolved in another substance (the solvent)

Solution (so-LU-shun) Mixture, the components of which are evenly distributed

Solvent (SOL-vent) Substance in which another substance (the solute) is dissolved

Spermatozoon (sper-mah-to-ZO-on) Male reproductive cell or gamete; pl., spermatozoa

Sphincter (SFINK-ter) Muscular ring that regulates the size of an opening

Sphygmomanometer (sfig-mo-mah-NOM-eh-ter) Device used to measure blood pressure

Spirillum (spi-RIL-um) Corkscrew or spiral-shaped bacterium; pl., spirilla

Spirochete (SPI-ro-kete) Spiral-shaped microorganism that moves in a waving and twisting motion

Spleen Lymphoid organ in the upper left region of the abdomen

Spore Resistant form of bacterium; reproductive cell in lower plants

Staphylococcus (staf-ih-lo-KOK-us) Round bacterium found in a cluster resembling a bunch of grapes; pl., staphylococci (staf-ih-lo-KOK-si)

Stasis (STA-sis) Stoppage in the normal flow of fluids, such as blood, lymph, urine, or contents of the digestive tract

STD See Sexually transmitted disease

Stenosis (sten-O-sis) Narrowing of a duct or canal

Sterilization (ster-ih-li-ZA-shun) Process of killing every living microorganism on or in an object; procedure that makes an individual incapable of reproduction

Steroid (STE-royd) Category of lipids that includes the hormones of the sex glands and the adrenal cortex

Stethoscope (STETH-o-skope) Instrument for conveying sounds from the patient's body to the examiner's ears

Stimulus (STIM-u-lus) Change in the external or internal environment that produces a response

Subcutaneous (sub-ku-TA-ne-us) Under the skin

Sudoriferous (su-do-RIF-er-us) Producing sweat; referring to the sweat glands

Sulcus (SUL-kus) Shallow groove, as between convolutions of the cerebral cortex; pl., sulci (SUL-si)

Superior (su-PE-re-or) Above; in a higher position

Superior vena cava (VE-nah KA-vah) Large vein that drains the upper part of the body and empties into the right atrium of the heart

Surfactant (sur-FAK-tant) Substance in the alveoli that prevents their collapse by reducing surface tension of the fluids within

Suspension (sus-PEN-shun) Mixture that will separate unless shaken

Suture (SU-chur) Type of joint in which bone surfaces are closely united, as in the skull; stitch used in surgery to bring parts together or to stitch parts together in surgery

Sympathetic nervous system Thoracolumbar division of the autonomic nervous system

Symptom (SIMP-tom) Evidence of disease noted by the patient; such evidence noted by an examiner is called a *sign* or an *objective symptom*

Synapse (SIN-aps) Junction between two neurons or between a neuron and an effector

Synarthrosis (sin-ar-THRO-sis) Immovable joint

Syndrome (SIN-drome) Group of symptoms characteristic of a disorder

Synovial (sin-O-ve-al) Pertaining to a thick lubricating fluid found in joints, bursae, and tendon sheaths; pertaining to a freely movable (diarthrotic) joint

System (SIS-tem) Group of organs functioning together for the same general purposes

Systemic (sis-TEM-ik) Referring to a generalized infection or condition

Systole (SIS-to-le) Contraction phase of the cardiac cycle; adj., systolic (sis-TOL-ik)

Tachycardia (tak-e-KAR-de-ah) Heart rate over 100 beats per minute

Target tissue Tissue that is capable of responding to a specific hormone

T cell Lymphocyte active in immunity that matures in the thymus gland; destroys foreign cells directly; T lymphocyte

Tendinitis (ten-din-I-tis) Inflammation of a tendon

Tendon (TEN-don) Cord of fibrous connective tissue that attaches a muscle to a bone

Testis (TES-tis) Male reproductive gland; pl., testes (TES-teze)

Testosterone (tes-TOS-ter-one) Male sex hormone produced in the testes; promotes the development of sperm cells and maintains secondary sex characteristics

Tetanus (TET-an-us) Constant contraction of a muscle; infectious disease caused by a bacterium (*Clostridium tetani*); lockjaw

Tetany (TET-an-e) Muscle spasms due to abnormal calcium metabolism, as in parathyroid deficiency

Thalamus (THAL-ah-mus) Region of the brain located in the diencephalon; chief relay center for sensory impulses traveling to the cerebral cortex

Thorax (THO-raks) Chest; adj., thoracic (tho-RAS-ik)

Thrombocyte (THROM-bo-site) Blood platelet; cell fragment that participates in clotting

Thrombocytopenia (throm-bo-si-to-PE-ne-ah) Deficiency of platelets in the blood

Thrombosis (throm-BO-sis) Condition of having a thrombus

Thrombus (THROM-bus) Blood clot within a vessel

Thymus (THI-mus) Endocrine gland in the upper portion of the chest; stimulates development of T cells

Thyroid (THI-royd) Endocrine gland in the neck

Thyroid-stimulating hormone (TSH) Hormone produced by the anterior pituitary that stimulates the thyroid gland

Thyroxine (thi-ROK-sin) Hormone produced by the thyroid gland; increases metabolic rate and needed for normal growth

Tissue Group of similar cells that performs a specialized function

Tonsil (TON-sil) Mass of lymphoid tissue in the region of the pharynx

Tonus (TO-nus) Also tone. Partially contracted state of muscle

Toxemia (tok-SE-me-ah) General toxic condition in which poisonous bacterial substances are absorbed into the bloodstream; condition caused by abnormal metabolism

Toxin (TOK-sin) Poison

Toxoid (TOK-soyd) Altered toxin used to produce active immunity

Trachea (TRA-ke-ah) Tube that extends from the larynx to the bronchi; windpipe

Tracheostomy (tra-ke-OS-to-me) Surgical opening into the trachea for the introduction of a tube through which a person may breathe

Tract Bundle of neuron fibers within the central nervous system

Trait Characteristic

Transplant Organ or tissue used to replace an injured or incompetent part

Trauma (TRAW-mah) Injury or wound

Tricuspid (tri-KUS-pid) **valve** Valve between the right atrium and right ventricle of the heart

TSH See Thyroid-stimulating hormone

Tympanic (tim-PAN-ik) **membrane** Membrane between the external and middle ear that transmits sound waves to the bones of the middle ear; eardrum

Ulcer (UL-ser) Sore or lesion associated with death and disintegration of tissue

Umbilical (um-BIL-ih-kal) **cord** Structure that connects the fetus with the placenta; contains vessels that carry blood between the fetus and placenta

Umbilicus (um-BIL-ih-kus) Small scar on the abdomen that marks the former attachment of the umbilical cord to the fetus; navel

Urea (u-RE-ah) Nitrogen waste product excreted in the urine; end product of protein metabolism

Ureter (U-re-ter) Tube that carries urine from the kidney to the urinary bladder

Urethra (u-RE-thrah) Tube that carries urine from the urinary bladder to the outside of the body

Urinalysis (u-rin-AL-ih-sis) Laboratory examination of the physical and chemical properties of urine

Urinary bladder Hollow organ that stores urine until it is eliminated

Urine (U-rin) Liquid waste excreted by the kidneys

Uterus (U-ter-us) Muscular, pear-shaped organ in the female pelvis within which the fetus develops during pregnancy

Uvea (U-ve-ah) Middle coat of the eye, including the choroid, iris, and ciliary body; vascular and pigmented structures of the eye

Uvula (U-vu-lah) Soft, fleshy, V-shaped mass that hangs from the soft palate

Vaccine (vak-SENE) Substance used to produce active immunity; usually a suspension of attenuated or killed pathogens given by inoculation to prevent a specific disease

Vagina (vah-JI-nah) Lower part of the birth canal that opens to the outside of the body; female organ of sexual intercourse

Valve Structure that prevents fluid from flowing backward, as in the heart, veins, and lymphatic vessels

Varicose (VAR-ih-kose) Pertaining to an enlarged and twisted vessel, as in varicose vein

Vas deferens (DEF-er-enz) Tube that carries sperm cells from the testis to the urethra; ductus deferens

Vasectomy (vah-SEK-to-me) Surgical removal of part or all of the ductus (vas) deferens; usually done on both sides to produce sterility

Vasoconstriction (vas-o-kon-STRIK-shun) Decrease in the diameter of a blood vessel

Vasodilation (vas-o-di-LA-shun) Increase in the diameter of a blood vessel

VD Venereal disease; see Sexually transmitted disease

Vein (vane) Vessel that carries blood toward the heart

Vena cava (VE-nah KA-vah) A large vein that carries blood into the right atrium of the heart; superior vena cava or inferior vena cava

Venereal (ve-NE-re-al) **disease (VD)** Disease acquired through sexual activity; sexually transmitted disease (STD)

Venous sinus (VE-nus SI-nus) Large channel that drains deoxygenated blood

Ventilation (ven-tih-LA-shun) Movement of air into and out of the lungs

Ventral (VEN-tral) Toward the front or belly surface; anterior

Ventricle (VEN-trih-kl) Cavity or chamber; one of the two lower chambers of the heart; one of the four chambers in the brain in which cerebrospinal fluid is produced; adj., ventricular (ven-TRIK-u-lar)

Venule (VEN-ule) Very small vein that collects blood from the capillaries

Vertebra (VER-teh-brah) A bone of the spinal column; pl., vertebrae (VER-teh-bre)

Vesicle (VES-ih-kl) Small sac or blister filled with fluid

Vestibule (VES-tih-bule) Part of the internal ear that contains receptors for the sense of equilibrium; any space at the entrance to a canal or organ

Vibrio (VIB-re-o) Slight curved or comma-shaped bacterium; pl., vibrios

Villi (VIL-li) Small finger-like projections from the surface of a membrane; projections in the lining of the small intestine through which digested food is absorbed; sing., villus

Virulence (VIR-u-lens) Power of an organism to overcome defenses of the host

Virus (VI-rus) Extremely small microorganism that can reproduce only within a living cell

Viscera (VIS-er-ah) Organs in the ventral body cavities, especially the abdominal organs

Vitamin (VI-tah-min) Organic compound needed in small amounts for health

Vitreous (VIT-re-us) **body** Soft, jellylike substance that fills the eyeball and holds the shape of the eye; vitreous humor

White matter Nervous tissue composed of myelinated fibers

X-ray Ray or radiation of extremely short wavelength that can penetrate opaque substances and affects photographic plates and fluorescent screens; an image made by means of such rays

Zygote (ZI-gote) Fertilized ovum; cell formed by the union of a sperm and an egg

Medical Terminology

Medical terminology, the special language of the health occupations, is based on an understanding of a relatively few basic elements. These elements—roots, prefixes, and suffixes—form the foundation of almost all medical terms. A useful way to familiarize yourself with each term is to learn to pronounce it correctly and say it aloud several times. Soon it will become an integral part of your vocabulary.

The foundation of a word is the word root. Examples of word roots are *abdomin-*, referring to the belly region; and *aden-*, pertaining to a gland. A word root is often followed by a vowel to facilitate pronunciation, as in *abdomino-* and *adeno-*. We then refer to it as a "combining form." The hyphen appended to a combining form indicates that it is not a complete word; if the hyphen precedes the combining form, then it commonly appears as the terminal element of the word ending, as in *-algia*, meaning "a painful condition."

A prefix is a part of a word that precedes the word root and changes its meaning. For example, the prefix *mal-* in *malunion* means "abnormal." A suffix, or word ending, is a part that follows the word root and adds to or changes its meaning. The suffix *-rrhea* means "profuse flow" or "discharge," as in *diarrhea*, a condition characterized by excessive discharge of liquid stools.

Many medical words are compound words; that is, they are made up of more than one root or combining form. Examples of such compound words are *erythrocyte* (red blood cell) and *hydrocele* (fluid-containing sac), and many more difficult words, such as *sternoclavicular* (indicating relations to both the sternum and the clavicle).

A general knowledge of language structure and spelling rules is also helpful in mastering medical terminology. For example, adjectives include words that end in *-al*, as in *sternal* (the noun is *sternum*), and words that end in *-ous*, as in *mucous* (the noun is *mucus*).

The following list includes some of the most commonly used word roots, combining forms, prefixes, and suffixes, as well as examples of their use.

a-, an- absent, deficient, lack of: *atrophy, anemia, anuria*

ab- away from: *abduction, aboral*

abdomin-, abdomino- belly or abdominal area: *abdominocentesis, abdominoscopy*

acou- hearing, sound: *acoustic*

acr-, acro- extreme ends of a part, especially of the extremities: *acromegaly, acromion*

actin-, actini-, actino- relation to raylike structures or, more commonly, to light or roentgen (x-) rays, or some other type of radiation: *actiniform, actinodermatitis*

ad- (sometimes converted to *ac-, af-, ag-, ap-, as-, at-,*) toward, added to, near: *adrenal, accretion, agglomerated, afferent*

aden-, adeno-, gland: *adenectomy, adenitis, adenocarcinoma*

-agogue inducing, leading, stimulating: *cholagogue, galactagogue*

alge-, algo-, algesi- pain: *algetic, algophobia, analgesic*

-algia pain, painful condition: *myalgia, neuralgia*

amb-, ambi-, ambo- both, on two sides: *ambidexterity, ambivalent*

ambly- dimness, dullness: *amblyopia*

angio- vessel: *angiogram, angiotensin*

ant-, anti- against; to prevent, suppress, or destroy: *antarthritic, antibiotic, anticoagulant*

ante- before, ahead of: *antenatal, antepartum*

antero- position ahead of or in front of (*i.e.,* anterior to) another part: *anterolateral, anteroventral*

arthr-, arthro- joint or articulation: *arthrolysis, arthrostomy, arthritis*

-ase enzyme: *lipase, protease*

-asis *see* -sis.

audio-   sound, hearing: *audiogenic, audiometry, audiovisual*
aur-   ear: *aural, auricle*
aut-, auto-   self: *autistic, autodigestion, autoimmune*

bi-   two, twice: *bifurcate, bisexual*
bio-   life, living organism: *biopsy, antibiotic*
blast-, blasto-, -blast   early stage, immature cell or bud: *blastula, blastophore, erythroblast*
bleph-, blephar-, blepharo-   eyelid, eyelash: *blepharism, blepharitis, blepharospasm*
brachi-   arm: *brachial, brachiocephalic, brachiotomy*
brachy-   short: *brachydactylia, brachyesophagus*
brady-   slow: *bradycardia*
bronch-   windpipe or other air tubes: *bronchiectasis, bronchoscope*
bucc-   cheek: *buccal*

carcin-   cancer: *carcinogenic, carcinoma*
cardi-, cardia-, cardio-   heart: *carditis, cardiac, cardiologist*
-cele   swelling; enlarged space or cavity: *cystocele, meningocele, rectocele*
centi-   relating to 100 (used in naming units of measurements): *centigrade, centimeter*
-centesis   performation, tapping: *amniocentesis, paracentesis*
cephal-, cephalo-   head: *cephalalgia, cephalopelvic*
cerebro-   brain: *cerebrospinal, cerebrum*
cervi-   neck: *cervical, cervix*
cheil-, cheilo-,   lips; brim or edge: *cheilitis, cheilosis*
cheir-, cheiro-   (also written **chir-, chiro-**) hand: *cheiralgia, cheiromegaly, chiropractic*
chol-, chole-, cholo-   bile, gall: *chologogue, cholecyst, cholecystitis, cholecystokinin*
chondr-, chondri-, chondrio-   cartilage: *chondric, chondrocyte, chondroma*
-cid, -cide   to cut, kill or destroy: *bactericidal, germicide, suicide*
circum-   around, surrounding: *circumorbital, circumrenal, circumduction*
-clast   break: *osteoclast*
colp-, colpo-   vagina: *colpectasia, colposcope, colpotomy*
contra-   opposed, against: *contraindication, contralateral*
cost-, costa-, costo   ribs: *intercostal, costosternal*
counter-   against, opposite to: *counteract, counterirritation, countertraction*
crani-, cranio-   skull: *cranium, craniotomy*
cry-, cryo-, crymo-   low temperature: *cryalgesia, cryogenic, cryotherapy*
crypt-, crypto-   hidden, concealed: *cryptic, cryptogenic, cryptorchidism*
cut-   skin: *subcutaneous*
cysti-, cysto-   sac, bladder: *cystitis, cystoscope*
cyt-, cyto-, -cyte   cell: *cytolytic, cytoplasm*

dactyl-, dactylo-   digits (usually fingers, but sometimes toes): *dactylitis, polydactyly*
de-   remove: *detoxify, dehydration*
dendr-   tree: *dendrite*
dento-, dent-, denti-   tooth: *dentition, dentin, dentifrice*
derm-, derma-, dermo-, dermat-, dermato-   skin: *dermatitis, dermatology; dermatosis*
di-, diplo-   twice, double: *dimorphism, diplopia*
dia-   through, between, across, apart: *diaphragm, diaphysis*
dis-   apart, away from: *disarticulation, distal*
dors-, dorsi-, dorso-   back (in the human, this combining form refers to the same regions as postero-): *dorsal, dorsiflexion, dorsonuchal*
-dynia   pain, tenderness: *myodynia, neurodynia*
dys-   disordered, difficult, painful: *dysentery, dysphagia, dyspnea*

-ectasis   expansion, dilation, stretching: *angiectasis, bronchiectasis*
ecto-   outside, external: *ectoderm, ectogenous*
-ectomy   surgical removal or destruction by other means: *appendectomy, thyroidectomy*
edem-   swelling: *edema*
-emia   blood: *glycemia, hyperemia*
encephal-, encephalo-   brain: *encephalitis, encephalogram*
end-, endo-   within, innermost: *endarterial, endocardium, endothelium*
enter-, entero-   intestine: *enteritis, enterocolitis*
epi-   on, upon: *epicardium, epidermis*
eryth-, erythro-   red: *erythema, erythrocyte*
-esthesia   sensation: *anesthesia, paresthesia*
eu-   well, normal, good: *euphoria, eupnea*
ex-, exo-   outside, out of, away from: *excretion, exocrine, exophthalmic*
extra-   beyond, outside of, in addition to: *extracellular, extrasystole, extravasation*

fasci-   fibrous connective tissue layers: *fascia, fascitis, fascicle*
-ferent   to bear, to carry: *afferent, efferent*
fibr-, fibro-   threadlike structures, fibers: *fibrillation, fibroblast, fibrositis*

galact-, galacta-, galacto-   milk: *galactemia, galactagogue, galactocele*
gastr-, gastro-   stomach: *gastritis, gastroenterostomy*
-gen   an agent that produces or originates: *allergen, pathogen, fibrinogen*
-genic   produced from, producing: *endogenic, pyogenic*
genito-   organs of reproduction: *genitoplasty, genitourinary*
gen-, geno-   a relationship to reproduction or sex: *genealogy, generate, genetic, genotype*
-geny   manner of origin, development or production: *ontogeny, progeny*

glio-, -glia gluey material; specifically, the connective tissue of the central nervous system: *glioma, neuroglia*
gloss-, glosso- tongue: *glossitis, glossopharyngeal*
gly-, glyco- sweet, relating to sugar: *glycemia, glycosuria*
-gnath, gnatho- related to the jaw: *prognathic, gnathoplasty*
gon- seed, knee: *gonad, gonarthritis*
-gram record, that which is recorded: *electrocardiogram, electroencephalogram*
-graph instrument for recording: *electrocardiograph, electroencephalograph*
gyn-, gyne-, gyneco-, gyno- female, woman: *gynecology, gynecomastia, gynoplasty*

hem-, hema-, hemato-, hemo- blood: *hematoma, hematuria, hemorrhage*
hemi- one half: *hemisphere, heminephrectomy, hemiplegia*
hepat-, hepato- liver: *hepatitis, hepatogenous*
heter-, hetero- other, different: *heterogeneous, heterosexual, heterochromia*
hist-, histo-, histio- tissue: *histology, histiocyte*
homeo-, homo- unchanging, the same: *hemeostasis, homosexual*
hydr-, hydro- water: *hydrolysis, hydrocephalus*
hyper- above, over, excessive: *hyperesthesia, hyperglycemia, hypertrophy*
hypo- deficient, below, beneath: *hypochondrium, hypodermic, hypogastrium*
hyster-, hystero- uterus: *hysterectomy*

-ia state of: *myopia, hypochondria, ischemia*
-iatrics, -trics medical specialty: *pediatrics, obstetrics*
idio- self, one's own, separate, distinct: *idiopathic, idiosyncrasy*
im-, in- in, into, lacking: *implantation, inanimate, infiltration*
infra- below: *infraspinous, infracortical*
inter- between: *intercostal, interstitial*
intra- within a part or structure: *intracranial, intracellular, intraocular*
-ism state of: *alcoholism, hyperthyroidism*
iso- equal: *isotonic, isometric*
-itis inflammation: *dermatitis, keratitis, neuritis*

juxta- next to: *juxtaglomerular*

karyo- nucleus: *karyotype, karyoplasm*
kerat-, kerato- cornea of the eye, certain horny tissues: *keratin, keratitis, keratoplasty*

lacri- tear: *lacrimal*
lact-, lacto- milk: *lactation, lactogenic*
later- side: *lateral*
leuk-, leuko- (also written as *leuc-, leuco-*) white: *leukocyte, leukoplakia*

lig- bind: *ligament, ligature*
lith-, litho- stones (calculi): *lithiasis, lithotripsy*
-logy, -ology study of: *physiology, gynecology*
lyso-, -lysis flowing, loosening, dissolution (dissolving of): *hemolysis, paralysis, lysosome*

macro- large, abnormal length: *macrophage, macroblast.* See also **-mega, megalo-**
mal- disordered, abnormal: *malnutrition, malocclusion, malunion*
malac-, malaco-, malacia- softening: *malacoma, malacosarcosis, osteomalacia*
mast-, masto- breast: *mastectomy, mastitis*
meg-, mega-, megal-, megalo- unusually or excessively large: *megacolon, megaloblast, megakaryocyte*
men-, meno- physiologic uterine bleeding: *menses, menorrhagia, menopause*
mening-, meningo- membranes covering the brain and spinal cord: *meningitis, meningocele*
mes-, mesa-, meso- middle, midline: *mesencephalon, mesoderm*
meta- change, beyond, after, over, near: *metabolism, metacarpus, metaplasia*
-meter measure: *hemocytometer, sphygmomanometer*
micro- very small: *microscope, microbiology, microabscess, micrometer*
mono- single: *monocyte, mononucleosis*
my-, myo- muscle: *myenteron, myocardium, myometrium*
myc-, mycet-, myco- fungi: *mycid, mycete, mycology, mycosis*
myel-, myelo- marrow (often used in reference to the spinal cord): *myeloid, myeloblast, osteomyelitis, poliomyelitis*

necr-, necro- death, corpse: *necrosis*
neo- new, strange: *neopathy, neoplasm*
neph-, nephro- kidney: *nephrectomy, nephron*
neur-, neuro- nerve: *neuron, neuralgia, neuroma*
noct-, nocti- night: *noctambulation, nocturia, noctiphobia*
nos-, noso- disease: *nosocomial, nosophobia*

ocul-, oculo- eye: *oculist, oculomotor, oculomycosis*
odont-, odonto- tooth, teeth: *odontalgia, orthodontics*
-oid likeness, resemblance: *lymphoid, myeloid*
olig-, oligo- few, a deficiency: *oligemia, oligospermia, oliguria*
-oma tumor, swelling: *hematoma, sarcoma*
onych-, onycho- nails: *paronychia, onychoma*
oo-, ovi-, ovo- ovum, egg: *oocyte, oviduct, ovoplasm* (do not confuse with **oophor-**)
oophor-, oophoro- ovary: *oophorectomy, oophoritis, oophorocystectomy.* See also **ovar-**
ophthalm-, ophthalmo- eye: *ophthalmia, ophthalmologist, ophthalmoscope*
orth-, ortho- straight, normal: *orthopedics, orthopnea, orthosis*

oscillo- to swing to and fro: *oscilloscope*

oss-, osseo-, ossi-, oste-, osteo- bone, bone tissue: *osseous, ossicle, osteocyte, osteomyelitis*

-ostomy creation of a mouth or opening by surgery: *colostomy, tracheostomy*

ot-, oto- ear: *otalgia, otitis, otomycosis*

-otomy cutting into: *phlebotomy, tracheotomy*

ovar-, ovario- ovary: *ovariectomy*. See also **oophor-**

ox-, -oxia pertaining to oxygen: *hypoxemia, hypoxia, anoxia*

para- near, beyond, apart from, beside: *paramedical, parametrium, parathyroid, parasagittal*

path-, patho-, -pathy disease, abnormal condition: *pathogen, pathognomonic, pathology, neuropathy*

ped-, pedia- child, foot: *pedialgia, pedophobia, pediatrician*

-penia lack of: *leukopenia, thrombocytopenia*

per- through, excessively: *percutaneous, perfusion*

peri- around: *pericardium, perichondrium*

-pexy fixation: *nephropexy, proctopexy*

phag-, phago- to eat, to ingest: *phage, phagocyte*

-phagia, -phagy eating, swallowing: *aphagia, dysphagia*

-phasia speech, ability to talk: *aphasia, dysphasia*

-phil, -philic to be fond of, to like (have an affinity for): *eosinophilia, hemophilia, hydrophilic*

phleb-, phlebo- vein: *phlebitis, phlebotomy*

phob-, -phobia fear, dread, abnormal aversion: *phobic, acrophobia, hydrophobia*

phot-, photo- light: *photoreceptor, photophobia*

pile-, pili-, pilo- hair, resembling hair: *pileous, piliation, pilonidal*

-plasty molding, surgical formation: *cystoplasty, gastroplasty, kineplasty*

pleur-, pleuro- side, rib, serous membrane covering the lung and lining the chest cavity: *pleurisy, pleurotomy*

-pnea air, breathing: *dyspnea, eupnea*

pneum-, pneuma-, pneumo-, pneumon-, pneumato- lung, air: *pneumectomy, pneumograph, pneumonia*

pod-, podo- foot: *podiatry, pododynia*

-poiesis making, forming: *erythropoiesis, hematopoiesis*

polio- gray: *polioencephalitis, poliomyelitis*

poly- many: *polyarthritis, polycystic, polycythemia*

post- behind, after, following: *postnatal, postocular, postpartum*

pre- before, ahead of: *precancerous, preclinical, prenatal*

presby- old age: *presbycusis, presbyopia*

pro- in front of, before: *prodromal, prosencephalon, prolapse, prothrombin*

proct-, procto- rectum: *proctitis, proctocele, proctologist*

pseud-, pseudo- false: *pseudoarthrosis, pseudostratified, pseudopod*

psych-, psycho- mind: *psychosomatic, psychotherapy*

-ptosis downward displacement, falling, prolapse: *blepharoptosis, enteroptosis, nephroptosis*

pulmo-, pulmono- lung: *pulmonic, pulmonology*

py-, pyo- pus: *pyuria, pyogenic, pyorrhea*

pyel-, pyelo- kidney pelvis: *pyelitis, pyelogram, pyelonephrosis*

rachi-, rachio- spine: *rachicentesis, rachischisis*

radio- emission of rays or radiation: *radioactive, radiography, radiology*

re- again, back: *reabsorption, reaction, regenerate*

ren- kidney: *renal, renopathy*

retro- backward, located behind: *retrocecal, retroperitoneal*

rheo- flow of matter, or of a current of electricity: *rheology, rheoscope*

rhin-, rhino- nose: *rhinitis, rhinoplasty*

-rrhage, -rrhagia excessive flow: *hemorrhage, menorrhagia*

-rrhaphy suturing of or sewing up of a gap or defect in a part: *enterorrhaphy, herniorrhaphy, perineorrhaply*

-rrhea flow, discharge: *diarrhea, gonorrhea, seborrhea*

salping-, salpingo- tube: *salpingitis, salpingoscopy*

scler-, sclero- hardness; *scleroderma, sclerosis*

scolio- twisted, crooked: *scoliosis, scoliosometer*

-scope instrument used to look into or examine a part: *bronchoscope, cystoscope, endoscope*

semi- mild, partial, half: *semipermeable, semicoma*

sep-, septic- poison, rot, decay: *sepsis, septicemia*

-sis condition or process, usually abnormal: *dermatosis, osteoporosis*

somat-, somato- body: *somatic, somatotype*

sono- sound: *sonogram, sonography*

splanchn-, splanchno- internal organs: *splanchnic, splanchnoptosis*

sta-, stat- stop, stand still, remain at rest: *stasis, static*

sten-, steno- contracted, narrowed: *stenosis*

sthen-, stheno-, -sthenia, -sthenic strength: *asthenic, calisthenics, neurasthenia*

stomato- mouth: *stomatitis*

sub- under, below, near, almost: *subclavian, subcutaneous, subluxation*

super- over, above, excessive: *superego, supernatant, superficial*

supra- location above or over: *supranasal, suprarenal*

sym-, syn- with, together: *symphysis, synapse*

syring-, syringo- fistula, tube, cavity: *syringectomy, syringomyelia*

tacho-, tachy- rapid, fast, swift: *tachycardia, tachypnea*

tars-, tarso- eyelid, foot: *tarsitis, tarsoplasty, tarsoptosis*

-taxia, -taxis order, arrangement: *ataxia, chemotaxis, thermotaxis*

tens- stretch, pull: *extension, tensor*

therm-, thermo-, -thermy heat: *thermalgesia, thermocautery, diathermy, thermometer*

-tomy incision of, cutting: *anatomy, arthrotomy, laparotomy*

tox-, toxic-, toxico-   poison: *toxemia, toxicology, toxicosis*

trache-, tracheo-   windpipe: *trachea, tracheitis, tracheotomy*

trans-   across, through, beyond: *transorbital, transpiration, transplant, transport*

tri-   three: *triad, triceps*

trich-, tricho-   hair: *trichiasis, trichocardia, trichosis*

-trophic, -trophy   nutrition (nurture): *atrophic, hypertrophy*

-tropic   turning around, influencing, changing: *adrenotropic, gonadotropic, thyrotropic*

ultra-   beyond or excessive: *ultrasound, ultraviolent, ultrastructure*

uni-   one: *unilateral, uniovular, unicellular*

-uria   urine: *glycosuria, hematuria, pyuria*

vas-, vaso-   vessel, duct: *vascular, vasectomy, vasodilation*

viscer-, viscero-   internal organs: *viscera, visceroptosis*

xero-   dryness: *xeroderma, xerophthalmia, xerosis*

Appendices

Unit	Abbreviation	Metric Equivalent	U.S. Equivalent
Units of length			
Kilometer	km	1000 meters	0.62 miles; 1.6 km/mile
Meter*	m	100 cm; 1000 mm	39.4 inches, 1.1 yards
Centimeter	cm	$1/100$ m; 0.01 m	0.39 inches; 2.5 cm/inch
Millimeter	mm	$1/1000$ m; 0.001 m	0.039 inches; 25 mm/inch
Micrometer	μm	$1/1000$ mm; 0.001 mm	
Units of weight			
Kilogram	kg	1000 g	2.2 lb
Gram*	g	1000 mg	0.035 oz.; 28.5 g/oz
Milligram	mg	$1/1000$ g; 0.001 g	
Microgram	μg	$1/1000$ mg; 0.001 mg	
Units of volume			
Liter*	L	1000 mL	1.06 qt
Deciliter	dL	$1/10$ L; 0.1 L	
Milliliter	mL	$1/1000$ L; 0.001 L	0.034 oz., 29.4 mL/oz
Microliter	μL	$1/1000$ mL; 0.001 mL	

*Basic unit

APPENDIX 2. Celsius–Fahrenheit Temperature Conversion Scale

Celsius to Fahrenheit	Fahrenheit to Celsius

Use the following formula to convert Celsius readings to Fahrenheit readings:

$$°F = 9/5°C + 32$$

For example, if the Celsius reading is 37°:

$°F = (9/5 \times 37) + 32$
$= 66.6 + 32$
$= 98.6°F$ (normal body temperature)

Use the following formula to convert Fahrenheit readings to Celsius readings:

$$°C = 5/9 (°F - 32)$$

For example, if the Fahrenheit reading is 68°:

$°C = 5/9 (68 - 32)$
$= 5/9 \times 36$
$= 20°C$ (a nice spring day)

temperature
conversion
scale

APPENDIX 3. Laboratory Tests

Table 1. Routine Urinalysis

Test	Normal Value	Clinical Significance
General characteristics and measurements		
Color	Pale yellow to amber	Color change can be due to concentration or dilution, drugs, metabolic or inflammatory disorders
Odor	Slightly aromatic	Foul odor typical of urinary tract infection, fruity odor in uncontrolled diabetes mellitus
Appearance (clarity)	Clear to slightly hazy	Cloudy urine occurs with infection or after refrigeration; may indicate presence of bacteria, cells, mucus, or crystals
Specific gravity	1.003–1.030 (first morning catch; routine is random)	Decreased in diabetes insipidus, acute renal failure, water intoxication; increased in liver disorders, heart failure, dehydration
pH	4.5–8.0	Acid urine accompanies acidosis, fever, high protein diet; alkaline urine in urinary tract infection, metabolic alkalosis, vegetarian diet
Chemical determinations		
Glucose	Negative	Glucose present in uncontrolled diabetes mellitus, steroid excess
Ketones	Negative	Present in diabetes mellitus and in starvation
Protein	Negative	Present in kidney disorders, such as glomerulonephritis, acute kidney failure
Bilirubin	Negative	Breakdown product of hemoglobin; present in liver disease or in bile blockage
Urobilinogen	0.2–1.0 Ehrlich units/dL	Breakdown product of bilirubin; increased in hemolytic anemias and in liver disease; remains negative in bile obstruction
Blood (occult)	Negative	Detects small amounts of blood cells, hemoglobin, or myoglobin; present in severe trauma, metabolic disorders, bladder infections
Nitrite	Negative	Product of bacterial breakdown of urine; positive result suggests urinary tract infection and needs to be followed up with a culture of the urine
Microscopic		
Red blood cells	0–3 per high-power field	Increased due to bleeding within the urinary tract from trauma, tumors, inflammation, or damage within the kidney
White blood cells	0–4 per high-power field	Increased in infection of the kidney or bladder
Renal epithelial cells	Occasional	Increased number indicates damage to kidney tubules
Casts	None	Hyaline casts normal; large number of abnormal casts indicates inflammation or a systemic disorder
Crystals	Present	Most are normal; may be acid or alkaline
Bacteria	Few	Increased in infection of urinary tract or contamination from infected genitalia
Others		Any yeasts, parasites, mucus, spermatozoa, or other microscopic findings would be reported here

Table 2. Complete Blood Count (CBC)

Test	Normal Value*	Clinical Significance
Red blood cell count (RBC)	Men: 4.2–5.4 million/μL Women: 3.6–5.0 million/μL	Decreased in anemia; increased in dehydration, polycythemia
Hemoglobin (HGB)	Men: 13.5–17.5 g/dL Women: 12–16 g/dL	Decreased in anemia, hemorrhage and hemolytic reactions; increased in dehydration, heart and lung disease
Hematocrit (HCT) or packed cell volume (PCV)	Men: 40%–50% Women: 37%–47%	Decreased in anemia; increased in polycythemia, dehydration
Red blood cell indices (examples)		These values, calculated from the RBC, HGB, and HCT, give information valuable in the diagnosis and classification of anemia
Mean corpuscular volume (MCV)	87–103 μL/red cell	Measures the average size or volume of each red blood cell: Small size (microcytic) in iron-deficiency anemia; large size (macrocytic) typical of pernicious anemia
Mean corpuscular hemoglobin (MCH)	26–34 pg/red cell	Measures the weight of hemoglobin per red blood cell; useful in differentiating types of anemia in a severely anemic patient
Mean corpuscular hemoglobin concentration (MCHC)	31–37 g/dL	Defines the volume of hemoglobin per red blood cell; used to determine the color or concentration of hemoglobin per red cell
White blood count (WBC)	5,000–10,000/μL	Increased in leukemia and in response to infection, inflammation, and dehydration; decreased in bone marrow suppression
Platelets	150,000–350,000/μL	Increased in many malignant disorders; decreased in disseminated intravascular coagulation (DIC) or toxic drug effects; spontaneous bleeding may occur at platelet counts below 20,000
Differential (Peripheral blood smear)		A stained slide of the blood is needed to perform the differential. The percentages of the different white cells are estimated, and the slide is microscopically checked for abnormal characteristics in WBCs, RBCs, and platelets.
White cells		
Segmented neutrophils (SEGs, POLYs)	40%–74%	Increased in bacterial infections; low numbers leave person very susceptible to infection
Immature neutrophils (BANDs)	0%–3%	Increase when neutrophil count increases
Lymphocytes (LYMPHs)	20%–40%	Increased in viral infections; low numbers leave a person dangerously susceptible to infection
Monocytes (MONOs)	2%–6%	Increased in specific infections
Eosinophils (EOs)	1%–4%	Increased in allergic disorders
Basophils (BASOs)	0.5%–1%	Increased in allergic disorders

*Values vary depending on instrumentation and type of test.

Table 3. Blood Chemistry Tests

Test	Normal Value	Clinical Significance
Basic panel: An overview of electrolytes, waste product management, and metabolism		
Blood urea nitrogen (BUN)	7–18 mg/dL	Increased in renal disease and dehydration; decreased in liver damage and malnutrition
Carbon dioxide (CO_2) (includes bicarbonate)	23–30 mmol/L	Useful to evaluate acid–base balance by measuring total carbon dioxide in the blood: Elevated in vomiting and pulmonary disease; decreased in diabetic acidosis, acute renal failure, and hyperventilation
Chloride (Cl)	98–106 mEq/L	Increased in dehydration, hyperventilation, and congestive heart failure; decreased in vomiting, diarrhea, and fever
Creatinine	0.6–1.2 mg/dL	Produced at a constant rate and excreted by the kidney; increased in kidney disease
Glucose	Fasting: 70–110 mg/dL Random: 85–125 mg/dL	Increased in diabetes and severe illness; decreased in insulin overdose or hypoglycemia
Potassium (K)	3.5–5 mEq/L	Increased in renal failure, extensive cell damage and acidosis; decreased in vomiting, diarrhea, and excess administration of diuretics or IV fluids
Sodium (Na)	101–111 mEq/L or 135–148 mEq/L (depending on test)	Increased in dehydration and diabetes insipidus; decreased in overload of IV fluids, burns, diarrhea, or vomiting
Additional blood chemistry tests		
Alanine aminotransferase (ALT)	Men: 7–24 U/L Women: 7–17 U/L	Used to diagnose and monitor treatment of liver disease and to monitor the effects of drugs on the liver; increased in myocardial infarction
Albumin	3.8–5.0 g/dL	Albumin holds water in blood; decreased in liver disease and kidney disease
Albumin–globulin ratio (A/G ratio)	Greater than 1	Low A/G ratio signifies a tendency for edema because globulin is less effective than albumin at holding water in the blood.
Alkaline phosphatase (ALP)	20–70 U/L (varies by method)	Enzyme of bone metabolism; increased in liver disease and metastatic bone disease
Amylase	21–160 U/L	Used to diagnose and monitor treatment of acute pancreatitis and to detect inflammation of the salivary glands
Aspartate aminotransferase (AST)	0–41 U/L (varies)	Enzyme present in tissues with high metabolic activity; increased in myocardial infarction and liver disease
Bilirubin, total	0.2–1.0 mg/dL	Breakdown product of hemoglobin from red blood cells; increased when excessive RBCs are being destroyed or in liver disease
Calcium (Ca)	8.8–10.0 mg/dL	Increased in excess parathyroid hormone production and in cancer; decreased in alkalosis, elevated phosphate in renal failure, and excess IV fluids
Cholesterol	120–220 mg/dL desirable range	Screening test used to evaluate risk of heart disease; levels of 200 or above indicate increased risk of heart disease and warrant further investigation

(continued)

Table 3. *(Continued)*

Test	Normal Value	Clinical Significance
Creatine phosphokinase (CPK or CK)	Men: 38–174 U/L Women: 96–140 U/L	Elevated enzyme level indicates myocardial infarction or damage to skeletal muscle. When elevated, specific fractions (isoenzymes) are tested for
Gamma glutamyl transferase (GGT)	Men: 6–26 U/L Women: 4–18 U/L	Used to diagnose liver disease and test for chronic alcoholism
Globulins	2.3–3.5 g/dL	Proteins active in immunity; help albumin keep water in blood
Iron, serum (Fe)	Men: 75–175 μg/dL Women: 65–165 μg/dL	Decreased in iron deficiency and anemia; increased in hemolytic conditions
High-density lipoproteins (HDLs)	Men: 30–70 mg/dL Women: 30–85 mg/dL	Used to evaluate the risk of heart disease
Lactic dehydrogenase (LDH or LD)	95–200 U/L (normal ranges vary greatly)	Enzyme released in many kinds of tissue damage; including myocardial infarction, pulmonary infarction, and liver disease
Lipase	4–24 U/L (varies with test)	Enzyme used to diagnose pancreatitis
Low-density lipoproteins (LDLs)	80–140 mg/dL	Used to evaluate the risk of heart disease
Magnesium (Mg)	1.3–2.1 mEq/L	Vital in neuromuscular function; decreased levels may occur in malnutrition, alcoholism, pancreatitis, diarrhea
Phosphorus (P) (inorganic)	2.7–4.5 mg/dL	Evaluated in response to calcium; main store is in bone: elevated in kidney disease; decreased in excess parathyroid hormone
Protein, total	6–8 g/dL	Increased in dehydration, multiple myeloma; decreased in kidney disease, liver disease, poor nutrition, severe burns, excessive bleeding
Serum glutamic oxalacetic transaminase (SGOT)		See Aspartate aminotransferase (AST)
Serum glutamic pyruvic transaminase (SGPT)		See alanine aminotransferase (ALT)
Thyroxin (T_4)	5–12.5 μg/dL (varies)	Screening test of thyroid function; increased in hyperthyroidism; decreased in myxedema and hypothyroidism
Thyroid stimulating hormone (TSH)	0.5–6 mIU/L	Produced by pituitary to cause thyroid gland to function; elevated when thyroid gland is not functioning
Triiodothyronine (T_3)	120–195 mg/dL	Elevated in specific types of hyperthyroidism
Triglycerides	Men: 40–160 mg/dL Women: 35–135 mg/dL	An indication of ability to metabolize fats; increased triglycerides and cholesterol indicate high risk of atherosclerosis
Uric acid	Men: 3.5–7.2 mg/dL Women: 2.6–6.0 mg/dL	Produced by breakdown of ingested purines in food and nucleic acids; elevated in kidney disease, gout, and leukemia

APPENDIX 4. Typical Disease Conditions and Causative Organisms

Table 1. Bacterial Diseases

Organism	Disease and Description
Cocci	
Neisseria gonorrhoeae (gonococcus)	Gonorrhea. Acute inflammation of mucous membranes of the reproductive and urinary tracts (with possible spread to the peritoneum in the female). Systemic infection may cause gonococcal arthritis and endocarditis. Organism also causes ophthalmia neonatorum, an eye inflammation of the newborn.
Neisseria meningitidis (meningococcus)	Epidemic meningitis. Inflammation of the membranes covering brain and spinal cord.
Staphylococcus aureus and other staphylococci	Boils, carbuncles, impetigo, osteomyelitis, staphylococcal pneumonia, cystitis, pyelonephritis, empyema, septicemia, toxic shock, and food poisoning. Strains resistant to antibiotics are a cause of infections originating in the hospital, such as wound infections.
Streptococcus pneumoniae (*Diplococcus pneumoniae*)	Pneumonia; inflammation of the alveoli, bronchioles, and bronchi; middle ear infections; meningitis. May be prevented by use of polyvalent pneumococcal vaccine.
Streptococcus pyogenes, Streptococcus hemolyticus, and other streptococci	Septicemia, septic sore throat, scarlet fever, puerperal sepsis, erysipelas, streptococcal pneumonia, rheumatic fever, subacute bacterial endocarditis, acute glomerulonephritis.
Bacilli	
Bordetella pertussis	Pertussis (whooping cough). Severe infection of the trachea and bronchi. The "whoop" is caused by the effort to recover breath after coughing. All children should be inoculated with appropriate vaccine.
Brucella abortus (and others)	Brucellosis, or undulant fever. Disease of animals such as cattle and goats transmitted to humans through unpasteurized dairy products or undercooked meat. Acute phase of fever and weight loss; chronic disease with abscess formation and depression.
Clostridium botulinum	Botulism. Very severe poisoning caused by eating food in which the organism has been allowed to grow and excrete its toxin. Can cause paralysis of the muscles and death from asphyxiation.
Clostridium perfringens	Gas gangrene. Acute wound infection. Organisms cause death of tissues accompanied by the generation of gas within them.
Clostridium tetani	Tetanus. Acute, often fatal poisoning caused by introduction of the organism into deep wounds. Characterized by severe muscular spasms. Also called *lockjaw.*
Corynebacterium diphtheriae	Diphtheria. Acute inflammation of the throat with the formation of a leathery membrane-like growth (pseudomembrane) that can obstruct air passages and cause death by asphyxiation. Toxin produced by this organism can damage heart, nerves, kidneys, and other organs. Disease preventable by appropriate vaccination.
Escherichia coli, Proteus bacilli, and other colon bacilli	Normal inhabitants of the colon, and usually harmless there. Cause of local and systemic infections, food poisoning, diarrhea (especially in children), septicemia, and septic shock. *E. coli* is a common hospital-acquired infection.
Francisella tularensis	Tularemia, or deer fly fever. Transmitted by contact with an infected animal or bite of a tick or fly. Symptoms are fever, ulceration of the skin, and enlarged lymph nodes.
Haemophilus influenzae type b (HIB)	Severe infections in children under 3 years of age. Causes meningitis, also epiglottitis, septicemia, pneumonia, pericarditis, and septic arthritis. A vaccine is available.
Helicobacter pylori	Acute inflammation of the stomach (gastritis), ulcers of the pyloric area of the stomach and of the duodenum.
Legionella pneumophila	Legionnaires' disease (pneumonia). Seen in localized epidemics, may be transmitted by air conditioning towers and by contaminated soil at excavation sites. Not spread person to person. Characterized by high fever, vomiting, diarrhea, cough, and bradycardia. Mild form of the disease called *Pontiac fever.*

(continued)

Table 1. *(Continued)*

Organism	Disease and Description
Mycobacterium leprae (Hansen's bacillus)	Leprosy. Chronic illness in which hard swellings occur under the skin, particularly of the face, causing a grotesque appearance. In one form of leprosy the nerves are affected, resulting in loss of sensation in the extremities.
Mycobacterium tuberculosis (tubercle bacillus)	Tuberculosis. Infectious disease in which the organism causes primary lesions called *tubercles.* These break down into cheeselike masses of tissue, a process known as *caseation.* Any body organ can be infected, but in adults the usual site is the lungs. Still one of the most widespread diseases in the world, tuberculosis is treated with chemotherapy; resistant strains of the bacillus have developed.
Pseudomonas aeruginosa	Ubiquitous organism is a frequent cause of wound and urinary infections in debilitated hospitalized patients. Often found in solutions that have been standing for long periods.
Salmonella typhi (and others)	Salmonellosis occurs as enterocolitis, bacteremia, localized infection, or typhoid. Depending on type, presenting symptoms may be fever, diarrhea, or abscesses; complications include intestinal perforation and endocarditis. Carried in water, milk, meat, and other food.
Shigella dysenteriae (and others)	A serious bacillary dysentery. Acute intestinal infection with diarrhea (sometimes bloody); may cause dehydration with electrolyte imbalance or septicemia. Transmitted through fecal–oral route or other poor sanitation.
Yersinia pestis (formerly called *Pasteurella pestis*)	Plague, the "black death" of the Middle Ages. Transmitted by fleas from infected rodents to humans. Symptoms of the most common form are swollen, infected lymph nodes or *buboes.* Another form may cause pneumonia. All forms may lead to a rapidly fatal septicemia.

Note: The following organisms are smaller than other bacteria and vary in shape. Like viruses, they grow within cells, but they differ from viruses in that they are affected by antibiotics.

Chlamydia oculogenitalis	Inclusion conjunctivitis, acute eye infection. Carried in genital organs, transmitted during birth or through water in inadequately chlorinated swimming pools.
Chlamydia psittaci	Psittacosis, also called *ornithosis.* Disease transmitted by various birds, including parrots, ducks, geese, and turkeys. Primary symptoms are chills, headache, and fever, more severe in older persons. The duration may be from 2 to 3 weeks, often with a long convalescence. Antibiotic drugs are effective remedies.
Chlamydia trachomatis	A sexually transmitted disease causing pelvic inflammatory disease and other infections of the reproductive tract. Also causes inclusion conjunctivitis, an acute eye infection, and trachoma, a chronic infection that is a common cause of blindness in underdeveloped areas of the world. Infection of the conjunctiva and cornea characterized by redness, pain, and lacrimation. Antibiotic therapy is effective if begun before there is scarring. The same organism causes lymphogranuloma venereum (LGV), a sexually transmitted disease characterized by swelling of inguinal lymph nodes and accompanied by signs of general infection.
Coxiella burnetti	Q fever. Infection transmitted from cattle, sheep, and goats to humans by contaminated dust and also carried by arthropods. Symptoms are fever, headache, chills, and pneumonitis. This disorder is almost never fatal.
Rickettsia prowazekii	Epidemic typhus. Transmitted to humans by lice; associated with poor hygiene and war. Main symptoms are headache, hypotension, delirium, and a red rash. Frequently fatal in older persons.
Rickettsia rickettssi	Rocky Mountain spotted fever. Tick-borne disease occurring throughout the United States. Symptoms are fever, muscle aches, and a rash that may progress to gangrene over bony prominences. The disease is rarely fatal.
Rickettsia typhi	Endemic or murine typhus. A milder disease transmitted to man from rats by fleas. Symptoms are fever, rash, headache, and cough. The disease is rarely fatal.

Table 1. *(Continued)*

Organism	Disease and Description
Curved rods	
Vibrio	
Vibrio cholerae (Vibrio comma)	Cholera. Acute infection of the intestine characterized by prolonged vomiting and diarrhea, leading to severe dehydration, electrolyte imbalance, and in some cases death.
Spirochetes	
Borrelia burgdorferi	Lyme disease, transmitted by the extremely small deer tick. Symptoms include rash, palsy, and joint inflammation, which may become chronic.
Borrelia recurrentis (and others)	Relapsing fever. Generalized infection in which attacks of fever alternate with periods of apparent recovery. Organisms spread by lice, ticks, and other insects.
Treponema pallidum	Syphilis. Infectious disease transmitted mainly by sexual intercourse. Untreated syphilis is seen in the following three stages: primary—formation of primary lesion (chancre); secondary—skin eruptions and infectious patches on mucous membranes; tertiary—development of generalized lesions (gummas) and destruction of tissues resulting in aneurysm, heart disease, and degenerative changes in brain, spinal cord and ganglia, meninges. Also a cause of intrauterine fetal death or stillbirth.
Treponema vincentii (formerly called *Borrelia vincentii*)	Vincent's disease (trench mouth). Infection of the mouth and throat accompanied by formation of a pseudomembrane, with ulceration.

Table 2. Fungal Diseases

Disease/Organism	Description
Actinomycosis	"Lumpy jaw," which occurs in cattle and humans. The organisms cause the formation of large masses of tissue, which are often accompanied by abscesses. The lungs and liver may be involved.
Blastomycosis (Blastomyces dermatitidis)	A general term for any infection caused by a yeast-like organism. There may be skin tumors and lesions in the lungs, bones, liver, spleen, and kidneys.
Candidiasis *(Candida albicans)*	An infection that can involve the skin and mucous membranes. May cause diaper rash, infection of the nail beds, and infection of the mucous membranes of the mouth (thrush), throat, and vagina.
Coccidioidomycosis *(Coccidioides immitis)*	Also called *San Joaquin Valley fever*, it is another systemic fungal disease. Because it often attacks the lungs, it may be mistaken for tuberculosis.
Histoplasmosis *(Histoplasma capsulatum)*	This fungus may cause a variety of disorders, ranging from mild respiratory symptoms or enlargement of liver, spleen, and lymph nodes to cavities in the lungs with symptoms similar to those of tuberculosis.
Ringworm tinea capitis tinea corporis tinea pedis	Common fungal infections of the skin, many of which cause blisters and scaling with discoloration of the affected areas. All are caused by similar organisms from a group of fungi called *dermatophytes*. They are easily transmitted from person to person or by contaminated articles.

Table 3. Viral Diseases

Organism	Disease and Description
Cytomegalovirus (CMV)	Common mild infection of the salivary glands. In an immunosuppressed person may cause infection of the retina, lung, and liver, ulceration of GI tract, and inflammation of the brain. Causes severe fetal or neonatal damage.
Hepatitis A virus (HAV)	Liver inflammation that is rarely fatal. Transmitted by fecal contamination. Does not become chronic or produce carrier state. Infection provides lifetime immunity. Vaccine now available.
Hepatitis B virus (HBV)	Liver inflammation. Transmitted by direct exchange of blood and body fluids. Can cause rapidly fatal disease or develop into chronic disease and carrier state. Risk of progress to liver cancer. Vaccine available.
Hepatitis C virus (HCV)	Liver inflammation. Known as *post-transfusion hepatitis*. Can cause rapidly fatal disease or develop into chronic state.
Hepatitis D virus (HDV)	Liver inflammation. Occurs as coinfection with hepatitis B. Responsible for half of cases of rapidly fatal liver failure as well as a high rate of chronic disease that progresses to death.
Hepatitis E virus (HEV)	Liver inflammation. Occurs in epidemics in certain areas of the world; resembles hepatitis A. Can be fatal in pregnant women.
Herpes simplex virus type 1	Cold sores or fever blisters that appear about the mouth and nose of persons with colds or other illnesses accompanied by fever.
Herpes simplex virus type 2	Genital herpes. Acute inflammatory disease of the genitalia, often recurring. A very common sexually transmitted disease.
Human immunodeficiency virus (HIV)	Acquired immunodeficiency syndrome (AIDS). Fatal disease which infects T lymphocytes of the immune system. Diagnosed by decline in specific (CD4) cells and presenting disease, including *Pneumocystis carinii* pneumonia, Kaposi's sarcoma, acute viral syndrome, persistent swelling of lymph nodes (lymphadenopathy), chronic diarrhea, and wasting syndrome. Spread through sexual contact and contact with contaminated blood and body fluids.
Human papillomavirus (HPV)	Genital warts (condylomata acuminata). Sexually transmitted warts of the genital and perianal area in men and women. Associated with certain types of cervical dysplasia and cancer.
Influenza virus	An epidemic viral infection, marked by chills, fever, muscular pains, and prostration. The most serious complication is bronchopneumonia caused by *Haemophilus influenzae* (a bacillus) or streptococci.
Mumps virus	Epidemic parotitis. Acute inflammation with swelling of the parotid salivary glands. Mumps can have many complications, such as orchitis (inflammation of the testes) in young men and meningitis in young children.
Poliovirus	Poliomyelitis. Acute viral infection that may attack the anterior horns of the spinal cord, resulting in paralysis of certain voluntary muscles. North and South America have been declared free of polio as a result of vaccination programs.
Rhabdovirus	Rabies. An acute, fatal disease transmitted to humans through the saliva of an infected animal. Rabies is characterized by violent muscular spasms induced by the slightest sensations. Because the swallowing of water causes spasms of the throat, the disease is also call *hydrophobia* ("fear of water"). The final stage of paralysis ends in death.
Rubella virus	German measles. A less severe form of measles, but especially dangerous during the first 3 months of pregnancy because the disease organism can cause heart defects, deafness, mental deficiency, and other permanent damage in the fetus.

(continued)

Table 3. *(Continued)*

Organism	Disease and Description
Rubeola virus	Measles. An acute respiratory inflammation followed by fever and a generalized skin rash. Patients are prone to the development of dangerous complications, such as bronchopneumonia and other secondary infections caused by staphylococci and streptococci.
Varicella-zoster herpesvirus	Chickenpox (varicella). A usually mild infection, almost completely confined to children, characterized by blister-like skin eruptions. Vaccine now available.
	Shingles (herpes zoster). A very painful eruption of blisters of the skin that follows the course of certain peripheral nerves. These blisters eventually dry up and form scabs that resemble shingles.
Other viral diseases	
Common cold (coryza)	A viral infection of the upper respiratory tract. Victims are highly susceptible to complications such as pneumonia and influenza.
Viral encephalitis	*Encephalitis* usually is understood to be any brain inflammation accompanied by degenerative tissue changes, and it can have many causes besides viruses. There are several forms of viral encephalitis (Western and Eastern epidemic, equine, St. Louis, Japanese B, etc.), some of which are known to be transmitted from birds and other animals to humans by insects, principally mosquitoes.
Viral pneumonia	Caused by a number of different viruses, such as the influenza and parainfluenza viruses, adenoviruses, and varicella viruses.

Table 4. Protozoan Diseases

Organism	Disease and Description
Amebae	
Entamoeba histolytica	Amebic dysentery. Severe ulceration of the wall of the large intestine caused by amebae. Acute diarrhea may be an important symptom. This organism also may cause liver abscesses.
Ciliates	
Balantidium coli	Gastrointestinal disturbances and ulcers of the colon.
Flagellates	
Giardia lamblia	Gastrointestinal disturbances.
Leishmania donovani (and others)	Kala-azar. In this disease there is enlargement of the liver and spleen, as well as skin lesions.
Trichomonas vaginalis	Inflammation and discharge from the vagina of the female. In the male it involves the urethra and causes painful urination.
Trypanosoma	African sleeping sickness. Disease begins with a high fever, followed by invasion of the brain and spinal cord by the organisms. Usually the disease ends with continued drowsiness, coma, and death.
Sporozoa	
Cryptosporidium	Cramps and diarrhea that can be long term and severe in persons with a weakened immune system, such as those with AIDS.
Plasmodium; varieties include *vivax, falciparum, malariae*	Malaria. Characterized by recurrent attacks of chills followed by high fever. Severe attacks of malaria can be fatal because of kidney failure, cerebral disorders, and other complications.
Pneumocystis carinii	Pneumonia. Opportunistic infection in persons with depressed immune systems.
Toxoplasma gondii	Toxoplasmosis. Common infectious disease transmitted by cats and raw meat. Mild forms cause fever and enlargement of lymph nodes. May cause fatal encephalitis in immunosuppressed persons. Infection of a pregnant woman is a cause of fetal stillbirth or congenital damage.

The letter *f* following a page number indicates a figure; *n* denotes a footnote; and *t* indicates tabular material.

A

A antigens, transfusion and, 207–8
abdomen
    frontal view of, 10*f*
    muscles of, 126–27, 127*f*
    quadrants of, 8, 11*f*
    regions of, 8
abdominal aorta, 241, 244
abdominal cavity, 8
    diagram of, 305*f*
    frontal view of, 10*f*
abdominopelvic cavity, 8
    frontal view of, 10*f*
abdominopelvic organs disorders,
    backache and, 111
abducens nerve, function of, 163
abduction, 107
abnormal blood pressure, 253
abnormal body temperature, 328–29
abnormal respiration, 294
abnormal uterine bleeding, 371
abnormal ventilation, effects of, 294
ABO blood group system, 210*t*
abortions, 387
abruptio placentae, 386
abscess, 205
absorption, 302, 316
accessory nerve, function of, 163
accessory organs
    of digestion, 302, 313*f*
        gallbladder, 315
        liver, 313–15
        pancreas, 315
    of reproductive system, 360
accommodation, 172
ACE inhibitors, 351
acetabulum, 97
acetaminophen, 182
acetylcholine (ACh), 139
ACh. *See* acetylcholine
Achilles tendon, 129
acid, 20, 21*f*
acid-base balance, 336
acid-fast organisms, 69
acid-fast stain, 69
acidity, pH scale measures degree of, 21*f*
acidosis, 336
acne, 80
acne vulgaris, 80
acoustic nerve, 163

acquired immunity, 274, 275–76
    artificial, 277–79
    natural, 276–77
acquired immunodeficiency syndrome
    (AIDS), 280. *See also* AIDS
acromegaly, 189
acromion, 94
acrosome, 363
ACTH. *See* adrenocorticotropic
    hormone
actin, 116
action potential, 116, 138
active acquired immunity, 274
active immunity, 276
active transport, 34–35
acute coryza, 295–96
acute disease, 57
acute glomerulonephritis, 350–51
acute pancreatitis, 315
acute renal failure, 351
Adam's apple, 286
Adams-Stokes syndrome, 230
adaptation, 180
Addison's disease, 192, 336
adduction, 107
adductor muscles, 128
adeno, 49
adenocarcinoma, 310
adenoids, 266
adenoma, 49
adenomatous goiter, 190
adenosine triphosphate (ATP), 5
ADH. *See* antidiuretic hormone
adhesion, 298
adipose capsule, 340
adipose tissue, 42
adrenal glands, 191–92
adrenal steroids, 195
adrenaline, 139
    effects of, 191
    heat production and, 325–26
    in treatment, 195
adrenergic system, 147
adrenocorticotropic hormone (ACTH),
    186, 189, 196
adult polycystic kidney, 395
adult spine, curves of, 97*f*
aerobic bacteria, 59
aerobic reactions, 322
afferent arteriole, 341–42

afferent nerves, 139
afferent neurons, 137–38
African sleeping sickness, 161
afterbirth, 383
Ag. *See* antigens
agglutination, 208
agglutinins, 207
agglutinogens, 207
aging
    disease and, 56
    effects of
        on hair, 79
        on the kidneys, 353
        on male reproduction, 363–64
        on muscles, 129
        on nails, 79
        on skin, 78–79
    skeletal changes and, 103–4
agranulocytes, 205
AIDS. *See also* HIV virus
    syphilis and, 365
    update on, 280
air, 292
air movement, 290–91, 292*t*
air passageways, 288–89, 289*f*
albinism, 396
albumin
    function of, 203
    in urine, 354
albuminuria, 354
alcohol, pregnancy and, 395
aldosterone, 192, 336
alimentary tract, 303
alkaline phosphatase, 214
alkalinity, measurement of, 21*f*
alkalosis, 336
allergens, 279
allergy, 279
alopecia, 81
alveoli, 289
Alzheimer's disease, 162
amebas, 62
ameboid motion, 205
amenorrhea, 370
amino acid compounds, 186
amino acids, 22, 323
    in plasma, 204
amniocentesis, 397
amniotic fluid, 382
amniotic sac, 382

amphiarthrosis, 105
amylase, 316
amyotrophic lateral sclerosis, 143
anabolic steroids, 120
anabolism, 5
    metabolism and, 322, 322–23
anaerobic bacteria, 59
anaerobic reactions, 322
anal canal, 311
analgesics, 182
anaphase, 32
anaphylactic shock, 255
anaphylaxis, treatment of, 195
anastomosis, 244, 246*f*
anatomic position, 5
anatomy, 4
androgens, 196
anemia, 210–11
    red cells and, 211–12
    renal failure and, 351
anesthetics, 182
aneurysm, 253
aneurysm of the aorta, 253
angina pectoris, 229
angioma, 49
angioplasty, 233
angiotensin, 343, 351
angiotensin I, 351
angiotensin II, 351
angiotensin-converting enzyme
        inhibitors, 351
anion, 19
ankle, bones of, 100
anorexia, 318
anorexia nervosa, 318
ANP. *See* atrial natriuretic peptide
ANS. *See* autonomic nervous system
antagonist, 120
anterior attachment, 94
anterior direction, 5, 6*f*
anterior fontanelle, 92–93, 93*f*
anterior lobe of pituitary gland, 186,
        188*f*
    hormones of, 188–89
anterior superior iliac spine, 97
anti-A serum, 208
antiarrhythmic agents, 231
anti-B serum, 208
antibiotics, 67–68
    allergic reactions to, 279
    resistance to, 68
antibodies, 207, 276
    B cells and, 275–76
    function of, 203
anticoagulants, 207
    in treatment of heart disease, 232
antidiuretic hormone (ADH), 186, 187,
        337
    concentration of urine and, 347
    function of, 189
antigen-antibody reaction, 276
antigens (Ag), 207
    acquired immunity and, 275
antihistamines, 279
antineoplastic agents, 68
    in treatment of cancer, 50
antipyretic drug, 328
antisepsis, 66–67

antiseptics, 66–67
antisera, 278–79
antiserum, 278
anti-snake bite sera, 279
antitoxins, 278
antivenins, 279
anus, 311
aorta, 223, 226*f*, 239, 239–40, 243*f*
    aneurysm of, 253
    branches of, 242*f*
    coarctation of, 228
    sections of, 240–41
aortic arch, 240
    branches of, 241
aortic bodies, 293–94
aortic valve, 223
aphasia, 162
aplastic anemia, 212
apnea, 294
aponeurosis, 119–20, 121*f*, 122*f*
appendicitis, 311
appendicular skeleton, 90
    bones of
        lower extremity, 97, 100, 102*f*
        pelvic girdle, 97
        shoulder girdle, 94, 98*f*
        upper extremity, 94, 96, 97
appendix, 311
appetite, 318
aqueous humor, 170
arachnoid, 153
areolar, 42
arm
    movement of, 126
    bones of, 94, 97
arrhythmia, 229
arterial arches, 244
arterial degeneration, 254*f*, 254–55
arterial obstruction, 256–57
arteries, 221, 238
    to brain, 245*f*
    coronary, 224*f*
    cross section of, 241*f*
    hardening of, 253
    systemic, 239–41, 243*f*, 244
    walls of, 239, 240*f*
arterioles, 238
arteriosclerosis, 254
arteriosclerosis obliterans, 256–57
arteriovenous shunt, 246*f*
arthritis, 108–9, 111
articular cartilage, 105
articulation, 104
artificial pacemakers, 232
artificial respiration, 299
artificially acquired immunity, 274,
        277–79
ascaris, 64
ascending aorta, 240
    branches of, 241
ascending colon, 311
ascites, 337
asepsis, 66
aseptic methods, 66–67, 67*f*
asexual reproduction, 360
aspirin, 182, 279
asthma, 279, 297
astigmatism, 175

atelectasis, 297
atherosclerosis, 228, 254
    coronary, 229*f*
    stages in, 254*f*
athlete's foot, 81
atlas, 93
atoms, 16, 17*f*
atonic intestinal muscle, 312
atopic dermatitis, 80
ATP. *See* adenosine triphosphate
atrial natriuretic peptide (ANP), 195
atrioventricular bundle, 224
atrioventricular node, 224
atrioventricular valves, 223
atrophy, 129, 396
attenuated organism, 277
auditory area, 156, 157–58
auditory nerve, 163
auditory tube, 177
auricle, 176
autoclave, 66
autoimmune diseases, 29
autoimmunity, 279
autolysis, 29
autonomic nervous system (ANS), 136,
        146*f*
    disorders of, 148
    functions of, 147–48
    heart rate and, 225, 227*f*
    parts of, 144, 146–47
AV. *See* atrioventricular
axial skeleton, 90
axillary, 244
axillary lymph nodes, 262, 264
axillary vessel, 245
axis, 93
axon fiber, 46
axons, 136
azygos vein, 246

B

B antigens, 207–8
B cells, 275–76
baby teeth, 304
bacilli, 60
back, deep muscles of, 127–29
backache, 111
bacteria
    aerobic, 59
    anaerobic, 59
    chlamydias, 61–62
    curved, 60–61, 61*f*
    rickettsias, 61
    rod-shaped, 60, 61*f*
    spherical, 60*f*, 60
bactericide, 66
bacteriology, 58
bacteriostasis, 66
bag of waters, 382
balance, 160
baldness, 81
ball-and-socket joint, 106
band cells, 205
barbiturates, 279
Bartholin's gland, 368
basal ganglia, 156
basal metabolism, 322

base, 20–21, 21*f*
basilar artery, 244
basilic vein, 245
basophils, 205
Bell's palsy, 164
bellyache, 129
benign tumors, 48, 49
    of female reproductive system, 371–72
beta-adrenergic blocking agents, 231
bicarbonate ion, 292
biceps brachii, 126
biceps femoris, 129
bile, production of, 314
binge-purge syndrome, 318
biopsy, in diagnosis of cancer, 50
births
    live, 387–88
    multiple, 384–85
birth control pill
    as contraception, 373*t*
    in treatment, 196
birth defects, as cause of disease, 56
bitter tastes, 179
bladder. *See* urinary bladder
bleeding, uterine, abnormal, 371
blood
    classification of, 202
    components, 208–10
    uses of, 208–10
    constituents of, 202
        formed elements, 204–7
        plasma, 202–4
    deoxygenated, 238, 238*f*
    description of, 202
    flow of, 250–51
    functions of, 202
    occult, 312
    oxygenated, 238, 238*f*
    return of, 251
    studies of
        blood cell counts, 214
        blood chemistry tests, 214–16
        blood slide, 214
        blood smear, 214
        bone marrow biopsy, 216
        coagulation studies, 216
        hematocrit, 213–14
        hemoglobin tests, 214
    in urine, 354
    uses of, 208–10
blood banks, 208
blood cell counts, 214
blood cells, 203*f*
blood chemistry tests, 214–16
blood clotting, 207, 207*f*
blood disorders, 210–13
blood groups, 207–8
    ABO system, 210*t*
blood plasma, 202–4
    body fluid and, 334
blood poisoning, 267–68
blood pressure, 252, 253
blood slide, 214
blood smear, 214
blood supply to myocardium, 223, 224*f*
blood typing, 209*f*
    transfusions and, 207–8
blood vessels, 238*f*

disorders involving, 253–57
    in fetus, 239
    functional classification of, 238–39
    structure of, 239, 240*f*
    of the urinary system, 341*f*
"blue baby"
    in tetralogy of Fallot, 228
body
    directions in, 5–7, 6*f*
    imaging of, 10
    introduction to, 3–11
    map of, 8
    planes of division in, 6–7, 7*f*
    of the sternum, 94
    studies of, 4
body cavities, 7–8, 9*f*
body defenses
    against disease, 272
    nonspecific, 272–74
body fluids
    compartments of, 334, 335*f*
    disorders involving, 336–37
    extracellular, 334
    fluid therapy, 337–38
    intake/output of water, 334
        acid-base balance, 336
        electrolytes and, 335–36
        sense of thirst, 334–35
    intracellular, 334
    replacement of, 335
    in special compartments, 334
body heat
    loss of, 326–27
    production of, 325–26
body mechanics, 122–23
body processes, 5
body systems, 4–5
body temperature, 325
    abnormal, 328–29
    heat loss, 326–27
    heat production, 325–26
    normal, 328
    regulation of, 327–28
        skin and, 77
boil, 81
bolus, 308
bone markings, 89–90
bone marrow, 86
    biopsy of, 216
    suppression of, 212–13
bones, 86
    of appendicular skeleton, 94, 96–97,
        98*f*, 99*f*, 100, 100*f*, 101*f*, 102*f*
    description of, 86
    disorders of, 102–3
    formation of, 93
    growth of, 86, 88
    long, 88*f*
    main functions of, 88–89
    repair of, 86, 88
    of right foot, 103*f*
    of right hand, 100*f*
    of shoulder girdle, 98*f*
    structure of, 86, 87*f*, 88*f*
    of thorax, 98*f*
    tumors of, 102–3
bony labyrinth, 177
booster shots, 278
botulism antitoxin, 279

Bowman's capsule, 341
boxer's muscle, 126
brachial artery, 244
    as pressure point for hemorrhage, 255
brachial plexus, 144
brachial vessel, 245
brachiocephalic trunk, 241
brachiocephalic veins, 245
bradycardia, 226
brain, 46
    arteries to, 245*f*
    base of, 164*f*
    coverings of, 152–53
    external surface of, 152*f*
    imaging of, 160–61
    main parts of, 152, 152*f*
    protective structures of, 152–55
    studies of, 160–61
    tumors of, 162
brain disorders, 161–62
brain stem, 152, 160
breasts
    cancer of, 372
    lactation and, 385
    section of, 386
breathing, 286
brittle bones, 396
broad ligaments, 366
bronchi, 288
bronchial tree, 289
bronchioles, 289
bronchitis, 297
bronchogenic carcinoma, 297
bronchopneumonia, 296
bronchoscope, 298
buboes, 266
buccinator, 123
buffer systems, 336
buffers, 21
building blocks, 22
bulbourethral gland, 362
bulimia, 318
bundle of His, 224
bunions, 130
burns, 80
bursae, 105
bursitis, 105, 130
bypass surgery, 232–33

C

calcaneus, 100
calcitonin, 190, 191, 336
calcium
    bone disorders and, 103
    functions of, 336
calculi, 352
calyces, 340
calyx, 340
cancer, 280
    of the breast, 372
    causes of, 35–36
    cells and, 35–36
    of the cervix, 372
    of the colon, 312–13
    definition of, 49
    diagnosis of, 50
    of the endometrium, 372

cancer *(continued)*
  of the liver, 314–15
  of the lungs, 297–98
  ovarian, 372
  of the prostate, 365
  of the rectum, 312–13
  of the skin, 81
  of the stomach, 310
  symptoms of, 49–50
  of the testes, 365
  treatment of, 50
Candida, 62
canine teeth, 304
capacity, lung, 291, 292*t*
capillaries, 238
  lymphatic, 262, 264
  network of, 246*f*
  peritubular, 342
  walls of, 239
  workings of, 248–50, 251*f*
capsules around organs, 42
carbon, 22
carbonic anhydrase, 292
carbuncle, 81
carcinogen, 36
carcinoma, 49
cardiac center of medulla oblongata, 160
cardiac cycle, 224, 225*f*
  normal EKG of, 232*f*
cardiac muscle, 45, 45*f*
cardiac muscle tissue, 221*f*
cardiac output, 224
cardiac reserve, 225
cardiac sphincter, 309
cardiac veins, 224*f*
cardiogenic shock, 255
cardiopulmonary resuscitation (CPR), 299
cardiovascular system, 263*f*
caries, 305
carotenemia, 78
carotid bodies, 293–94
carpal bones, 97, 100*f*
carpal tunnel syndrome, 130
carrier(s)
  of genes, 392
  for molecules, 34
cartilaginous joint, 105
casts in urine, 354
catabolism, 5
  metabolism and, 322
catalysts, 23
cataracts, 175, 395
catheterization, 231
cation, 19
caudal direction in body, 5, 6*f*
cavernous sinuses, 248
CBC. *See* complete blood count
CD4 receptor, 280
CDC. *See* Centers for Disease Control and Prevention
cecum, 311
celiac trunk, 241, 244
cell(s), 4
  cancer and, 35–36
  columnar, 40
  cuboidal, 40
  division of, 30, 32, 32*f*

energy obtained by, 322
functions of
  active transport, 34–35
  cell division, 30, 32, 32*f*
  diffusion, 33, 33*f*, 34*f*
  DNA, 29–30, 31*f*
  filtration, 33–34, 35*f*
  movement of substances across cell membrane, 32–35
  osmosis, 33, 35
  phagocytosis, 35
  pinocytosis, 35
  protein synthesis, 29
  RNA and, 30
germ, 360
multinucleated, 119
  osmosis and, 35, 36*f*
  pancreatic, 192, 192*f*
  squamous, 40
  structure of, 26, 28*f*, 30*t*
  study of, 26, 27*f*
cell division, 30, 32, 32*f*
cell junctions, 29
cell membrane
  description of, 26, 28*f*
  movement of substances across, 32–33
    by active transport, 34–35
    by diffusion, 33, 33*f*, 34*f*
    by filtration, 33–34, 35*f*
    by osmosis, 33
cell receptors, 147
cell-mediated immunity, 275
cellular respiration, 286, 322
Celsius thermometer, 328
centers of medulla oblongata, 160
Centers for Disease Control and Prevention (CDC), 57
centigrade scale, 328
centimeters, 11*f*
central canal, 142
central nervous system (CNS), 136
central neuron, 140
central sulcus, 156
centrifuge, 210
centrioles, 26
cephalic vein, 245
Cephalosporium, 67
cerebellum, 152, 160
cerebral aqueduct, 155
cerebral cortex
  functional areas of, 158*f*
  functions of, 156
  motor areas of, 157*f*
cerebral hemispheres, 152, 156
cerebral hemorrhage, 161
cerebral palsy, 162
cerebrospinal fluid (CSF), 143, 154
  flow of, 154*f*
cerebrospinal meningitis, 153
cerebrovascular accident (CVA), 161
cerebrum, 152
  lobes and sulci of, 152*f*
cerumen, 76, 177
ceruminous glands, 76, 177
cervical nodes, 264
cervical plexus, 144
cervical vertebrae, 93
cervix, 366
  cancer of, 372

cesarean section, 386
chambers of the heart, 220–21
change of life. *See* menopause
charley horse, 129–30
chemical barriers, 272
chemical bonds
  covalent, 20
  electrolytes, 19–20
  ionic, 19
chemical elements, 18*t*
chemicals
  as cause of cancer, 36
  as cause of disease, 56
chemistry, 16
  enzymes, 23, 23*f*
  of living matter, 22–23
  organic compounds, 22
chemoreceptors, 293–94
chemotherapy, 67
  antibiotics, 67–68
  antineoplastic agents, 68
  in treatment of cancer, 50
Cheyne-Stokes respiration, 294
chicken pox vaccine, 278
childbirth, 383–85
Chlamydia trachomatis, 173, 374
chlamydias, 61
chloride, functions of, 336
cholecystitis, 315
cholecystokinin, 317
cholelithiasis, 315
cholinergic system, 147
chondroma, 49
chordae tendineae, 223
choriocarcinoma, 388
chorionic villus sampling, 397
choroid, 170
choroid plexus, 154
  flow of cerebrospinal fluid from, 154*f*
chromosomal errors, 397
chromosomes, 29
  distribution of, 392–93
  male, normal, 398*f*
chronic bronchitis, 297
chronic disease, 57
chronic obstructive pulmonary disease (COPD), 297
chronic renal failure, 351
chyle, 262–63, 316
chyme, 310
cigarette smoking, pregnancy and, 395
cilia, 29, 48, 62
ciliary body, 170
ciliary glands, 76
ciliated epithelium, 289*f*
ciliates, 62
circle of Willis, 244, 245*f*
circulation
  fetal, 381*f*
  heart and, 220
  physiology of, 248–51
circulatory system, 4
circumduction, 107
cirrhosis of the liver, 314
cisterna chyli, 262
clavicle, 94
cleft palate, 102
clitoris, 368

closed fracture, 103
Clostridium botulinum, 66
clotting disorders, 211, 213
clotting factors, 203
clubfoot, 395
CNS. *See* central nervous system
coagulation
    platelets and, 206
    studies on, 216
coarctation of aorta, 228
cocci, 60
coccyx, 94
cochlea, 177
cochlear duct, 177
cochlear nerve, 177
cold, 168
    in treatment of pain, 182
colic, 129, 312
collagen, 42, 44, 86
collarbone, 94
colloidal suspension, 19
colon, 311
    cancer of, 312–13
colostrum, 385
columnar cells, 40
comminuted fracture, 103
common acne, 80
common bile duct, 314
common carotid artery, 241, 244
    as pressure point for hemorrhage, 255
common cold, 295–96
common hepatic duct, 314
common iliac arteries, 244
communicable disease, 57
communication areas of cerebral cortex,
    156–58
compact bone, 86
    composition of, 88*f*
complement, 276
    function of, 203
complete blood count (CBC), 214
complete proteins, 323
compound microscope. *See* optical mi-
    croscope
compounds, 17
computed tomography (CT)
    in diagnosis of cancer, 50
    of the brain, 160–61
computed tomography (CT) scanner,
    161*f*
conchae, 286
conduction, 143, 326
conduction system of the heart, 224–25,
    226*f*
conductive hearing loss, 179
condyloid joint, 106
cones, 170
confluence of sinuses, 248
congenital, definition of, 395
congenital birth defects, disease and, 56
congenital versus hereditary diseases,
    394–95
congenital heart disease, 227, 227–28
conjunctiva, 172–73
conjunctivitis, 173
connective tissue, 40, 42, 43*f,* 44
constipation, 312
contamination of food, 66

contraception, 376
    methods of, 373*t*
contractility, 116
contractions, 119
    of smooth muscles in wall of blood
        vessel, 207
control center of cell, 26
convection, 326
convergent strabismus, 175
convulsion, 129
coordination of voluntary muscles, 160
COPD. *See* chronic obstructive pul-
    monary disease
coracoid process, 94
corium, 74
cornea, 170
coronal plane, 6–7, 7*f*
coronary arteries, 224*f*
coronary atherosclerosis, 229*f*
coronary bypass surgery, 232–33
coronary heart disease, 227, 228–29, 229*f*
coronary occlusion, 228
coronary sinus, 223, 246, 248
coronary thrombosis, 228
corpus, 363
    uteral, 366
corpus callosum, 156
corpus cavernosum, 363
corpus luteum, 366
corpus spongiosum, 363
corpuscles, 202
cortex
    of adrenal glands, 191
        hormones from, 192
    renal, 340
cortisol, 192
coryza, 295
costal cartilages, 94
counseling
    genetic, 396–97
    of prospective parents, 397
counter-current mechanism, 347
covalent bonds, 20
Cowper's gland, 362
CPK. *See* creatine phosphokinase
CPR, 299
cramps, muscle, 130
cranial bones, 91*f*
cranial cavity, 8
cranial direction in body, 5, 6*f*
cranial nerves, 136
    disorders involving, 163–64
    functions of, 163
    location of, 163, 164*f*
    names of, 163
cranial venous sinuses, 248, 249*f*
craniosacral areas, 147
cranium, 90–91
creatine phosphokinase (CPK), 214
creatinine, 214, 354
creatinism, treatment of, 195
crest, as bone marking, 90
cretinism, 190
crisis, fever and, 328
crossbridges between filaments, 116–17
cryoprecipitate, 210
cryptorchidism, 364–65
cryptosporidium, 63

crystalline lens, 170
C-section, 386
CSF. *See* cerebrospinal fluid
CT. *See* computed tomography
cuboidal cells, 40
cul-de-sac, 367
curvatures of the spine, 103, 104*f*
curved rod bacteria, 60–61, 61*f*
Cushing's syndrome, 192
cutaneous, definition of, 74
cutaneous membrane, 48
CVA. *See* cerebrovascular accident
cyanosis, 228, 294
cystic duct, 315
cystic fibrosis, 396
    as inheritable disease, 395
    treatment of, 396
cystitis, 353
cysts, sebaceous, 76
cytoplasm, 26
cytotoxic T cells, functions of, 275

D

D antigen, 208
dander, 279
dark adaptation, 170
DCT. *See* distal convoluted tubule
deafness, 178–79, 395
deaths, fetal, 387–88
deciduous teeth, 304
decubitus ulcers, 81
deep fascia, 42, 116
deep lymphatic vessels, 262
deep muscles of the back, 127–29
deep veins, 245
defecation, 311
deficiency anemia, 212
degeneration, 56
    arterial, 254*f,* 254–55
    macular, 175
degenerative joint disease (DJD), 108–9
degenerative processes disease and, 56
deglutition, 307
dehydration, 337
    renal failure and, 351
delivery, 383
deltoid, 126
dendrite fiber, 46
dendrites, 136
deoxygenated blood, 238, 238*f*
deoxyribonucleic acid (DNA), 29–30,
    392
depressions, as bone markings, 90
dermatitis
    atopic, 80
    stasis, 256
dermatomes, 145
dermatosis, 80
dermis, 74
descending colon, 311
desmosomes, 29
detached retina, 175
detoxification, 314
deviated septum, 295
dextrose, in fluid therapy, 337
diabetes, types of, 192–93
diabetes insipidus, 189

diabetes mellitus, 192–93
  glucose in urine and, 354
  as inheritable disease, 395
diabetic retinopathy, 175
diagnosis of disease, 57
dialysis, 352
dialysis machines, 352
diaphragm, 8, 126, 127f, 290
diaphysis, 86
diarrhea, 312
diarthrosis, 105
diastole, 223
diastolic pressure, 252
DIC. *See* disseminated intravascular
    coagulation
diencephalon, 152, 159
diet, cancer and, 36
differential white count, 214
diffusion, 33, 291
  of gas molecules in lungs, 293f
  of gaseous molecules throughout
    given space, 33f
  of solid in liquid, 34f
  through semipermeable membrane,
    34
diffusion of gases, respiration and, 286
digestion, 302
  accessory organs of, 313f, 313–15
  control of, 316–17
  hormones active in, 317t
  process of, 315–16
    absorption, 316
    appetite, 317–18
    hunger, 317–18
digestive juices, 317t
digestive system, 4, 302, 306f
  accessory organs, 313f
    gallbladder, 315
    liver, 313–15
    pancreas, 315
  digestive tract, 302–3
    esophagus, 308
    large intestine, 311–12
    mouth, 303–7, 307f
    pharynx, 307–8
    small intestine, 310–11
    stomach, 309f, 309–10
    wall of, 302, 303f, 304f
  substances eliminated by, 340
digestive tract, 302, 302–3, 306f
  esophagus, 308
  large intestine, 311–12
  mouth, 303–7, 307f
  of newborn baby, 385
  pharynx, 307–8
  salivary glands, 307
  small intestine, 310–11
  stomach, 309f, 309–10
  surface area, 311
  teeth, diseases of, 305–7
  transport, 311
  wall of, 302, 303f, 304f
digitalis, 231
dilation, 172
diphtheria antitoxin, 278
diplococci, 60, 60f
discoloration of skin, 78
disease terminology, 57

disease-producing organisms, 56
diseases. *See also specific diseases*; disorders
  acute, 57
  body defenses against, 272
  chronic, 57
  communicable, 57
  congenital versus hereditary, 394–95
  definition of, 4, 56
  diagnosis of, 57
  genetic, 393–96
    examples of, 395–96
    treatment/prevention of, 396–97,
      399
  immune deficiency and, 280
  of the liver, 314–15
  membranes and, 48
  of mouth/teeth, 305–7
  nosocomial, susceptibility to, 68
  prevention of, 57–58
  prognosis of, 57
  of skin, 80–81
  study of, 56–58
  subacute, 57
  treatment of, 57
  venereal, 365
  of vertebrae, backache and, 111
disinfectants, 66
disinfection, 66
dislocations, 108
disorders. *See also specific disorders*;
    diseases
  of autonomic nervous system, 148
  of the blood, 210–13
  of the bones, 102–3
  of the brain, 161–62
  of the ear, 177–79
  of the eyes, 173–75
  of female reproductive system, 370–72,
    374
  genetic, 396
  of the immune system, 279–80
  of the intestines, 312–13
  involving blood vessels, 253–57
  involving body fluids, 336–37
  involving the bladder, 353
  involving the spinal cord, 143
  involving the stomach, 310
  of the joints, 108–9, 111
  of the kidneys, 350–52
  of male reproductive system, 364–65
  menstrual, 370–71
  muscular, 129–30
  postpartum, 388
  of pregnancy, 385–88
  of the spinal nerves, 144
  of the ureters, 352–53
  of the urethra, 353
  of urinary system, 350–53
dissect, 4
disseminated intravascular coagulation
    (DIC), 213
distal convoluted tubule (DCT), 342
distal direction in body, 5, 6f
distraction techniques, in treatment of
    pain, 182
divergent strabismus, 175
diverticulitis, 312
DJD. *See* degenerative joint disease

DNA, 29–30, 392
DNA molecule, basic structure of, 31f
dominant genes, 392
dopamine, 156
dorsal cavity, 8
dorsal direction in body, 5, 6f
dorsal horn, 142
dorsal root, 144
dorsal root ganglion, 144
dorsalis pedis, 244
dorsiflexion, 108
dose of pathogens, infection and, 272
double helix, 30
dowager's hump, 129
Down syndrome, 395
DPT, 277
drugs
  administration of, to spinal cord, 143
  allergic reactions to, 279
  genetic defects and, 395
  in treatment
    of pain, 182
    of tuberculosis, 296–97
DTP, 277
duct system, of male reproductive system,
    360–61
ductus arteriosus, 239
ductus deferens, 360
duodenal ulcers, 310
duodenum, 309f, 310
  microscopic view of, 304f
dura mater, 153
dural sinuses, 153
dust cells, 267
dysmenorrhea, 370
dyspnea, 294

E

ear, 175–76, 176f
  disorders of, 177–79
  external, 176, 176–77
  internal, 176, 177, 178f
  middle, 176, 177
ECG. *See* electrocardiograph
echocardiography, 231
eclampsia, 387
ectoderm layer of tissue, 383
ectopic, 385
ectopic pregnancy, 385
eczema, 80
edema, 336–37
  renal failure and, 351
EEG. *See* electroencephalograph
effector, 138, 140
efferent arteriole, 342
efferent nerves, 139
efferent neurons, 138
effusion, 296, 337
ejaculation, 363
ejaculatory duct, 360
ejection of milk, 385
EKG. *See* electrocardiograph
elbow
  right, lateral view of, 99f
  student's, 130
elderly, heart in, 230
electrocardiogram (ECG), 20

electrocardiograph (ECG)
  in heart studies, 231, 232*f*
  showing cardiac cycle, 232*f*
electroencephalogram (EEG), 20
electroencephalograph (EEG), of the
    brain, 161
electrolytes, 19–20, 336
  functions of, 335–36
  in plasma, 204
  in the urine, 354
electron microscope, 27*f*
electrons, 16, 17*f*
electrophoresis, 214
elements, 16–17
  trace, 22
elephantiasis, 65, 268
embolus, 256
embryo, 380, 382*f*
embryology, 380
emphysema, 297
emulsify, 314
encephalitis, 161
encephalon, 161
endemic disease, 57
endocarditis, 227
endocardium, 220
endocrine glands
  adrenal, 191–92
  description of, 4–5, 41
  hormones from, 186–95, 194*t*
  pancreas, 192–93
  parathyroid, 191
  pituitary, 186–89, 188*f*
  prostaglandins, 195
  secretions of, 41–42
  sex, 193
  thyroid, 189–91, 190*f*, 191*f*
endocrine system, 4–5
  glands and hormones of, 186–95
endoderm layer of tissue, 383
endolymph, 177
endometrium, 366
  cancer of, 372
endoplasmic reticulum (ER), 26
end-organ, 168
endorphins, 182
endoscope, 313
endoscopy, 313
endosteum, 86
endothelium, 239
  of lymphatic capillaries, 262
energy
  muscles and, 117, 119
  obtained from food, 322
enteritis, 312
Enterobius vermicularis, 64
enucleation, 173–74
enveloping membrane, skin as, 74
environmental agents, genetic defects
    and, 395
enzyme action, 23*f*
enzymes, 23
eosinophils, 205
epicardium, 220
epidemic disease, 57
epidermis, 74
epidermophytosis, 81
epididymis, 360

epidural hematoma, 154
epigastric region, 8, 10*f*
epiglottis, 288, 308
epilepsy, 162
epinephrine, 139
  effects of, 191
  heat production and, 325–26
  in treatment, 195
epiphysis, 86
episiotomy, 384
epistaxis, 295
epithelial membranes, 46
epithelial tissue, 40, 40–42, 41*f*
epithelium
  ciliated, 289*f*
  transitional, 40
  types of, 41*f*
EPO. *See* erythropoietin
equilibrium, 168
equipment, for respiratory tract
    treatments, 298–99
ER. *See* endoplasmic reticulum
erection, 362
eruption, 78
erythema, 78
erythroblastosis fetalis, 208
erythrocytes, 204
  function of, 202
erythropoietin (EPO), 204
  function of, 193, 195
  production of, 343
Escherichia coli, 66
esophageal hiatus, 308
esophagus, 241, 303, 308
essential amino acids, 323
essential hypertension, 253
estrogen(s)
  functions of, 193
  menstrual cycle and, 368
ethmoid bone, 91
etiology, 57
eustachian tube, 177
evaporation, 327
eversion, 108
excoriation, 78
excretory system, 4, 340
exercise, metabolism and, 119
exhalation, 290, 291*f*
exocrine glands, 40, 41
exophthalmic goiter, 190
extension, 107
extensor carpi muscles, 126
extensor digitorum muscles, 126
external abdominal oblique muscle,
    126–27
external auditory canal, 176–77
external auditory meatus, 176–77
external carotid artery, 244
external ear, 176, 176–77
external genitalia
  female, 370*f*
  male, 363
external iliac artery, 244
external secretions of exocrine glands,
    40
extracellular fluid, 5, 334
extremities, framework for, 90
extrinsic muscles, 123, 170

exudate, 296
eye(s), 169*f*
  cataracts, 175
  coats of, 168, 170
  conjunctiva, 172–73
  disorders of, 173–75
  glaucoma, 175
  infections, 173
  injuries, 173–74
  lacrimal apparatus, 172–73, 173*f*
  muscles of, 170, 172
  nerve supply to, 172
  pathway of light rays, 170
  protection of, 168
  retinal disorders, 175
  visual defects, 174*f*, 174–75
eye disorders, 173–75
eyeball. *See* eye
eyeteeth, 304

F

facial artery, as pressure point, 255
facial nerve, function of, 163
facial portion of the skull, 91–92
fallopian tubes, 366
false ribs, 94
familial genetic diseases, 395
family history, genetic counseling and,
    396–97
fascia, 42
fascicles, 116
fat(s)
  emulsification of, 314
  saturated, 324
  unsaturated, 325
febrile, 328
feces, 311
female genitalia, external, 370*f*
female pelvis, bones of, 101*f*
female perineum, muscles of, 128*f*
female reproductive system, 367*f*, 369*f*
  contraception, 373*t*, 376
  disorders of, 370–72, 374
  greater vestibular glands, 368
  menopause, 374–76
  menstrual cycle, 368–69
  ovaries, 365–66
  oviducts, 366
  perineum, 368
  uterus, 366
  vagina, 366–68
  vulva, 368
femoral artery, 244
  as pressure point, 255
femoral lymphatic vessels, 262
femoral vessel, 245
femur, 97, 102*f*
fertility drugs, multiple births and, 385
fertilization
  pregnancy and, 380
  in vitro, 364
fetal alcohol syndrome, 395
fetal circulation, 381*f*
fetal deaths, 387–88
fetus, 380, 382
  blood vessels in, 239
  development of embryo into, 382*f*

fetus *(continued)*
  pregnant uterus with, 384*f*
  viable, 387
fever, 328
  as defense against disease, 273
fiberoptic endoscope, 313
fibrillation, 229
fibrils, 159
fibrin, 206, 207
fibrinogen, 206
fibroid tumor, 49
fibroids, 371–72
fibromyositis, 129–30
fibrositis, 129–30
fibrous connective tissue, 42, 44
fibrous joint, 104–5
fibula, 100
fight-or-flight hormones, 191
fight-or-flight response, 147
filariae, 268
filariasis, 65
filtration, 33–34, 35*f*
  in formation of urine, 346*f*
  glomerular, 343, 345, 347
fimbriae, 366
finger, extra, 395
first aid, hemorrhage and, 255
first stage of parturition, 383
fissures, 78, 179
flaccid constipation, 312
flagella, 59, 62
flagellates, 62
flagellum, 29
flatfoot, 130
flatulence, 310
flatus, 310
flatworms, 65
flexion, 107
flexor carpi muscles, 126
flexor digitorum muscles, 126
floating ribs, 94
floppy baby syndrome, 396
flow murmur, 227
flu. *See* influenza
fluid compartments, 334, 335*f*
fluid therapy, 337–38
fluids
  extracellular, 5
  intracellular, 5
  replacement of, 335
flukes, 65
fluoroscope, 231
flutter, 229
follicle, 76
follicle-stimulating hormone (FSH), 363
  function of, 189
  menstrual cycle and, 368–69
food
  contamination of, 66
  energy obtained from, 322
  as good as medicine, 325
  guide to daily choices of, 326*f*
foot
  bones of, 100, 103*f*
  movement of, 129
  structure of, 100
foramen, as bone marking, 90
foramen magnum, 91

foramen ovale, 239
foramina, 90, 155
forearm
  bones of, 96
  movement of, 126
formed elements of blood
  classification of, 202
  erythrocytes, 204
  leukocytes, 204–6, 205*t*
  origin of, 206–7
  platelets, 206
fornices, 366
fossa, as bone marking, 90
fourth stage of parturition, 383
fovea centralis, 172
fractures
  aging and, 103–4
  types of, 105*f*
fraternal twins, 385
freckles, 74
free nerve endings, 181
frontal bone of the cranium, 90–91
frontal lobe, 156
frontal plane, 6–7, 7*f*
frontal sinuses, 90–91
frostbite, 329
FSH. *See* follicle-stimulating hormone
functional murmur, 227
fundus, 366
fungi, 62
furuncle, 81

G

gallbladder, 315
gametes, 360
gamma globulin, 276
ganglion, 144
gangrene, arterial degeneration and, 255
gap junctions, intercellular, 29
garbage disposal, microbial control and, 66
gas exchanges, 291
gas transport, 291–93
gastric juice, 309
  components of, 309–10
gastric ulcers, 310
gastric vein, 248
gastrin, digestion and, 316–17
gastritis, 310
gastrocnemius muscle, 129
gastroenteritis, 312
gastrointestinal (GI) tract, 303
gene mutation, 393
general senses, 168, 179–82, 181*f*
general sensory impulses, of cranial nerves, 163
generalized peritonitis, 302
genes, 30, 392, 393
genetic counseling, 396–97
genetic diseases, 393–96
  treatment/prevention of, 396–97, 399
genetic disorders, correction of, 396
genetic identification, 79
genital herpes, 365
genital warts, 374
genitalia, external

female, 370*f*
  male, 363
genitourinary system, male, 361*f*
germ, 58
germ cells, 360
germ layers of tissue, 383
German measles (rubella), 395
German measles vaccine, 277
germicide, 66
gestation, 380
GH. *See* growth hormone (GH)
GH-IH. *See* growth hormone-inhibiting hormone
GH-RH. *See* growth hormone-releasing hormone
GI. *See* gastrointestinal (GI) tract
gigantism, 189
gingivitis, 306
glands, 40–42
  ceruminous, 76
  ciliary, 76
  endocrine, 4–5, 41
    hormones from, 186–95
  exocrine, 40
  mammary, 76
    lactation and, 385
  mucous-producing, 362
    female, 368
  parotid, 307
  pituitary, 188*f*
    hormones from, 186–89
  salivary, 307
  sebaceous, 76
  sublingual, 307
  submandibular, 307
  submaxillary, 307
  sudoriferous, 75
  sweat, 75–76
glaucoma, 175
glenoid cavity, 94
glia, 46
glial cells, 46
gliding joint, 106
gliomas, 49, 162
glomerular filtrate, 347
glomerular filtration, 343, 345, 347
glomeruli, 341
glomerulonephritis, acute, 350–51
glomerulus, 341
glossopharyngeal nerve, 163
glottis, 288, 289*f*
glucagon, 192
glucocorticoids, 192
glucose, 322
  in plasma, 203–4
  in urine, 354
glucose tolerance test, 215
glucosuria, 347
gluteus maximus, 128
gluteus medius, 128
glycerine, 22
glycerol, 22
glycogen, 22, 117
  metabolism and, 322
  storage of, 313
glycosuria, 354
goiter, 190
Golgi apparatus, 26

gonadotropins, 189
gonads, 360
gooseflesh, 46
gout, 111
Graafian follicles, 366, 368*f*
grafting. *See* transplantation
gram, 10–11
gram negative, 68–69
gram positive, 68
gram stain, 68
granulocytes, 205
Graves' disease, 190
gray matter, 46, 137
great vessels, 221*f*
greater curvature, 309
greater omentum, 302
greater trochanter, 97
greater vestibular glands, 368
greenstick fracture, 103
gristle, 44
growth hormone (GH), 189
    function of, 188
    in treatment, 195
growth hormone-inhibiting hormone
        (GH-IH), 189
growth hormone-releasing hormone
        (GH-RH), 189
gustatory center, 179
gyri, 156

H

habits, disease and, 56
Haemophilus influenzae type B (HIB), in-
        oculation against, 277
hair, 76, 79
hamstring muscles, 129
hand(s)
    bones of, 100*f*
    movement of, 126
handwashing, 67
hard connective tissue, 44
hardening of the arteries, 253
Haversian canal, 86
hay fever, 279, 297
HCG. *See* human chorionic go-
        nadotropin (HCG)
HCl. *See* hydrochloric acid
HDL. *See* high-density lipoprotein
head
    as bone marking, 89
    framework of, 90–92
    frontal section of top of, 153*f*
    muscles of, 122*f*, 123
health, public, 65–66
hearing, 168
    loss of, 179
heart
    a double pump, 222*f*
    circulation and, 220
    conduction system of, 224–25, 226*f*
    description of, 220
    in elderly, 230
    function of, 223–27, 225*f*, 227*f*
    great vessels and, 221*f*
    nervous stimulation of, 227*f*
    pumping cycle of, 225*f*
    return of blood to, 251

rheumatic fever and, 228
    structure of, 220–21, 221*f*, 222*f*, 223,
        223*f*, 224*f*
    studies of, 231
    valves of, 223*f*
    wall of, layers of, 220
    work of, 223–24, 225*f*
heart ailments, 230–31
heart block, 230
heart disease
    classification of, 227
    treatment of
        artificial pacemakers, 232
        medications, 231–32
        surgery, 232–33
heart failure, 227, 230
heart rates, 224, 226
    control of, 225, 227*f*
heart rhythm, abnormalities of, 229–30
heart sounds, 226–27
heart surgery, 232–33
heartburn, 310
heat, 168, 182
heat cramps, 329
heat exhaustion, 329
heat loss, 326–27
heat production, 325–26
heat stroke, 329
heel bone, 100
Helicobacter pylori, 310
helminthology, 63
helminths, 63
helper T cells, 275
hemapheresis, 210
hematocrit, 213–14
hematuria, 354
hemiplegia, 161–62
hemiplegic, 162
hemocytometer, 214, 215*f*
hemoglobin, 204
    anemia and, 211–12
    gas transport and, 291–92
hemoglobin tests, 214
hemolytic anemia, 211
hemolytic disease of the newborn, 208
hemolyze, 35
hemolyzed cells, 207
hemophilia, 213
hemorrhage, 207
    pressure points for, 255
hemorrhoids, 256, 311
hemostasis, 207
hemothorax, 298
hepat, 313
hepatic artery, 244
hepatic metastases, 314–15
hepatic portal system, 248, 250*f*
hepatic portal vein, 246
hepatic veins, 246
hepatitis, 314
hepatitis B, 314
hepatitis B immune globulin, 279
hepatitis B virus, vaccine against, 277–78
hereditary, 395
hereditary versus congenital diseases,
        394–95
hereditary genetic diseases, 395
hereditary traits, 393

heredity, 392
    cancer and, 36
    disease and, 56
hernia, 365
herpes, genital, 365
herpes simplex, 81
herpes zoster, 81, 144
hiatal hernia, 308
HIB. *See Haemophilus influenzae* type B
high-density lipoprotein (HDL), 230
hilum, 288
hilus, 264, 288
hinge joint, 106
hip joint and pelvis, 102*f*
hip replacement, 111
hippocampus, 159
histamine, inflammation and, 273
histology, 40
HIV, 280. *See also* AIDS
    lymphadenopathy and, 268
hives, 81
Hodgkin's disease, 268
holes, as bone markings, 90
homeostasis, 5
hookworms, 65
horizontal plane, 7, 7*f*
hormonal control of male reproduction,
        363–64
hormones, 41
    from adrenal cortex, 192
    from adrenal medulla, 191
    from anterior lobe, 188–89
    definition of, 186
    description of, 4–5
    digestion and, 317*t*
    from endocrine glands, 186–95,
        194*t*
    heat production and, 325–26
    menstrual cycle and, 372*f*
    from pituitary gland, 186–89
    stress and, 196
    treatment and, 195–96
hormone-secreting organs, 187*f*
horns, 155
host, 58
hot flashes, 375–76
housemaid's knee, 130
HPV. *See* human papilloma virus
human body. *See* body
human chorionic gonadotropin (HCG),
        380
Human Genome Project, 392
human immune globulin, 279
human immunodeficiency virus (HIV),
        280
human papilloma virus (HPV), 374
humerus, 94, 97
humoral immunity, 275–76
hunchback, 103
hunger, 317–18
Huntington's disease, 395
hyaline membrane disease, 297
hydrocephalus, 155
hydrochloric acid (HCl), 309
hydrocortisone, 192
hydrolysis, 315
hydronephrosis, 351
hydroxycholecalciferol, 191

hymen, 367
hyoid bone, 92
hyperglycemia, 214–15
hyperkalemia, 337
hypernatremia, 337
hyperopia, 174
hyperpnea, 294
hypersensitivity, 279
hypertension, 253
    heart failure and, 230
    pregnancy-induced, 387
    renal failure and, 351
hyperthyroidism, 190
hypertonic solution, 35
hyperventilation, 294
hypochondriac regions, 8, 10f
hypogastric region, 8, 10f
hypoglossal nerve, 163
hypokalemia, 337
hyponatremia, 337
hypophysis gland, 186
hypospadias, 353
hypotension, 253
hypothalamus, 159
    male reproduction and, 363
    pituitary gland and, 187
    relationships of, 159
    temperature regulation and, 327–28
hypothermia, 329
hypothyroidism, 190
hypotonic solution, 35
hypoventilation, 294
hypovolemic shock, 255
hypoxemia, 294
hypoxia, 294
hysterectomy, 372

I

ICSH. *See* interstitial cell-stimulating
    hormone
IDDM. *See* insulin-dependent diabetes
    mellitus
identical twins, 385
idiopathic, 57
ileocecal valve, 311
ileum, 310
iliac arteries, subdivisions of, 244
iliac crest, 97
iliac regions, 8, 10f
iliac veins, 246
iliac vessel, 245
iliopsoas, 128
ilium, 97
illness, preexisting, disease and, 56
imaging, 10
immature infant, 387
immune deficiency diseases, 280
immune globulin, 279
immune serum, 278
immune surveillance, cancer and, 280
immune system, 272
    disorders of, 279–80
immunity, 274
    acquired, 275–76
    active, 276
    artificially acquired, 277–79
    categories of, 274

cell-mediated, 275
    humoral, 275–76
    inborn, 274–75
    individual, 275
    naturally acquired, 276–77
    passive, 277
    racial, 274–75
    species, 274
    types of, 274f
immunization, 277, 278
immunoglobulins (Ig), 275
    antibodies and, 276
immunotherapy, cancer and, 50, 280
impacted fracture, 103
impetigo, 81
implantation, 380
in vitro fertilization, 364
inborn immunity, 274–75
inches, comparison of centimeters
    and, 11f
incidence of disease, 57
incisors, 304
inclusion conjunctivitis, 173
incus, 177
individual immunity, 275
induced abortions, 387
infant skull, 92–93, 93f
infant spine, curves of, 97f
infants, premature, 387
infarct, 228
infections
    definition of, 58
    of the eye, 173
    of female reproductive system, 374
        pathway of, 375f
    inflammation and, 273
    of male reproductive organs, 365
    modes of, 58
    occurrence of, 272
    of respiratory tract, 295
    puerperal, 388
infectious arthritis, 109, 111
infectious mononucleosis, 268
inferior direction in body, 5, 6f
inferior mesenteric artery, 244
inferior mesenteric vein, 248
inferior nasal conchae, 92
inferior vena cava, 246
infertility, 364
    in females, 374
infestation, 63
inflammation
    as body defense against disease, 273
    intestinal, 312
    of the meninges, 153
inflammatory exudate, 273
inflammatory reaction, 273
influenza, 296
infundibulum, 186
ingestion, 303
inguinal canal, 360
inguinal hernia, 365
inguinal nodes, 262, 266
inguinal regions, 8, 10f
inhalation, 290, 291f
injuries to the eye, 173–74
insertion, 120
insula, 156

insulin
    function of, 192
    in treatment, 195
insulin-dependent diabetes mellitus
    (IDDM), 192–93
intake of water, 334
integument, 4, 74
integumentary system, 4
    description of, 74
    structure of, 74–75, 75f
    substances eliminated by, 340
interatrial septum, 220
interbrain, 159
intercalated, 220
intercalated disks, 220
intercellular material, 42, 86
intercostal arteries, 241
intercostal muscles, 126
intercostal spaces, 94
interferon, 273–74
interleukins, 275
internal abdominal oblique muscle,
    126–27
internal capsule, 156
internal carotid artery, 244
internal ear, 176, 177, 178f
internal iliac artery, 244
internal secretions of endocrine glands,
    41–42
internal sphincter muscle, 349
interphase, in cell division, 30, 32
interstitial cells, 360
interstitial cell-stimulating hormone
    (ICSH), 363
    function of, 189
interstitial cystitis, 353
interstitial fluid, 334
interventricular septum, 220
intervertebral disks, backache and, 111
intervertebral foramina, 93
intestinal disorders, 312–13
intestinal roundworms, 64
intestinal tract, body fluids and, 334
intestines
    large, 311–12
    small, 310–11
intoxication, water, 337
intracellular fluid, 5, 334
intrinsic factor, 212
intrinsic muscles, 123, 170
inversion, 108
involuntary muscle, 45
involuntary nervous system, 136
ion, 19
ionic bonds, 19, 19f, 20f
ionizing radiation, cancer and, 36
iris, 170
iron deficiency anemia, 212
irritability, 116
irritation, physical, as cause of cancer, 36
ischemia, 228
    arterial degeneration and, 254
ischial bursitis, 130
ischial spine, 97
ischial tuberosity, 97
ischium, 97
islets, 192
islets of Langerhans, 192

isometric contractions, 119
isotonic contractions, 119
isotonic solutions, 35
isotopes, 21
isthmus, 190
itching of hives, 81

J

jaundice, 78, 314
jejunum, 310
JG apparatus, 342
joint(s)
    definition of, 104
    disorders of, 108–9, 111
    synovial, 105–8
        movements at, 109–10*f*
        table of, 106*t*
        types of, 104–5, 106*f*
joint capsule of connective tissue, 105
joint cavity, 105
joint replacement, 111
jugular veins, 245
juvenile diabetes. *See* insulin-dependent
    diabetes mellitus
juxtaglomerular apparatus, 342
    structure of, 345*f*

K

kalium, 337
Kaposi's sarcoma, 280
karyon, 397
karyotype, 397, 398*f*
keloids, 44
keratin, skin and, 74
ketoacidosis, 336
ketones in urine, 354
kidney disorders, 350–52
kidney stones, 352
kidney transplants, 352
kidneys, 340, 341*f*
    adult polycystic kidney, 395
    aging and, 353
    body fluids and, 334
    functions of, 342–43
        acid-base balance and, 336
        transport maximum, 347
    internal structure of, 343*f*
    location of, 340
    microscopic view of, 345*f*
    polycystic, 352
    renal physiology
        concentration of urine, 347, 349
        glomerular filtration, 343, 345, 347,
          349
        tubular reabsorption, 347
        tubular secretion, 347
    structure of, 340–42
    tumors of, 351–52
killer T cells, 275
Klinefelter's syndrome, treatment of,
    397, 399
knee joint, 106
    sagittal section of, 107*f*
    x-ray of, 108*f*
knee replacement, 111
kneecap, 100

knees, housemaid's, 130
Kupffer's cells, 267
kyphosis, 103

L

labia, 368
labor, 383
laceration, 78
lacrimal apparatus, 172–73, 173*f*
lacrimal bones, 92
lacrimal gland, 172
lactase, digestion and, 316
lactation
    disturbances in, 387
    mammary glands and, 385
lacteals, 262, 316
lactic acid, muscle fatigue and, 117,
    119
lactic dehydrogenase (LDH), 214
lactiferous ducts, 385
large intestine, 303, 311–12
laryngeal pharynx, 286
larynx, 190*f*, 286, 288
laser, 50
lateral direction in body, 5, 6*f*
lateral lobes of thyroid gland, 189–90
lateral malleolus, 100
lateral sinuses, 248
lateral sulcus, 156
latissimus dorsi, 126
LDH. *See* lactic dehydrogenase
LDL. *See* low-density lipoprotein
learning process, memory and, 158–59
left atrioventricular valve, 223
left atrium, 220
left bundle branches, 224
left cerebral hemisphere, 152
left common carotid artery, 241
left coronary arteries, 223, 241
left gastric artery, 244
left heart, 220
left lower quadrant (LLQ), 8, 11*f*
left scapula, 99*f*
left subclavian artery, 241
left upper quadrant (LUQ), 8, 11*f*
left ventricle, 220–21
Legionella pneumophila, 296
legs
    bones of, 100
    movement of, 128–29
lens, 170
LES. *See* lower esophageal sphincter
lesser curvature, 309
lesser omentum, 302
lesser trochanter, 100
letdown of milk, 385
leukemia, 211, 213
leukocytes, 204–6, 205*t*
    function of, 202
leukocytosis, 214
leukopenia, 212, 214
leukoplakia, 307
leukorrhea, 370
levator ani muscle, 127
levator palpebrae superioris, 123
levers, body mechanics and, 122–23
levodopa, 162

LH. *See* luteinizing hormone
LH surge, menstrual cycle and, 368
ligaments, 42
    of hip joint and pelvis, 102*f*
    of synovial joints, 105
light rays, pathway of, 170
limb buds, 380
limbic system, 159–60
linea alba, 127
lingual tonsils, 266
lipase, digestion and, 316
lipids, 22
    in plasma, 204
lipoma, 49
lipoproteins, 230
liquid ventilation, 298
liter, 11
lithotripsy, 352
lithotriptor, 352
live births, 387–88
liver, 313*f*
    diseases of, 314–15
    functions of, 313–14
living conditions, disease and, 56
LLQ. *See* left lower quadrant
lobar pneumonia, 296
lobes, 156
local infection, 58
localized peritonitis, 302
lockjaw, 191
long bones, 86
    structure of, 88*f*
longitudinal fissure, 152
long-term memory, 158
loop of Henle, 342, 348*f*
lordosis, 103
low-density lipoprotein (LDL), 230
lower esophageal sphincter (LES), 309
lower extremities
    bones of, 97, 100, 101*f*
    muscles of, 127–29
lower jaw bone, 91
lower limb, bones of, 97, 100
lower vertebrae, backache and, 111
lumbago, 129–30
lumbar arteries, 244
lumbar curve, excessive, 103
lumbar puncture, 143
lumbar regions, 8, 10*f*
lumbar veins, 246
lumbar vertebrae, 93–94
lumbosacral plexus, 144
lumbrosacral joint, backache and, 111
lung(s), 286, 289–90
    body fluids and, 334
    cancer of, 297–98
    diffusion of gas molecules in, 293*f*
    root of, 290
lung capacity, 291, 292*t*
lung cavities, 290
lung tissue, 290*f*
lung volumes, 292*t*
LUQ. *See* left upper quadrant
luteinizing hormone (LH), 363
    function of, 189
    menstrual cycle and, 368–69
lymph, 262
    body fluid and, 334

lymph *(continued)*
  movement of, 264
lymph nodes, 262, 264, 266
  structure of, 266*f*
lymphadenitis, 268
lymphadenopathy, 268
lymphangitis, 267–68
lymphatic system, 262, 265*f*
  cancer and, 268
  capillaries, 262, 264
  disorders of, 267–68
  movement of lymph, 264
  nodes, 264, 266
  in relation to cardiovascular system, 262*f*
  spleen, 267
  thoracic duct, 262–63
  thymus, 266–67
  tissue, 264
  tonsils, 266
  vessels, 262–64, 265*f*
lymphocytes, 205
lymphocytic leukemia, 213
lymphoid tissue, 264
  disorders of, 267–68
  nodes, 264, 266
  spleen, 267
  thymus, 266–67
  tonsils, 266
lymphoma, 49, 268
Lymphosarcoma, 268
lysis, 328
lysosomes, 26, 29

M
macrophages, 205, 267
  acquired immunity and, 275
macular degeneration, 175
macules, 78
magnetic resonance imaging (MRI), 162
  of the brain, 161
  in diagnosis of cancer, 50
maintenance of balance, 160
maintenance of muscle tone, 160
male chromosomes, 398*f*
male genitourinary system, 361*f*
male pelvis, bones of, 101*f*
male reproductive system, 360–63
  disorders of, 364–65
  reproduction, 363–64
malignant tumor, 48, 49
  of female reproductive system, 372, 374
malleolus, 100
malleus, 177
malnutrition, 324
  disease and, 56
maltase, digestion and, 316
mammary glands, 76
  lactation and, 385
mammary lymphatic vessels, 262
mammogram, 372
mandible, 91
manubrium, 94
maple syrup urine disease, 397
marrow, 86
masseter muscles, 123

mastectomy, 372
master gland, 186–87
mastication, 307
mastitis, 387
mastoid process, 91
mastoid sinuses, 91
maxillae, 91–92
maxillary sinus, 92
measles vaccine, 277
meatus, as bone marking, 90
mechanical barriers, disease and, 272
mechanoreceptors, 168
media, 69, 170
medial direction in body, 5, 6*f*
medial malleolus, 100
median cubital vein, 245
mediastinum, 290
medical procedures, involving spinal cord, 143
medical treatment, of genetic diseases, 397, 399
medications, in heart disease, 231–32
medicine, food as good as, 325
medulla
  of adrenal glands, 191
  renal, 340
medulla oblongata, 152, 160
megakaryocytes, 206
meiosis, 360, 392–93
melanin, 74, 78
melanoma, 49
melatonin, 193
membranes
  cell, 26, 28*f*
    movement of substances across, 32–35, 33*f*, 34*f*, 35*f*
  cutaneous, 48
  disease and, 48
  enveloping, 74
  epithelial, 46
  mucous, 46, 48
  plasma, 26, 28*f*
  semipermeable, 32
    diffusion through, 34*f*
  serous, 46, 47–48
  synovial, 48
membranous canals, 177
membranous cochlea, 177
memory, 156
  learning process and, 158–59
  T cells, 275
meninges, 42, 152–53, 153*f*
  inflammation of, 153
  trauma to, 154
meningitis, 153
meningococcus, 153
menopause, 374–76
menstrual cycle, 368–69, 371*f*
  disorders of, 370–71
  hormones in, 372*f*
menstrual flow, 368
mesenteric arches, 244
mesenteric nodes, 266
mesentery, 302
mesocolon, 302
mesoderm layer of tissue, 383
mesothelium, 47
messenger RNA, 30

metabolic rate, 322
metabolism, 322–23
  purpose of, 5
metacarpal bones, 97
metal atom, 16
metaphase, in cell division, 32
metaphase spread of male chromosomes, 398*f*
metaphase stage, 397
metastasis, 49
metatarsal bones, 100
meter, 10
metric system, 8–11, 11*f*
microbes, 58
  public health and, 65–66
microbial control
  aseptic methods for, 66–67, 67*f*
  microbes and public health, 65–66
  spread of microorganisms, 65
  universal precautions, 67
microbiology
  chemistry in, 16
  definition of, 58
micrometer, 26
microorganisms, 58–59, 59*f*
  bacteria, 59–62
    curved, 61*f*
    rod-shaped, 61*f*
    spherical, 60*f*
  disease and, 58
  fungi, 62
  modes of transmission for, 58
  portal of entry/exit, 58
  protozoa, 62–63
    pathogenic, 63*f*
  spread of, 65
  viruses, 62
microscopes, comparison of, 27*f*
microvilli, 311
micturition, 349–50
midbrain, 152, 160
middle ear, 176, 177
midsagittal section, 6
migraine headache, 279
mild shock, 255
milk pasteurization, 66
mineralocorticoids, 192
minerals, 323, 323*t*
miscarriage, 387
mitochondria, 26
mitosis, 30
  stages of, 32*f*
mitral valve, 223
mixed cranial nerves, 163
mixed nerves, 139
mixtures, 18–19
MMR vaccine, 277
modified radical mastectomy, 372
molars, 304, 308*f*
molds, 62
molecules
  DNA, basic structure of, 31*f*
  of water, 17, 18*f*
monocyte-macrophage system, 267
monocytes, 205
mononuclear phagocyte system, 267
monosaccharides, 22
morphine, 182

mother, during pregnancy, 383
motor areas of cerebral cortex, 157*f*, 158
motor impulses, conduction of, 143
motor nerves, 139
motor neuron, 138, 140
   diagram of, 138*f*
mouth, 307*f*
   digestive functions of, 303
   diseases of, 305–7
   salivary glands, 307
   teeth, 304
movement, muscular
   of the foot, 129
   of forearm/hand, 126
   mechanics of, 119–20, 120*f*, 121*f*, 122*f*, 122–23
   of shoulder/arm, 126
   of thigh/leg, 128–29
MRI. *See* magnetic resonance imaging
mucosa, 46
mucous membranes, 46, 48,
   of digestive tract wall, 302
mucus, 40, 46
mucus-producing glands
   of female reproductive system, 368
   of male reproductive system, 362
multifactorial inheritance, 393
multinucleated cells, 119
multiple births, 384–85
multiple sclerosis, 143
mumps, 307
mumps vaccine, 277
murmurs, 226–27
muscle(s)
   of abdomen/pelvis, 126–27, 128*f*
   in action, 116–17, 118*f*
   aging and, 129
   of the back, 127–29
   contractions of, 119
   disorders of, 129–30
   energy and, 117, 119
   extrinsic, 123
   of the eye, 170, 172
   of female perineum, 128*f*
   of the head, 122*f*, 123
   intrinsic, 123
   involuntary, 45
   of lower extremities, 127–29
   masseter, 123
   movement of, 120, 122
      mechanics of, 119–20, 120*f*, 121*f*, 122*f*, 122–23
   naming of, 123
   of the neck, 123
   orbicularis, 123
   of respiration, 126
   review of, 125*t*
   skeletal
      attachments of, 119–20, 120*f*, 121*f*, 122*f*
      characteristics of, 116–17, 118*f*, 119
      structure of, 117*f*
   sternocleidomastoid, 123
   striated, 119
   superficial
      anterior view of, 121*f*
      posterior view of, 124*f*

temporal, 123
trapezius, 126
trumpeter's, 123
of the trunk, 126
of upper extremities, 126
visceral, 45–46
voluntary, 45, 116
   coordination of, 160
muscle cramps, 130
muscle fibers, 44–45, 116, 116*f*
muscle sheaths, 42
muscle tissue, 40, 44–46, 45*f*
   cardiac, 45, 45*f*, 221*f*
   classification of, 45–46
   description of, 116
   skeletal, 45, 45*f*
   smooth, 45*f*, 45–46
muscle tone, 119, 160
muscular dystrophy, 129
muscular system, 4
mutagenic agent, 393
mutation(s)
   of cells, 35
   genetic, 393
myalgia, 129–30
myasthenia gravis, 129
Mycobacterium tuberculosis, 296
mycology, 58
mycotic infections, 62
myelin, 46, 136
myelin sheath, 139*f*
myelogenous leukemia, 213
myeloid tissue, 213
myocarditis, 227
myocardium, 45
   blood supply to, 223, 224*f*
   description of, 220
   special features of, 220
myoglobin, 117
myomas, 49, 371–72
myopia, 174–75
myosin, 116
myositis, 129
myringotomy, 178
myxedema, 190, 195

N

nail root, 76
nails, 76
   aging and, 79
narcotics, 182
nasal bones, 92
nasal cavities, 286
   disorders of, 294–95
nasal polyps, 294–95
nasal septum, 286, 295
nasolacrimal duct, 172
nasopharynx, 286
natrium, 337
natural killer cells, 273
naturally acquired immunity, 274, 276–77
nausea, 310
   during pregnancy, 383
navel, 383
neck, muscles of, 123
necrosis, 254

negative feedback, 5
Neisseria, 109
neoplasms, 48, 56
neoplastic blood disease, 213
neoplastic disease, of blood-forming organs, 211
nephrons, 340, 343*f*, 344*f*
nerve impulse, 138
nerves, 46, 139–40
   cranial, 163–64, 164*f*
   to the eye, 172
   spinal, 144
nervous system, 136
   anatomic division of, 137*f*
   autonomic nervous system, 146*f*
      disorders of, 148
      functions of, 147–48, 148*t*
      parts of, 144, 146–47
   brain, 152–155, 153*f*, 154*f*
      cranial nerves, 163–64, 164*f*
      disorders of, 161–62
      divisions of, 156–60, 157*f*, 158*f*, 159*f*
      external surface of, 152*f*
      studies of, 160–61, 161*f*
      ventricles of, 155*f*
   description of, 4
   neurons, 136–38, 138*f*, 139*f*
      nerve impulse, 138
      nerves, 139–40
      reflex arc, 140–41
      synapse, 138–39, 140*f*, 141*f*
   spinal cord
      cerebrospinal fluid, 154, 154*f*
      coverings of, 152–53
      disorders involving, 143
      functions of, 143
      location of, 141
      medical procedures involving, 143
      structure of, 141–42, 142*f*
   spinal nerves, 144
nervous tissue, 40, 46, 47*f*
neuralgia, 164
neurilemma, 136–37
neuritis, 144
neurofibromatosis, 396
neuroglia cells, 46
neuroma, 49
neuromuscular junction, 116, 118*f*
neurons, 46
   description of, 136
   functions of, 136–38, 141*f*
      nerve impulse, 138
      nerves, 139–40
      reflex arc, 140–41
      synapse, 138–39, 140*f*, 141*f*
   for receiving impulses from special sense organs, 171*f*
   stimulation of, 141*f*
neurotransmitters, 116, 138
neutrons, 16, 17*f*
neutrophils, 205
nevus, 49
newborn baby, digestive tract of, 385
NIDDM. *See* non-insulin-dependent diabetes mellitus
night blindness, 170
nitrogen, 22
nitrogen fixation, 58

nitrogenous waste products, 354
nitroglycerin, 231
nodes, 136, 224, 262, 264, 266, 266*f*
nodular goiter, 190
nodule, 78
non-insulin-dependent diabetes mellitus
       (NIDDM), 193
nonmetal atom, 16
non-narcotic analgesics, 182
non-rapid eye movement (NREM), 159
nonspecific body defenses, 272–74
noradrenaline, 139
norepinephrine, 139, 191
normal body temperature, 328
normal saline solution, 35
nosebleed, 295
nosocomial diseases, susceptibility to, 68
NREM. *See* non-rapid eye movement
nuclei of medulla oblongata, 160
nucleic acids, 29
nucleolus, 26
nucleotides, 29
nucleus, 26
nurse cells, 363
nutrients, metabolism and, 322
nutrition, 323–25
nutritional anemia, 211–12

O

obesity, 327
obligate intracellular parasites, 61
obstetrics, use of ultrasound in, 386, 387
obstruction, arterial, 256–57
obturator foramina, 97
occipital bone, 91
occipital lobe, 156
occult blood, 312
occupation, disease and, 56
oculomotor nerve, 163
offspring, distribution of chromosomes
       to, 392–93
olecranon, 96
olfactory area, 156
olfactory epithelium, 179
olfactory nerve, 163
oligospermia, 364
open fracture, 103
ophthalmia neonatorum, 173
ophthalmic branch of trigeminal nerve,
       172
ophthalmic veins, 248
ophthalmologist, 173
opportunistic infection, 58
optic disk, 172
optic nerve, 163, 172
optical microscope, comparison of, 27*f*
oral cavity, 303–7, 307*f*
oral contraceptives, 196
orbicularis muscles, 123
orbicularis oculi, 123
orbicularis oris, 123
orbitals, 16
orchitis, 365
organ of Corti, 177
organelles, 26, 28*f*
organic compounds, 22
organic murmur, 227

organisms, 58
   attenuated, 277
   as cause of disease, 56
organs
   definition of, 4
   hormone-secreting, 187*f*
   skin as, 74
   of taste and smell, 180*f*
   of urinary system, 340
origin, 120
oropharynx, 286
orthopnea, 294
os coxae, 97
oscilloscope, 231
osmosis, 33, 36*f*
   effect of, on cells, 35, 36*f*
osmotic pressure, 33
osseous tissue, 44, 86
ossicles, 92, 177
ossification, 93
osteitis deformans, 103
osteoarthritis, 108
osteoblasts, 86
osteoclasts, 86
osteocytes, 86
osteogenesis imperfecta, 396
osteoma, 49
osteomyelitis, 102
osteoporosis, 102
otitis media, 177–78
otosclerosis, 179
output of water, 334
ova, 360, 365
oval window, 177
ovarian arteries, 244
ovarian cancer, 372
ovarian follicles, 366
ovarian veins, 246
ovaries, 360, 365–66
   microscopic view of, 368*f*
oviducts, 366
ovulation, 366
oxidation, metabolism and, 322
oxygen debt, 119
oxygen therapy, 298
oxygenated blood, 238, 238*f*
oxytocin, 187, 385
   function of, 189
   in treatment, 196

P

pacemakers, 224, 232
pain, 168
   referred, 182
   sense of, 181–82
palate, cleft, 102
palatine bones, 92
palatine tonsils, 266
pancreas, 192, 192*f*, 315
pancreatic cells, 192*f*
pancreatic vein, 248
pandemic disease, 57
Pap smear, 372
Papanicolaou test, 372
papilloma, 49
papules, 78
paranasal sinuses, 91

paraplegia, 143
parasite, 58
parasitic worms, 63–65, 64*f*
parasitology, 58
parasympathetic nervous system, 136,
       146, 146*f*
   effects of, 148*t*
parathyroid glands, 191, 191*f*
parathyroid hormone, 336
parathyroid hormone (PTH), 191
parents, prospective, counseling of, 397
parietal bones, 91
parietal layer, 47
parietal peritoneum, 302
parietal pleura, 290
Parkinson's disease, 162
parotid glands, 307
parotitis, 307
partition of the hearts, 220
parturition, stages of, 383
passive acquired immunity, 274
passive immunity, 277
passive immunization, 278
pasteurization of milk, 66
patella, 100
pathogenic microorganisms, 58
pathogenic protozoa, 63*f*
pathogens, 58
   laboratory identification of, 68–69
pathology, 4
pathophysiology, 56
PCT. *See* proximal convoluted tubule
PCW. *See* pulmonary capillary wedge
pectoralis major, 126
pedigree, 397
pelvic cavity, 8, 10*f*
pelvic girdle, bones of, 97, 101*f*
pelvic inflammatory disease (PID), 374
pelvis
   female, 101*f*
   hip joint and, 102*f*
   male, 101*f*
   muscles of, 126–27, 128*f*
penicillin, allergic reactions to, 279
Penicillium, 67
penis, 362–63
   cross section of, 363
   urination and, 350
pepsin, 310
peptic ulcer, 310
pericarditis, 227
pericardium, 47, 220
perichondrium, 44
perilymph, 177
perineum, 127, 128*f*, 368
periodontitis, 306
periosteum, 42, 86
peripheral nervous system (PNS), 136
peristalsis, 302
peritoneum, 47, 302, 305*f*
   pathway of infection to, 375*f*
peritonitis, 48, 302
peritubular capillaries, 342
permanent teeth, 304
pernicious anemia, 212
peroneus longus muscle, 129
pertussis, inoculation against, 277
PET. *See* positron emission tomography
pH scale, 21*f*

phagocytosis, 205, 206*f*
  as defense against disease, 273
  description of, 35
phalanx(ges)
  of fingers, 97, 100*f*
  of toes, 100, 103*f*
pharmacology, chemistry in, 16
pharyngeal tonsil, 266
pharyngitis, 295
pharynx, 177, 286, 303, 307–8
phenylalanine, 395
phenylketonuria (PKU), 395
  treatment of, 397
phimosis, 365
phlebitis, 256
phosphate, 336
phospholipids, 22
phrenic arteries, 244
phrenic nerve, 293
physical agents, disease and, 56
physical exposure, disease and, 56
physical irritation, cancer and, 36
physiology, 4
pia mater, 153
PID. *See* pelvic inflammatory disease
pigment of skin, 78
pilcae circulares, 311
piles, 256
pineal body, hormone from, 193
pinkeye, 173
pinna, 176
pinocytosis, 35
pinworms, 64
pituitary gland, 159, 188*f*
  hormones from, 186–89
  tumors of, 189
pivot joint, 106
PKU. *See* phenylketonuria
placenta, 380, 381*f*
  development/functions of, 380
  hormones from, 195
placenta previa, 385–86
placental abruption, 386
planes of division, 6–7, 7*f*
plantar flexion, 108
plaque, 254
plasma, 202
  body fluids and, 334
plasma cells, 205–6
  acquired immunity and, 275
plasma membrane, 26, 28*f*
plasmapheresis, 210
plasmodium, 62–63
platelet plug, 207
platelets, 202, 206
pleura(e), 47, 290
pleural space, 290
  disorders involving, 298
pleurisy, 298
plexuses, 144
PMNs. *See* polymorphs
PMS. *See* premenstrual syndrome
pneumocystis carinii, 63
Pneumocystis carinii, 280
pneumonectomy, 297
pneumonia, 296
pneumothorax, 298
PNS. *See* peripheral nervous system (PNS)

poisons. *See also* toxins
poliomyelitis, 143
polycystic kidney, 352
polycythemia, 214
polycythemia vera, 214
polydactyly, 395
polydipsia, 335
polymorphs, 205
polyps, 294–95
pons, 152, 160
popliteal artery, 244
portal of entry, infection and, 272
portal system, 187
position, 168
  sense of, 182
positron emission tomography (PET), 161
posterior direction in body, 5, 6*f*
posterior fornix, 366
posterior lobe, of pituitary gland, 186, 188*f*
  hormones of, 189
postpartum disorders, 388
postsynaptic cell, 138
potassium
  functions of, 336
  imbalance of, 337
potential spaces, 47
Pott's disease, 111
precapillary sphincter, 251
predisposing causes of disease, 56
predisposition, infection and, 272
preeclampsia, 387
preexisting illness, disease and, 56
pregnancy
  disorders of, 385–88
  ectopic, 385
  embryo, 380
  fertilization, 380
  fetus, 380, 382
  mother during, 383
  origin of tissues, 383
  placenta, 380
  start of, 380
  toxemia of, 387
pregnancy-induced hypertension, 387
pregnant uterus, with intact fetus, 384*f*
premature beat, 226
premature infant, 387
premenstrual syndrome (PMS), 370–71
premenstrual tension, 370–71
presbycusis, 179
presbyopia, 172
pressure, 168
  sense of, 179
pressure points for hemorrhage, 255
preterm infants, 387
prevention
  of disease, 57–58
  of food contamination, 66
  of genetic diseases, 396–97, 399
  of heart ailments, 230–31
prime mover, 120
PRL. *See* prolactin
process, as bone marking, 90
procoagulants, 207
progesterone
  function of, 193
  menstrual cycle and, 368
prognosis, 57

progressive muscular atrophies, 396
projections, as bone marking, 89–90
prolactin (PRL), 385
  function of, 189
pronation, 107
prophase, in cell division, 32
proprioceptors, 182
prospective parents, counseling of, 397
prostaglandins, hormones from, 195
prostate gland, 361–62
  cancer of, 365
protection, blood and, 202
proteins, 323
  in plasma, 202–3
  synthesis of, 29
prothrombin, 207
protons, 16, 17*f*
protozoa, 62–63
  inflamation of the brain and, 161
  pathogenic, 63*f*
protozoology, 58
proximal bone, 94, 97
proximal convoluted tubule (PCT), 342
proximal direction in body, 5, 6*f*
pruritus, 81
pseudostratified tissue, 288
psoriasis, 81
psychogenic influences, disease and, 56
PTH. *See* parathyroid hormone
ptosis, 172
pubis, 97
public health, microbes and, 65–66
puerperal infections, 388
pulmonary capillary wedge (PCW) pressure, 253
pulmonary circuit, 220, 238
pulmonary edema, 336
pulmonary embolism, 256
pulmonary valve, 223
pulmonary ventilation, 290, 291*f*
  respiration and, 286
pulmonic valve, 223
pulse, 252
pump failure, 255
pumping cycle of the heart, 225*f*
pupil, 170
purification of water supply, 66
Purkinje fibers, 224
pus, 205
  inflammation and, 273
pustules, 78
pyelonephritis, 351
pyloric sphincter, 309
pyloric stenosis, 310
pylorus, 309
pyogenic bacteria, 102
pyramids, 340
pyrogens, 328
pyuria, 354

Q

quadrants, of abdomen, 8, 11*f*
quadriceps femoris, 128
quinidine, 231

R

rabies antiserum, 279
rabies vaccine, 278

racial immunity, 274–75
radial artery, 244
radiation, 326
    as cause of cancer, 36
radioactive, 21
radioactivity, 21–22
radiography, 10
radius, 96
rapid eye movement (REM), 159
rash, 78
RDS. *See* respiratory distress syndrome
reabsorption, tubular, 347
receptors, 138, 140, 168
    for deep touch, 179
    dendrites as, 136
    hormones and, 186
recessive genes, 392
rectum, 311
    cancer of, 312–13
rectus abdominis, 127
rectus femoris, 128
red blood cells
    anemia and, 211
    impaired production of, 211–12
    loss/destruction of, 211
red marrow, 86
referred pain, 182
reflex activities, 143
reflex arc, 140–41, 142*f*
reflexes, 143
refraction, 170
regulation, blood and, 202
rejection syndrome, 280–81
relaxation, pain and, 182
releasing hormones, 187
REM. *See* rapid eye movement
renal arteries, 244, 340
renal colic, 352
renal cortex, 340
renal failure, 351
renal medulla, 340
renal pelvis, 340
renal physiology
    concentration of urine, 347, 349
    glomerular filtration, 343, 345, 347
    tubular reabsorption, 347
    tubular secretion, 347
renal ptosis, 352
renal veins, 246, 340
renin, 253, 343
reproductive system, 5
    female, 367*f*, 369*f*
        contraception, 376
        disorders of, 370–72, 374
        greater vestibular glands, 368
        menopause, 374–76
        menstrual cycle, 368–69
        ovaries, 365–66
        oviducts, 366
        perineum, 368
        uterus, 366
        vagina, 366–68
        vulva, 368
    male
        aging and, 363–64
        disorders of, 364–65
        duct system, 360–61
        hormonal control of reproduction, 363–64

mucus-producing glands, 362
        penis, 362–63, 363*f*
        prostate gland, 361–62
        semen, formation of, 361–62
        seminal vesicles, 361
        spermatozoa, 363, 364*f*
        testes, 360, 362*f*
        urethra, 362
resistance to antibiotics, 68
resorption, 86
respiration
    abnormal, 294
    acid-base balance and, 336
    artificial, 299
    cellular, 286
    definition of, 286
    muscles of, 126
    phases of, 286
    physiology of
        air movement, 290–91, 292*t*
        gas exchanges, 291, 293*f*
        gas transport, 291–93
        pulmonary ventilation, 290, 291*f*
        respiratory rates, 294
    regulation of, 293–94
respiratory center, 160
respiratory distress syndrome (RDS), 297
respiratory passageways, ciliated
        epithelium of, 289*f*
respiratory rates, 294
respiratory system, 287*f*
    bronchi, 288–89
    description of, 4
    disorders of
        asthma, 297
        atelectasis, 297
        bronchitis, 297
        common cold, 295–96
        deviated septum, 295
        emphysema, 297
        hayfever, 297
        hemothorax, 298
        infection, 295–96
        influenza, 296
        lung cancer, 297–98
        nasal polyps, 294–95
        nosebleed, 295
        pleurisy, 298
        pneumonia, 296
        pneumothorax, 298
        sinusitis, 294–95
        tuberculosis, 296–97
    larynx, 286, 288, 288*f*
    lining of air passageways, 288–89, 289*f*
    lungs, 289–90, 290*f*, 293*f*
    nasal cavities, 286
    pharynx, 286
    substances eliminated by, 340
    trachea, 288
    windpipe, 288
respiratory tract disorders treatment for,
        298–99
reticular formation, 159
reticulocytes, 211
reticuloendothelial system, 267
retina, 170
    eye disorders involving, 175
retinal detachment, 175

retinopathy, diabetic, 175
retroperitoneal space, 340
Rh factor, 208
    transfusion and, 207–8
Rh negative, 208
Rh positive, 208
rheumatic endocarditis, 228
rheumatic fever, 228
rheumatic heart disease, 227
rheumatism, 129
rheumatoid arthritis, 108–9
    membranes and, 48
RhoGAM, 279
ribonucleic acid (RNA), 29
ribosomes, 26
ribs, 94, 94, 94
rickets, 103
rickettsias, 61
right atrioventricular valve, 221, 223
right atrium, 220
right bundle branches, 224
right cerebral hemisphere, 152
right common carotid artery, 241
right coronary arteries, 223, 241
right elbow, 99*f*
right foot, bones of, 103*f*
right hand, bones of, 100*f*
right heart, 220
right lower quadrant (RLQ), 8, 11*f*
right lymphatic duct, 262
right subclavian artery, 241
right upper quadrant (RUQ), 8, 11*f*
right ventricle, 220
rigor mortis, 119
Ringer's lactate solution, 337
RLQ. *See* right lower quadrant
RNA, 29, 30
rods, 170
rod-shaped bacteria, 60, 61*f*
root of lung, 290
rotation, 107
rotator cuff, 126
roundworms, 64–65
RU 486
    contraception and, 376
rubella (German measles), 395
rubella vaccine, 277
rubeola vaccine, 277
rugae, 309
    of the urinary bladder, 349
rupture, 365
RUQ. *See* right upper quadrant

S

SA. *See* sinoatrial
sacral vertebrae, 94
sacrospinalis muscle, 127
sacrum, 94
SAD. *See* seasonal affective disorder
        (SAD)
saddle joint, 106
sagittal plane, 6, 7*f*
saline solution, 337
saliva, 307
salivary amylase, 307
salivary glands, 307
salpingitis, 374

salt, 20
salty tastes, 179
saphenous veins, 245
sarcoma, 49
sartorius muscle, 128
saturated fats, 324
scab, 77
scanning electron microscope, 26
scapula, 94, 99*f*
scapular spine, 94
scar tissue, 42, 44
Schwann cells, 136
sciatic nerve, 144
sciatica, 144
sclera, 168, 170
scleroderma, 81
sclerosis, 143
scoliosis, 103
scrotum, 360
seasonal affective disorder (SAD), 195
seat worm. *See* pinworm
sebaceous cysts, 76
sebaceous glands, 76
sebum, 76
second stage of parturition, 383
secondary sex characteristics, 193, 360
secretin, digestion and, 317
secretion, tubular, 347
seizure, 129
selectively permeable, 35
semen, formation of, 361
semicircular canals, 177
semilunar valves, 221
semimembranosus, 129
seminal vesicles, 360, 361
seminiferous tubules, 360
semipermeable membrane, 32
    diffusion through, 34*f*
semitendinosus, 129
sense of pain, 181–82
sense of position, 182
sense of pressure, 179
sense of smell, 179
sense of taste, 179
sense of temperature, 180
sense of thirst, 334–35
sense of touch, 180–81
senses, 168
    general, 179–82
    special, 179, 180*f*
sensitized T lymphocytes, 275
sensitized tissues, 279
sensorineural hearing loss, 179
sensory area, 156
sensory impulses, 143
sensory mechanisms, 168
sensory nerves, 139
sensory neurons, 137–38, 140
sensory system
    the ear, 175–79, 176*f*, 177–79, 178*f*
    the eye, 168, 169*f*, 170, 171*f*, 172–75, 173*f*, 174*f*
    general senses, 179–82
    senses/sensory mechanisms, 168
    special sense organs, 179, 180*f*
sepsis, 66
septic arthritis, 109, 111
septic shock, 255

septicemia, 267–68
septum, 220
Sequential Multiple Analyzer (SMA), 214
serosa, 46
serous membranes, 46, 47–48, 302
serratus anterior, 126
sertoli cells, 363
serum, 207
serum albumin, 338
serum hepatitis, 314
serum sickness, 278
sesamoid bone, 100
severe shock, 256
sewage, microbial control and, 66
sex
    determination of, 393, 394*f*
    disease and, 56
sex characteristics, secondary, 360
sex glands, 193
sex hormones, 192
sex-linked traits, 393
sexual reproduction, 360
sexually transmitted disease (STD), 374
shin bone, 100
shingles, 81, 144
shock, 255–56
short-term memory, 158
shoulder, movement of, 126
shoulder blade, 94, 99*f*
shoulder girdle, bones of, 94, 98*f*
sick headache, 279
sickle cell anemia, 211, 212*f*, 395
sickle cell crisis, 211
sickle cell disease, 211
sickle cell trait, 211
sigmoid colon, 311
sigmoidoscope, 312
signs of disease, 57
simple, definition of, 40
simple goiter, 190
simple squamous epithelium, 40
sinoatrial node, 224
sinus, 246
    as bone marking, 90
sinus arrhythmia, 226
sinus rhythm, 225
sinuses, 286
    dural, 153
sinusitis, 294–95
sinusoids, 248
skeletal changes, aging and, 103–4
skeletal muscle(s), 45, 45*f*
    of the abdomen/pelvis, 126–27, 127*f*, 128*f*
    action, 116–117, 118*f*
    attachments of, 119–20, 120*f*, 121*f*, 122*f*
    of the back, 127
    contractions, 119 sliding filament mechanism of, 118*f*
    exercise and, 119
    energy and, 117, 119
    fibers of, 116*f*
    of the foot, 129
    of the forearm/hand, 126
    of the head, 122*f*, 123

of the neck, 123
    review of, 125*t*
    of the shoulder/arm, 126
    structure, 116, 116*f*, 117*f*
    of the thigh/leg, 128–29
    of the trunk, 126, 127*f*
    of upper extremities, 126
skeletal system, 4
skeleton, 4, 86, 87*f*
    divisions of, 90
        bones of appendicular skeleton, 94, 96–97, 98*f*, 99*f*, 100, 100*f*, 101*f*, 102*f*
        framework of the head, 89*f*, 90*f*, 90–92, 91*f*, 92*f*
        infant skull, 92–93, 93*f*
        trunk, 93–94, 95*f*, 96*f*, 97*f*, 98*f*
skin
    aging and, 78–79
    appendages of, 75–76, 76*f*
    body fluids and, 334
    cancer of, 81
    care of, 79–80
    classification of, 74
    cross section of, 75*f*
    deeper injuries to, 78
    discoloration of, 78
    diseases of, 80–81
    functions of, 76–77
    observation of, 77–79
    pigment of, 78
    surface lesions of, 78
skin lesions, 78, 79*f*
skull, 89*f*, 90
    base of, 91*f*
    from below, 90*f*
    facial portion of, 91–92
    from front and left, 89*f*
    of infant, 92–93, 93*f*
    internal view of, 92*f*
SLE. *See* systemic lupus erythematosus
sleep, 159
sleeping sickness, 161
slow channel calcium blockers, 231–32
SMA. *See* Sequential Multiple Analyzer
small intestine, 303, 310–11
    wall of, 303*f*
smallpox, inoculation against, 277
smell, 168
    organs of, 180*f*
    sense of, 179
smooth muscle, 45*f*, 45–46
    of digestive tract wall, 302
sodium
    functions of, 335
    imbalance of, 337
soft connective tissue, 42
soft palate, 308
solute, 18
solutions, 18, 35
solvent, 18
    universal, 17–18
somatic motor impulses, 163
somatic nervous system, 136
somatostatin, 189
somatotropin, 189
    function of, 188
sonogram, 387

sore throat, 295
sour tastes, 179
spasm, 129
spastic constipation, 312
special senses, 168
  organs of, 179
  impulses from
    of cranial nerves, 163
    neurons for recieving of, 171*f*
species immunity, 274
specific body defenses, 272
specific gravity of urine, 354
specificity, 274
sperm, 360
  course of, 361*f*
sperm cells, 360
spermacide, 373*t*
spermatic arteries. *See* testicular arteries
spermatic cord, 360
spermatozoa, 360, 363, 364*f*
sphenoid bone, 91
spherical bacteria, 60, 60*f*
spherical cells, 60
sphincter, 309
sphygmomanometer, 252
spinal accessory nerve, 163
spinal canal, 8
spinal cord, 46
  coverings of, 152–53
  cross section of, 142*f*
  disorders involving, 143
  functions of, 143
  location of, 141
  medical procedures involving, 143
  structure of, 141–42, 142*f*
spinal curves, abnormalities of, 103, 104*f*
spinal meningitis, 153
spinal nerves, 136, 144
spinal reflex, 143
spine
  as bone marking, 90
  curves of, 97*f*
  of scapula, 94
  tuberculosis of, 111
spinous process, 93
spiral fracture, 104
spirillum, 60
spirochetes, 60
spleen, 267
splenectomy, 267
splenic anemia, 268
splenic artery, 244
splenic vein, 248
splenomegaly, 268
spongy bone, 86
spontaneous abortions, 387
spores, 59
sporozoa, 62
sprains, 108, 129
squamous cells, 40
staghorn calculi, 352
staging, 50
staining techniques, 68–69
stains, 26, 68
stapes, 177
staphylococci, 60, 60*f*
Staphylococcus, 109
Staphylococcus aureus, 66

stasis, 256
stasis dermatitis, 256
STD. *See* sexually transmitted disease
stem cells, 206
sterility, 364
sterilization, 66
sternal angle, 94
sternal puncture, 216
sternocleidomastoid muscle, 123
sternomastoid, 123
sternum, 94
steroids, 186
  anabolic, 120
stethoscope, 231
stillbirth, 387
stimuli, 136
stimulus, 168
stomach, 303, 309*f*, 309–10
  disorders involving, 310
stomach cancer, 310
stool, 311
strabismus, 175
straddle injuries, 353
strains, 129
  backache and, 111
strata of epidermis, 74
stratified, 40
stratified squamous epithelium, 40
stratum corneum, 74
stratum germinativum, 74
streptococci, 60, 60*f*
Streptococcus, 109
stress, hormones and, 196
stretch reflex, 143
striated muscle, 119
strictures, ureteral, 352
stroke, 161
stroke volume, 224
student's elbow, 130
subacute disease, 57
subclavian artery, 255
subclavian vessel, 245
subcutaneous layer, 74
subdeltoid bursitis, 130
subdural hematoma, 154
sublingual glands, 307
submandibular glands, 307
submaxillary glands, 307
submucosa, 302
sucrase, digestion and, 316
suction apparatus, 298
sudoriferous glands, 75
suffocation, 294
sulci, 156
sunburn, 80
sunstroke, 329
superficial fascia, 42, 74
superficial lymphatic vessels, 262
superficial muscles
  anterior view of, 121*f*
  posterior view of, 124*f*
superficial veins, 244–45, 247*f*
superior direction in body, 5, 6*f*
superior mesenteric artery, 244
superior mesenteric vein, 248
superior sagittal sinus, 248
superior vena cava, 245–46
supination, 107
suppressor T cells, 275

suprarenal arteries, 244
suprarenal glands, 191–92
suprarenal veins, 246
surface area of digestive tract, 311
surfactant, 290
surgery, in treatment of heart disease, 232–33
surgical obstetrical perineum, 384
surgical transplantation, 233
suspensions, 19
suspensory ligaments, 172
suture, 105
swayback, 103
sweat glands, 75–76
sweet tastes, 179
sympathetic chains, 147
sympathetic nervous system, 136, 146, 146*f*
  effects of, 148*t*
symphysis pubis, 97
symptoms of disease, 57
synapse, 138–39, 141*f*
  close-up view of, 140*f*
synaptic cleft, 138
synarthrosis, 105
syndrome of disease, 57
synovial fluid, 105
synovial joints
  function of, 107–8
  movements at, 109–10*f*
  structure of, 105
  types of, 105–6
synovial membranes, 48
syphilis, 365
systemic arteries
  names of, 239–41, 242*f*, 243*f*, 244, 245*f*, 246*f*
  subdivisions of, 244
systemic circuit, 220, 239
systemic infection, 58
systemic lupus erythematosus (SLE), 48
systemic veins, 244–46, 247*f*, 248
systems, 4
systole, 223
systolic pressure, 252

T

T cells, 193
  acquired immunity and, 275
  functions of, 275
  thymosin and, 193, 267
tachycardia, 226
tactile corpuscles, 180
tailor's muscle, 128
talipes, 395
tanning, 78
target tissue, 186
tarsal bones, 100
taste, 168
  organs of, 180*f*
  sense of, 179
taste buds, 179, 304
TB. *See* tuberculosis
teeth, 304, 307*f*
  diseases of, 305–7
  molars, 308*f*
telophase, in cell division, 32
temperatures

outside, body temperature and, 328–29
sense of, 180
temporal artery, hemorrhage and, 255
temporal bones, 91
temporal lobe, 156
temporal muscles, 123
tendinitis, 130
tendons, 42, 119, 120*f*
tenosynovitis, 130
terminal bronchioles, 289
testes, 360, 362*f*
testicular arteries, 244
testicular cancer, 365
testicular veins, 246
testosterone, 193, 360
tetanus, 191
tetanus immune globulin, 278–79
tetany, 191
tetralogy of Fallot, 228
thalamus, 159, 159*f*
"the pill". *See* oral contraceptives
therapeutic abortions, 387
therapy for disease, 57
thiamine, 144
thigh, movement of, 128–29
third stage of parturition, 383
thirst, sense of, 334–35
thoracic aorta, 240, 241
thoracic cavity, 8, 10*f*
thoracic curve, exaggeration of, 103
thoracic duct, 262–64
thoracic vertebrae, 93
thoracolumbar area, 147
thorax, 93
    bones of, 94, 98*f*
throat, 307–8
thrombin, 207
thrombocytes, 202
thrombocytopenia, 213
thrombolytic drugs, 229
thrombophlebitis, 256
thromboplastin, 207
thrombosis, 229*f*
thrombus, 228
thrush, 62
thymosin, 193
    T lymphocytes and, 267
thymus gland, 266–67
    hormones from, 193
thyroid function, 190–91
thyroid gland, 189–91, 190*f*, 191*f*
thyroid hormones, 195–96
thyroid storm, 190
thyroid-stimulating hormone (TSH), 188
thyrotropin, 187, 188
thyroxine hormone, 190
    heat production and, 325–26
    metabolic rate and, 322
tibia, 100
tibial artery, 244
tibial lymphatic vessels, 262
tibialis anterior muscle, 129
tic douloureux, 164
tight junction, intercellular, 29
tinea capitis, 62
tinea corporis, 62

tissue fluid, 248
tissue macrophage system, 267
tissue repair, 42, 44
tissues
    adipose, 42
    classification of, 40
    connective, 40, 42, 43*f*, 44
    definition of, 4
    description of, 40
    epithelial, 40, 40–42, 41*f*
    lymphoid, 264, 266–67
    muscle, 40, 44–46, 45*f*
    nervous, 40, 46, 47*f*
    origin of, 383
    osseous, 44
T-lymphocytes. *See* T cells
toe dancer's muscle, 129
toes, phalanges of, 100
tongue, taste map of, 179
tonsils, 266, 307*f*
tonus, 119
torticollis, 123
touch, 168
    sense of, 180–81
toxemia of pregnancy, 387
toxic shock syndrome (TSS), 374
toxins
    infection and, 272
    produced by bacteria, 59
toxoid, 277
trace elements, 22, 323
trachea, 190*f*, 286, 288
tracheobronchial nodes, 264, 266
tracheostomy, 298–99
tracheostomy tube, 298–99
trachoma, 173
tracts, 139
traits, 393
transducer, 387
transfusions, 207–8
transitional epithelium, 40
transmission electron microscope, 26, 27*f*
transplants
    of heart, 233
    of kidneys, 352
    rejection syndrome and, 280–81
transport
    in digestive tract, 311
    of oxygen, 286
transportation, blood and, 202
transverse colon, 311
transverse plane, 7, 7*f*
transverse sinuses, 248
transversus abdominis muscle, 126–27
trapezius muscle, 126
trauma
    to the meninges, 154
    skin lesions and, 78
treatment
    for cancer, 50
    of cystic fibrosis, 396
    of disease, 57
    of genetic diseases, 396–97, 399
    of heart disease
        artificial pacemakers, 232
        medications, 231–32
        surgery, 232–33
    hormones and, 195–96

of osteoporosis, 102
of respiratory disorders, 298–99
of tuberculosis, 296–97
Treponema pallidum, 365
triceps brachii, 126
trichina, 65
trichinosis, 65
tricuspid valve, 221, 223
trigeminal nerve
    function of, 163
    ophthalmic branch of, 172
trigeminal neuralgia, 164
trigone, 349
trimesters, in fetal development, 382
trochlea, 97
trochlear nerve, 163
trochlear notch, 97
true ribs, 94
trumpeter's muscle, 123
trunk
    framework of, 93–94, 96–97, 100
    muscles of, 126
trypanosomes, 62, 161
trypsin, digestion and, 316
TSH. *See* thyroid-stimulating hormone (TSH)
TSS. *See* toxic shock syndrome
tubal ectopic pregnancy, 385
tuberculosis (TB), 296–97
    of the spine, 111
tubular reabsorption, 347
tubular secretion, 347
tumors, 48–49
    benign, 49
        of female reproductive system, 371–72
    in bone tissue, 102–3
    of the brain, 162
    of the kidneys, 351–52
    of male reproductive organs, 365
    malignant, 49
        of female reproductive system, 372, 374
    of pituitary gland, 189
    of the urinary bladder, 353
twins, types of, 385
tympanic membrane, 177
tympanostomy tube, 178
type I diabetes. *See* insulin-dependent - diabetes mellitus
type II diabetes. *See* non-insulin-dependent diabetes mellitus

U

UA. *See* urinalysis
ulcers
    decubitus, 81
    duodenal, 310
    gastric, 310
    peptic, 310
    skin, 78
    stasis dermatitis and, 256
ulna, 96
ulnar artery, 244
ultrasonography, 387
ultrasound
    in diagnosis of cancer, 50
    in heart studies, 231

ultrasound *(continued)*
  in obstetrics, 386, 387
ultrasound cardiography, 231
umbilical cord, 380
umbilical region, 8, 10*f*
umbilicus, 383
universal precautions, 67
universal solvent, 17–18
unsaturated fats, 325
upper extremities
  bones of, 94, 96–97
  muscles of, 126
upper limb, bones of, 94, 96, 97
upper respiratory infection (URI), 295
urea, 342
  synthesis of, 313
uremia, 351
ureters, 340, 341*f*, 349
  disorders of, 352–53
urethra, 340, 341*f*, 350
  disorders of, 353
  male, 362
urethral meatus, 350
urethritis, 353
URI. *See* upper respiratory infection
urinalysis (UA), 354
urinary bladder, 340, 341*f*, 349–50
  characteristics of, 349
  disorders involving, 353
  interior of, 350*f*
  tumors of, 353
  urination and, 349–50
urinary system, 340, 341*f*
  bladder, 349–50, 350*f*
  description of, 4
  disorders of, 350–53
  kidneys
    functions of, 342–43
    location of, 340
    renal physiology, 343, 345, 347,
      349
    structure of, 340–42
  organs of, 340
  pathway of infection to, 375*f*
  substances eliminated by, 340
  ureters, 349
  urethra, 350
urination, 349–50
urine
  concentration of, 347, 349
  constituents of
    abnormal, 354
    normal, 353–54
  filtration in formation of, 346*f*
urticaria, 81, 279
uterine bleeding, abnormal, 371
uterine tubes, 366
uterus, 366
  pregnant, with intact fetus, 384*f*
uvula, 308

V

vaccination, 277
vaccines, 277–78
vagina, 366, 366–68
vagus nerve, 163

valve replacement, 233
valves of the heart, 221, 223, 223*f*
varicella vaccine, 278
varicose veins, 256, 256*f*
varix, 256
vas deferens, 360
vascular system
  blood pressure, 252–53
  blood vessels, 238*f*, 238–39, 240*f*, 241*f*
  circulation, 248–51, 251*f*
  pulse, 252
  systemic arteries, 239–41, 242*f*, 243*f*,
    244, 245*f*, 246*f*
  systemic veins, 244–46, 247*f*, 248, 249*f*,
    250*f*
vasectomy, 364
vasoconstriction, 207, 251
vasodilation, 119, 251
vasomotor activities, 251
vasomotor center, 160, 251
vastus intermedius, 128
vastus lateralis, 128
vastus medialis, 128
VD. *See* sexually transmitted disease
veins, 220, 238
  cardiac, 224*f*
  cross section of, 241*f*
  superficial, 244–45, 247*f*
  systemic, 244–46, 247*f*, 248
  walls of, 239, 240*f*
venereal diseases, 365
venous sinuses, 246, 248, 249*f*
ventilation
  abnormal, 294
  liquid, 298
  pulmonary, 286, 290, 291*f*
ventral cavity, 8
ventral direction in body, 5, 6*f*
ventral horn, 142
ventral root, 144
ventricles, 154–55, 155*f*
venules, 238, 239
vermiform, 311
vermiform appendix, 311
vernix caseosa, 76, 382
vertebrae, 93, 93–94
  backache and, 111
  from above, 96*f*
vertebral column, 93
  front view of, 96*f*
  lateral curvature of, 103
  side view of, 95*f*
very-low-density lipoprotein (VLDL),
  230
vesicles, 78
vestibular nerve, 177
vestibule, 177
vestibulocochlear nerve, 163
viable fetus, 387
vibrio, 60
villi, 311, 380
Vincent's disease, 306
virology, 58
virulence, infection and, 272
viruses, 62
  as cause of cancer, 36
  inflamation of the brain and, 161

viscera, 45
visceral layer, 47
visceral motor impulses, 163
visceral muscles, 45–46
visceral nervous system, 136
visceral peritoneum, 302
visceral pleura, 290
viscosity, 252
vision, 168
visual areas of cerebral cortex, 156,
  158
visual defects, 174*f*, 174–75
vitamin D, 77
vitamins, 323, 324*t*
vitreous body, 170
VLDL. *See* very-low-density lipoprotein
vocal cords, 286, 289*f*
volar arch, 244
voluntary muscles, 45, 116
  coordination of, 160
voluntary nervous system, 136
vomer, 92
vomiting, 310
  during pregnancy, 383
vulva, 368

W

warts, genital, 374
water, 17
  intake/output of, 334
  molecule of, 18*f*
  properties of, 17–18
water intoxication, 337
water supply, purification of, 66
wax, 177
white blood cells, 204–6, 205*t*
  in urine, 354
white fibers, 137
white matter, 137
white of the eye, 170
Wilson's disease, 397
window of the eye, 170
windpipe, 288
wisdom teeth, 304
worms, 63–65
wound healing, 44
Wright's stain, 214
wryneck, 123

X

xiphoid process, 94
x-linked traits, 393
x-ray of knee joint, 108*f*

Y

yeasts, 62
yellow marrow, 86
yellow pigment in the urine, 354

Z

zygomatic bones, 92
zygote, 380

Please remember that this is a library book,
and that it belongs only temporarily to each
person who uses it. Be considerate. Do
not write in this, or any, library book.